Studies in Computational Intelligence

Volume 628

Series editor

Janusz Kacprzyk, Polish Academy of Sciences, Warsaw, Poland
e-mail: kacprzyk@ibspan.waw.pl

About this Series

The series "Studies in Computational Intelligence" (SCI) publishes new developments and advances in the various areas of computational intelligence—quickly and with a high quality. The intent is to cover the theory, applications, and design methods of computational intelligence, as embedded in the fields of engineering, computer science, physics and life sciences, as well as the methodologies behind them. The series contains monographs, lecture notes and edited volumes in computational intelligence spanning the areas of neural networks, connectionist systems, genetic algorithms, evolutionary computation, artificial intelligence, cellular automata, self-organizing systems, soft computing, fuzzy systems, and hybrid intelligent systems. Of particular value to both the contributors and the readership are the short publication timeframe and the worldwide distribution, which enable both wide and rapid dissemination of research output.

More information about this series at http://www.springer.com/series/7092

Subana Shanmuganathan
Sandhya Samarasinghe
Editors

Artificial Neural Network Modelling

 Springer

Editors
Subana Shanmuganathan
School of Computer and Mathematical
  Sciences
Auckland University of Technology
Auckland
New Zealand

Sandhya Samarasinghe
Department of Informatics and Enabling
  Technologies
Lincoln University
Christchurch
New Zealand

Studies in Computational Intelligence
ISBN 978-3-319-80363-0   ISBN 978-3-319-28495-8 (eBook)
DOI 10.1007/978-3-319-28495-8

Printed on acid-free paper

This Springer imprint is published by SpringerNature
The registered company is Springer International Publishing AG Switzerland

Contents

Artificial Neural Network Modelling: An Introduction

Subana Shanmuganathan

Abstract While scientists from different disciplines, such as neuroscience, medicine and high performance computing, eagerly attempt to understand how the human brain functioning happens, Knowledge Engineers in computing have been successful in making use of the brain models thus far discovered to introduce heuristics into computational algorithmic modelling. Gaining further understanding on human brain/nerve cell anatomy, structure, and how the human brain functions, is described to be significant especially, to devise treatments for presently described as incurable brain and nervous system related diseases, such as Alzheimer's and epilepsy. Despite some major breakthroughs seen over the last few decades neuroanatomists and neurobiologists of the medical world are yet to understand how we humans think, learn and remember, and how our cognition and behaviour are linked. In this context, the chapter outlines the most recent human brain research initiatives following which early Artificial Neural Network (ANN) architectures, components, related terms and hybrids are elaborated.

1 Introduction

Neuroanatomists and Neurobiologists of the medical world are yet to discover the exact structure and the real processing that takes place in human nerve cells and to biologically model the human brain. This is despite the breakthroughs made by research ever since the human beings themselves began wondering how their own thinking ability happens. More recently, there has been some major initiatives with unprecedented funding, that emphasise the drive, to accelerate research into unlocking the mysteries of human brain's unique functioning. One among such big funding projects is the Human Brain Project (HBP) initiated in 2013. The HBP is a European Commission Future and Emerging Technologies Flagship that aims to

S. Shanmuganathan (✉)
Auckland University of Technology, Auckland, New Zealand
e-mail: Subana@gmail.com

© Springer International Publishing Switzerland 2016
S. Shanmuganathan and S. Samarasinghe (eds.), *Artificial Neural Network Modelling*, Studies in Computational Intelligence 628,
DOI 10.1007/978-3-319-28495-8_1

1

understand what makes the brain unique, the basic mechanisms behind cognition and behaviour, how to objectively diagnose brain diseases, and to build new technologies inspired by how the brain computes. There are 13 subprojects (SPs) within this ten-year one-billion pound HBP programme. The scientists involved in the HBP accept that the current computer technology is insufficient to simulate complex brain functioning. However, they are hopeful of having sufficiently powerful supercomputers to begin the first draft simulation of the human brain within a decade. It is surprising that despite the remarkable and ground breaking innovations achieved in computing leading to transformations never seen in human development, even the modern day's most powerful computers still struggle to do things that humans find instinctive. "Even very young babies can recognise their mothers but programming a computer to recognise a particular person is possible but very hard." [1]. Hence, SP9 scientists of HBP are working on developing "neuromorphic computers-machines" that can learn in a similar manner to how the brain functions. The other major impediment in this regard is the humongous amount of data that will be produced, which is anticipated to require massive amount of computing memory. Currently, HBP scientists of The SpiNNaker project at the University of Manchester are building a model, which will mimic 1 % of brain function. Unlocking brain functioning secrets in this manner is anticipated to yield major benefits in information technology as well. The advent of neuromorphic computers and knowledge could lead to the production of computer chips with specialised cognitive skills that truly mimic those of the human brain, such as the ability to analyse crowds, or decision-making on large and complex datasets. These digital brains should also allow researchers to compare healthy and diseased brains within computer models [2].

Meanwhile, across the Atlantic, the unveiling of Brain Research Through Advancing Innovative Neurotechnologies—or BRAIN in the USA by President Obama took place in 2013 [3]. This was announced to keep up with the brain research initiated in Europe. The BRAIN project was said to begin in 2014 and be carried out by both public and private-sector scientists to map the human brain. The President announced an initial $100 m investment to shed light on how the brain works and to provide insight into diseases such as Alzheimer's, Parkinson's, epilepsy and many more. At the White House inauguration, President Obama said: "There is this enormous mystery waiting to be unlocked, and the BRAIN initiative will change that by giving scientists the tools they need to get a dynamic picture of the brain in action and to better understand how we think and learn and remember. And that knowledge will be transformative." In addition, the US President as well pointed out a lack of research in this regard, "As humans we can identify galaxies light years away, we can study particles smaller than the atom, but we still haven't unlocked the mystery of the 3 lb of matter that sits between our ears," [3].

With that introduction to contemporary research initiatives to unlock unique human brain functioning, Sect. 2 looks at the early brain models in knowledge engineering following which initial ANN models and their architectures are elaborated. In the final section some modern day ANN hybrids are outlined.

2 Early Brain Models in Knowledge Engineering

Using the Brain models developed based on our understanding thus far made on human like thinking, researchers in "Knowledge Engineering" continue to introduce functional models simulating the heuristic ability that is still considered as a unique characteristics of human intelligence. "We [Knowledge Engineers] are surprisingly flexible in processing information in the real world..." [4]. As reiterated by HRP as well as the US brain research initiatives of this decade, the discovery of actual processing in the human brain (consisting of 10^{11} neurons, participating in perhaps 10^{15} interconnections over transmission paths) seems to be very unlikely to be made in the near future. Nevertheless, the functional models of the knowledge engineers and combinations of these models have been put into successful use in knowledge representation and processing, and are known as Artificial Intelligence (AI) in computing. In the last few decades, there has been considerable research carried out with an appreciable amount of success in using knowledge-based systems for solving problems those needed heuristics.

Brain functions are mainly processed in the form of algorithms suggested John Holland (1975) [5], who was the first to compare the heuristic methods for problem solving with nature's evolution process using genetic approaches and genetic algorithms. Genetic algorithms solve complex combinational and organizational problems with many variants i.e., genes, chromosomes, population, mutation. In [6], it is explained that the brain is capable of acquiring information-processing algorithms automatically. This kind of elucidation not only forms a basis for understanding the growth of the brain and the factors needed for mental growth, but also enables us to develop novel information processing methods. Expert systems are an instance of rule-based expressions of knowledge, represented in the conditional mathematical form of "if and then" causal relationships.

In the last few decades, the performance of conventional computing has been growing spectacularly [1]. The reasons for this have been; the falling cost of large data storage devices, the increasing ease of collecting data over networks, the development of robust and efficient machine learning algorithms to process this data along with the falling cost of computational power. They have indeed enabled the use of computationally intensive methods for data analysis [7]. The field of "data mining" also called as "knowledge discovery" is one among them that has already produced practical applications in many areas. i.e., analysing medical outcomes, predicting customer purchase behaviour, predicting the personal interests of Web users, optimising manufacturing process, predicting trends in stock markets, financial analysis and sales in real estate investment appraisal of land properties, most of them using past observational data.

"Traditionally, human experts have derived their knowledge that is described as explicit from their own personal observations and experience. With advancing computer technology, automated knowledge discovery has become an important AI research topic, as well as practical business application in an increasing number of organisations..." [8]. Knowledge discovery has been identified as a method of learning the implicit knowledge that is defined as previously unknown, non-trivial knowledge hidden in the past data or observations.

Above all, our expectations from computers have been growing. "In 40 years' time people will be used to using conscious computers and you wouldn't buy a computer unless it was conscious...." [9] as envisioned by Aleksander in 1999.

3 Artificial Neural Networks and Their Components

An Artificial Neural Network (ANN) in simple terms is a biologically inspired computational model, which consists of processing elements (called neurons), and connections between them with coefficients (weights) bound to the connections. These connections constitute the neuronal structure and attached to this structure are training and recall algorithms. Neural networks are called the connectionist models because of the connections found between the neurons [10].

Deboeck and Kohonen [11] described Neural networks (NNs) as a collection of mathematical techniques that can be used for signal processing, forecasting and clustering and termed it as non-linear, multi-layered, parallel regression techniques. It is further stated that neural network modelling is like fitting a line, plane or hyper plane through a set of data points. A line, plane or hyper plane can be fitted through any data set to define the relationships that may exist between (what the user chooses to be) the inputs and the outputs; or it can be fitted for identifying a representation of the data on a smaller scale.

The first definition describes the ANN from its similarities to the human brain like functioning (Fig. 1) and the latter (Kohonen) in an application perspective.

It is truly accepted inclusive of recent brain research initiatives that the human brain is much more complicated as many of its cognitive functions are still unknown. However, the following are the main characteristics considered and described as common functions in real and artificial networks:

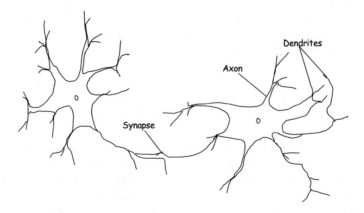

Fig. 1 Biological neuron

1. Learning and adaptation
2. Generalisation
3. Massive parallelism
4. Robustness
5. Associative storage of information
6. Spatiotemporal information processing.

Intrigued by the potentials of the ANNs, professionals from almost all fields are finding methods by way of creating new models using all possible combinations of symbolic and sub-symbolic paradigms, many of them with Fuzzy techniques to suit a variety of applications within their own disciplines.

McCulloch and Pitts were the first to introduce a mathematical model of a neuron in 1943. They continued their work [12] and explored network paradigms for pattern recognition despite rotation angle, translation, and scale factor related issues. Most of their work involved simple neuron model and these network systems were generally referred to as perceptrons.

The perceptron model of McCulloch and Pitts [12] created using Pitts and McCulloch [13] neuron is presented in Fig. 2. The Σ unit multiplies each input x by a weight w, and sums the weighted inputs. If this sum is greater than a predefined threshold, the output is one, otherwise zero. In general, they consist of a single layer. In 1958, using this neuron model of Pitts and McCulloch [13], Rosenblatt made a network with the aim of modelling the visual perception phenomena. In the 1960s, these perceptrons created a great interest and in 1962, Rosenblatt proved a theorem about perceptron learning. He showed that a perceptron could learn anything that it could represent. Consequently Widrow [14, 15], Widrow and Angell [16], and Widrow and Hoff [17] demonstrated convincing models. The whole world was exploring the potential of these perceptrons. But eventually as these single layer systems found to fail at certain simple learning tasks, researchers lost interest in ANNs. Consequently, Minsky [18] proved that single layer perceptrons had severe restrictions on their ability to represent and learn [18]. He further doubted a learning algorithm could be found for multi-layer neural networks. This caused almost an eclipse to ANN research and in turn made the researchers to develop symbolic AI methods and systems i.e., Expert Systems.

Later, from 1977 onwards, new connectionist models were introduced. Such as associative memories [19, 20], multi-layer perceptron (MLP) and back propagation learning algorithm [21, 22]; adaptive resonance theory (ART) [23, 24],

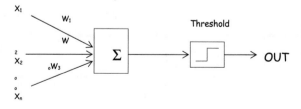

Fig. 2 Perceptron neuron of Pitts and McCulloch [12]

self-organising networks [25] and more. These new connectionist models drew the interest of many more researchers into sub-symbolic systems and as a result many more networks have been designed and used since then: i.e., Bi-directional associative memory introduced by [26], radial basis function by [27], probabilistic RAM neural networks of [28, 29], fuzzy neurons and fuzzy networks presented by [30, 31], oscillatory neurons and oscillatory neural networks [32–34] and many more. Based on these different neuron and network models enormous applications have been developed and successfully used in invariably all disciplines.

ANNs are increasingly being used across a variety of application areas where imprecise data or complex attribute relationships exist that are difficult to quantify using traditional analytical methods [10]. Research elaborated in the following chapters of this book show the more recent trends in ANN applications and the success achieved in using them.

The following are the parameters that describe a neuron based on Fig. 3.

1 Input connections (inputs): $x1, x2, \ldots, x_n$. There are weights bound to the input connections: $w1, w2, \ldots, w_n$. One input to the neuron, called the bias has a constant value of 1 and is usually represented as a separate input, let's refer to as $x0$, but for simplicity it is treated here just as an input, clamped to a constant value.

2 Input functions f: Calculates the aggregated net input signal to the neuron
   $u = f(x, w)$,
   where x and w are the input and weight vectors correspondingly;
   f is usually the summation function; $u = \Sigma i = 1, nx_i.w_i$

3 An activation (signal) function s calculates the activation level of the neuron

$$a = s(u).$$

4 An output function calculates the output signal value emitted through the output (the axon) of the neuron; $o = g(a)$; the output signal is usually assumed to be equal to the activation level of the neuron, that is, o = a.

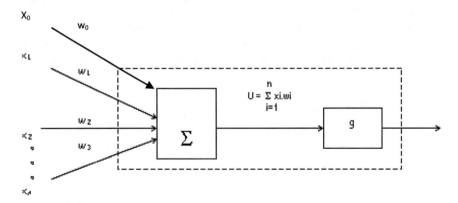

Fig. 3 A model of an artificial neuron

Artificial neural networks are usually defined by the following four parameters:

1 Type of neuron (or nodes as the neural network resembles a graph) i.e., Perceptron Pitts and McCulloch [13], Fuzzy neuron Yamakawa [30]
2 Connectionist architecture: The organisation of the connections between the neurons is described as the architecture. The connections between the neurons define the topology of the ANN. i.e., fully connected, partially connected (Fig. 4).
Connectionist architecture can also be distinguished depending on the number of input and output neurons and the layers of neurons used

   (a) Autoassociative: Input neurons are the output neurons i.e., Hopfield network
   (b) Heteroassociative: There are separate input neurons and output neurons i.e., Multi-layer perceptron (MLP), Kohonen network.
   Furthermore, depending on the connections back from the output to the input neurons, two different kinds of architectures are determined:
   (a) Feedforward architecture: There are no connections back from the output neurons to the input neurons. The network does not remember of its previous output values and the activation states of its neurons.
   (b) Feedback architecture: There are connections back from the output neurons to the input neurons and as such the network holds in memory of its previous states and the next state depends on current input signals and the previous states of the network. i.e., Hopfield network.

3 Learning algorithm: is the algorithm, which trains the networks. Lots of research have been carried out in trying various phenomena and it gives the researchers an enormous amount of flexibility and opportunity for innovation and discussing the whole set of Learning algorithms is far beyond the scope of this chapter.

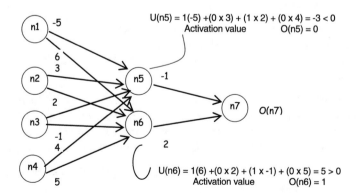

Fig. 4 A simple neural network with four input nodes (with an input vector 1, 0, 1, 0), two intermediate, and one output node. The connection weights are shown, presumably as a result of training

Nevertheless, the learning algorithms so far used are currently classified into three groups.

(a) Supervised learning: The training examples consist of input vectors x and the desired output vectors y and training is performed until the neural network "learns" to associate each input vector x to its corresponding output vector y (approximate a function $y = f(x)$). It encodes the example in its internal structure.

(b) Unsupervised learning: Only input vectors x are supplied and the neural network learns some internal feature of the whole set of all the input vectors presented to it. Contemporary unsupervised algorithms are further divided into two (i) noncompetitive and (ii) competitive.

(c) Reinforcement learning: Also referred to as reward penalty learning. The input vector is presented and the neural network is allowed to calculate the corresponding output and if it is good then the existing connection weights are increased (rewarded), otherwise the connection weights involved are decreased (punished).

4   Recall algorithm: By which learned knowledge is extracted from the network.

The following are the contemporary applications of ANN in general:

1   Function approximation, when a set of data is presented.
2   Pattern association.
3   Data clustering, categorisation, and conceptualisation.
4   Learning statistical parameters.
5   Accumulating knowledge through training.
6   "Extracting" knowledge through analysis of the connection weights.
7   Inserting knowledge in a neural network structure for the purpose of approximate reasoning.

The problem solving process using the neural networks actually consists of two major phases and they are:

(i) Training phase: During this phase the network is trained with training examples and the rules are inserted in its structure.
(ii) Recall phase: When new data is fed to the trained network the recall algorithm is used to calculate the results.

The problem solving process is described as mapping of problem domain, problem knowledge and solution space into the network's input state space, synaptic weights space and the output space respectively. Based on recent studies construction of a neural network could be broken into the following steps [10]:

(1) Problem identification: What is the generic problem and what kind of knowledge is available?
(2) Choosing an appropriate neural network model for solving the problem.
(3) Preparing data for training the network, which process may include statistical analysis, discretisation, and normalisation.

(4) Training a network, if data for training is available. This step may include creating a learning environment in which neural networks are "pupils".
(5) Testing the generalisation ability of the trained neural network and validating the results.

In recent years, neural networks have been considered to be universal function approximators. They are model free estimators [35]. Without knowing the type of the function, it is possible to approximate the function. However, the difficult part of it is, how to choose the best neural network architecture. i.e., to choose the neural network with the smallest approximation error. In order to understand further, one should look into the structure of the networks that have evolved over the years and the ones that are currently in use (Fig. 5).

The perceptron network proposed by Rosenblatt [36], one of the first network models, made using the neuron model of McCulloch and Pitts [13], was used to model the visual perception phenomena. The neurons used in the perceptron have a simple summation input function and a hard-limited threshold activation function or linear threshold activation function. The input values are in general real numbers and the outputs are binary. The connection structure of this perceptron is feed forward and three-layered. The first layer is a buffer, in which the sensory data is stored. The second layer is called the 'feature layer' and the elements of the first layer are either fully or partially connected to the second layer. The neurons from the second layer are fully connected to neurons in the output layer, which are also referred to as the "perceptron layer". The weights between the buffer and the feature layer are generally fixed and due to this reason, perceptrons are sometimes called as "single layer" networks.

A perceptron learns only when it misclassifies an input vector from the training example. i.e., if the desired output is 1 and the value produced by the network is 0, then the weights of this output neuron is increased and vice versa.

Widrow and Hoff [37] proposed another formula for calculating the output error during training: $Err_j = y_j - \Sigma w_{ij} x_i$. This learning rule was used in a neural machine called ADALINE (adaptive linear neuron).

Fig. 5 A simple two-input, one output perceptron and a bias

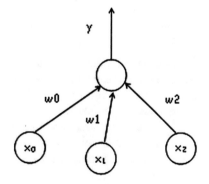

The recall procedure of the perceptron simply calculates the outputs for a given input vector using the standard summation thresholding formula and can be defined as:

$$U_j = \sum (X_i \cdot w_{ij}) \text{ for } i = (1, 2, \ldots, n) \text{ for } j = (1, 2, \ldots, m)$$

where
   U_j is the net input signal to each output neuron j
   $X_0 = 0$ is the bias,
   X = input feature vector
This perceptron can only solve problems that are linearly separable, hence could be used only to solve such examples. Still it is used due to its simplicity in structure, architecture and unconditional convergence when linearly separable classes are considered.

In order to overcome the linear separability limitations of the perceptrons, Multi-layer Perceptrons (MLPs) were introduced. A MLP consists of an input layer, at least one intermediate layer or "hidden layer" and one output layer. The individual neurons of layers are either fully or partially connected to the neurons of the next layers depending on the type and architecture of the network.

The MLPs were actually put into use only after the development of learning algorithms for multi-layer networks. i.e., back propagation algorithm [21, 22]. The neurons in the MLP have continuous valued inputs and outputs, summation input function and non-linear activation function. A MLP with one hidden layer can approximate any continuous function to any desired accuracy, subject to a sufficient number of hidden nodes. Finding the optimal number of hidden nodes for different kinds of problems has been tried out by research work and the following are considered to be the latest techniques in finding the optimal number of hidden nodes.

1 Growing neural networks: Training starts with the small number of hidden nodes and depending on the error calculated the number of the hidden nodes might increase during the training procedure.
2 The weak connections and the neurons connected by weak connections are removed from the network during the training procedure. After removing the redundant connections and nodes the whole network is trained and the remaining connections take the functions of the pruned ones. Pruning may be implemented through learning-with-forgetting methods.

Growing and pruning could be applied to input neurons hence, the whole network could be made dynamic according to the information held in the network or to be more precise according to the requirement of the nodes needed to hold the information in the data set.

Ever since the introduction of MLPs, research with diversified approaches has been conducted to find out the best network architecture, the network paradigm with the smallest approximation error for different kind of problems. Such research conducted by different interested groups and teams has proved that certain classes of network paradigms be best used to solve particular set of problems. One such approach is the use of Self-Organizing Maps for "data mining" purposes in "knowledge discovery".

4   Knowledge Extraction from ANNs

In the past decade, another important AI topic has emerged in the form of knowledge extraction using trained neural networks. Knowledge processing is performed in a "black box" approach with the trained neural network models. Some of ANN's black box related issues are discussed in Chap. "Order in the Black Box: Consistency and Robustness of Hidden Neuron Activation of Feed Forward Neural Networks and its Use in Efficient Optimization of Network Structure". Meanwhile, the following approaches are currently used in order to extract or interpret the symbolic knowledge encoded in the structure of trained network models [8]:

(a) Decompositional: Each neuron is examined and the knowledge extracted at this level is then combined to form the knowledge base of the entire network.
(b) Pedagogical: Only the network input/output behaviour is observed, viewed as a learning task, in which the target concept is the function, computed by the network.

Today artificial neural networks have been recognized in the commercial sphere too as a powerful solution for building models of systems or subjects you are interested in, just with the data you have without knowing what's happening internally. This is not possible with the conventional computing and currently there are plenty of areas where ANN applications are commercially available i.e., Classification, Business, Engineering, Security, Medicine, Science, Modelling, Forecasting and Novelty detection to name a few.

AI not only tends to replace the human brain in some representational form, it also provides the ability to overcome the limitations faced in conventional (sequential) computing and in [10] Kasabov (1995) classifies the main paradigms adapted to achieve AI as the following:

1 Symbolic—Based on the theory of physical symbolic systems proposed by Newel and Simon [38] symbolic AI is further classified into two and they are:

(i) A set of elements (or symbols) which can be used to construct more complicated elements or structures and
(ii) A set of processes and rules which, when applied to symbols and structures, produce new structures.
    Symbolic AI systems in the recent past have been associated with two issues, namely, representation and processing (reasoning). The currently

developed symbolic systems or models solve AI problems without following the way, how humans think, but produce similar results. They have been very effective in solving problems that can be represented exactly and precisely and have been successfully applied in natural language processing, expert systems, machine learning, and modelling cognitive processes. At the same time, symbolic AI systems have very limited power in handling inexact, uncertain, corrupted, imprecise or ambiguous information (Fig. 6).

2 Subsymbolic—Based on Neurocomputing explained by Smolenski [39] Subsymbolic, intelligent behaviour is performed at a subsymbolic level, which is higher than the neuronal level in the brain but different from the symbolic paradigms. Knowledge processing is carried out by changing states of networks constructed of small elements called neurons that are similar to the biological neurons. A neuron or a collection of neurons could be used to represent a micro feature of a concept or an object. It has been shown that it is possible to design an intelligent system that achieves the proper global behaviour even though all the components of the system are simple and operate on purely local information. ANNs, also referred to as connectionist models, are made possible by subsymbolic paradigms and have produced good results especially, in the last two decades. ANN applications i.e., pattern recognition, image and speech processing, have produced significant progress.

Increasingly, Fuzzy systems are used to handle inexact data and knowledge in expert systems. Fuzzy systems are actually rule-based expert systems based on fuzzy rules and fuzzy inference. They are powerful in using inexact, subjective, ambiguous, data and vague knowledge elements. Many automatic systems (i.e., automatic washing machines, automatic camera focusing, control of transmission are a few among the many applications) are currently in the market. Fuzzy systems can represent symbolic knowledge and also use numerical representation similar to the subsymbolic systems.

Fig. 6 Usability of different methods for knowledge engineering and problem solving depending on the availability of data and expertise on a problem based on [10] p. 67

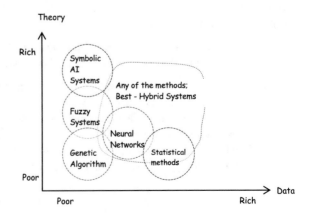

Symbolic and subsymbolic models could interact in the following ways in knowledge processing:

1 Developed and used separately and alternatively.
2 Hybrid systems with both symbolic and subsymbolic systems.
3 Subsymbolic systems could be used to model pure symbolic systems.

With that introduction to ANN initiatives, architectures, their components and hybrid systems, the remaining chapters of the book look at more recent ANN applications in a range of problem domains and are presented under three categories, namely, (1) Networks, structure optimisation and robustness (2) Advances in modelling biological and environmental systems, and (3) Advances in modelling social and economic systems.

References

1. S. Furber, Billion pound brain project under way. BBC by Fergus Walsh, 2 Oct 2013, http://www.bbc.co.uk/news/health-24428162. Accessed 18 Oct 2015
2. H. Markram, Billion pound brain project under way. BBC by Fergus Walsh, 7 Oct 2013, http://www.bbc.co.uk/news/health-24428162. Accessed 18 Oct 2015
3. B. Obama, Obama proposes brain mapping project. BBC, 2 April 2013, http://www.bbc.co.uk/news/science-environment-22007007. Accessed 18 Oct 2015
4. S.-i. Amari, Foreword, in *Foundations of Neural Networks, Fuzzy systems and Knowledge Engineering*, ed. by N.K. Kasabov (USA, Bradford, 1995), p. xi
5. J. Holland, Adaptation in Natural and Artificial Systems. The University of Michigan, 1975
6. G. Matsumato, The Brain can Acquire its Algorithm in a Self-Organized Fashion, in *Proceedings of the ICONIP/ANZIIS/ANNES'99 International Workshop*, Perth, Australia, 22–23 Nov 1999, p. 116
7. T.M. Mitchell, Machine learning and Data Mining. *Communications of the ACM,* Nov 1999, vol 42, No. 11
8. L. Fu, Knowledge discovery basedon neural networks. *Communications of the ACM, Nov 1999,* vol. 42. No. 11, pp. 47–50
9. I. Aleksander, I compute therefore I am. London BBC Science and Technology, 1999, http://news.bbc.co.uk/2/hi/science/nature/166370.stm. Accessed 27 Aug 2015
10. N.K. Kasabov, *Foundations of Neural Networks*, A Bradford Book, (The MIT Press, Fuzzy systems and Knowledge Engineering, Cambridge, Massachusetts, London, England, 1995), 581 pp
11. T. Kohonene, G. Deboeck, *Visual Explorations in Finance with Self-organizing Maps*, (Springer, London, 1998), 258 pp
12. W. Pitts, W.S. McCulloch, How we know univesrsals, The perception of auditory and visual forms. Bull. Math. Biophys. **9**(3), 127–147 (1947)
13. W.S. McCulloch, W.H. Pitts, A logical calculus of the ideas immanent in nervous activity. Bull. Math. Biophys. **5**, 115–133 (1943)
14. B. Widrow, Generalization and information storage in networks of Adaline 'Neurons', in *Self-Organizing Systems, symposium proceedings*, ed. by M.C. Yovitz, G.T. Jacobi, G. Goldstein (Spartan Books, Washington, DC, 1962), pp. 435–461
15. B. Widrow, Review of "Adaptive Systems of Logic Network and Binary Memories", Trans. Electron. Comput.: IEEE J. Aleksander **EC16**(5), 710–711, (1967)

16. B. Widrow, J.B. Angell, Reliable, trainable networks for computing and control. Aerosp. Eng. **21**(9), 78–123 (1962)
17. B. Widrow, M.E. Hoff, Associative Storage and Retrieval of Digital Information in Networks of Adaptive `Neurons'. Biol. Prototypes Synth. Syst. **1**, 160 (1962)
18. M.L. Minsky, S. Papert, *Perceptrons: An Introduction to Computational Geometry* (MIT Press, Cambridge, MA, 1969)
19. S. Amari, Neural theory of association and concept formation. Biol. Cybern. **26**, 175–185 (1977)
20. J. Hopfield, Neural networks and physical systems with emergent collective computational abilities. Proc. Natl. Acad. Sci. U.S.A. **79**, 2554–2558 (1982)
21. D.E. Rumelhart, G. Hinton, R. Williams, Learning internal representation by error propagation, in D.E. Rumelhart, J.L. McClelland, PDP Research Group (eds), *Parallel Distributed Processing Exploration in the Microstructure of Cognition: Foundations*, vol. 1 (MIT Press, Cambridge, MA, 1986) p. 1
22. P. Werbos, Backpropagation through time: What it does and how to do it., Proc. IEEE **87**, 10 (1990)
23. G.A. Carpenter, S. Grossberg, ART 2: Stable self-organization of pattern recognition codes for analog input patterns. Appl. Opt. **26**, 4919–4930 (1987)
24. G.A. Carpenter, S. Grossberg, A massively parallel architecture for a self-organizing neural pattern recognition machine. Comput. Vision, Graph., Image Proc. **37**, 54–115 (1987)
25. T. Kohonen, Self-organized formation of topologically correct feature maps. Biol. Cybern. **43**, 59–69 (1982)
26. B. Kosko, Bidirectional associative memories. IEEE Trans. Syst., Man Cybern. **18**, 49–60 (1988)
27. T. Moody, C. Darken, Fast learning in networks of locally tuned processing units. Neural Comput. **1**, 281–294 (1989)
28. J. Taylor, C. Mannion, *New Developments in Neural Computing* (Adam Hilger, Bristol, England 1989)
29. I. Aleksander, Neural Computing Architectures. The Design of Brain-like Machines, in I. Aleksander, H. Morton (eds.), (MIT Press, Cambridge, MA, 1990) (An Introduction to Neural Computing, London, Chapman & Hall., 1989)
30. T. Yamakawa, Pattern recognition hardware system employing a fuzzy neuron, in *Proceedings of the International Conference on Fuzzy Logic and Neural Networks,* Iizuka, Japan, July 1990, pp. 943–948
31. T. Furuhashi, T. Hasegawa, S. Horikawa et al., An adaptive fuzzy controller using fuzzy neural, in *networks. In: Proceedings of Fifth International Fuzzy Systems Association World Congress, IEEE* (1993), pp. 769–772
32. W. Freeman, C. Skarda, Spatial EEG patterns, non-linear dynamics and perception: The neo-Sherringtonian view. Brain Res. Rev. **10**(10), 147–175 (1985)
33. K. Kaneko, Clustering, coding, switching, hierarchical ordering, and control in network of chaotic elements. Physica **41D**, 137–172 (1990)
34. R. Borisyuk, A. Kirillov, Bifurcation analysis of neural network model. Biol. Cybern. **66**, 319–325 (1992)
35. B. Kosko, *Neural Networks and Fuzzy Systems: A Dynamical Approach to Machine Intelligence* (Englewood Cliffs, NJ, Prentice-Hall, 1992)
36. F. Rosenblatt, The perceptron: A probabilistic model for information storage and organization in the Brain. Psychol. Rev. **65**, 386–408 (1958)
37. B. Widrow, M.E. Hoff, Adaptive switching circuits. IRE WESCON Convention Rec. N.Y. **4**, 96–104 (1960)
38. A. Newell, H.A. Simon, *Human Problem Solving, Englewood Cliffs* (Prentice Hall, NJ, 1972)
39. P. Smolenski, Tensor product variable binding and the representation of symbolic structures in connectionist systems. Artif. Intell. **46**, 159–216 (1990)

Order in the Black Box: Consistency and Robustness of Hidden Neuron Activation of Feed Forward Neural Networks and Its Use in Efficient Optimization of Network Structure

Sandhya Samarasinghe

Abstract Neural networks are widely used for nonlinear pattern recognition and regression. However, they are considered as black boxes due to lack of transparency of internal workings and lack of direct relevance of its structure to the problem being addressed making it difficult to gain insights. Furthermore, structure of a neural network requires optimization which is still a challenge. Many existing structure optimization approaches require either extensive multi-stage pruning or setting subjective thresholds for pruning parameters. The knowledge of any internal consistency in the behavior of neurons could help develop simpler, systematic and more efficient approaches to optimise network structure. This chapter addresses in detail the issue of internal consistency in relation to redundancy and robustness of network structure of feed forward networks (3-layer) that are widely used for nonlinear regression. It first investigates if there is a recognizable consistency in neuron activation patterns under all conditions of network operation such as noise and initial weights. If such consistency exists, it points to a recognizable optimum network structure for given data. The results show that such pattern does exist and it is most clearly evident not at the level of hidden neuron activation but hidden neuron input to the output neuron (i.e., weighted hidden neuron activation). It is shown that when a network has more than the optimum number of hidden neurons, the redundant neurons form clearly distinguishable correlated patterns of their weighted outputs. This correlation structure is exploited to extract the required number of neurons using correlation distance based self organising maps that are clustered using Ward clustering that optimally cluster correlated weighted hidden neuron activity patterns without any user defined criteria or thresholds, thus automatically optimizing network structure in one step. The number of Ward clusters on the SOM is the required optimum number of neurons. The SOM/Ward based optimum network is compared with that obtained using two documented pruning methods: optimal brain damage and variance nullity measure to show the efficacy of

S. Samarasinghe (✉)
Integrated Systems Modelling Group, Lincoln University, Christchurch, New Zealand
e-mail: sandhya.samarasinghe@lincoln.ac.nz

© Springer International Publishing Switzerland 2016
S. Shanmuganathan and S. Samarasinghe (eds.), *Artificial Neural Network Modelling*, Studies in Computational Intelligence 628,
DOI 10.1007/978-3-319-28495-8_2

15

the correlation approach in providing equivalent results. Also, the robustness of the network with optimum structure is tested against perturbation of weights and confidence intervals for weights are illustrated. Finally, the approach is tested on two practical problems involving a breast cancer diagnostic system and river flow forecasting.

Keywords Feed-forward neural networks · Structure optimization · Correlated neuron activity · Self organizing maps · Ward clustering

1 Introduction

Feed forward neural networks are the most powerful and most popular neural network for nonlinear regression [1]. A neural network with enough parameters can approximate any nonlinear function to any degree of accuracy due to the collective operation of flexible nonlinear transfer functions in the network. However, neural networks are still treated as black boxes due to lack of transparency in the internal operation of networks. Since a neural network typically is a highly nonlinear function consisting of a number of elementary functions, it is difficult to summarize the relationship between the dependent and independent variables in a way similar to, for instance, statistical regression where the relationships are expressed in a simple and meaningful way that builds confidence in the model. In these statistical models, coefficients or model parameters can be tested for significance and indicate directly the strength of relationships in the phenomena being modeled. Although neural networks are used extensively and they can provide very accurate predictions, without internal transparency, it is not easy to ensure that a network has captured all the essential relationships in the data in the simplest possible structure in classification or function approximation. Therefore, it is vital for the advancement of these networks that their internal structure is studied systematically and thoroughly. Furthermore, the validity and accuracy of phenomena they represent need thorough assessment. Additionally, any consistency in the activation of neurons can reveal possibilities for efficient optimization of the structure of neural networks.

2 Objectives

The goal of this Chapter is to address in detail the issue of internal consistency in relation to robustness of network structure of feed forward (multiplayer perceptron) networks. Specifically, it has the following objectives:

- To investigate if there is a recognizable pattern of activation of neurons that reveals the required complexity and is invariable under all conditions of network operation such as noise and initial weights.
- To investigate the possibility of efficient optimization of structure (i.e., by pruning) based on internal consistency of neuron activations in comparison to existing structure optimization methods, such as, optimal brain damage and variance nullity measure.
- Apply the above structure optimization approach to multi-dimensional data and practical real-life problems to test its efficacy.
- To test the robustness of a network with the optimum structure against perturbation of weights and develop confidence intervals for weights.

3 Background

Feed forward networks have been applied extensively in many fields. However, little effort has gone into systematic investigation of parameters or weights of neural networks and their inter-relationships. Much effort has been expended on resolving bias variance dilemma (under- or over- fitting) [2] and pruning networks [1, 3–9]. In these approaches, the objective is to obtain the optimum or best possible model that provides the greatest accuracy based on either the magnitude of weights or sensitivity.

A network that under-fits, lacks nonlinear processing power and can be easily corrected by adding more hidden neurons. Over-fitting is more complex and occurs when the network has too much flexibility. Two popular methods for resolving over-fitting are early stopping (or stopped search) and regularization (or weight decay) [10]. In early stopping, a network with larger than optimum structure is trained and excessive growth of its weights is prevented by stopping training early at the point where the mean square error on an independent test set reaches a minimum. Regularization is a method proposed to keep the weights from getting large by minimizing the sum of square weights in the error criterion along with the sum of square error. Pruning methods such as optimal brain damage [5–7] and variance nullity measure [8] make use of this knowledge to remove less important weights. However, they do not reveal if there is a pattern to the formation of weights in networks in general and if they are internally consistent, unique, and robust.

Aires et al. [11–13] in addressing the complexity of internal structure of networks have shown that de-correlated inputs (and outputs) result in networks that are smaller in size and simpler to optimize. This was confirmed by Warner and Prasad [14]. In addressing uncertainty of network output, Rivals and Personnaz [15] constructed confidence intervals for neural networks based on least squares estimation. However, these studies do not address the relationships of weights within a network due to sub-optimum network complexity and uncertainty of response of the simplest structure.

Teoh et al. [16] proposes singular value decomposition (SVD) of hidden neuron activation to determine correlated neuron activations in order to optimize network structure. It is a step toward meaningful investigation into hidden neuron activation space; however, as authors point out, the method requires heuristic judgment in determining the optimum number of neurons. Furthermore, our research, as will be presented in this chapter, revealed that the most meaningful patterns are found not in the hidden neuron activation space but in the weighted hidden neuron activation feeding the output neuron. Xian et al. [17] used an approach based on the knowledge of the shape of the target function to optimize network structure, which is only possible for 2- or 3-dimensional data as target function shape cannot be ascertained easily for high-dimensional data. Genetic and evolutionary algorithms [18, 19] have also been used for identifying network structure, but they typically involve time consuming search in large areas in the weight space and rely on minimum insight from the operation of a network compared to other approaches to network structure optimisation.

In this Chapter, a systematic and rigorous investigation of the internal structure and weight formation of feed forward networks is conducted in detail to find out if there is a coherent pattern to weights formation that reveals the optimum structure of a network that can be easily extracted based on such knowledge. We also greatly expand our previous work presented in [20] for structure optimization.

4 Methodology

A one-dimensional nonlinear function shown in Fig. 1a (solid line) is selected for simplicity of study and interpretation of the formation of weights in detail. This has the form

$$t = \begin{cases} 0.3\,Sin\,x & If\,x<0 \\ Sin\,x & otherwise \end{cases} \tag{1}$$

A total of 45 observations were extracted from this function depicted by the solid line in Fig. 1 and these were modified further by adding a random noise generated

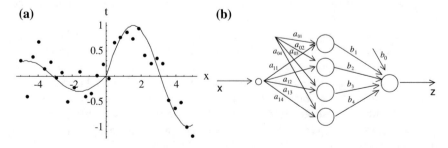

Fig. 1 Data and the model: **a** Target data generator and noisy data (random sample 1) generated from it and **b** network with redundant neurons [1]

from a Gaussian distribution with 0 mean and standard deviation of 0.25 as depicted by dots in Fig. 1. It is worth noting that the size of the extracted data set was purposely kept relatively small with a reasonably large amount of noise to approximate a real situation and to test the robustness of networks rigorously. Also, the fact that the target data generator is known helps assess how the network approaches the data generator through the cloud of rather noisy data.

This data pattern requires 2 neurons to model the regions of inflection. A larger network of 4 hidden neurons, as shown in Fig. 1b is, therefore, used for the purpose of investigation. In this network, the hidden neuron activation functions are logistic, output neuron is linear, the bias and input-hidden layer neuron weights are depicted by a_{0j} and a_{ij}, respectively, and hidden-output weights and the corresponding bias are denoted by b_j and b_0, respectively. The network is extensively studied in the following sections for patterns of hidden neuron activation as well as robustness of activation patterns and its potential for structure optimization.

5   Consistency of Network Weights

5.1   Consistency with Respect to Initial Weights

It is desirable that there is just one minimal and consistent set of weights that produces the global minimum error on the error surface and that the network reaches that global optimum regardless of the initial conditions. The data set was randomly divided into 3 sets: training, test and validation, each consisting of 15 observations. The network was trained with the training set based on Levenberg-Marquardt method [1] on Neural Networks for Mathematica [21] and test set was used to prevent over-fitting based on early stopping. (Levenberg-Marquardt is a second order error minimization method that uses the gradient as well as the curvature of the error surface in weight adaptation).

Since the network has excessive weights, it is expected that it will experience over-fitting unless training is stopped early. The performance of the optimum network (validation root mean square error RMSE = 0.318) obtained from early stopping is shown in Fig. 2a (solid line) along with the target pattern (dashed line) and training data. It shows that the network generalizes well. The performance of the over-fitted network that underwent complete training until the training error reached a minimum is illustrated in Fig. 2b. Here, the network fits noise as well due to too much flexibility resulting in over-fitting caused by large magnitude weights. The network has 13 weights and their updates during the first 10 epochs are shown in Fig. 2c. Over-fitting sets in at epoch 2. Two weights that increase drastically are two hidden-output weights.

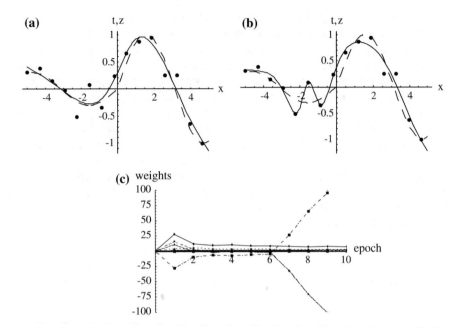

Fig. 2 Network performance: **a** Optimum network performance (*solid line*) plotted with the target data generator (*dashed line*) and training data, **b** over-fitted network performance and **c** Evolution of weights during training [1]

The experiment was repeated twice more with two different random weight initializations. The optimum networks for these two cases produced similar outputs to that shown in Fig. 2a with validation RMSE of 0.364 and 0.299, respectively. However, the first case produced over-fitting with complete training similar to the first weight initialization but the other did not although there were excessive weights. A closer inspection of the evolution of weights for the latter non-overfitting case revealed that 4 of the 13 weights in fact grew to large values after reaching the optimum similar to that shown in Fig. 2c. However, these appear to have pushed the network to operate in the saturated region of the activation functions thereby not affecting the network performance.

Are the optimum weights in these three cases similar? Figure 3a shows the 8 input-hidden neuron weights (a_{01}, a_{11}, a_{02}, a_{12}, a_{03}, a_{13}, and a_{04}, a_{14}) denoted by 1, 2, 3, 4, 5, 6, 7 and 8, respectively, for the three weight initializations and Fig. 3b shows the 5 hidden-output weights (b_0, b_1, b_2, b_3, b_4) denoted by 1, 2, 3, 4 and 5, respectively. These show that the values for an individual weight as well as the overall pattern across all the weights for the three weight initializations are generally dissimilar. In Fig. 3, the network that did not over-fit was for initialization 3.

Fig. 3 Parallel plots of weights for 3 weight initializations: **a** input-hidden neuron weights and **b** hidden-output neuron weights

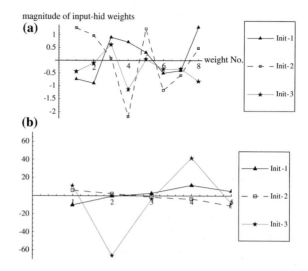

5.2 Consistency of Weights with Respect to Random Sampling

A good model must be robust against chance variations in representative samples extracted from the same population. The effect of random sampling on weight structure was tested by training the same network as in Fig. 1b along with the very first set of random initial weights (Init-1 in the previous section) on two more random data sets extracted from the original target function in Fig. 1a. These new datasets are labeled random samples 2 and 3 and the original sample used in the previous section is labeled random sample 1. The optimum network output for sample 2 was similar to that for sample 1 shown in Fig. 2a and had a validation RMSE of 0.270 and produced over-fitting with complete training. Results for sample 3 were interesting in that there was no over-fitting at all with complete training and the weight evolution for this case revealed that weight remained constant after reaching the optimum. The validation RMSE for this case was 0.32. This is natural control of over-fitting by the data and as Siestma and Dow [22] also illustrated, training with properly distributed noise can improve generalization ability of networks.

In order to find out if there is a pattern to the final structure of optimum weights, these are plotted (along with those for the very initial random sample 1 used in the previous section) for comparison in Fig. 4a, b. Here, the weights for random sample 3 stand out in Fig. 4b. Comparison of Figs. 3 and 4 indicate that there is no consistency in the network at this level. However, both non-over-fitted networks- one in Fig. 4b (sample 3) and the other in Fig. 3b (Init-3)-have similar hidden-output weight patterns.

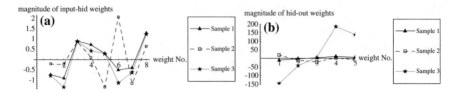

Fig. 4 Final optimum weight structure after training with three random samples: **a** input-hidden weights and **b** hidden-output neuron weights

6 Consistency of Hidden Neuron Activation

Since the weights in Figs. 3 and 4 do not provide any clues as to the existence of an internally consistent pattern, we next explore hidden neuron activation. Activation y_j for each neuron j is a nonlinear transformation of the weighted sum of inputs:

$$y_j = f(a_{0j} + a_{1j}x) = \frac{1}{1 + e^{-(a_{0j} + a_{1j}x)}} \qquad (2)$$

The hidden neuron activations for the previous three cases of weight initialization are shown in Fig. 5 as a function of the input x. The Figure reveals an interesting effect. Although actual weights are not identical for the three cases, hidden neuron activations follow some identifiable patterns. For example, the first two cases that produced over-fitting, have a similar pattern of hidden neuron activation whereas the third case that did not produce over-fitting has a unique pattern with only partial similarity to the previous two cases [1].

In the first two cases of weight initialization, early stopping was required to prevent over-fitting. For these, all four activation functions are strongly active in the range of inputs, as indicated by their slopes at the boundary point where the activation y is equal to 0.5, and an external measure is required to suppress their activity. In the third case with no over-fitting however, two of the neurons (solid lines) have low activity and these do not contribute greatly to the output.

Careful observation of Fig. 5 reveals that in the first two cases, there appear to be two sets of neurons each consisting of two neurons of similar activity that could

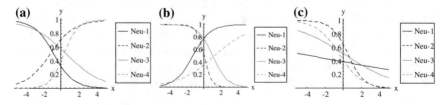

Fig. 5 Activation functions for the 4 hidden neurons for the three weight initializations for random sample 1: **a** Init-1, **b** Init-2 and **c** Init-3 (no-over-fitting)

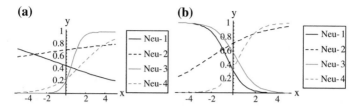

Fig. 6 Hidden neuron function for random samples: **a** sample 2, **b** sample 3 (no over-fitting)

intuitively be interpreted as evidence for redundancy. However, in the third case where there was no over-fitting, there still seem to be two sets of neurons with similar activation but the situation is not so clear. For the case of random sampling, hidden neuron activation patterns for the two networks created from random samples 2 and 3 are shown in Fig. 6. (The patterns for random sample 1 are shown in Fig. 5a).

The activation pattern for sample 3 in Fig. 6b did not produce over-fitting; however, interestingly, this pattern is similar to that for sample 1 which resulted in over-fitting (Fig. 5a, b). Furthermore, now there is a totally new activation pattern for random sample 2 (Fig. 6a) with activation of 3 neurons having a positive slope and one having a negative slope.

Thus, for the six trials (3 weight initializations and 3 random samples), there are 3 distinct activation patterns. In two trial cases, networks did not over-fit but their activation patterns are dissimilar. As for the search for internally consistent weights or activations, there is still ambiguity.

7 Consistency of Activation of Persistent Neurons

It is known that the target data generator used in this study (Fig. 1) requires 2 hidden neurons. The experiment so far indicates that in some cases, there is a persistent 2-neuron structure, such as that in Figs. 5a, b and 6b. In order to confirm if the ones that are persistent point to the optimum, a network with two hidden neurons was tested on the random data sample 1 and with 3 random weight initializations. None of the networks over-fitted even after full training, as expected, since this is the optimum number of neurons. The network produces a closer agreement with the target function and data, similar to that shown in Fig. 2a. However, the hidden neuron functions for the 3 trials produced 3 quite different patterns as shown in Fig. 7.

Figure 7 displays the main features of all previous networks, e.g., activation functions can be all positive, all negative or a combination of positive and negative slopes- and still produce the optimum final output. The figure shows that the optimum network does not have a unique pattern of activation of neurons and still produces the correct output. In order to test the optimality of the two-neuron

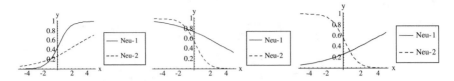

Fig. 7 Hidden neuron activations for the optimum 2-neuron network for 3 random weight initializations for random sample 1

network, another neuron was added and the network trained. This case resulted in a mild form of over-fitting and required early stopping to stop training at the optimum weights. Hidden neuron activation patterns were generally similar to Fig. 5a, b (i.e., 2 functions had negative slope and one had positive slope).

Since the results so far has not yet pointed to a structure that is internally consistent and robust, we next explore hidden neuron contribution to output generation.

8 Internal Consistency of Hidden Neuron Contribution to Output

Contribution of each neuron j to output generation is its weighted activation:

$$y_{weighted_j} = y_j b_j \tag{3}$$

where y_j is output of hidden neuron j and b_j is the corresponding weight linking neuron j with the output. Returning to our original 4-neuron network, these weighted activation patterns for the first three random weight initializations are presented in Fig. 8.

The plots in Fig. 8 reveal a pattern that is consistent. In each plot, there is one dominant weighted activation pattern with a negative slope. In Fig. 8b, c, the other

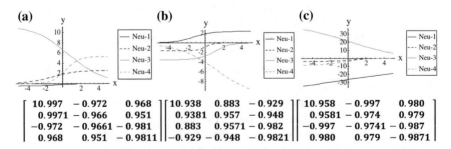

Fig. 8 Weighted hidden neuron activation and correlation matrices for the three random weight initializations for random sample 1: **a** Initialization 1, **b** Initialization 2, and **c** Initialization 3

three patterns are almost parallel to each other. The patterns that are parallel indicate redundancy. In Fig. 8a also, this pattern is obvious but to a lesser extent. These observations for the weighted hidden neuron activation are quite convincing and persistent compared to those observed for the hidden neuron activation in Figs. 5, 6 and 7.

The activation patterns that are parallel can be identified by their strong correlation. This way, it should be possible to eliminate redundant neurons. The correlation matrix for the 3 sets of weighted hidden neuron activation plots are presented below each figure in Fig. 8. The correlation matrices for Fig. 8a, c clearly indicate that neurons 1, 2 and 4 are very highly correlated and all these are inversely correlated with neuron 3 activity. In matrix for Fig. 8b, neurons 1, 2, and 3 are highly correlated and they are inversely correlated with neuron 4 activity. The fact that the correlation coefficients are strong indicate consistency and resilience of the activation patterns. Highly correlated patterns can be replaced by a single representative, leaving two neurons for the optimum network as required. Furthermore, correlation confirms that the optimum network has one neuron with positive weighted activation and another with negative activation for all 3 weight initializations.

The weighted activation patterns and correlations for the network trained with different random samples (samples 2 and 3) are shown in Fig. 9. Results for sample 1 is in Fig. 8a.

Analogous to Fig. 8, highly correlated structure of weighted hidden neuron activation patterns for random samples 2 and 3 is evidenced in Fig. 9 where the left image indicates that neurons 1 and 3 as well as neurons 2 and 4 are highly correlated in an opposite sense. The right image indicates that neurons 1, 2, and 4 are highly correlated with each other and inversely correlated with neuron 3. By replacing the correlated neurons with a representative, an optimum 2-neuron structure is obtained for both these cases.

The above experiment was conducted for 5- and 3-neuron networks with early stopping and results are presented in Fig. 10.

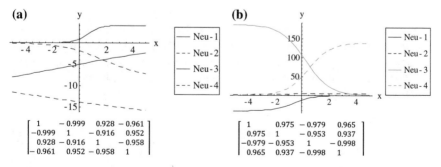

Fig. 9 Weighted hidden neuron activation for the random data samples 2 and 3 and corresponding correlation matrices

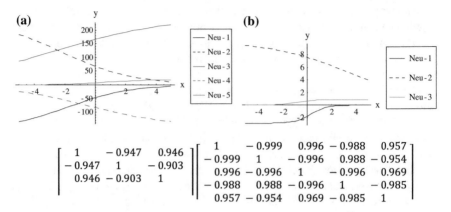

Fig. 10 Weighted hidden neuron activation of a 3-neuron **a** and 5-neuron **b** networks and corresponding correlation matrices

Figure 10 illustrate convincingly that the redundant neurons can be identified by their high correlation. By removing redundant neurons, both networks are left with 2 (optimum number) of neurons.

In summary, the network attains a consistent pattern of weighted hidden neuron activation for this data regardless of the number of hidden neurons, initial weights and random data samples. By replacing highly correlated neurons with similar sign (+ or −) with a single representative, the optimum structure for this example can be obtained with certainty. In what follows, the robustness of weights and hidden neuron activation patterns is further investigated by examining the results obtained from regularization.

9 Internal Structure of Weights Obtained from Regularization

Regularization is another method used to reduce the complexity of a network directly by penalizing excessive weight growth [10]. The amount of regularization is controlled by the parameter δ shown in Eq. 4 where MSE is the mean square error and w_j is a weight in the total set of m weights in the network. In regularization, W is minimized during training.

$$W = MSE + \delta \sum_{j=1}^{m} w_j^2 \tag{4}$$

The user must find the optimum regularization parameter through trial and error. Too large a parameter exerts too much control on weight growth and too small a value allows too much growth. In this investigation, the original four-neuron

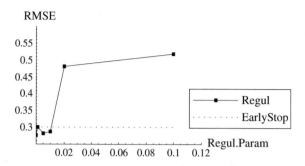

Fig. 11 Comparison of accuracy of networks with weights optimized from regularization and early stopping [1]

network in Fig. 1b with the first set of random initial weights used earlier (Init-1 in Fig. 3) was trained on random sample 1 shown in Fig. 1a. Three regularization parameters, 0.0001, 0.001 and 0.02, were tested. For the first two values, weights initially grew and then dropped to optimum values and from then on they remained constant. For these, the network followed the target pattern closely but the smallest regularization parameter of 0.0001 resulted in the smallest validation MSE of 0.276 which is smaller than that obtained from early stopping (MSE = 0.318). With the parameter value of 0.001, MSE is similar to that obtained from early stopping. For the largest chosen parameter of 0.02, however, weights are controlled too much and therefore, they are not allowed to reach optimum values. In this case, the network performance was very poor.

The experiment was continued further and Fig. 11 shows RMSE for various values of regularization parameter. The horizontal line indicates the RMSE obtained from early stopping. The figure indicates that for this example, regularization can produce networks with greater accuracy than early stopping. However, considering the trial and error nature of regularization, early stopping is efficient. Furthermore, Fig. 11 highlights the sensitivity of RMSE to regularization parameter beyond a certain value.

9.1 Consistency of Weighted Hidden Neuron Activation of Networks Obtained from Regularization

The hidden neuron activations for the two regularization parameters (0.0001 and 0.001) that produced smaller than or similar validation MSE to early stopping are plotted in Fig. 12. They illustrate again that the correlation structure as well as the 2-neuron optimum structure identified in the previous investigations remain persistent.

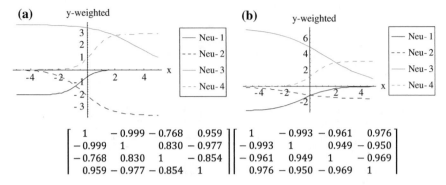

Fig. 12 Weighted hidden neuron activations for regularization parameters 0.0001 **a** and 0.001 **b** and corresponding correlation matrices

10   Identification of Correlated Weighted Hidden Neuron Activations Using Self Organizing Maps

Previous sections demonstrated that the redundant neurons can be identified by the correlation of their weighted neuron activations. It is useful, if these can be identified automatically. In this section, SOM is used to efficiently cluster similar activation patterns. Such approach would be especially useful for larger networks. An input vector to SOM contains weighted hidden neuron activation ($b_j y_j$) for each neuron over the input data. Input vectors were normalized to 0 mean and unit standard deviation and the correlation distance was used as the distance measure [1]. The normalized activation patterns for each network presented so far were mapped to a 2-dimensional SOM [23] (4 neuron map) and the most efficient number of clusters was determined by the Ward clustering [24] of SOM neurons. Ward is an efficient statistical clustering method suitable and effective for relatively small datasets. Figure 13 presents the results for only two networks, one with 4 neurons that was depicted as random weight initialization 2 in Fig. 8b and the other with 5 neurons presented in Fig. 10b. Maps were trained very quickly with default parameter settings of the program [21] indicating the efficiency of clustering highly correlated patterns.

The top two images in Fig. 13 are graphs of Ward likelihood index (vertical axis) against likely number of clusters. The higher the index, more likely that the corresponding number of clusters is the optimum. These images reveal that undoubtedly there are two clusters of activation patterns. The index for other possible cluster sizes is almost zero, which increases the confidence in the two-cluster structure. The bottom images of Fig. 13 show these two clusters on the corresponding SOMs. Here, the two clusters are depicted by brown and black colors, respectively. For example, in the bottom left image, one clusters has 3 correlated patterns distributed in two neurons and the other cluster has one pattern, whereas, in the bottom right image, depicting a 5 neuron network, 2 patterns are grouped into the cluster depicted by the

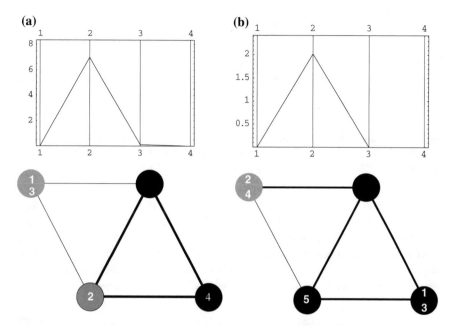

Fig. 13 Self organizing map/Ward clustering of correlated weighted hidden neuron activation patterns. **a** 4-neuron network and **b** 5 neuron network

top left neuron and the other three patterns are spread among the bottom two neurons that form the second cluster. The optimum network structure is obtained by selecting one representative from each cluster and then retraining the network. Similar two-cluster neuron maps were found for all the networks presented previously revealing that, for this example, networks maintain consistent patterns in their internal structure at the level of weighted hidden neuron activation feeding the output neuron. Importantly, the network structure is optimized in one iteration of clustering correlated hidden neuron activation patterns.

11 Ability of the Network to Capture Intrinsic Characteristics of the Data Generating Process

A good model not only should follow the target data but also must capture the underlying characteristics of the data generating process. These can be represented by first and higher order derivative of the generating process. When a network model converges towards the target function, all the derivatives of the network must also converge towards the derivatives of the underlying target function [25, 26]. A new network with two hidden neurons (7 weights in total) was trained and its weighted hidden neuron activation patterns are shown in Fig. 14a that highlights the features already described. The network model is given in Eq. 5 and its first and

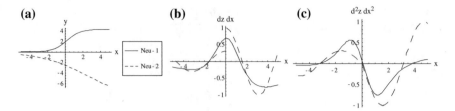

Fig. 14 Optimum network characteristics: **a** weighted hidden neuron activations **b** and **c** gradient and second derivative (*solid line*), respectively, of network function superimposed on the corresponding values for the target data generator function (*dashed line*)

second derivatives are superimposed on those of the target function in Fig. 14b, c. They show that, notwithstanding the small sample size and large noise, both derivatives follow the corresponding trends in the target function reasonably well indicating that the trained network is a true representation of the data generating process and can be used for gaining further insight into the process such as sensitivities and errors as well as for further ascertaining the robustness of the weights.

$$z = 0.91 + \frac{4.34}{1 + 1.25e^{-1.27x}} - \frac{9.23}{1 + 2.65e^{-0.37x}} \tag{5}$$

12 Comparison of Correlation Approach with Other Network Pruning Methods

Since it is clear in all the previous experiments that redundant neurons in too flexible networks form highly correlated weighted hidden neuron activation patterns, it is interesting to find out if other pruning methods identify the neurons with the most consistent patterns and prune the redundant ones. A notable feature of the commonly used pruning methods is that they optimize the structure iteratively and require a certain amount of heuristic judgment. In what follows, two pruning methods, Optimal Brain Damage (OBD) [5–7] and Variance Nullity measure (VN) [8] are implemented and compared with the proposed correlation method.

12.1 Network Pruning with Optimum Brain Damage (OBD)

In OBD [5–7], weights that are not important for input-output mapping are found and removed. This is based on a saliency measure of a weight, as given in Eq. 6, that is an indication of the cost of setting it to zero. The larger the s_i, the greater the influence of w_i on error. It is computed from the Hessian (H) which is the matrix containing the second derivative of error with respect to a pair of weights in the network. This matrix is used in error minimization and weight update by the Levenberg Marquardt method [1]. Since Hessian is nonlocal and computationally intensive, an approximation is used by utilizing only the diagonal entries (H_{ii}) of the Hessian matrix.

$$s_i = H_{ii}\, w_i^2 / 2 \tag{6}$$

The original 4 neuron network (Fig. 1b) with the first set of initial weights (Init-1) that was trained using a regularization parameter of 0.0001 on the random sample 1 (Fig. 1a) was pruned using saliency measure of the weights. The network was pruned in stages. In the first stage, 5 (or 40 %) of the 13 weights were pruned and the reduced network that retained neurons 2, 3, and 4 was re-trained. The network was further subjected to pruning in the next stage and 2 more weights were removed resulting in a total removal of 7 or (54 %) of the weights from the original network. What remained were neurons 2 and 3 with bias on neuron 2 eliminated leaving 6 weights in the optimum network. The 2 weighted hidden neuron activations for the retrained network are plotted in Fig. 15 by a solid line and a dashed line. (The other set will be discussed shortly). These resemble those of neuron 2 and 3 of the full network in Fig. 8a indicating that the OBD has identified and removed the redundant neurons.

The network performs similarly to that shown in Fig. 2a. Any further pruning resulted in severe loss of accuracy and therefore, the above network was the optimum network obtained from OBD. The output z of the pruned network is [1]

$$z = -5.36 + \frac{2.63}{1 + e^{-1.84x}} + \frac{5.86}{1 + 0.345\, e^{0.399x}} \tag{7}$$

Equations 5 and 7 are not identical as the network obtained from the proposed correlation method has all 7 weights associated with the hidden and output neurons whereas the one from OBD has only 6 weights. This also reveals that the network can still have redundant bias weights. If a set of weights that are invariant is desired, these redundant weights can be pruned with potentially one extra weight pruning step applied to the trained network with the optimum number of neurons.

Fig. 15 Weighted hidden neuron activation patterns for networks pruned by OBD and Variance nullity

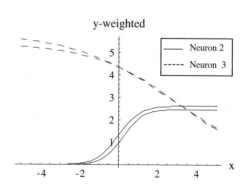

12.2 Network Pruning Based on Variance of Network Sensitivity

In this section, the same original network used in the previous section is pruned using a very different method- variance analysis of the sensitivity of the output of the network to perturbation of weights- as proposed by Engelbrecht [8]. Variance nullity (VN) measure tests whether the variance of the sensitivity of network output over all input-output patterns is significantly different from zero. It is based on a hypothesis test using χ^2 (chi square) distribution to test statistically if the parameter should be pruned. If the sensitivity with respect to a parameter is denoted by S_θ, then the variance of the sensitivity for N patterns can be expressed as

$$\sigma^2_{S_\theta} = \frac{\sum_{i=1}^{N} (S_{\theta i} - \mu_{S_\theta})^2}{N} \tag{8}$$

where $\mu_{S\theta}$ is the mean sensitivity. This is used to obtain an expression for a variance nullity measure $\gamma_{s\theta}$, that indicates the relevance of a parameter as

$$\gamma_{s\theta} = \frac{(N-1)\sigma^2_{s\theta}}{\sigma^2_0} \tag{9}$$

where σ^2_0 is a value close to zero. The hypothesis that the variance is close to zero is tested for each parameter θ with the null and alternative hypotheses of

$$
\begin{aligned}
H_0 &: \sigma^2_{s\theta} = \sigma^2_0 \\
H_1 &: \sigma^2_{s\theta} < \sigma^2_0
\end{aligned} \tag{10}
$$

Under the null hypothesis, $\gamma_{s\theta}$ follows a χ^2 (N-1) where N-1 is the degree of freedom. A parameter is removed if the alternative hypothesis is accepted with the condition $\gamma_{s\theta} \leq \gamma_c$ where γ_c is a critical χ^2 value obtained from $\gamma_c = \chi^2_{N-1,(1-\alpha/2)}$. The α is the level of significance which specifies the acceptable level of incorrectly rejecting null hypothesis. Smaller values result in a stricter pruning algorithm.

The success of the algorithm depends on the value chosen for σ^2_0. If it is too small, no parameter is pruned. If too large, even relevant parameters will be pruned. Thus, some trial and error is necessary. The method was applied to the original 4-neuron network described in this chapter with an initial value of 0.01 for σ^2_0 at 0.05 significance level. Only one weight was targeted for pruning. When it was increased to 0.1, all six weights associated with neurons 1 and 4 became targets for pruning leaving those for neurons 2 and 3, that are the required neurons, with all 7 corresponding weights. This outcome, in terms of exactly which neurons remain, is similar to that obtained from OBD in the previous section and in both these cases, variance nullity and OBD, considerable subjective judgment is required in setting

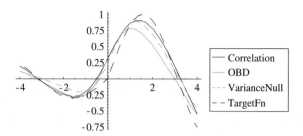

Fig. 16 The pruned network based on correlated activity patterns plotted with networks pruned by OBD and variance nullity and the target function

up the parameters. The weighted hidden neuron activations for the retrained network obtained from variance nullity method are plotted in Fig. 15b indicating that the two methods produce almost identical activation patterns. The corresponding plot for the network obtained from the correlation method was shown in Fig. 14a. Here the shape of the patterns are the same as those for VN and OBD based networks; however, since the initial weights are different in this case, one activation pattern has shifted vertically but this effect is offset by the larger bias weight (0.91 in Eq. 5) on the output neuron when the final output is produced. The important point is that the slopes and trends of the patterns are identical for the 3 methods.

The final network output from variance nullity based weight pruning is

$$z = -5.09 + \frac{2.48}{1 + 1.57e^{-1.99x}} + \frac{5.38}{1 + 0.235\,e^{0.465x}} \tag{11}$$

which is very similar to that obtained from OBD (Eq. 7). Reason why Eqs. 7 and 11 are similar is that both of them were retrained with the same set of original initial weights that remained on the network after pruning. In the network obtained from the correlation method, new initial weights were used as only one representative from each cluster was used. The three network outputs are superimposed on the target function in Fig. 16.

Figure 16 reveals that the performance of the two full networks with 7 weights obtained from the proposed method and variance nullity method is closer to the target function than that with 6 weights obtained from OBD. The validation RMSE from the proposed correlation method, variance nullity and OBD were, 0.272, 0.272 and 0.285, respectively. However, the proposed method is more efficient and does not require heuristic judgment as in OBD and Variance Nullity. This point applies to other past approaches for structure optimization, such as singular value decomposition as well. The validation RMSE for the correlation based network (0.272) is slightly smaller than that for the best full networks obtained from regularization (0.276) and early stopping (0.318).

13 Robustness and Uncertainty of Networks

13.1 Robustness

The fact that the optimum networks do not match the target pattern perfectly is due to the large noise deliberately added to the data generated from the true function. The noise allows a number of possible outputs that follow the target function closely. Smaller the noise, the tighter the band around the target function within which output of various optimum networks can lie. In order to test this interval, optimum weights were perturbed by randomly adding noise from a Gaussian distribution with 0 mean and standard deviations of 0.01, 0.05, 0.1, to 0.2. Thus there were 4 sets of weights. The network output for these 4 sets of weights showed that the weights are robust against variations up to 0.1 standard deviation which is equivalent to ±30 % random perturbation of the weights. The 0.2 standard deviation representing ±60 % random perturbations was detrimental to the network performance (see p. 238 of [1]).

13.2 Confidence Interval for Weights

Since the weights are robust against perturbation of at least up to ±30 %, confidence intervals for weights were developed for a noise level of ±15 % (noise standard deviation of 0.05). Ten sets of weights, each representing a network, were drawn by superimposing noise on the optimum weights of the network obtained from the proposed approach based on correlation of weighted hidden neuron activation. Confidence intervals were constructed using methods of statistical inference based on sampling distribution as:

$$(1 - \alpha)CI = \overline{w} \pm t_{\alpha,n-1} \frac{s_w}{\sqrt{n}} \tag{12}$$

where $\overline{w}$ is the mean value of a weight, s_w is the standard deviation of that weight, and n is the sample size. In this case, we have 10 observations. The $t_{\alpha,n-1}$ is the t-value from the t-distribution for (1-α) confidence level and degree of freedom (dof) of n-1. The 95 %confidence intervals were constructed for each of the 7 weights and the resulting 95 % Confidence Intervals (CIs) for the network performance are plotted in Fig. 17a with the two solid lines depicting upper and lower limits. In this figure, the smaller dashed line represents the mean and larger dashed line is the target function.

In order to assess all the models developed so far, network outputs from 4 random weight initializations using the proposed method involving correlation of weighted hidden neuron activations were superimposed along with outputs from OBD and variance nullity (6 curves altogether) on the above confidence interval

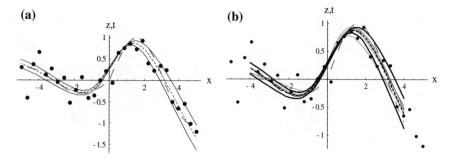

Fig. 17 Confidence Interval (CI) bands and comparison of optimum networks: **a** Mean and 95 % confidence interval limits for the correlation based network and **b** optimum network performance from the 3 methods superimposed on the CIs

plots containing the original data. The results are shown in Fig. 17b that illustrates that that all models are within the confidence limits for the entire range of the data and covers most of the target function (larger dashed line). The target function is the combination of two functions (Eq. 1 and also see Fig. 14b) and all networks experience difficulty in the region near the axes origin where the two functions merge.

13.3   2-D Function Approximation

In this section, the correlation of weighted hidden neuron activation is tested on a two-dimensional problem. The function from which 120 data vectors were generated is shown in Fig. 18a. The network was trained with 15 hidden neurons with sigmoid functions and linear output function. Training was done with Levenberg Marquardt method with early stopping to optimise the network. The optimum network output is shown in Fig. 18b. After training, weighted hidden neuron activations were analysed and the correlation matrix is given in Fig. 18c.

The weighted hidden neuron activations were projected onto a 16-neuron SOM and trained SOM weights were clustered with Ward clustering. Figure 19a shows the Ward index plot which clearly indicates 7 clusters as the optimum. The SOM clustered into 7 groups are shown in Fig. 19b.

A new network was trained with 7 hidden neurons and results identical to Fig. 18b was found confirming that the optimum number of hidden neurons is 7. In order to test further, individual networks were trained with hidden neuron numbers increasing from 1 to 10 with a number of weight initializations. Root Mean Square Error (RMSE) plot for these cases are shown in Fig. 20 which clearly indicates that the 7 neurons do provide the minimum error.

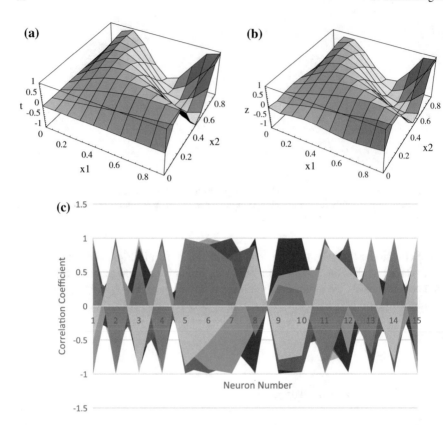

Fig. 18 a Two-dimensional target function, **b** and network prediction and **c** correlation of weighted activation of the 15 hidden neurons in the network (colours visually display the general character and strength of correlations across neurons)

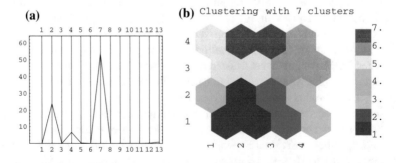

Fig. 19 a Ward index against number of clusters; **b** SOM clustered into optimum number of 7 clusters: *yellow* (neurons 1, 6, 7), *red* (5, 14), *brown* (3, 9, 10), *cyan* (4, 13), *pale blue* (2, 11), *dark blue* (8, 15) and *green* (12)

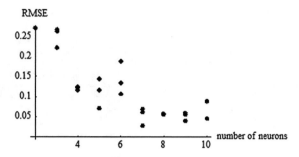

Fig. 20 RMSE for increasing number of hidden neurons trained for a number of random weight initialisations

14 Optimising a Network for Practical Real-Life Problems

14.1 Breast Cancer Classification

Correlation of weighted hidden neuron networks were also tested on a real world problem of breast cancer classification. In this study, a total of 99: 46 malignant and 53 benign, samples obtained from The Digital Database for Breast Ultrasound Image (DDBUI) was used to develop a feed forward network for breast cancer classification. Using a preprocessing and feature selection approach, 6 features were selected with the ability to discriminate between cancer and healthy cases. These were: depth-width ratio of the dense mass and its shape and margin, blood flow, age and a newly identified effective feature called central regularity degree (CRD) that explicitly incorporates irregularity of the mass that has been known to be indicative of malignancy [27].

Networks were developed with sigmoid hidden and output neurons on 70 % and 30 % training and testing data, respectively, and trained with Levenberg Marquardt method and early stopping. First a network with a large number of hidden neurons was developed and then the number of neurons were decreased gradually, every time comparing results with previous results. It turned out that 15 hidden neurons provide optimum results: Training (100 % Sensitivity, Specificity and Accuracy) and Testing (100, 90.9, 95.4 %, respectively, for the above measure). Then we tested the clustering of weighted hidden neuron activation approach on the best network using SOM topology with 20 neurons. The trained SOM results are shown in Fig. 21 where several individual SOM neurons represent a number hidden neurons as indicated by Fig. 21a—hidden neuron groups (4, 7, 10), (6, 12) and (14, 15) each share an SOM neuron. Other neurons are each represented by an individual SOM neuron. The U-matrix in Fig. 21b shows further similarity among the nodes. For example, neurons (8, 1, 9) were found close to each other (blue colour on the top right corner of the map) and were considered as one cluster by Ward clustering that divided the SOM into 9 clusters suggesting that 9 neurons should adequately model the data.

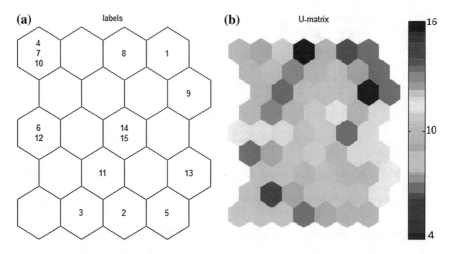

Fig. 21 Twenty neuron SOM representing hidden neuron activation patterns. **a** Distribution of 15 neurons over the map; **b** U-matrix for the 15 hidden neurons (*blue colour* indicates similarity/closeness)

To test this finding, a 9 neuron network was trained and tested on the same data sets as before and the network accuracy (95.4 %), sensitivity (100 %) and specificity (90.9 %) were found to be the same as those for the 15 neuron network. Thus the SOM/Ward reduced the number of neurons without any comprise on the network performance confirming that the redundant weighted hidden neuron activations do form correlated clusters and the number of these clusters indicate the required number of neurons.

14.2 River Flow Forecasting

The efficacy of the correlation method was tested in another complex real world problem of river flow forecasting for a multi-basin river system in New Zealand and the detail results were presented in [28]. Forecasting river flows are very complicated due to the effect of daily, monthly, seasonal and yearly variability of the contributing factors of rainfall and temperature etc. The inputs, selected from an extensive feature selection process, were: previous month's flow, current temperature and a river basin geometric factor and the output was current month's flow. The data were divided into training (70 %) and calibration (30 %) with 1079 and 269 data points, respectively, and validation set with 63 data points. In the original study, it was found that 70 hidden neurons (logistic activation) and one linear output neuron provided the optimum results.

To test this result, a network with 100 hidden neurons with logistic function was trained and the weighted hidden activations of 100 neurons were projected onto a 100 neuron square SOM. Results showed that the 100 patterns were projected onto 59 SOM neurons and the Ward method further clustered these neurons indicating that 2 and 3 neurons provided the highest Ward Likelihood Index, which is much smaller than the original optimum of 70 neurons found by trial and error. A 2-neuron network was trained and the results (training $R^2 = 0.88$; Validation $R^2 = 0.71$) were similar to original results [28]. Results for 59 and 70 neurons were similar.

15   Summary and Conclusions

This chapter presented the results from a systematic investigation of the internal consistency and robustness of feed forward (multi-layer perception) networks. It demonstrated that weighted hidden neuron activations feeding the output neuron display meaningful and consistent patterns that are highly correlated for redundant neurons. By representing each correlated group with one neuron, the optimum structure of the network is obtained. Furthermore, the chapter illustrated that the correlated activation patterns can be mapped on to a self organizing map (SOM) where Ward clustering convincingly revealed the required number of clusters. The chapter also compared the proposed method with two pruning approaches from literature: Optimal Brian Damage (OBD) and Variance Nullity (VN) and demonstrated the efficacy of the proposed correlation based method. A clear advantage of the correlation method is that it does not require heuristic judgment in selecting parameters for optimizing the network as in other methods. Another advantage is that network is optimized in one step of clustering correlated weighted hidden neuron activation patterns thus minimizing the time and effort spent on structure optimization. Yet another advantage is that as the redundant neurons are highly correlated, they cluster easily on the SOM with default network learning parameters and Ward clustering automatically produces the required optimum number of neurons. This chapter used a one-dimensional problem to allow the presentation of a thorough assessment of various modeling issues deemed important and demonstrated that the insights gained are relevant to larger problems as well by successfully applying the concept to multi-dimensional and complex real world problems. These demonstrated that the approach is robust to initial weights, random samplings and for networks with logistic activation function and either linear or logistic output neuron activation function. In future, it will be useful to test the validity of the method for other activation functions and networks.

Appendix: Algorithm for Optimising Hidden Layer of MLP Based on SOM/Ward Clustering of Correlated Weighted Hidden Neuron Outputs

I. Train an MLP with a relatively larger number of hidden neurons

1. For input vector X, the weighted input u_j and output y_j of hidden neuron j are:

$$u_j = a_{0j} + \sum_{i=1}^{n} a_{ij} x_i$$

$$y_j = f(u_j)$$

where a_{oj} is bias weight and a_{ij} are input-hidden neuron weights. f is transfer function.

2. The net input v_k and output z_k of output neuron k are:

$$v_k = b_{0k} + \sum_{j=1}^{m} b_{jk} y_j$$

$$z_k = f(v_k)$$

where b_{ok} is bias weight and b_{jk} are hidden-output weights.

3. Mean Square error (MSE) for the whole data set is:

$$MSE = \frac{1}{2N} \left[\sum_{i=1}^{N} (t_i - z_i)^2 \right]$$

where t is target and N is the sample size.

4. Weights are updated using a chosen method of least square error minimisation, such as Levenberg Marquardt method:

$$w_m = w_{m-1} - \varepsilon R d_m$$

where d_m is sum of error gradient of weight w for epoch m, R is inverse of curvature, and ε is learning rate.

5. Repeat the process 1 to 4 until minimum MSE is reached using training, calibration (testing) and validation data sets.

II. SOM clustering of weighted hidden neuron outputs

Inputs to SOM
An input vector X_j into SOM is:

$$X_j = y_j b_j;$$

where y_j is output of hidden neuron j and b_j is its weight to output neuron in MLP. Length n of the vector X_j is equal to the number of samples in the original dataset.

Normalise X_j to unit length

SOM training

1. *Projecting weighted output of hidden neurons onto a Self Organising Map*:

$$u_j = \sum_{i=1}^{n} w_{ij} x_i$$

where u_j is output of SOM neuron j and w_{ij} is its weight with input component x_i

2. *Winner selection*: Select winner neuron based on the minimum correlation distance between an input vector and SOM neuron weight vectors (same as Euclidean distance for normalised input vectors)

$$d_j = \mathbf{x} - \mathbf{w}_j$$
$$\sqrt{\sum_{i}^{n} (x_i - w_{ij})^2}$$

3. *Update of weights of winner and neighbours at iteration t*:
Select neighbourhood function $NS(d, t)$ (such as Gaussian) and learning rate function $\beta(t)$ (such as exponential or linear) where d is distance from winner to a neighbour neuron and t is iteration.

$$\mathbf{w}_j(t) = \mathbf{w}_j(t-1) + \beta(t)NS(d,t)[\mathbf{x}(t) - \mathbf{w}_j(t-1)]$$

4. *Repeat the process until mean distance D between weights W_i and inputs x_n is minimum.*

$$D = \sum_{i=1}^{k} \sum_{n \in c_i} (\mathbf{x}_n - \mathbf{w}_i)^2$$

where k is number of SOM neurons and c_i is the cluster of inputs represented by neuron i

III. **Clustering of SOM neurons**

Ward method minimizes the within group sum of squares distance as a result of joining two possible (hypothetical) clusters. The within group sum of squares is the sum of square distance between all objects in the cluster and its centroid. Two clusters that produce the least sum of square distance are merged in each step of clustering. This distance measure is called the Ward distance (d_{ward}) and is expressed as:

$$d_{wand} = \frac{(n_r^* n_s)}{(n_r + n_s)} \|x_r - x_s\|^2$$

where x_r and x_s are the centre of gravity of two clusters. n_r and n_s are the number of data points in the two clusters.

The centre of gravity of the two merged clusters $x_{r(new)}$ is calculated as:

$$x_{r(new)} = \frac{1}{n_r + n_s} \left(n_r^* x_r + n_s^* x_s \right)$$

The likelihood of various numbers of clusters is determined by *WardIndex* as:

$$WardIndex = \frac{1}{NC} \left(\frac{d_t - d_{t-1}}{d_{t-1} - d_{t-2}} \right) = \frac{1}{NC} \left(\frac{\Delta d_t}{\Delta d_{t-1}} \right)$$

where d_t is the distance between centres of two clusters to be merged at current step and d_{t-1} and d_{t-2} are such distances in the previous two steps. NC is the number of clusters left.

The numbers of clusters with the highest *WardIndex* is selected as the optimum.

IV. Optimum number of hidden neurons in MLP

The optimum number of hidden neurons in the original MLP is equal to this optimum number of clusters on the SOM.

Train an MLP with the above selected optimum number of hidden neurons.

References

1. S. Samarasinghe, *Neural Networks for Applied Sciences and Engineering-From Fundamentals to Complex Pattern Recognition* (CRC Press, 2006)
2. C. Bishop, *Neural Networks for Pattern Recognition* (Clarendon Press, Oxford, UK, 1996)
3. S. Haykin, *Neural Networks: A comprehensive Foundation*, 2nd edn. (Prentice Hall Inc, New Jersey, USA, 1999)
4. R. Reed, Pruning algorithms-A survey. IEEE Trans. Neural Networks **4**, 740–747 (1993)
5. Y. Le Cun, J.S. Denker, S.A. Solla, Optimal brain damage, in *Advances in Neural Information Processing (2)*, ed. by D.S. Touretzky (1990), pp. 598–605
6. B. Hassibi, D.G. Stork, G.J. Wolff, Optimal brain surgeon and general network pruning. *IEEE International Conference on Neural Networks*, vol. 1, (San Francisco, 1992), pp. 293–298
7. B. Hassibi, D.G. Stork, Second-order derivatives for network pruning: Optimal brain surgeon, in *Advances in Neural Information Processing Systems*, vol. 5, ed. by C. Lee Giles, S. J. Hanson, J.D. Cowan, (1993), pp. 164–171
8. A.P. Engelbrecht, A new pruning heuristic based on variance analysis of sensitivity information. IEEE Trans. Neural Networks **12**(6), 1386–1399 (2001)

9. K. Hagiwara, Regularization learning, early stopping and biased estimator. Neurocomputing **48**, 937–955 (2002)
10. M. Hagiwara, Removal of hidden units and weights for backpropagation networks. Proc. Int. Joint Conf. Neural Networks **1**, 351–354 (1993)
11. F. Aires, Neural network uncertainty assessment using Bayesian statistics with application to remote sensing: 1. Network weights. J. Geophys. Res. **109**, D10303 (2004). doi:10.1029/2003JD004173
12. F. Aires, Neural network uncertainty assessment using Bayesian statistics with application to remote sensing: 2. Output Error. J. Geophys. Res. **109**, D10304 (2004). doi:10.1029/2003JD004174
13. F. Aires, Neural network uncertainty assessment using Bayesian statistics with application to remote sensing: 3. Network Jacobians. J. Geophys. Res. **109**, D10305 (2004). doi:10.1029/2003JD004175
14. K. Warne, G. Prasad, S. Rezvani, L. Maguire, Statistical computational intelligence techniques for inferential model development: A comparative evaluation and novel proposition for fusion. Eng. Appl. Artif. Intell. **17**, 871–885 (2004)
15. I. Rivals, L. Personnaz, Construction of Confidence Intervals for neural networks based on least squares estimation. Neural Networks **13**, 463–484 (2000)
16. E.J. Teoh, K.C. Tan, C. Xiang, Estimating the number of hidden neurons in a feed forward network using the singular value decomposition IEEE Trans. Neural Networks **17**(6), (2006)
17. C. Xian, S.Q. Ding, T.H. Lee, Geometrical interpretation and architecture selection of MLP, IEEE Trans. Neural Networks **16**(1), (2005)
18. P.A. Castillo, J. Carpio, J.J. Merelo, V. Rivas, G. Romero, A. Prieto, Evolving multilayer perceptrons. Neural Process. Lett. **12**(2), 115–127 (2000)
19. X. Yao, Evolutionary artificial neural networks. Proc. IEEE **87**(9), 1423–1447 (1999)
20. S. Samarasinghe, Optimum Structure of Feed Forward Neural Networks by SOM Clustering of Neuron Activations. *Proceedings of the International Modelling and Simulation Congress (MODSM)* (2007)
21. Neural Networks for Mathematica, (Wolfram Research, Inc. USA, 2002)
22. J. Sietsma, R.J.F. Dow, Creating artificial neural networks that generalize. Neural Networks **4**(1), 67–77 (1991)
23. Machine learning framework for Mathematica. 2002 Uni software Plus. www.unisoftwareplus.com
24. J.H. Ward Jr, Hierarchical grouping to optimize an objective function. J. Am Stat. Assoc. **58**, 236–244 (1963)
25. K. Hornik, M. Stinchcombe, H. White, Universal approximation of an unknown mapping and its derivatives using multi-layer feedforard networks. Neural Networks **3**, 551–560 (1990)
26. A.R. Gallant, H. White, On learning the derivative of an unknown mapping with multilayer feedforward networks. Neural Networks **5**, 129–138 (1992)
27. A. Al-yousef, S. Samarasinghe, Ultrasound based computer aided diagnosis of breast cancer: Evaluation of a new feature of mass central regularity degree. *Proceedings of the International Modelling and Simulation Congress (MODSM)* (2011)
28. S. Samarasinghe, Hydrocomplexity: New Tools for Solving Wicked Water Problems Hydrocomplexité: Nouveaux outils pour solutionner des problèmes de l'eau complexes *(IAHS Publ.* 338) (2010)

Artificial Neural Networks as Models of Robustness in Development and Regeneration: Stability of Memory During Morphological Remodeling

Jennifer Hammelman, Daniel Lobo and Michael Levin

Abstract Artificial neural networks are both a well-established tool in machine learning and a mathematical model of distributed information processing. Developmental and regenerative biology is in desperate need of conceptual models to explain how some species retain memories despite drastic reorganization, remodeling, or regeneration of the brain. Here, we formalize a method of artificial neural network perturbation and quantitatively analyze memory persistence during different types of topology change. We introduce this system as a computational model of the complex information processing mechanisms that allow memories to persist during significant cellular and morphological turnover in the brain. We found that perturbations in artificial neural networks have a general negative effect on the preservation of memory, but that the removal of neurons with different firing patterns can effectively minimize this memory loss. The training algorithms employed and the difficulty of the pattern recognition problem tested are key factors determining the impact of perturbations. The results show that certain perturbations, such as neuron splitting and scaling, can achieve memory persistence by functional recovery of lost patterning information. The study of models integrating both growth and reduction, combined with distributed information processing is an essential first step for a computational theory of pattern formation, plasticity, and robustness.

J. Hammelman · M. Levin (✉)
Biology Department, School of Arts and Science, Tufts University,
200 Boston Avenue, Suite 4600, Medford, MA 02155, USA
e-mail: michael.levin@tufts.edu

D. Lobo
Department of Biological Sciences, University of Maryland, Baltimore County,
1000 Hilltop Circle, Baltimore, MD 21250, USA

© Springer International Publishing Switzerland 2016
S. Shanmuganathan and S. Samarasinghe (eds.), *Artificial Neural
Network Modelling*, Studies in Computational Intelligence 628,
DOI 10.1007/978-3-319-28495-8_3

1   Introduction

The animal brain is widely considered to be the material substrate of memory. It may thus be expected that maintenance of complex memories requires stability of brain structure. Remarkably, several studies in a number of animal species revealed that this is not the case [1]. Planarian flatworms are able to recall learned information after their entire brain has been amputated and regenerated (suggesting an imprinting process by the somatic tissues onto the newly developing brain) [2]. Insects that metamorphose into butterflies or beetles still remember information learned as larvae, despite the fact that this process involves an almost complete rewiring of their brain [3]. Mammals are not exempt from this, as illustrated by ground squirrels that retain memories after winter hibernation has drastically pruned and rebuilt large regions of the brain [4]. How can memories persist in the face of reorganization of the recording medium? Specifically, it is unknown how bio-electric networks (whether neural or non-neural) can stably store information despite cellular turnover and changes of connectivity required for regenerative repair and remodeling. The answer to this question would have huge implications not only for regenerative neuromedicine (e.g., cognitive consequences of stem cell implants into the adult brain), but also for the fundamental understanding of how propositional content is encoded in properties of living tissues. Despite the importance of this question, and on-going research into the molecular mechanisms of cellular turnover in the brain, no conceptual models have been developed to probe the robustness of information in remodeling cell networks.

Artificial neural networks are well-established quantitative models of cognition that show promise to study the patterning capabilities of a dynamic system that is robust to components that move, proliferate, and die. While the learning performance properties of ANNs have been widely studied [5, 6], much less attention has been paid to the consequences of topology change that would mimic cellular turnover in vivo. Thus, a quantitative study is needed for the understanding of memory properties of neural networks under dynamic rearrangements of topology. We sought to develop a proof-of-concept for analysis of models that subsume both, structural change (remodeling) and functional (cognitive) performance. Here, we introduce perturbation methods that mimic aspects of the dynamic cellular regeneration observed in vivo. Using feed-forward single hidden layer networks trained with different backpropagation methods on pattern-recognition learning tasks, we assayed performance (memory persistence) after node removal, addition, and connection-blocking perturbations. These data demonstrate how artificial neural networks can be used as models for cognitive robustness and plasticity in developmental and regenerative biology.

2 Development, Regeneration, and Artificial Neural Networks

2.1 Biological Relevance: Memory Is Retained During Remodeling

It was once a long-standing central dogma of neurobiology that the central nervous system was incapable of accommodating neuron growth and death. This theory has since been overturned [7–9], which opens up a new field of study to understand how individual neurons and their environment contribute to the overall plasticity and growth of the brain [10, 11]. Much evidence now suggests that biological neural networks in many organisms, including mammals, have the capability of pruning, growing, and altering connectivity [12, 13]. Some species of insects undergo significant remodeling of the central nervous system and brain during metamorphosis from the larval to adult form, yet have been shown to maintain stored memories. Drosophila and manduca larvae have an ability to remember aversive associative learning tasks pairing electric shock with odor into adulthood [3, 14]. Weevils and wasps demonstrate a preference for the odors of their larval environments, supporting a long-standing hypothesis in entomology known as the Hopkins 'host-selection principle' that beetles and other insects favor their larval environment in selection of their adult homes [15, 16]. Planarian flatworms are another animal that demonstrate persistent memory after undergoing extensive remodeling. These flatworms have an incredible regeneration capacity in that they can regrow a fully functioning central nervous system and brain after head amputation [17]. In studies of memory persistence after head regeneration, it was discovered that flatworms trained on learning tasks before amputation require significantly less trials to re-learn the same task (the "savings" paradigm) than amputated but previously untrained animals [2, 18].

While amphibians, zebrafish, and planaria are capable of complete central nervous system regeneration, it appears that most mammals and birds are more limited in neurogenesis to specific regions of the brain [12]. Human neuronal progenitors have already proven to be a promising candidate as a treatment of traumatic injury, shown to successfully integrate and aid functional recovery in mice [19, 20]. Conversely, elevated neurogenesis in the dentate gyrus region of the hippocampus is a cause of forgetfulness in adulthood and infancy [21]. Neuronal death is also implicated as a primary cause of neurodegenerative diseases, especially diseases like amyotrophic lateral sclerosis, where motor neurons weaken and die slowly over time, and in traumatic brain injury, where delayed neuron death occurs in selective cell regions [22]. Yet cells also die during initial neurogenesis in the development of the nervous system as a pruning mechanism for neurons that are poorly wired or functionally redundant [23]. Understanding and leveraging the neural plasticity of animals capable of full brain repair is the first step for finding a treatment of traumatic brain injury: replacing or inducing formation of lost neurons and allowing the network to dictate their differentiation and functional response, which will contribute to innovative solutions for treatment of traumatic brain injury and neurological disorders [24].

Despite the rapid move toward stem cell therapy for degenerative disease and brain damage, it is still completely unknown what the cognitive consequences will be for an adult patient with decades of memories when the brain is engrafted with descendants of naïve stem cells. The studies of morphogenetic remodeling mechanisms and of memory and behavioral performance have not been integrated. Indeed, there are no established platforms for computational modeling of behavior during brain remodeling. It is thus imperative to begin to formalize the understanding of what happens to memories and behavioral programs when cells are added (proliferation), removed (apoptosis/necrosis), relocated (migration), and re-wired (synaptic plasticity).

2.2 Traditional and Adaptive Artificial Neural Network Models

Artificial neural networks (ANN) are a computational model of the neuronal connectivity in the brain. The feed-forward network is one of the most common neural network architectures in which connections between neurons are directed and going only in a global forward direction, avoiding the formation of feed-back loops. The output of a neuron can be calculated as a dot product of the input vector, x, and connection weight vector, w, plus some bias value, b, which acts as the neurons' firing threshold [25]. The output of the neuron is generally modulated by some activation function that is nonlinear, usually a sigmoid, such that the neural network is able to learn patterns that are not linearly separable [26, 27].

While most of the work to-date has focused on ANNs with constant (fixed) topologies, a few studies have exampled dynamical changes to ANN structure. Artificial neural network growth and pruning techniques have been primarily addressed as a part of training to avoid over-fitting (network is too large) or under-fitting (network is too small) [28]. The majority of techniques developed are meant as a pre-training network optimization or an intermediate step in training the network and therefore the immediate effects of perturbation on memory persistence after all training has been completed have not been studied [29–31]. While some studies of artificial neural network architecture have attempted to draw connections between such computational methods and their biological implications [28, 31], none have formally attempted to test perturbation methods prior to re-training the network.

The combination of genetic algorithms and artificial neural networks has given rise to a series of studies related to evolving adaptive and developmental neural networks [32–36]. These and other works studying neural-like cellular models of development [37, 38] suggest that artificial evolution has the capacity to produce robust information processing systems. Compositional pattern producing networks (CPPNs) are an example of a developmental artificial neural network model evolved by genetic algorithms [33, 35], and incorporate protein diffusion along with excitatory proteins into an artificial tissue model. This model was later extended with the neuroevolution of augmented topologies algorithm, NEAT, [39] which provides a method for augmenting artificial neural network topologies to become incrementally complex throughout genetic algorithm evolution [32]. Most recently, the CPPN-NEAT method was used to evolve 3-dimensional structures, a promising avenue for the study of morphogenesis [36].

Significant questions remain open with respect to the degree of robustness of memory after ANN structure change. Moreover, it is not known what value assignment policies for new cells (and the neighbors of dying cells) in a network optimally preserve function during topological change. Here, we approach these questions with a novel quantitative analysis of the behavioral performance of an ANN model subjected to cellular removal and proliferation.

3 Formalizing Artificial Neural Network Cellular Perturbation

3.1 Experimental Methods

We utilized a simple ANN topology for our primary experiment: a feed-forward single hidden layer neural network (Fig. 1a). A tangent sigmoid transfer function was used as activation function for all neurons. All networks had 2 input neurons and a hidden layer of a size of either 10 neurons or 20 neurons. We chose not to optimize the hidden layer size in order to compare the perturbations across the two training algorithms (likely with different optimal network sizes). The MATLAB Neural Network Toolbox [40] provided the implementations for the artificial neural networks and related functions for training and testing them.

We performed the training of the networks with two different backpropagation training algorithms: Levenberg-Marquardt and resilient back propagation. Both algorithms use the output error of the training network to adjust the internal weights, starting from the neurons in the output layer and propagating backwards through the rest of the network [41]. The Levenberg-Marquardt training algorithm is an optimization of the traditional backpropagation algorithm that approximates the computation of the Hessian matrix using a Jacobian matrix [42]. The resilient backpropagation algorithm is a modified backpropagation algorithm that was designed to combat the minimization of extreme changes due to taking the derivative of the tangent sigmoid transfer function [43].

We evaluated the networks using three benchmark data sets of varying complexity (two spirals and half-kernel). The input was defined as the x and y coordinates for a data point, and the output as a value (color classification) for the point (Fig. 1).

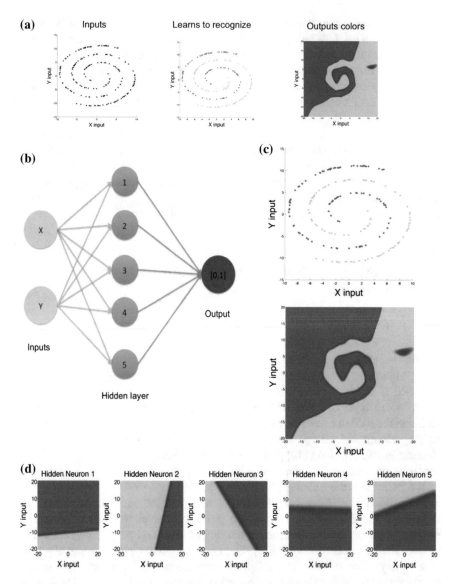

Fig. 1 Basic model structure and qualitative methods for studying artificial neural networks. **a** Pattern recognition goals for network to discriminate between two spirals; area of uncertainty can be seen from output on entire input space. **b** Feedforward single hidden layer neural network with 5 hidden neurons. **c** The artificial neural network output can be visualized over the training data set (spiral *x*, *y* points) or mapped over the entire input space to show a global categorization. **d** The output of each hidden neuron can be visualized to determine when individual neurons are firing (*purple*) and not-firing (*blue*). Visualizing the output of the network shows how its memory is stored

The quantitative error of the network was computed using the mean squared error function, which is commonly used to quantitatively test errors in artificial neural networks as well as other machine learning algorithms. Given n inputs for the actual outputs x and expected outputs r, the equation calculates the mean difference between the expected and actual outputs over the input set (1).

$$\frac{1}{n}\sum_{t=1}^{n}(r_t - x_t)^2 \tag{1}$$

3.2 Perturbation Methods

We implemented and tested seven perturbation methods on artificial neural networks, falling under the categories of neuron removal, addition, and connection blocking. The merging and splitting techniques, used to implement cell proliferation and programmed cell death (apoptosis) were adapted from a neuron pruning and growing method in an iterative training algorithm [30]. The seven perturbation methods implemented are the following:

- Remove—set the input and output weights and the bias of a specific neuron to 0 (Fig. 2a).
- Merge—merge two neurons, a and b, into neuron c by combining the weights of the input layer and output layer (2a, 2b). Adapted from a pruning method for artificial neural network size optimization [30]

$$\text{Input layer} \quad \vec{w}_c = \frac{\vec{w}_a + \vec{w}_b}{2} \tag{2a}$$

$$\text{Output layer} \quad \vec{w}_c = \vec{w}_a + \vec{w}_b \tag{2b}$$

- Merge insignificant with correlated neuron—select the least significant neuron, where significance is computed as the standard deviation of the hidden neurons' output over the training set (3), and select the neuron that is most correlated to as the neuron that has the smallest mean squared error to the insignificant neuron and merge as in 2.

$$\text{Significance} \quad s_a = \sigma\left(o\left(\vec{w}_a\right)\right) \tag{3}$$

- Add randomly—add a neuron with random weights (Fig. 2b).

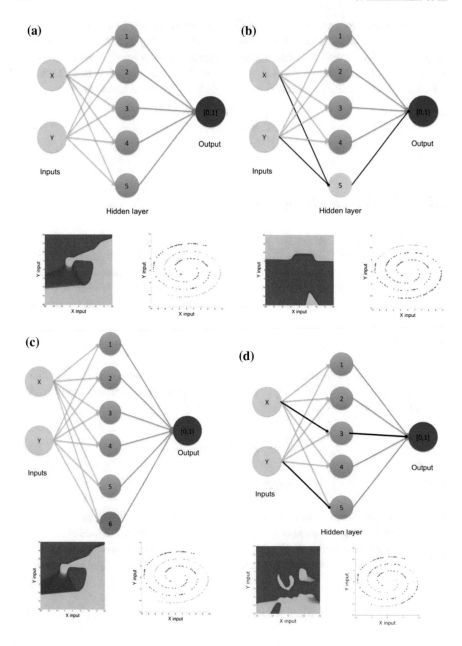

◀ **Fig. 2** Effects of modification of topology by three connections. **a** Trained artificial neural network before any perturbation has learned the two-spiral pattern with some area of uncertainty where it cannot discriminate between the two spirals. **b** Removing hidden neuron five from the artificial neural network by blocking its two input and one output connections causes perturbation of the network output, and loss of memory of the learned pattern. **c** Addition of hidden neuron six to the artificial neural network post-training by addition of two input and one output connections causes perturbation of the network output, and an alteration of the learned pattern. In comparison to the original network (a), the addition of the neuron has improved performance. The neuron addition post-training is not universally beneficial, sometimes causing loss of learned patterning information. **d** Connection blocking between input neurons and hidden neurons (X to 3 and Y to 5) and hidden neurons and output neuron (3 to output) causes loss of memory of correct patterning information

- Split—splitting a neuron involves modifying the weights of the original neuron a and new neuron b so that the two new neurons contain the same number of connections as the parent neuron, w_a (4a, 4b). The parameter α represents a random "mutation" value that for the scope of this study was a normally distributed random number [30].

$$\text{Old neuron} \quad \vec{w}'_a = (1+\alpha)\vec{w}_a \tag{4a}$$

$$\text{New neuron} \quad \vec{w}_b = -\alpha\vec{w}_a \tag{4b}$$

- Split insignificant—select the least significant hidden neuron (3) and split it (4a, 4b).
- Block connections—set the connection between neuron a and neuron b to be 0 (5), this can be done between both the input to hidden layer connections and the hidden layer to output connections (Fig. 2c).

$$w_{ab} = 0 \tag{5}$$

4 Results

4.1 Robustness of Feedforward Single Hidden Layer Artificial Neural Networks

We assayed the performance of two different artificial neural networks, one trained with resilient backpropagation and one with Levenberg-Marquardt training algorithm, on two different pattern recognition problems and seven types of perturbation to determine how topological changes impacted neural network memory. The results show how the memory effects of even slight changes to network topology by perturbing only 3 connections could be quite severe for any of the perturbations: neuron removal, addition, or connection blocking (Fig. 2).

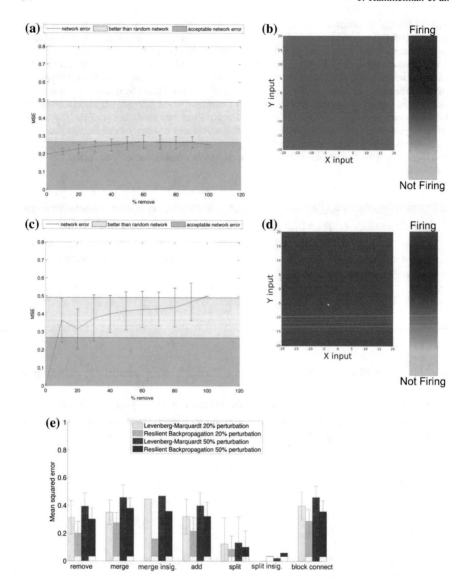

◄ **Fig. 3** Testing resilience as a function of percent perturbation for different training algorithms. **a** Network trained with resilient backpropagation loses little patterning information when neurons are removed using the simple remove neuron perturbation method. The network ends with <0.3 mean-squared error (MSE) since **b** the output neuron, due to the intermediate bias value characteristic of the Resilient Backpropagation training method, is always firing with an intermediate value for all inputs of the binary spiral classifications after all hidden neurons are removed. **c** Levenberg-Marquardt trained network reaches 0.3 mean squared error before 10 % perturbation and quickly approaches the average error of a random artificial neural network, since **d** the output neuron is constantly firing with a high value and hence classifying every point into one of the spirals because of the extreme bias values characteristic of the Levenberg-Marquardt training method. For both a and c, random network error was calculated by randomly assigning weights to neurons that were within the range (min and max with standard deviation) of a trained network and computed the error for the training problem. The networks are single hidden layer network (10 neurons in the hidden layer such that 10 % removal is equivalent to the removal of just one neuron). Performance was measured by the mean squared error (MSE). **e** Networks trained with the Levenberg-Marquardt algorithm (*yellow* 20 % manipulation; *red* 50 % manipulation) lose patterning information more quickly than those trained with resilient backpropagation (*green* 20 % manipulation; *blue* 50 % manipulation) for all seven perturbation methods. Starting error for the network is in white. One network trained with each algorithm was used, with each perturbation (i.e. remove 20 %) performed randomly 250 times. Performance was measured by the mean squared error w/standard deviation. Overall, while networks trained by either training algorithm are negatively affected by perturbation, the networks trained with resilient backpropagation have greater memory persistence that those trained with the Levenberg-Marquardt training algorithm

Neuron removal or blocking of 3 network connections caused almost a complete pattern degradation (Fig. 2a, c) while the effect of neuron addition was less severe (Fig. 2b). In order to determine how network performance degrades with increasing perturbation of cells, we trained a network and then progressively removed individual nodes at random, while testing its performance (Fig. 3). Patterning memory in the feedforward artificial neural networks was overall quickly affected by all seven perturbation methods, with removal and connection blocking being more traumatic to memory than neuron addition, but all having severe memory degradation at 50 % perturbation (Fig. 3e). Most perturbations reach above the threshold for network error at 20 % perturbation and certainly by 50 % perturbation except for neuron splitting, which was nonlinear in that further cellular perturbation did not always lead to a greater network error.

4.2 Differential Robustness of Training Algorithms

It was discovered that the resilient backpropagation algorithm overall was more robust to all types of perturbation than the Levenberg-Marquardt training algorithm (Fig. 3). This was particularly noticeable for removal and merging methods; the perturbation of resilient backpropagation networks had a slower effect on the network error, whereas the Levenberg-Marquardt networks lost almost all patterning capability after the removal of just one neuron (Fig. 3). One component that appeared to contribute to this robustness was the training of the bias values of the

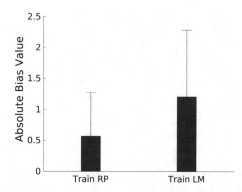

Fig. 4 Comparing absolute bias values of output neurons between training algorithms. The output bias values were compared for the resilient backpropagation and Levenberg-Marquardt training algorithms on 10 hidden neuron feed-forward artificial neural networks learning the two-spiral problem. We sampled the output values of 100 networks for each training algorithm, and took the mean and standard deviation of the bias value of their output neurons. Resilient backpropagation output neurons on average have a significantly lower absolute value (less extreme firing) than the output neurons of Levenberg-Marquardt trained networks. Using a two sample t-test with unequal variance we confirmed that the two training algorithms had significantly different bias values for their output neurons (p < 0.05e-4)

output neurons. This was then confirmed by training 100 networks using the two different training algorithms on the two spiral problem and for each network recording the value for its output bias neuron (Fig. 4). Using a two-tailed t-test with unequal variance, resilient backpropagation was found to have significantly lower bias values for its output neurons than Levenberg-Marquardt for output neurons (p < 0.05e-04).

Seemingly, the more extreme bias values of the output neurons from Levenberg-Marquardt training appeared to cause the network to quickly lose functionality, reaching the average error of a random artificial neural network. In contrast, artificial neural networks trained with resilient backpropagation never reached the error level of a random network, even after perturbing all hidden and input neurons, because of the output neuron biases. It is biologically plausible that bioelectrical networks with extreme differences in membrane potential would be less robust, as the loss of one neuron would significantly impact the response of its neighbors. This idea was further reinforced by the impact showed by random values of a mutation (scaling) parameter in the splitting perturbation methods. In future work, this mutation parameter will be tested to determine an optimal mutation value (or pattern) for splitting neurons in the network.

4.3 Resilience to Cell Death Depends on the Difficulty of the Problem

In order to determine if networks trained with easier or harder patterning problems were more robust to perturbation, we ran an experiment that determined which 20 % of the neurons in an artificial neural network gave the best memory persistence (least mean squared error) over the output, and then compared that value to the average performance of all of the possible combinations of 20 % neurons removal. The results showed that the two-spiral problem, a notoriously difficult machine-learning problem, trained with resilient backpropagation had the least difference in performance between removal of the best 20 % and the average 20 % (Fig. 5). In contrast, the half-kernel problem with the resilient backpropagation

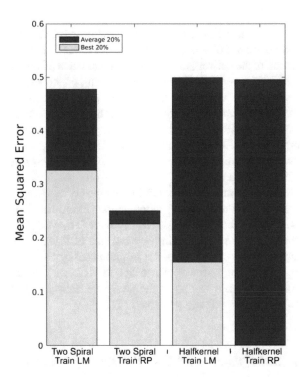

Fig. 5 Comparing average to optimal memory performance of 20 % neuron removal for different training problems and algorithms. Best performance compared to average performance for removal of 20 % of neurons for single hidden layer (20 neurons—remove 4 neurons) artificial neural networks trained on different data sets (two-spiral and half-kernel) and with different training algorithms: resilient backpropagation (RP); Levenberg-Marquardt (LM). Performance was measured as the mean squared error. Network trained with the half-kernel data set using resilient backpropagation had a low error for the best performance, indicating a high memory persistence with optimal neuron removal. Overall, the training algorithm had an impact on average resilience and the patterning problem had an impact on optimal resilience

training algorithm had the greatest difference between the best performance with 20 % neuron removal and average removal (Fig. 5). As the two-spiral problem has a higher degree of nonlinearity in its patterning, intuitively our results confirm that most if not all of the networks' hidden neurons are essential to the ability to learn the problem. The same is not true for the half-kernel pattern recognition problem because it is more linear and thus requires less neurons (in fact may contain redundant neurons) that can be pruned without compromising memory. In fact, it is likely that these excess nodes in the network learning the half-kernel problem were overgeneralizing, a common concern in ANN optimization literature [25].

4.4 Balancing Node Removal May Increase Memory Persistence

Using the same network trained using resilient backpropagation on the half-kernel problem with the optimal 20 % neuron removal, we looked at the firing output for each hidden neuron in the network to determine if there was a pattern of what neurons were found to be best to remove. By viewing the firing patterns of the hidden layer neurons, we were able to understand the network architecture and how the firing pattern plays a role in memory persistence. Interestingly, this network trained with resilient backpropagation for half-kernel was the only network tested in this study that contained neurons that fired weakly (neuron three and five; Fig. 6), and only one of the two weakly firing neurons was removed in the 20 % best performance case. The other weakly firing neuron (neuron five) was not part of the 20 % removed.

Overall, there was no immediate pattern in the types of neurons that were best to be removed from the artificial neural networks: we found that it was not best to always remove neurons that had consistent behavior (always firing or always not firing) but rather it seemed best to remove a balance of firing and non-firing neurons. This suggests that static neurons are in fact necessary for overall patterning recognition capability. The removal of neurons to "balance" the network may be comparable to the allometric scaling of the planarian flatworm, in which the proportions of the worm are perfectly preserved [44].

4.5 Functional Recovery of Lost Patterning Information

There were a few occurrences of a striking functional recovery of seemingly lost patterning information that was seen when using the splitting neuron and the splitting insignificant neuron perturbation methods (Fig. 7). In the splitting insignificant method, the neurons chose to split were the same, but the mutation parameter (4) varied, which, for extreme mutation parameters, may have caused the

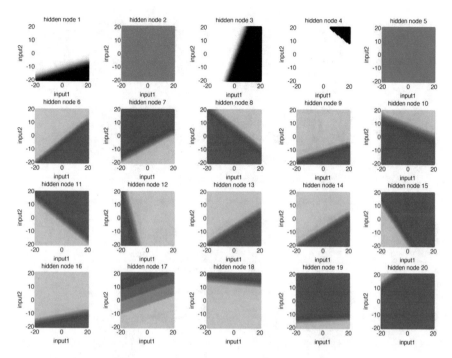

Fig. 6 Identifying patterns of best neurons to remove from network with memory persistence to 20 % removal. Individual neuron output for each of the 20 hidden neurons in a single layer feedforward artificial neural network trained using resilient backpropagation on the half-kernel data set showing optimal 20 % neuron removal. Grayscale neurons (neurons numbered 1–4) are removed. For colored neurons, *purple* indicates where the neuron fires over the input space and *blue* indicates where it is not firing. For grayscale (dead) neurons, *white* to *black* indicated firing level, with *white* being firing and *black* being not firing. There is not one consistent type of neuron that is removed, but rather four neurons that fire at different places in the input space. This implies that balancing neuron removal may be a method for maintaining memory persistence

extreme imbalance showed by the network (loss of all functional information). If this was followed by a more reasonable mutation parameter, this could cause a scaling of the network (changing the firing rates of all added neurons from extreme values to more balanced ones) that appeared to recover the patterning information. While the original intent was to study the impact of neuron splitting, these results suggest that neuron scaling may be very important to the recovery of lost patterning information. This may have implications for cancer and neurological diseases, where perhaps the bioelectrical properties of newly introduced cells can be adjusted to cause a certain alteration in the overall bioelectrical network and produce a functional recovery of the correct pattern.

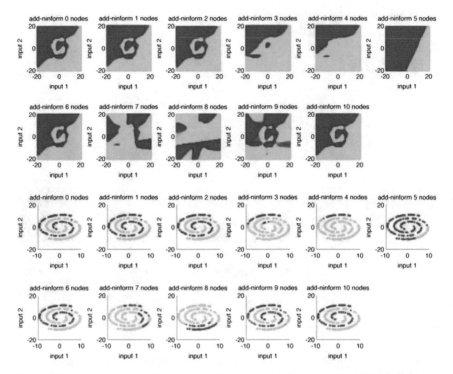

Fig. 7 Evidence of functional recovery of lost patterning information while splitting of insignificant neurons. Output over the entire input space (*top*) and training data coordinates (*bottom*) for each of 10 sequential perturbations in a single hidden-layer feed-forward artificial neural network. The network was initially trained with Levenberg-Marqardt training algorithm on the two-spiral dataset. After training, neurons were sequentially added using the splitting insignificant method. Functional information was lost after the addition of three neurons, but recovered after the addition of six neurons and nine neurons. This implies that addition of neurons using information from old neurons in the network may be a means of resilience towards topological perturbations

5   Discussion

We analyzed a computational model of memory in a neural network subject to morphological plasticity. We found that:

- Morphological perturbations to artificial neural networks corrupt learned patterns with a majority being outside of acceptable network error at 20 % perturbation, excluding neuron splitting as it remains in the acceptable error range after both 20 and 50 % perturbation.
- Resilient backpropagation-trained networks are more robust than the Levenberg-Marquardt trained networks to morphological change.
- Networks learning harder patterning problems are less capable of robustness after cell death.

- Selective cell death of neurons that balance firing patterns may increase memory persistence.
- Re-scaling of sequentially split neurons can facilitate recovery of lost patterning information in the network.

Our data represent a proof-of-principle of using computational models to understand the stability of memory in nervous systems undergoing significant remodeling. The findings also have implications for understanding the robustness of artificial cognitive devices that will increasingly be called upon to function despite damage or reconfiguration (e.g., in space exploration). Two major lines of inquiry represent the most important next steps. First is the modification of our paradigm to match specific instances of biological neural remodeling. We chose the training paradigm (e.g., backpropagation) to maximize tractability and connection with the existing body of work in the field of ANNs. Future work must extend this analysis to networks with greater biological realism and training mechanisms that are plausibly operational in real biological brains [45].

A second major area for investigation, on-going in our lab, is the extension of models used in computational neuroscience to understand goal-directed activity of non-neural tissues. The ability of some organisms to regenerate complex body organs [46] and adjust their structure toward the correct anatomy despite unpredictable perturbations [47, 48] reveals the existence of robust information-processing algorithms executed by somatic tissues that must make decisions guiding their growth and form toward adaptive outcomes. The body has an incredible capacity to learn and remember its structure—a capacity that is currently an untapped resource in biomedicine due to our limited understanding of the mechanisms of control of this complex system. Extensive data now demonstrate that all cells, not just neurons, communicate via electrical signals and neurotransmitters in making decisions that control morphogenesis [49–52]. Thus, we have conjectured that the ability to restore a target shape after injury or deformation may represent a kind of memory, and could be tractably modeled by techniques currently used to understand memory in the nervous system [52–55]. Mathematical models of such learning and memory, and particularly those inspired by cognitive science, may hold the key to understanding the storage and access of somatic cell memory. This is a top-down, information-centered approach which complements the current paradigm's near-exclusive focus on bottom-up models of molecular interaction pathways.

Simply the ability to manipulate somatic memory would hold great potential for biomedicine in treatment of traumatic injuries and birth defects, because it may help avoid the complexity explosion that hampers efforts to control complex shapes by tweaking specific molecular activities. It is possible that the mechanisms implicated in the massively distributed computing mechanisms of the brain are also present in the body, especially given that somatic cells contain the same signaling components that are essential for neuronal information processing and that neuronal communication likely evolved from more primitive somatic signaling [56–59]. If so, the mathematical formalizations employed in cognitive science may serve to further our understanding of cellular coordination towards complex pattern formation in development and regeneration. In particular, artificial neural networks and self-organizing maps are known for creating attractor states from which stable patterns may arise [60, 61]. Basins of attraction are gaining popularity in developmental biology to explain the course of gene expression via gene regulatory networks [62–64], but can also explain the collective phenomena of stable morphologies [54].

The planarian flatworm is an ideal model system for investigating such work due to its highly regenerative nature and demonstrated morphological plasticity [65–68]. Planarian flatworms have the capability of extending their genetically encoded morphologies to a number of additional stable states, including double and quadruple head morphologies [69]. The question that remains is how and why such stable states exist and if it is a property unique to regenerative organisms or if it is innate to all stem cell populations. In future work, we hope to expand our proposed method for artificial neural network perturbation to more complex network types such as deep multilayer perceptrons and recurrent neural networks which are known to have a more complex "memory" [28] as well as unsupervised networks (not trained with output criteria) such as the self-organizing map which are known to mimic the organization of like cells into clusters that share recognition properties, much like biological neural networks [70]. Additionally, future experiments will employ data that is related to biological shape to draw further connections between pattern recognition and target morphology. The work presented in this paper is the first to attempt to formalize perturbation methods for artificial neural networks in order to determine their use as a biological model. Evidence of functional recovery and sensitivity to the method of perturbation implies that artificial neural networks are capable of memory persistence and can serve as a model of a dynamic information processing system.

For biologists, important advances in neuroscience and the manipulation of somatic target morphology for regenerative medicine await a better understanding of memory robustness in dynamic cellular networks [54]. For computer scientists, understanding the mechanism of cell memory is an interesting question as well as a practical one: such a system could be the first step toward designing self-building and regenerative machinery. It is a question that has to be answered with care, as the balance between robustness and adaptability is a long-standing challenge in biology and computer science. As a mature model of cognitive information processing, artificial neural networks may be able to help solve the great mystery of how the body remembers and inform our understanding of development and regeneration in the process.

Acknowledgments We thank the Levin lab, Francisco J. Vico, and many others in the community for helpful discussions at the intersection of neuroscience and developmental biology. This work was supported by NSF (subaward #CBET-0939511 via EBICS at MIT), the G. Harold and Leila Y. Mathers Charitable, and Templeton World Charity Foundations.

References

1. D. Blackiston, T. Shomrat, M. Levin, The stability of memories during brain remodeling: a perspective, *Communicative and Integrative Biology* (In Press, 2015)
2. J.V. McConnell, A.L. Jacobson, D.P. Kimble, The effects of regeneration upon retention of a conditioned response in the planarian. J. Comp. Physiol. Psychol. **52**, 1–5 (1959)
3. D.J. Blackiston, E. Silva Casey, M.R. Weiss, Retention of memory through metamorphosis: can a moth remember what it learned as a caterpillar?. PLoS ONE **3**, e1736 (2008)
4. J.M. Mateo, Self-referent phenotype matching and long-term maintenance of kin recognition. Anim. Behav. **80**, 929–935 (2010)
5. K.J. Anil, Artificial neural networks: a tutorial (1996), pp. 31–44, http://doi. ieeecomputersociety.org/10.1109/2.485891
6. H. White, *Artificial Neural Networks: Approximation and Learning Theory* (Blackwell Publishers, Inc., 1992)
7. P. Arlotta, B. Berninger, Brains in metamorphosis: reprogramming cell identity within the central nervous system. Curr. Opin. Neurobiol. **27**, 208–214 (2014)
8. D.A. Berg, L. Belnoue, H. Song, A. Simon, Neurotransmitter-mediated control of neurogenesis in the adult vertebrate brain. Development **140**, 2548–2561 (2013)
9. M. Koehl, D.N. Abrous, A new chapter in the field of memory: adult hippocampal neurogenesis. Eur. J. Neurosci. **33**, 1101–1114 (2011)
10. W. Deng, J.B. Aimone, F.H. Gage, New neurons and new memories: how does adult hippocampal neurogenesis affect learning and memory? Nat. Rev. Neurosci. **11**, 339–350 (2010)
11. Y. Kitabatake, K.A. Sailor, G.L. Ming, H. Song, Adult neurogenesis and hippocampal memory function: new cells, more plasticity, new memories?. Neurosurg. Clin. North Am. **18**, 105–13 (2007)
12. S. Couillard-Despres, B. Iglseder, L. Aigner, Neurogenesis, cellular plasticity and cognition: the impact of stem cells in the adult and aging brain–a mini-review. Gerontology **57**, 559–564 (2011)
13. C. Wiltrout, B. Lang, Y. Yan, R.J. Dempsey, R. Vemuganti, Repairing brain after stroke: A review on post-ischemic neurogenesis. Mech. Neurodegeneration **50**, 1028–1041 (2007)
14. T. Tully, V. Cambiazo, L. Kruse, Memory through metamorphosis in normal and mutant *Drosophila*. J. Neurosci. **14**, 68–74 (1994)
15. M. Gandolfi, L. Mattiacci, S. Dorn, Preimaginal learning determines adult response to chemical stimuli in a parasitic wasp. Proc. R. Soc. Lond. Ser. B-Biol. Sci. **270**, 2623–2629 (2003)
16. K. Rietdorf, J.L.M. Steidle, Was Hopkins right? Influence of larval and early adult experience on the olfactory response in the granary weevil *Sitophilus granarius* (Coleoptera, Curculionidae). Physiological Entomol. **27**, 223–227 (2002)
17. K. Agata, Y. Umesono, Brain regeneration from pluripotent stem cells in planarian. Philos. Trans. R. Soc. Lond. B Biol. Sci. **363**, 2071–2078 (2008)
18. T. Shomrat, M. Levin, An automated training paradigm reveals long-term memory in planarians and its persistence through head regeneration. J. Expe. Biol. **216**, 3799–3810 (2013)
19. R.A. Fricker, M.K. Carpenter, C. Winkler, C. Greco, M.A. Gates, A. Björklund, Site-specific migration and neuronal differentiation of human neural progenitor cells after transplantation in the Adult Rat Brain. J. Neurosci. **19**, 5990–6005 (1999)
20. A. Wennersten, X. Meijer, S. Holmin, L. Wahlberg, T. Mathiesen, Proliferation, migration, and differentiation of human neural stem/progenitor cells after transplantation into a rat model of traumatic brain injury. J. Neurosurg. **100**, 88–96 (2004)
21. K.G. Akers, A. Martinez-Canabal, L. Restivo, A.P. Yiu, A. De Cristofaro, H.-L. Hsiang et al., Hippocampal neurogenesis regulates forgetting during adulthood and infancy. Science **344**, 598–602 (2014)

22. L.J. Martin, Neuronal death in amyotrophic lateral sclerosis is apoptosis: possible contribution of a programmed cell death mechanism. J. Neuropathol. Exp. Neurol. **58**, 459–471 (1999)
23. W.M. Cowan, J.W. Fawcett, D.D. O'Leary, B.B. Stanfield, Regressive events in neurogenesis. Science **225**, 1258–1265 (1984)
24. Y. Xiong, A. Mahmood, M. Chopp, Angiogenesis, neurogenesis and brain recovery of function following injury. Curr. Opin. Investig. Drugs (Lond., Engl.: 2000) **11**, 298–308 (2010)
25. I.A. Basheer, M. Hajmeer, Artificial neural networks: fundamentals, computing, design, and application. J. Microbiol. Methods **43**, 3–31 (2000)
26. M. Anthony, P.L. Bartlett, *Neural Network Learning: Theoretical Foundations* (Cambridge University Press, 2009)
27. S. Haykin, *Neural Networks: A Comprehensive Foundation* (Prentice Hall PTR, 1998)
28. A.M. Hermundstad, K.S. Brown, D.S. Bassett, J.M. Carlson, Learning, memory, and the role of neural network architecture. PLoS Comput. Biol. **7**, e1002063 (2011)
29. P.G. Benardos, G.C. Vosniakos, Optimizing feedforward artificial neural network architecture. Eng. Appl. Artif. Intell. **20**, 365–382 (2007)
30. M.M. Islam, M.A. Sattar, M.F. Amin, K. Murase, A new adaptive strategy for pruning and adding hidden neurons during training artificial neural networks, in *Intelligent Data Engineering and Automated Learning—IDEAL 2008*, vol. 5326, ed. by C. Fyfe, D. Kim, S.-Y. Lee, H. Yin (Springer Berlin Heidelberg, 2008), pp. 40–48
31. Y. Lecun, J.S. Denker, S.A. Solla, *Optimal Brain Damage*, pp. 598–605
32. K.O. Stanley, R. Miikkulainen, Efficient reinforcement learning through evolving neural network topologies. Network (Phenotype) **1**, 3 (1996)
33. A.N. Hampton, C. Adami, Evolution of robust developmental neural networks. Proc. Artif. Life **9**, 438–443 (2004)
34. J.F. Miller, Evolving developmental programs for adaptation, morphogenesis, and self-repair, in *Advances in Artificial Life* (Springer, 2003), pp. 256–265
35. J.C. Astor, C. Adami, A developmental model for the evolution of artificial neural networks. Artif. Life **6**, 189–218 (2000)
36. J.E. Auerbach, J.C. Bongard, Evolving CPPNs to grow three-dimensional physical structures, in *Proceedings of the 12th Annual Conference on GENETIC and Evolutionary Computation* (2010), pp. 627–634
37. N. Bessonov, M. Levin, N. Morozova, N. Reinberg, A. Tosenberger, V. Volpert, On a model of pattern regeneration based on cell memory. PLoS ONE **10**, e0118091 (2015)
38. U. Yerushalmi, M. Teicher, Evolving synaptic plasticity with an evolutionary cellular development model. PLoS ONE **3**, e3697 (2008)
39. K.O. Stanley, Compositional pattern producing networks: A novel abstraction of development. Genet. Program Evolvable Mach. **8**, 131–162 (2007)
40. M. a. N. N. T. R., *Natick* (The MathWorks, Inc., Massachusetts, United States, 2012)
41. R. Hecht-Nielsen, Theory of the backpropagation neural network, in *Neural Networks, 1989. IJCNN., International Joint Conference on*, vil. 1 (1989), pp. 593–605
42. D. Marquardt, An algorithm for least-squares estimation of nonlinear parameters. J. Soc. Ind. Appl. Math. **11**, 431–441 (1963)
43. M. Riedmiller, H. Braun, A direct adaptive method for faster backpropagation learning: the RPROP algorithm, in *IEEE International Conference on Neural Networks*, pp. 586–591
44. N.J. Oviedo, P.A. Newmark, A. Sánchez Alvarado, Allometric scaling and proportion regulation in the freshwater planarian Schmidtea mediterranea. Dev. Dyn. **226**, 326–333 (2003)
45. G. Deco, E.T. Rolls, L. Albantakis, R. Romo, Brain mechanisms for perceptual and reward-related decision-making. Prog. Neurobiol. **103**, 194–213 (2013)
46. K.D. Birnbaum, A.S. Alvarado, Slicing across kingdoms: regeneration in plants and animals. Cell **132**, 697–710 (2008)
47. D. Lobo, M. Solano, G.A. Bubenik, M. Levin, A linear-encoding model explains the variability of the target morphology in regeneration. J. R. Soc., Interface/R. Soc. **11**, 20130918 (2014)

48. L.N. Vandenberg, D.S. Adams, M. Levin, Normalized shape and location of perturbed craniofacial structures in the Xenopus tadpole reveal an innate ability to achieve correct morphology. Dev. Dyn. **241**, 863–878 (2012)
49. J. Mustard, M. Levin, Bioelectrical mechanisms for programming growth and form: taming physiological networks for soft body robotics, Soft Rob. **1**, 169–191 (2014)
50. A. Tseng, M. Levin, Cracking the bioelectric code: Probing endogenous ionic controls of pattern formation. Commun. Integr. Biol. **6**, 1–8 (2013)
51. M. Levin, C.G. Stevenson, Regulation of cell behavior and tissue patterning by bioelectrical signals: challenges and opportunities for biomedical engineering. Annu. Rev. Biomed. Eng. **14**, 295–323 (2012)
52. M. Levin, Molecular bioelectricity: how endogenous voltage potentials control cell behavior and instruct pattern regulation in vivo. Mol. Biol. Cell **25**, 3835–3850 (2014)
53. M. Levin, Reprogramming cells and tissue patterning via bioelectrical pathways: molecular mechanisms and biomedical opportunities. Wiley Interdisc. Rev.: Syst. Biol. Med. **5**, 657–676 (2013)
54. M. Levin, Morphogenetic fields in embryogenesis, regeneration, and cancer: non-local control of complex patterning. Bio Syst. **109**, 243–261 (2012)
55. M. Levin, Endogenous bioelectrical networks store non-genetic patterning information during development and regeneration. J. Physiol. **592**, 2295–2305 (2014)
56. F. Keijzer, M. van Duijn, P. Lyon, What nervous systems do: early evolution, input-output, and the skin brain thesis. Adapt. Behav. **21**, 67–85 (2013)
57. N.D. Holland, Early central nervous system evolution: an era of skin brains? Nat. Rev. Neurosci. **4**, 617–627 (2003)
58. G.A. Buznikov, Y.B. Shmukler, Possible role of "prenervous" neurotransmitters in cellular interactions of early embryogenesis: a hypothesis. Neurochem. Res. **6**, 55–68 (1981)
59. G. Buznikov, Y. Shmukler, J. Lauder, From oocyte to neuron: do neurotransmitters function in the same way throughout development? Cell. Mol. Neurobiol. **16**, 537–559 (1996)
60. H. Yan, L. Zhao, L. Hu, X. Wang, E. Wang, J. Wang, Nonequilibrium landscape theory of neural networks. Proc. Natl. Acad. Sci. **110**, E4185–E4194 (2013)
61. K. Friston, B. SenGupta, G. Auletta, Cognitive dynamics: From attractors to active inference. Proc. IEEE **102**, 427–445 (2014)
62. S. Bhattacharya, Q. Zhang, M. Andersen, A deterministic map of Waddington's epigenetic landscape for cell fate specification. BMC Syst. Biol. **5**, 85 (2011)
63. B.D. MacArthur, A. Ma'ayan, I. Lemischka, Toward stem cell systems biology: from molecules to networks and landscapes. Cold Spring Harb. Symp.Quant. Biol. 2008, p. sqb. 2008.73. 061
64. S. Huang, The molecular and mathematical basis of Waddington's epigenetic landscape: A framework for post-Darwinian biology? BioEssays **34**, 149–157 (2012)
65. E. Aboukhatwa, A. Aboobaker, An introduction to planarians and their stem cells," in *eLS*, ed (John Wiley and Sons, Ltd, 2015)
66. D. Lobo, W.S. Beane, M. Levin, Modeling planarian regeneration: a primer for reverse-engineering the worm. PLoS Comput. Biol. **8**, e1002481 (2012)
67. E. Saló, K. Agata, Planarian regeneration: a classic topic claiming new attention. Int. J. Dev. Biol. **56**, 1–4 (2012)
68. P.W. Reddien, A. Sanchez Alvarado, Fundamentals of planarian regeneration. Annu. Rev. Cell Dev. Biol. **20**, 725–57 (2004)
69. N.J. Oviedo, P. Morokuma, P. Walentek, I. Kema, M.B. Gu, J.-M. Ahn et al., Long-range neural and gap junction protein-mediated cues control polarity during planarian regeneration. Dev. Biol. **339**, 188–199 (2010)
70. T. Kohonen, Self-organized formation of topologically correct feature maps. Biol. Cybern. **43**, 59–69 (1982)

A Structure Optimization Algorithm of Neural Networks for Pattern Learning from Educational Data

Jie Yang, Jun Ma and Sarah K. Howard

Abstract Digital technology integration is recognized as an important component in education reformation. Learning patterns of educators' and students' perceptions of, beliefs about and experiences in using digital technologies through self-reported questionnaire data is straightforward but difficult, due to the huge-volume, diversified and uncertain data. This chapter demonstrates the use of fuzzy concept representation and neural network to identify unique patterns via questionnaire questions. Fuzzy concept representation is used to quantify survey response and reform response using linguistic expression; while neural network is trained to learn the complex pattern among questionnaire data. Furthermore, to improve the learning performance of the neural network, a novel structure optimization algorithm based on sparse representation is introduced. The proposed algorithm minimizes the residual output error by selecting important neuron connection (weights) from the original structure. The efficiency of the proposed work is evaluated using a state-level student survey. Experimental results show that the proposed algorithm performs favorably compared to traditional approaches.

Keywords Neural networks · Sparse representation · Single measurement vector · Structure optimization · Education data · Fuzzy representation

J. Yang (✉) · J. Ma
SMART Infrastructure Facility Faculty of Engineering and Information Sciences, University of Wollongong, Wollongong, NSW 2522, Australia
e-mail: jiey@uow.edu.au

J. Ma
e-mail: junm@uow.edu.au

S.K. Howard
School of Education Faculty of Social Science, University of Wollongong, Wollongong, NSW 2522, Australia
e-mail: sahoward@uow.edu.au

© Springer International Publishing Switzerland 2016
S. Shanmuganathan and S. Samarasinghe (eds.), *Artificial Neural Network Modelling*, Studies in Computational Intelligence 628,
DOI 10.1007/978-3-319-28495-8_4

67

1 Introduction

Digital technology integration in schools has, as yet, resulted in relatively limited teacher and student engagement with new and sophisticated ways of learning supported through information and communication technologies (ICTs). One possible reason is that educational research as a field has struggled to grasp the complexity or dynamic nature of technology adoption and integration in the classroom. A large amount of student and teacher self-reported questionnaire therefore has been designed for modeling purpose. Relevant data has been collected and studied on perceptions of, beliefs about and experiences using digital technologies in learning and teaching.

However, there are still some issues remain in terms of understanding and predicting students' learning performance as a result of technology integration in the classroom using survey data. The reasons are summarized as follows: firstly, the survey or self-reported questionnaire data is heterogeneous, as it is often in categorical form or expressed in free text. Missing data is also very common due to reporters' unwillingness and incautious. Secondly, much of this data is subject to participant bias and self-selection. Therefore, survey data is always skewed and imbalanced. In addition, due to the difference in perceptions of a survey topic from survey designers and responders, collected data is full of inconsistent expressions and varied semantics. Last, with the exponential accumulation of survey data, the challenges associated with data volume (numbers of years, students, teachers, even parents) and veracity (language uncertainty) continue to multiply. This leads to a complex data mining processing, which is difficult to address using traditional linear analysis methods.

Neural networks (NNs) have increasingly found their wide applications in many fields. Given training samples, NNs are used to establish the potential model between the observed input and output samples. Because of their capability of learning complex models from data, NNs have been applied to statistical modeling and decision-making in many areas [1, 2]. Inspired by existing work, we consider utilizing NNs for modeling educational questionnaire data. The performance of the NNs modeling, however, depends on the network structure, which must be built ahead by fixing the number of hidden neurons and layers. Given the same network training algorithm for the same questionnaire dataset, different network structures commonly lead to varied performance. A larger structure, for example, may converge quickly to a local optimum but exhibit poor generalization capability due to over-fitting; on the other hand, a smaller structure may require more time to approach the proper fit so that it becomes ineffective in real questionnaire data application. Hence, designing advanced algorithms to optimize network structures is critical for learning complex patterns from educational questionnaire data.

There is not a theoretical formula giving clear insight for how to choose the optimal network structure. The structure is typically decided by trial-and-error experiments or cross validation, but this process is computationally demanding. In this chapter, we propose a structure optimization algorithm, termed *sparse weight optimization* (SWO), by selecting important weights from the initial network. The proposed algorithm benefits from the model of *sparse signal representation* by considering the structure of a neural network as a sparse signal, and then the structure optimization is cast as finding the sparse representation for the network structure. More precisely, the proposed algorithm adopts an *initialization-then-selection* procedure. In the *initialization* step, an overlarge network is initialized. In the *selection* step, the proposed algorithm iteratively selects weights that minimize the residual error. The selection step is repeated until the termination condition is met. To this end, the proposed algorithm is able to adjust the network structure while training the neural network.

The remainder of the chapter is organized as follows. Section 2 presents the problem with modeling the self-reported questionnaire data. Section 3 present a brief review of existing structure optimization algorithms for NNs and the model of sparse representation, respectively. Section 4 details the proposed algorithm, by analyzing the relationship between structure optimization and sparse representation. Section 5 introduces how to process education data using the fuzzy concept representation. The implementation and experimental results are discussed in Sect. 6, followed by concluding remarks in Sect. 7.

2 Background

In recent decades, young people are believed to be more confident using technology than older generations in education. The belief that young people can easily adopt digital technologies has influenced how the public and educational systems think about technology integration and learning. To better support young people, schools need to fundamentally change and become more technology-driven, collaborative and student-centered. Research now has started to unpack how students use technology in schools and what this means for teaching and learning [3, 4].

Data mining technique, as an advanced tool, is therefore applied in educational domain [5, 6]. The educational data mining approach has mainly focused on higher education using data from learning management systems and intelligent tutoring systems. The benefit is that validation of models is simple and researchers can easily build on special knowledge.

Nevertheless, less work has been conducted at the secondary school level, where there is also a large number of publicly available datasets (e.g. national surveys of teacher satisfaction and work, Organization for Economic Co-operation and Development (OECD) adult skills and student assessment, etc.). Meanwhile, most of these datasets are self-reported attitudinal data, such as teacher experience and students perceptions, which presents a number of challenges in educational data mining. For instance, schools are by nature heterogeneous populations. To compound this, one of the most common issues is missing data, what it means and how this should be addressed. Furthermore, much of this data is subject to participant bias and self-selection. These traits of the data increase heterogeneity in the sample and complicate data mining analysis. In the following section, we introduce the neural networks technology to identify the complex pattern from the educational survey data.

3 Related Work

Neural networks have been inspired by biological observations of the human brain function [7]. Over the past several decades, neural networks have evolved into powerful computation systems, which are able to learn complex nonlinear input-output relationship from data. There are two general types of network topology: *feedback* or *recurrent neural networks* and *feed-forward networks*. In recurrent networks, the connections between neurons can form directed cycles. Feed-forward neural networks, on the other hand, process the information only in the forward direction, from the input layer to the output layer. In this chapter, we focus on feed-forward networks. A multilayer perceptron (MLP) or multilayer feed-forward neural network consists of a set of input neurons, one or more hidden layers of perceptrons, and an output layer of linear or sigmoid type neurons. Each neuron is connected to all neurons in the succeeding layer, while signals only propagate through the network in the feed-forward direction.

3.1 Structure Optimization Algorithms for NNs

Structure optimization is critical for improving the generality of feed-forward net-works, which aims to balance the trade-off between the computational complexity and the generalization ability of a neural network. The trial-and-error or cross validation methods are used to search for the optimal structure in the past, which is computational costly. To automatically optimize the network structure, more sys-tematic methods are developed, which can be broadly categorized as follows:

- *Network pruning algorithm*, which is based on the saliency analysis on each network element (such as weights) in a multiple-iteration procedure. At each iteration, the significant weights which minimize the training error are main-tained; while those with the least contribution are removed. Some typical weight-based pruning algorithms include optimal brain surgeon (OBS) [8] and optimal brain damage (OBD) [9];
- *Network construction algorithm*, which begins with a small network and incrementally adds hidden neurons during the training process. The incremental step is repeated until a stopping criterion is met. Some construction methods are proposed in [10–12];

- *Hybrid construction-pruning algorithm,* which combines the pruning and the constructive strategies. Generally speaking, the hybrid construction-pruning algorithm starts with a small structure and dynamically adding a hidden node (neuron) if its significance is larger than a certain threshold or deleting a node on the contrary during the training process. Some typical work can be found in [13, 14].

Overall, the traditional structure optimization methods are characterized by different advantages: the pruning-based algorithms can converge quickly from the initial larger network; the construction algorithms are less sensitive to the initial settings; and the hybrid algorithms combines the merits of both.

3.2 Sparse Representation

Sparse representation arises from single processing. It is based on the assumption that a synthetic signal can be decomposed into a linear combination of a few elementary signals. By sparse representation, the majority information conveyed by the target signal can be represented by only a few non-zero elements. The model of the sparse representation has attracted a great amount of research effort, resulting in many exciting applications [15, 16].

The *single measurement vector* (SMV) model is one particular type of sparse representation, which is to recover a one-dimensional signal $x \in \mathbb{R}^N$ from few linear measurements $y \in \mathbb{R}^M$ [17, 18]. The mathematical model can then be expressed as

$$\min S(x) \text{ subject to } y = \mathcal{D}x, \tag{1}$$

where $S(x)$ denotes a sparsity measure, and $\mathcal{D} \in \mathbb{R}^{M \times N}$ is known as the *dictionary*. A dictionary *atom* is one column vector from the dictionary. The SMV model is also illustrated in Fig. 1 [19].

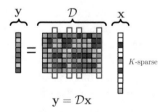

$$y = \mathcal{D}x$$

Fig. 1 The SMV model aims to minimize a vector's sparsity. In the sparse signal $x \in R^N$, only K entries are non-zero. Thus, the signal is K-sparse. The rectangular areas from the dictionary D are associated with non-zero coefficients from x

One simple strategy for solving Eq. (1) is to minimize the l_0-norm of x, i.e., $S(x) = \|x\|_0$. The l_0-norm is the cardinality or number of non-zero elements in x. Thus Eq. (1) can be also rewritten as

$$\min \|x\|_0 \text{ subject to } y = \mathcal{D}x. \tag{2}$$

Various algorithms have been developed for finding the sparse solution of the SMV model, including: non-convex local optimization algorithms and greedy algorithms. Non-convex local optimization algorithms solve the SMV model using self-weighted minimum norm, in which the Lagrange multiplier method is employed. Greedy algorithms, on the other hand, measure the similarity between the residual error and the dictionary atoms, and then select the atom that minimizes the residual error at each iteration. Typical greedy algorithms include matching pursuit (MP) [17], and orthogonal matching pursuit (OMP) algorithm [18].

4   Sparse Weight Optimization Algorithm

In this section, the sparse representation-based algorithm is proposed for optimizing a network structure, in which the structure optimization is formulated as an SMV model. We first introduce the definition of sparse rank which will be used later to measure the complexity or sparsity of the network structure.

Definition 1 *Given a matrix X, its sparse I rank $\mathcal{R}_1(X)$ is the largest number of non-zero elements in any column of X.*

Next, let us consider a three-layer fully-connected network with one hidden layer, an input layer and an output layer. Figure 2 shows the general structure of this fully connected network. On the left hand side, the input layer receives signals from the external environment. In the middle is the hidden layer, which receives signals from the input layer and sends its output signals to the output layer. The output layer processes the signals received from the hidden layer and produces the network response.

Suppose the initial network structure consists of Q inputs, N hidden neurons and M outputs. Let $P = [p_1; p_2; \ldots; p_L]$ be a matrix containing L training samples and $Y = [y_1; y_2; \ldots; y_L]$ be the desired output matrix. Each raw of P or Y represents one input or output sample; each pair of (p_i, y_i) forms an input-output observation.

Moreover, let $X \in \mathbb{R}^{L \times N}$ denote the output matrix of the hidden layer, in which the i-th column represents the output from the i-th hidden neuron, $i \in \{1, \ldots, N\}$, and $Z \in \mathbb{R}^{L \times M}$ denote neural network output matrix corresponding to the input matrix P. The output of the hidden layer can be expressed as

$$X = f(PV + B_1), \tag{3}$$

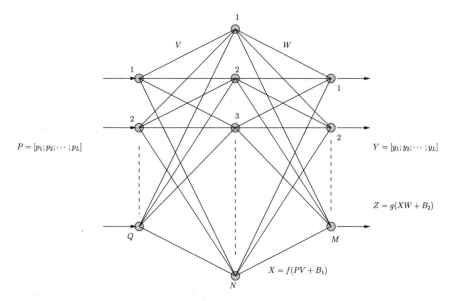

Fig. 2 An initial NN structure

where $f(\cdot)$ is the activation function, $V \in \mathbb{R}^{Q \times N}$ is the weight matrix between the input and hidden layers, and B_1 is the bias matrix having the bias vector as its columns. The final output of the network is given by

$$Z = g(XW + B_2), \tag{4}$$

where $g(\cdot)$ is the activation function, $W \in \mathbb{R}^{N \times M}$ comprises the weight vectors between the hidden and output layers, and B_2 is the bias matrix of the output layer with columns containing the bias vector.

Without loss of generality, we further assume that $f(\cdot)$ and $g(\cdot)$ are linear activation functions. Note that if $f(\cdot)$ and $g(\cdot)$ are invertible, we can transform the output neurons to linear units by applying the inverse function $f^{-1}(\cdot)$ or $g^{-1}(\cdot)$. In this case, the actual outputs of the hidden layer and of the final network can simplified as:

$$X = PV + B_1, \; Z = XW + B_2. \tag{5}$$

The desired output matrix Y can then be rewritten as:

$$Y = Z + E = XW + B_2 + E, \tag{6}$$

where $E = [e_1; e_2; \ldots; e_L]$ is the network error matrix; e_i is the error between the actual output z_i and the desired output y_i.

To optimize the structure of the NN, we need to remove some links between the input and hidden layers and between the hidden and output layers. Note that removing those links is equivalent to setting the weights of those links to zero. Hence, only non-zero weights will be kept in the network. Comparing the expressions of X, Z, Y in Eqs. (5) and (6), we can see that if we consider the weight matrices V or W as target signals, then finding an optimized structure is cast as finding the sparse representation of it.

Note that each single element in the matrix V (Eq. 3) is associated with one weight in-between the input and hidden neuron. For instance, the element $v_{i,j}$ from V in the i-th row and the j-th column represents the weight connecting the i-th input neuron and the j-th hidden neuron. Therefore, removing this weight connection is equivalent to setting $v_{i,j}$ to the zero value; on the other hand, those selected weights are given non-zero values. Similarly, to select the weight from the i-th hidden neuron to the j-th output neuron, we need to set the element $w_{i,j}$ in the matrix W as non-zero.

To this end, to optimize the structure of a NN, i.e., to select important weights while reducing the total number of weights in the original network is equivalent to minimizing the number of non-zero elements in V or W, or the minimization of $\mathcal{R}_1(V)$ and $\mathcal{R}_1(W)$. That is, given a trained network, the model of weight selection can be cast as follows:

$$\begin{cases} \min \mathcal{R}_1(V) \text{ subject to } X - B_1 = PV. \\ \min \mathcal{R}_1(W) \text{ subject to } Y - B_2 = XW. \end{cases} \tag{7}$$

The optimization to Eq. (7) can be calculated using the sparse representations for the Multiple-SMVs model simultaneously. We then propose a sparse algorithm, termed *SMVSI*, to solve the $\mathcal{R}_1(*)$ minimization, which is summarized in Algorithm 1.

---

input : The signal matrix $Y \in \mathbb{R}^{L \times M}$, the dictionary $\mathcal{D}$, and the maximal iteration T.
output: A sparse matrix X.

for $i \leftarrow 1$ **to** M **do**
 | *execute traditional SMV algorithm using* $(y_i, \mathcal{D})$ *with* T *iterations and obtain* x_i;
 | Replace the i-th column of X using x_i;
end

Algorithm 1: The SMVSI algorithm for the R_1 (*) minimization.

Remark 1 A variety of algorithms can be applied to SMVSI, such as greedy algorithms or non-convex local optimization algorithms. Herein, we apply greedy algorithms algorithm (such as MP or OMP) to solve the SMV model. The reason is that we are able to control the number of non-zero elements. Greedy algorithms start from an empty set and add one new atom iteratively. If the algorithm halts at the T-th iteration, there will be T non-zero element in the sparse solution. Therefore, when the greedy algorithm is applied in SMVSI, each column of V or W will have exactly T non-zero elements at the T-th iteration.

Overall, the sparse weight optimization (SWO) algorithm is proposed by an initialization-then-selection procedure. In the initialization step, an overlarge network is initialized. Then for the selection step, significant weights are selected to minimize the residual error. The SWO algorithm is summarized in Algorithm 2.

> **input** : The trained network and the dimension of the input Q.
> **output**: A pruned network.
>
> *Compute* $X_0 = PV + B_1$;
> **for** $i \leftarrow 1$ **to** Q **do**
> $V_i \leftarrow$ SMVSI($(X_{i-1} - B_1), P, i$);
>   Re-calculate $X_i = PV_i + B_1$;
>   **for** $j \leftarrow 1$ **to** N **do**
> $W_i \leftarrow$ SMVSI($(Y - B_2), X_i, j$);
>   *Evaluate the updated network using validation data set*;
>   **if** *the predefined termination condition is met* **then**
>   break;
>   **end**
>   **end**
> **end**
> *Take the network with the lowest error on the validation set as the final network*;

Algorithm 2: The proposed SWO algorithm for network structure optimization.

Remark 2 The termination criterion used in [13] is employed to halt the pruning process when the validation error keeps increasing for T successive times, where T is a parameter specified by the user. Apparently, the decision is made based on the assumption that such an increase error from the validation set indicates the beginning of the overfitting.

Remark 3 The computational complexity of weight-based SWO algorithm is analyzed hereafter. Note that the performance of conventional methods depends on the size of the original network as well as the remaining weights. For instance, traditional pruning methods including OBD and OBS remove weights iteratively, which lead to costly computation particularly for a large structure. On the other hand, the SWO algorithm focuses on the remaining weights rather than the eliminated ones, which speeds up the pruning process. Let L, N, N_w^*, and N_w denote the number of training samples, initial hidden neurons, initial weights, and remaining weights, respectively. Table 1 summaries the computational complexity of the proposed algorithm with existing methods in terms of the number of floating operations (flops).

Table 1 Comparison of computational complexity of different pruning algorithms

Algorithm	Computation cost	Flops
OBD	$O(LN)$	$O(N_w^* - N_w)$
OBS	$O(LN^2)$	$O(N_w^* - N_w)$
SWO	$O(\log(L)N)$	$O(N_w)$

5   Educational Dataset and Preprocessing

The dataset used in this study is taken from a state-level student and teacher survey as part of the federal Digital Education Revolution (DER) initiative in Australia. Data were collected by the Department of Education and Communities of New South Wales (NSW DEC), Australia. The DER was a federally funded program, which aimed to provide all secondary (year 9–12) students and teachers with information and communication technologies (ICTs) in various programs. The NSW government chose a one-to-one laptop program and provided laptops to all secondary teachers and year 9 students between 2009 and 2013. The NSW DEC collected the data over 4 years (2010–2013) through online questionnaires and school cases studies among all public secondary schools across the state. The student questionnaires covered five main subscales: school engagement, computer/laptop usage in and out school, learning experiences and beliefs, beliefs about the importance of ICTs in school subjects, and intentions after high school.

This study uses data from the 2012 year 9 student questionnaire. Of the approximately 80,000 year 9 students in NSW secondary schools, 27.2 % (21,800) students completed a two-part questionnaire. In the collected data, 12,978 students completed Part A and 8,817 students completed Part B. This work is based on Part B, i.e. the 8,817 student questionnaires.

The student questionnaire Part B contains 31 questions with a total of 147 items covering the five subscales. For example, the subscale "students' computer/laptop usage in and out school" includes items on frequency of use, confidence using, and computer/laptop related tasks and activities. Each question item contains a few options or a self-statement. Students either chose an option or answer using their language or leave it blank. Considering the categorical attributes, missing data and free expression as well as imbalanced features of the data, we adopt the fuzzy concept representation in data preprocessing step.

Fuzzy set was presented 50 years ago as a tool to express concepts without clear boundary such as "young age" and "high income" [20–22]. In collected educational data, many attributes can be summarized or aggregated using fuzzy concept to make the semantic of those data is much clear and understandable [23]. For example, a question in a state-level survey is about the ICT usage frequency of students. Provided response options include such as "once a week", "2-3 times a month", "once a month", "once a term", and "1-3 times a year". Obviously, these options are not easy to understand and compare; however, our experience and knowledge tell us that the options "once a week" and "2-3 times a month" have the similar meanings and so as to options "once a term" and "1-3 times a year". Therefore, it is possible to use fuzzy concept representation to summaries and aggregate information in the collected data for better analysis.

Generally, a fuzzy concept can be expressed by a fuzzy set F which is defined on a domain of discourse X and assigns a value from the real interval [0, 1] to each element of X as its membership degree with respect to the fuzzy set. As an example, we can define the fuzzy concept "frequent user" of ICT technology in teaching as:

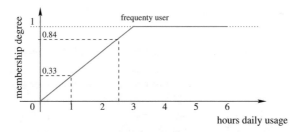

Fig. 3 Definition of fuzzy concept "frequent user" example

let X be the possible hours a user using a kind of ICT technology, say 0–6 h; then for each value between 0 and 10, we can assign a membership degree following below formula:

$$\mu(x) = \begin{cases} \frac{x}{3}, & x \leq 3 \\ 1.0, & 3 \leq x \leq 6. \end{cases} \quad (8)$$

We can draw the fuzzy set in Fig. 3 and claim that a user who spends 1 h daily on using ICT technology will be treated as "a frequent user with membership degree 0.33". The more time a user spends on ICT usage, the higher membership degree of being a "frequent user".

Note that the definition of a fuzzy concept is closely related to the context of discussion, hence the definition of the same fuzzy concept may vary in different situation. For instance, "frequent user" of ICT usage could be defined on a daily usage basis and can also be defined on a weekly, even monthly, usage basis. In this study, we define fuzzy concepts on top of a range of scores subjected to the question and the response options provided. Following the example of "frequent user" of ICT use, we firstly assign a score to each option, then collect all these scores as the domain X, and finally define fuzzy concepts such as "frequent user", "infrequent user" and "occasion user" on those scores. By using this method, we can easily summarize and aggregate information from the raw data.

6 Results and Analysis

In this section, we employ the DER dataset to investigate the efficiency of the proposed SWO algorithm for modelling purpose. The evaluation criterion is presented in the Sect. 6.1. The aims of our experiments are as follows:

- To understand the impact of initial network conditions on the SWO performance.
- To compare the proposed algorithm with conventional work for data modelling.

6.1 Experimental Setup

This experiment selects eight questions from the total 31 questions in the DER survey data, which include computer-efficacy, ICT engagement, learning preferences, learning beliefs, ICT and learning performance, school engagement, teacher directed ICT use frequency, and ICT importance in subject areas. Fourteen factors are further extracted from selected questions based on the main topics of their items (shown in Table 2), and the fuzzy representation is then applied to quantify these factors.

From the viewpoint of classification, the ICT engagement is regarded as the output while other 13 factors are cast as input attributes. For simplification, in this chapter we consider two typical modes, i.e. "positive" or "negative", for ICT engagement classification.

The DER dataset is further partitioned into three subsets: a training set, a validation set, and a test set. The training set is used to train and optimize the network

Table 2 Factors and their descriptions from the DER dataset

No	Factor	Description
1	Computer-efficacy (productivity)	Covers 6 productive tasks/activities; each task has 3 meaningful responses
2	Computer-efficacy (processing)	Covers 2 processing tasks/activities; each task has 3 meaningful responses
3	Computer-efficacy (creating)	Covers 2 creative tasks/activities; each task has 3 meaningful responses
4	ICT engagement	Covers 4 general engagement items; each item has 4 meaningful responses
5	Learning preferences (direct learning)	Covers 1 preferred item with 4 meaningful responses
6	Learning preferences (self-paced learning)	Covers 1 preferred item with 4 meaningful responses
7	Learning preferences (collaborative learning)	Covers 1 preferred item with 4 meaningful responses
8	Learning beliefs (self-learning)	Covers 1 preferred item with 4 meaningful responses
9	Learning beliefs (collaborative learning)	Covers 1 preferred item with 4 meaningful responses
10	Learning beliefs (instructed learning)	Covers 1 preferred item with 4 meaningful responses
11	ICT and learning performance	Covers 5 performance descriptions; each description has 4 meaningful responses
12	School engagement	Covers 5 questions items; each item has 4 responses
13	Teacher directed ICT use frequency	10 scales of frequency
14	ICT importance in subject areas	Covers 7 school subjects; each subject has 3 meaningful responses

architecture. The validation set is used for the stopping criterion and the test set is used for evaluation of the generalization ability of the pruned network. The size of the training, validation and test sets in all cases is 50, 25, and 25 %, respectively.

At the same time, a fully-connected feed-forward network is employed with an input layer, one hidden layer, and an output layer. The activation function of the hidden layer is set to tangent sigmoid function. The output neuron uses the linear function as activation function. The bias vector is implemented as an incoming connection from the particular bias node with the input value of 1. The network is initialized with random weights in the range [−0.1, 0.1], and it is trained with the resilient back-propagation algorithm (RPROP) [24]. The training parameters are set as follows:

- the maximum number of training iterations is 500;
- the minimum performance gradient is 10^{-6}; and
- the learning rate is 0.01.

The network training terminates when either the maximum number of iterations is reached or the performance gradient falls below 10^{-6}. Furthermore, the generalization performance is evaluated using the classification accuracy for the classification problem.

6.2 Performance Analysis

In this section, we investigate the effect of the initial network size on the performance of the SWO algorithm. A large network is more prone to converging to a local minimum than a smaller network. However, the over-sized structure will lead to the overfitting to the training samples and greatly affect the performance of the subsequent optimization. The purpose is then to find out the robustness of the proposed algorithm to various network sizes. To do that, initial networks with different numbers of hidden neurons are considered, i.e., N is set to 32 and 128, respectively. The initial networks are then trained by the RPROP method with training parameters as presented in Sect. 6.1.

Table 3 presents the average results for SWO over 30 independent runs. As observed, the performance across different initial network sizes is comparable; that is, the SWO algorithm yields similar generalization performance regardless of initial network sizes. The difference of average classification rate over all data sets between N = 32 and N = 128 is less than 1.03 %.

Table 3 Comparison with different sizes of the initial networks using the SWO algorithm

Data set	Classification accuracy (%)		Remaining weights	
	N = 32	N = 128	N = 32	N = 128
DER	94.50	95.53	42.33	51.07

The performance is evaluated using the classification accuracy on the test set

In terms of the final network structure, different initial structure lead to similar number of remaining weights. For instance, the SWO algorithm selects 42.33 weights when N = 32 compared to that of 51.07 from N = 128. This is because the SWO algorithm selects network weights instead of eliminating the redundant weights. As a result, the proposed algorithm achieves the similar number of weights for the remaining structure. This proves the robustness of the proposed SWO method to the initial network size. Based on the results presented in Table 3, we can conclude that the proposed SWO algorithm is less sensitive to changes in the original network size, which leads to similar generalization performance.

6.3 Comparison with Existing Works

In this section, the proposed SWO algorithm is compared with conventional structure optimization algorithms, such as OBS [8], OBD [9], and SBELM [25] algorithms. All the networks are initialized with 128 hidden nodes and the RPROP algorithm is used to train the initial network for OBS and OBD, while the random initialization approach is applied to SBELM. Meanwhile, the comparison between the proposed algorithm with others, such as standard constructive (SCA) and hybrid constructive-prune (AMGA) [12] algorithms are also implemented. In SCA and AMGA, a new hidden neuron will be added into the network if and only if the training error does not reduce after 20 epochs. These new neurons are randomly connected to the existing network. Table 4 reports the average classification error for various algorithms over 30 independent runs. For remaining structures, the results are for residual numbers of hidden neurons, while the numbers inside the bracket represent the number of remaining weights.

As seen from the above simulations, we find that the SWO algorithm outperforms the OBS and OBD methods by not only resulting in smaller size of the network structures, but also achieves a much better generalization ability. Compared to SCA and AMGA, although the SWO method performs worse by remaining bigger network structure, the proposed approach achieves better classification accuracy. Furthermore, the proposed approach is found competitive with the SBELM. In terms of the network size, SWO requires extra more 0.5 hidden neurons, while the average classification accuracy is slightly worse than AMGA (0.14 %).

Table 4 Summary of the performance of structure optimization methods using DER data

Algorithm	Classification accuracy (%)	Remaining structures
OBS	89.68	126 (1236.0)
OBD	85.68	126 (1245.0)
SBELM	95.67	5.8 (80.66)
SCA	91.44	3.5 (49.0)
AMGA	94.6	5.9 (77.31)
SWO	95.53	6.3 (61.07)

7 Conclusion and Future Work

Digital technology integration is one of main focuses in education reformation. Due to complexity of education system and practice, analyzing and understanding the perceptions and beliefs of technology integration becomes an important research area. Although huge amount of educational data is available, effectively analyzing those data is still a big challenge to researchers due to the heterogeneous features and the large amount of data. In this study, the fuzzy representation and neural network algorithm are employed to investigate potential patterns in educational data.

The fuzzy concept representation is applied for the data preprocessing to extract meaningful information from raw samples. We also presented a novel structure optimization algorithm for training neural networks, termed SWO, which is characterized by the model of sparse representation. The proposed algorithm is able to optimize the network structure while reducing the training error simultaneously.

The proposed work was evaluated on the DER dataset for classification purpose. We first consider the impact of initial network on the SWO performance. Experimental results show the robustness of the proposed method to the initial network; that is, the initial network size has limited effect on the performance of the proposed algorithm. Furthermore, a variety of optimization algorithms, such as pruning, constructive, and hybrid methods, have been introduced to compare with the proposed algorithm. A detailed investigation of the results has shown that the proposed algorithm performs reasonably well by eliminating average 95.31 % of the original networks structure, which also leads to a good generalization performance.

References

1. J. Chen, S. Lin, A neural network approach-decision neural network DNN for preference assessment. IEEE Trans. Syst. Man Cybern. Part C Appl. Rev. **34**(2), 219–225 (2004)
2. A. Zaknich, Introduction to the modified probabilistic neural network for general signal processing applications. IEEE Trans. Signal Process. **46**(7), 1980–1990 (1998)
3. P. Thompson, The digital natives as learners: technology use patterns and approaches to learning. Comput. Educ. **65**, 1233 (2013)
4. A. Margaryan, A. Littlejohn, G. Vojt, Are digital natives a myth or reality? University students use of digital technologies. Comput. Educ. **56**, 429440 (2011)
5. Y.K. Baker, R. S, The state of educational data mining in 2009: a review and future visions. J. Educ. Data Mining **1**, 317 (2009)
6. C. Romero, M.-I. Lpez, Luna, S. Comput. Educ. **68**, 458472 (2013)
7. S. Haykin, *Neural networks: A Comprehensive Foundation* (Prentice-Hall, Inc., New Jersey, 1999)
8. B. Hassibi, D.G. Stork, second-order derivatives for network pruning: optimal brain surgeon. Adv. Neural Inf. Process. Syst. **5**, 164–171 (1993)
9. Y.L. Cun, J. Denker, S. Solla, Optimal brain damage. Adv. Neural Inf. Process. Syst **2**, 598–605 (1990)

10. C. Kai, A. Noureldin, N. El-Sheimy, Constructive neural-networks-based MEMS/GPS integration scheme. IEEE Trans. Aerosp. Electron. Syst. **44**(2), 582–594 (2008)
11. H. Zheng, H. Wang, F. Azuaje, Improving pattern discovery and visualization of sage data through poisson-based self-adaptive neural networks. IEEE Trans. Inf Technol. Biomed. **12**(4), 459–469 (2008)
12. M. Islam, M. Sattar, M. Amin, X. Yao, K. Murase, A new constructive algorithm for architectural and functional adaptation of artificial neural networks. IEEE Trans. Syst. Man Cybern. B Cybern. **39**(6), 1590–1605 (2009)
13. M. Islam, A. Sattar, F. Amin, X. Yao, K. Murase, A new adaptive merging and growing algorithm for designing artificial neural networks. IEE Proc. Syst. Man Cybern. Part B: Cybern. **39**(3), 705–722 (2009)
14. M. Bortman, M. Aladjem, A growing and pruning method for radial basis function networks. IEEE Trans. Neural Netw. **20**(6), 1039–1045 (2009)
15. J. Haupt, R. Nowak, Signal reconstruction from noisy random projections. IEEE Trans. Inf. Theory **52**(9), 4036–4048 (2006)
16. H. Rauhut, K. Schnass, P. Vandergheynst, Compressed sensing and redundant dictionaries. IEEE Trans. Inf. Theory **54**(5), 2210–2219 (2008)
17. S. Mallat, Z. Zhang, Matching pursuits with time-frequency dictionaries. IEEE Trans. Signal Process. **41**(12), 3397–3415 (1993)
18. J. Tropp, A. Gilbert, Signal recovery from random measurements via orthogonal matching pursuit. IEEE Trans. Inf. Theory **53**(12), 4655–4666 (2007)
19. E. Candes, J. Romberg, T. Tao, Robust uncertainty principles: exact signal reconstruction from highly incomplete frequency information. IEEE Trans. Inf. Theory **52**(2), 489–509 (2006)
20. G.J. Klir, B. Yuan, *Fuzzy Sets and Fuzzy Logic: Theory and Applications* (Prentice-Hall, NJ, USA, 1995)
21. L.A. Zadeh, Fuzzy sets. Inf. Control **8**, 338–353 (1965)
22. H. Bustince, F. Herrera, J. Montero (Eds.), Fuzzy Sets and Their Extensions: Representation, Aggregation and Models, vol. 220 *Studies in Fuzziness and Soft Computing* (Springer, Berlin, 2008)
23. C. Fourali, Using fuzzy logic in educational measurement: the case of portfolio assessment. Eval. Res. Educ. **11**(3), 129 (1997)
24. M. Riedmiller, H. Braun, A direct adaptive method for faster backpropagation learning: the RPROP algorithm, in *IEEE International Joint Conference on Neural Networks* (1993), pp. 586–591
25. J. Luo, C.-M. Vong, P.-K. Wong, Sparse bayesian extreme learning machine for multiclassification. IEEE Trans. Neural Netw. Learn. Syst. **25**(4), 836–843 (2014)

Stochastic Neural Networks for Modelling Random Processes from Observed Data

Hong Ling, Sandhya Samarasinghe and Don Kulasiri

Abstract Most Artificial Neural Networks that are widely used today focus on approximating deterministic input-output mapping of nonlinear phenomena, and therefore, they can be well trained to represent the average behaviour of a nonlinear system. However, most natural phenomena are not only nonlinear but also highly variable. Deterministic neural networks do not adequately represent the variability observed in the natural settings of a system and therefore cannot capture the complexity of the whole system behaviour that is characterised by noise. This chapter implements a class of neural networks named Stochastic Neural Networks (SNNs) to simulate internal stochastic properties of natural and biological systems. Developing a suitable mathematical model for SNNs is based on the canonical representation of stochastic processes by means of Karhunen-Loève Theorem. In the implementation of this mathematical formulation for modelling nonlinear random processes from observed data, SNN is represented as a network of embedded deterministic neural networks, each representing a significant eigenfunction characterised by data, juxtaposed with random noise represented by White noise characterised by the corresponding eigenvalues. Two successful examples, including one from biology, are presented in the chapter to confirm the validity of the proposed SNN. Furthermore, analysis of internal working of SNNs provides an in-depth view of how SNNs work giving meaningful insights.

Keywords Stochastic neural networks · Random processes · Karhunen-Loève Theorem · White noise · Biological and environmental systems

1 Introduction

Most environmental and biological phenomena, such as cell signalling pathways, gene regulatory networks, underground water flow and pollution, blood flow through capillaries and properties of wood, exhibit variability which is not amenable

H. Ling · S. Samarasinghe (✉) · D. Kulasiri
Integrated Systems Modelling Group, Lincoln University, Christchurch,
New Zealand
e-mail: sandhya.samarasinghe@lincoln.ac.uz

© Springer International Publishing Switzerland 2016
S. Shanmuganathan and S. Samarasinghe (eds.), *Artificial Neural Network Modelling*, Studies in Computational Intelligence 628,
DOI 10.1007/978-3-319-28495-8_5

to be simulated realistically using deterministic approaches including deterministic Artificial Neural Networks (ANNs). The reason is that these ANNs do not adequately represent the variability which is observed in a systems' natural settings as well as do not capture the complexity of the whole system behaviour. These systems need to be considered as a class of stochastic processes with arbitrarily inherent nature for modeling the spatial and temporal behavior of the system. In the past, some mathematical models based on stochastic calculus along with stochastic differential equations have been established to simulate these particular cases of environmental, biological and natural systems. Some stochastic mathematical developments have been demonstrated earlier in the works of Bear [2] and Wiest et al. [18]. They focus on characterizing the flow within porous media for underground water flow such as a model for contamination flow in soil or underground aquifers. Furthermore, Kulasiri and Verwoerd [8] developed a stochastic mathematical model for the spatial and temporal solution of contaminant transport flow in a porous medium. However, a serious problem with these models is the difficulty in solving them analytically or numerically.

ANNs are another approach used to model some natural and biological systems on the basis of mimicking the information processing methods in the human brain. ANNs have a high capability in approximating input-output mappings that are complex and nonlinear to arbitrary degree of precision [14]. The incremental learning approaches used in ANNs make it possible for them to approximate all internal parameters iteratively. These capabilities of neural networks make them suitable to address some of the problems related to stochastic models and develop neural networks that approximate random processes. However, most widely used ANNs only focus on approximating deterministic input-output mappings although they generally operate in a stochastic environment where all signals could be inherently stochastic. Thus, there is a need to develop neural networks with the ability to learn stochastic processes or represent stochastic systems. There are some successful examples that demonstrate that SNNs can model natural, industrial and biological systems. For instance, a stochastic neural network has been created for modelling transportation systems in Italy [13], for fast identification of spatiotemporal sequences [1] and for generating multiple spectrum compatible accelerograms [9]. However, the development of these SNNs is not based on the theory of stochastic process. Truchetti [3, 17] proposed a new class of SNNs as a universal approximator of stochastic process. He [17] presented theoretical developments and a brief demonstration on the development of SNNs for a limited number of cases involving random functions. However, they have not used SNNs to model natural, biological and environmental systems for which the underlying random function must be extracted from observed data.

As presented by Truchetti [3, 17], there are two different approaches to incorporate stochastic properties into a network: Brownian motion and White noise, which are two fundamental stochastic processes represented by zero-mean Gaussian distributions. Furthermore, Brownian motion is used to simulate continuous stochastic processes whereas White noise is used to simulate discrete stochastic processes. Turchetti [3, 17] treats in detail the development of SNNs to simulate stochastic processes characterized by known random functions by means of Brownian motion. However, White noise is more appropriate for simulating real-world stochastic processes for the reason that for most real stochastic processes only some finite number of realizations (data) can be collected from the system as the governing stochastic functions are unknown. These collected realizations only record values at discrete times or locations of stochastic processes.

The goals of the research described in this chapter are to explore a method for developing SNNs to approximate real natural and biological stochastic processes based on White noise, implement SNNs for representing such natural stochastic systems, and provide an in-depth view of the network outcomes and internal workings of the networks in order to gain insights into networks and the processes they characterise. The chapter is organized as follows: Sect. 2 gives a brief review of related mathematical background in order to explore a suitable mathematical model for SNNs. Section 3 discusses how to develop SNNs based on a set of realizations of a stochastic process. Section 4 focuses on the implementation of the proposed neural networks and analysis of internal workings of SNNs. In Sect. 5, a discussion on some possible further work for improving the capability of the proposed stochastic neural network is presented.

2   A Brief Review of Related Mathematical Background

2.1   Stochastic Processes

In physical terms, a stochastic process may be regarded as a set of values obtained from a single experiment to observe the temporal evolutions of a stochastic variable which does not have a unique value for a corresponding time. This means that a different set of observations can be obtained when repeating the same experiment. Mathematically, a stochastic process is a collection of random variables over the time parameter t $(t \in T)$. These random variables can be considered as a function which maps a probability space (Ω, S, P) to real numbers, where Ω means a sample space, which is a collection of all possible values of the experiment, and we use ω to label each possible value of the experiment; S is a σ-algebra[1] of sets which is a nonempty collection of subsets of a sample space Ω [8, 12, 17] and each element of the σ-algebra S is regarded as an event; the sample space Ω and the σ-algebra S consisting of a measure space (Ω, S) and P, a probability measure, is a function which represents a probability of mapping each event from S to real numbers. Now let us consider a stochastic process $\xi(t) = \xi(t, \omega)$ or $\{\xi(t), t \in T\}$ as random

---

[1]A family S of subsets of Ω is a σ-algebra if:

1. the space of elementary events Ω is in S;
2. if a subset E of Ω is in S then so is the complement E^c of E;
3. if a countable number of subsets of Ω is in S then so is their union [8, 12, 17].

variables $\xi(\omega)$ which is defined on a fixed probability space (Ω, S, P) for each time t $(t \in T)$. If $\omega \in \Omega$ is kept constant, it can be shown as a function of time parameter t such as $\xi(\cdot, \omega) = \{\xi(t, \omega), t \in T\}$ and this function is called a realisation or trajectory of the stochastic process.

2.2 Canonical Representation of Stochastic Processes

Use of mathematical models for the representation of stochastic processes is a focus in the theory of stochastic processes. Canonical representation of stochastic processes has played an important role in simulating stochastic processes by means of mathematical models. The aim of canonical representation is to display a complex stochastic process using the sum of elementary stochastic functions, such as Brownian motion and White noise [3, 17]. A more general definition of canonical representation of a stochastic process is a linear combination of non-random functions and zero-mean random variables. Furthermore, canonical representation of the covariance function of stochastic processes is a key part in defining and developing stochastic neural networks.

Generally, an elementary stochastic function can be defined as

$$\xi(t) = \zeta \, \varphi(t), \tag{1}$$

where $\xi(t)$ is a stochastic process, ζ is an ordinary zero-mean random variable and $\varphi(t)$ is non-random function. As a consequence, the random property of the process depends on the random coefficient ζ while the evolution of the process is associated with the function $\varphi(t)$. Thus, a stochastic process can be regarded as the sum of elementary stochastic functions generally shown for M elementary functions as

$$\xi(t) = \sum_{i=1}^{M} \zeta_i \, \varphi_i(t) \tag{2}$$

In representing a stochastic process in a mathematical model, the main focus is to define the canonical representation of the covariance function of the stochastic process. Based on (2), canonical representation of the covariance function of the stochastic process is

$$B(t, s) = E\{\xi(t) \, \overline{\xi(s)}\} = \sum_{i=1}^{M} \varphi_i(t) \, \overline{\varphi_i(s)} \, E\{\zeta_i^2\} + \sum_{i \neq j} \varphi_i(t) \, \overline{\varphi_j(s)} \, E\{\zeta_i \, \zeta_j\} \tag{3}$$

where $B\{t, s\}$ is the covariance function of the stochastic process; $E\{\bullet\}$ represents the expectation of a random variable ζ; the bar denotes the conjugate complex quantity.

If random variables ζ_i and ζ_j are chosen as orthogonal random variables, they satisfy the following condition:

$$E\{\zeta_i \, \overline{\zeta_j}\} = 0, \quad \text{for } i \neq j \tag{4}$$

Therefore, canonical representation of the covariance function of stochastic processes (3) can display the process as a linear combination of orthogonal random variables as

$$B(t, s) = E\{\xi(t) \, \overline{\xi(s)}\} = \sum_{i=1}^{M} \varphi_i(t) \, \overline{\varphi_i(s)} \, E\{\zeta_i^2\} \, . \tag{5}$$

In summary, a complex stochastic process can be regarded as the summation of the product of deterministic functions and orthogonal random variables where the latter satisfy the condition in (4). Furthermore, properties of Brownian motion and White noise completely satisfy the definition of orthogonal random variables, and furthermore, they can be easily simulated in mathematical software. Therefore, we can use these two processes to represent properties of the orthogonal random variables. However, only a finite number of realisations can be collected from a real stochastic process without the knowledge of the governing stochastic function. Therefore, the development of a stochastic process based on these limited realistic data is a focus of this chapter. In order to achieve this objective, it is necessary to understand canonical representation of stochastic processes by means of Karhunen-Loève (KL) theorem that allows decomposition of the covariance matrix obtained from data.

2.3 Canonical Representation of Stochastic Processes by Means of Karhunen-Loève Theorem

This section provides fundamental developments on the canonical representation of stochastic processes by means of the Karhunen-Loève (KL) theorem. The KL theorem is given a central role in exploring a suitable mathematical model for SNNs [17]. The KL expansion is a representation of a stochastic process as a finite linear combination of orthogonal functions determined by the covariance function of the data collected from the random processes [6]. By definition, a stochastic process can be expanded as the following function (6) and the covariance function (7):

$$\xi(t) = \sum_{\lambda \in \Lambda} \zeta(\lambda) \, \varphi(t, \lambda) \, ; \tag{6}$$

$$B(t, s) = \sum_{\lambda \in \Lambda} \lambda \, \varphi(t, \lambda) \, \overline{\varphi(s, \lambda)} \, , \tag{7}$$

where $\zeta(\lambda)$ is an orthogonal sequence of random variables and the variance of $\zeta(\lambda)$ is equal to the eigenvalues (λ) of the covariance function of the stochastic process $(E\{|\zeta(\lambda)|^2\} = \lambda)$. $\varphi(t, \lambda)$ are the eigenfunctions of the covariance function.

If the set λ, $\lambda \in \Lambda \subset R$ could be decomposed into some discrete values λ_k in the real axis and the number of λ_k depends on the number of eigenvalues of the covariance of a stochastic process, (6) and (7) can be rewritten as,

$$\xi(t) = \sum_{k=1}^{n} \zeta(\lambda_k) \, \varphi(t, \lambda_k);$$ (8)

$$B(t, s) = \sum_{k=1}^{n} \lambda_k \varphi(t, \lambda_k) \overline{\varphi(s, \lambda_k)},$$ (9)

where the variance of $\zeta(\lambda_k)$ is equal to the eigenvalues (λ_k) of the covariance matrix of the stochastic process $E\{|\zeta(\lambda_k)|^2\} = \lambda_k$. Consequently, a stochastic process can be viewed as the summation of the product of eigenfunctions (deterministic or nonrandom) and their corresponding orthogonal random variables (random noise).

Based on the canonical representation of stochastic processes by means of the KL expansion in (8) and (9), it is possible to create a feasible method to develop a stochastic neural network for simulating a real stochastic process. According to the KL theorem (9), the covariance of the measured realisations can be decomposed into eigenvalues and corresponding eigenvectors using singular value decomposition method or Principal Component Analysis (PCA) using statistical or mathematical software. Applying these eigenvectors to original data, a set of discrete values of the corresponding eigenfunction are obtained. Since eigenfunction $\varphi(t, \lambda_k)$ are deterministic functions, they can be modeled by deterministic neural networks. These networks can be linearly combined with noise $\zeta(\lambda_k)$ as in (8) to develop stochastic neural networks. The process of development of SNNs based on a set of collected data (realisations) is described in Sect. 3.

3   Methods

Modeling stochastic processes using SNNs involves the following:

1. Find a Canonical representation of the stochastic process by means of Brownian or White noise;
2. Create deterministic input-output mappings from stochastic process based on KL expansion and develop deterministic neural networks;
3. Develop SNNs by adding White noise into each of the developed deterministic neural networks and assembling them.

The most important step in developing SNNs is exploring deterministic input-output mappings from a stochastic process. The purpose of this step is to develop deterministic neural networks for defining values of weights and parameters of SNNs. The following section discusses each step in the development of SNNs.

3.1 Creating Input-Output Mappings

A stochastic process can also be viewed as a set or a bundle of realizations in finite domain. Furthermore, only finite number of realizations can be collected from a real stochastic process. For example, Fig. 1 contains six realizations from the Sine function with random noise, which is used to illustrate model development.

In Fig. 1, all realizations represent the behavior of the same Sine function but they also represent random fluctuations. It is easy to see that the randomness becomes an inherent characteristic of this stochastic process. Following is a discussion on how to create deterministic input-output mappings using the KL theorem for the purpose of generating data for developing deterministic neural networks for representing eigenfunctions.

Suppose that Fig. 1 contains a bundle of realizations collected at discrete times of a stochastic process $\xi(t)$ and that $\xi^k(t)$ denotes the kth realization. For each realization, there are n different discrete values corresponding to each discrete time t and the value of n depends on the time interval Δt as well as the total time T for the whole realization ($n = \frac{T}{\Delta t}$). First task in the process is to establish the covariance matrix of this stochastic process. If we define that $\xi(t)$ at each discrete time t is viewed as an input variable for the covariance matrix, the values of each realization $\xi^k(t)$ at time t will be viewed as an element of this input variable. So we create n input variables on the dataset by vector $\xi(t) = \{\xi(t_1), \xi(t_2), \ldots \xi(t_n)\}$ where $\xi(t_i)$ contains all values of realizations at ith discrete time i.e., $\xi(t_i) = \{\xi^1(t_i), \xi^2(t_i), \ldots \xi^k(t_i)\}$. In this vector representation, the mean and variance of all realizations at a particular discrete time t_i and the covariance of all realizations between any two different discrete times t_i and t_j can be efficiently calculated by using the following equations [14]:

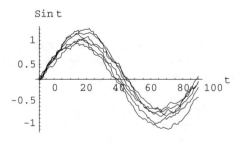

Fig. 1 A set of realizations from the Sine function with noise

$$E[\xi(t)] = \frac{1}{K}\sum_{k=1}^{K}\xi^{k}(t),\tag{10}$$

$$COV = \frac{1}{K-1}\sum_{k=1}^{K}(\xi^{k}(t) - E[\xi(t)])(\xi^{k}(t) - E[\xi(t)])^{T}.\tag{11}$$

In (10), $E[\xi(t)] = \{E[\xi(t_1)], E[\xi(t_2)], E[\xi(t_3)], \ldots E[\xi(t_n)]\}$ is a vector which contains all mean or expected values of realizations at each discrete time t_i and K is the number of realizations in the dataset. In (11), COV is the covariance matrix which contains all variances and covariances. COV is a symmetric matrix of size $n \times n$ where n is the number of time intervals over the whole time domain. The diagonal of COV represent variances and off-diagonals represent covariance between any two different discrete times.

According to the KL theorem, the COV matrix can be decomposed into a new matrix with new scaled variables based on the eigenvalue decomposition method. In this new matrix, all variables are independent of each other and all variables have their own variance. Therefore, the covariance between any two new variables is equal to zero. The COV matrix can be represented by using the KL theorem as

$$COV = \sum_{j=1}^{n}\lambda_j\varphi_j(t)\overline{\varphi_j(t)},\tag{12}$$

where n represents the total number of variables in the new matrix; λ_j represents the variance of the jth rescaled variable and the collection of λ_j values are the eigenvalues of the COV matrix; and $\varphi_j(t)$ are eigenvectors of the COV matrix. The number of eigenvectors depends on the number of discrete time intervals. It is easy to obtain eigenvalues and eigenvectors of the COV matrix using mathematical or statistical software.

Based on the KL theorem, a stochastic process can be represented by the function

$$\xi(t) = \sum_{j=1}^{n}\varphi_j(t)\,\zeta(\lambda_j),\tag{13}$$

where the set of $\varphi_j(t)$ are the eigenfunctions of the COV matrix; $\zeta(\lambda_j)$ is a stochastic measure defined on a second order random field [6]. The property of this stochastic measure $\zeta(\lambda_j)$ depends on its mean and variance. It is not possible to simulate this stochastic measure because of its randomness. Therefore, a number of deterministic neural networks is first developed to simulate eigenfunctions of the COV matrix from the rescaled data generated by projecting the original data onto the eigenvectors. These data represent eigenfunction values at discrete time points. A separate neural network is developed to simulate each eigenfunction based on the data for time (input) and eigenfunction value (output). The number of deterministic neural networks is decided by the number of eigenvalues which play a significant role according to the KL representation of these real realisations. Next, the stochastic measure is represented by White noise and is embedded into each of the neural networks that are then assembled into a SNN. White noise $\zeta(\lambda_j)$ is an element of a Gaussian distribution with zero mean and variance σ^2 which in this case is equal to the eigenvalue λ_j. In summary, the input and output mappings of deterministic neural networks for data collected from a real stochastic environment involve discrete times and corresponding eigenfunction values determined from the decomposition of the COV matrix.

3.2 Modelling Deterministic Neural Networks

After defining the input-output mapping of deterministic neural networks, the next step is to develop and model a suitable neural network to represent or mimic the patterns in each eigenfunction. Some of the main deterministic neural networks for function approximation include: Multilayer Perceptron Networks [4, 5, 7, 14], Radial Basis Function Neural Networks [11] and Approximate Identity Neural Networks [3]. All of them have powerful capability in approximating arbitrary deterministic input-output mapping. In this chapter, a series of Approximate Identity Neural Networks (AINNs) are used to learn these significant eigenfunctions whose values at discrete time points are obtained from the KL expansion of the COV matrix. Turchetti [3, 17] highlighted the efficacy of these networks.

In modelling with AINNs, three factors require attention: the number of neurons needed, the structure of networks and the learning algorithm.

3.2.1 Number of Neurons Required

The number of neurons in a deterministic neural networks is case-dependent. Generally, the number of neurons is adjusted during training until the network output converges on the actual output based on a least square error minimization approach. A neural network with an optimum number of neurons will reach the desired minimum error level more quickly than other networks with more complex structure. A trial and error process was used here for this purpose which worked out well.

3.2.2 The Structure of Networks

The proposed deterministic neural network is an AINN. The approximate function in neurons is the AI function given by

$$\omega(x) = \tanh\left(\frac{v(x-\vartheta)+\sigma}{2}\right) - \tanh\left(\frac{v(x-\vartheta)-\sigma}{2}\right), \tag{14}$$

where v defines the sharpness of the function, ϑ defines the centre of symmetry and the value of σ defines the position of the maximum of the function.

In order to approximate each significant eigenfunction based on AI function (14), it is necessary to develop one-dimensional AINNs. Now let us assume the input vector set $x = [t]$ and let the following function define the AINN for eigenfunction $\varphi_n(t)$:

$$\varphi_n(t) = \sum_{i=1}^{j} a_i \, \omega_i(t)$$

$$= \sum_{i=1}^{j} a_i(\tanh\left(\frac{v_i(t-\vartheta_i)+\sigma_i}{2}\right) - \tanh\left(\frac{v_i(t-\vartheta_i)-\sigma_i}{2}\right)) \tag{15}$$

where a_i displays weights of the ith neuron; j represents the total number of neurons in this AINN and $\omega(t, \sigma_i)$ are AI functions. Figure 2 displays the architecture of a deterministic neural network with one-dimensional input case. $\varphi_n(t)$ is the output of the deterministic neural network given in (15). There are as many networks as there are significant eigenfunctions.

3.2.3   The Learning Algorithm

The gradient descent algorithm in batch mode was used as the learning algorithm in this case to update weights and other parameters of networks. These are a_i, v_i ϑ_i and σ_i. It minimizes the network's global error between the actual network outputs and

Fig. 2 Architecture of one-dimensional AINN given by [14]

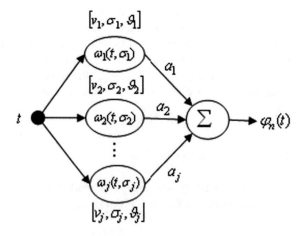

their corresponding desired outputs based on gradient descent error minimisation. The network's global error between the network output and actual output is

$$E = \frac{1}{2N} \sum_i^N (T_i - Z_i)^2 \tag{16}$$

where T_i and Z_i are the actual output and the network output, respectively, for the ith training pattern, and N is the total number of training patterns.

The method of modifying a weight or a parameter is the same for all weights and parameters. The change to a single weight of a connection between neuron j and neuron i in the AINN based on batch learning can be defined as

$$\Delta w_{ji} = \eta \sum_{p=1}^k \left(\frac{dE}{dw_{ji}} \right)_p \tag{17}$$

where η is the learning rate with a constant value. It controls the step size and the speed of weight adjustments. k is the total number of epochs (batches) and p is batch counter. Within brackets in Eq. 17 is the gradient of error with respect to the parameter to be adjusted for epoch p. The process of training is repeated until the weights and the parameters of the network converge on acceptably optimal values.

3.3 Modelling Stochastic Neural Networks

After the training of deterministic neural networks is completed, the adjustment of weights of the stochastic neural network has been completed. Next step is to obtain stochastic properties of the network by adding White noise processes into deterministic neural networks as shown in Fig. 3. From the KL expansion of the covariance matrix (13) for a stochastic process, it can be seen that the whole stochastic process can be regarded as a linear combination of the product of these independent eigenfunctions ($\varphi(t)$) and their corresponding stochastic measure $\zeta(\lambda)$ defined on the second order field. For these stochastic measures, the mean is equal to zero and variances are equal to the corresponding eigenvalues. Therefore, White

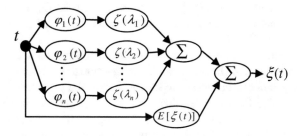

Fig. 3 The structure of stochastic neural network given in Eq. 18 [10]

noise process can be used to achieve the stochastic behaviour of the corresponding networks because White noise processes have the same attributes of these stochastic measures [3, 8, 17]. As a result, the function of this type of neural networks can be written as

$$\xi(t) = E[\xi(t)] + \sum_i \varphi_i(t)\zeta(\lambda_i) = E[\xi(t)] + \sum_i \sum_j a_{ij}\omega_{ij}(t)\zeta(\lambda_i) \qquad (18)$$

where $E[\xi(t)]$ is the expected value of the stochastic process at each discrete time t; $\varphi_i(t)$ is the output of the ith AINN; $\zeta(\lambda_j)$ is a White noise process; $\omega_{ij}(\bullet)$ is AI functions; a_{ij} is the weight of the jth neuron of the ith AINN; i is the number of eigenfunctions from KL expansion of the covariance matrix (i is also the number of neurons in the SNN (i.e., the number of AINNs)) and j is the number of neurons in each AINN. Figure 3 shows the structure of SNNs based on eigenfunctions and their corresponding White noise.

This stochastic neural network based on White noise was successfully applied to model two stochastic processes, a detail treatment of which is given in the next section.

4   Results and Discussion

Based on the mathematical developments of the stochastic neural network, some realistic examples are used to explain in detail steps of development of deterministic neural networks as well as stochastic neural networks.

4.1   Sine Function with Random Noise

The first example involves using a stochastic neural network to simulate the stochastic Sine Function model. For this purpose, the six realisations extracted over 100 time steps from the stochastic Sine Function model shown in Fig. 1 was used as the data set. The first step in developing a stochastic neural network is to calculate the covariance matrix for these six realisations in order to create input-output mappings for the networks using the KL theorem. We used Eq. 11 to approximate the covariance matrix and the behaviour of the covariance matrix for the six realisations is shown in Fig. 4.

According to the KL theorem, the covariance matrix can be decomposed into a series of eigenvalues and the corresponding eigenfunctions. Figure 5 shows only the first few eigenvalues in the KL representation of the covariance matrix as the others are zero. The figure shows that only four eigenvalues are significant and together they capture the total variance in the original data. Therefore, only four significant eigenvalues as well as their corresponding eigenfunctions are relevant to

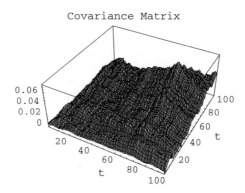

Fig. 4 The covariance matrix of six realisations extracted from the stochastic Sine Function model

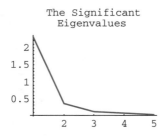

Fig. 5 The significant eigenvalues in the KL representation of the covariance matrix

network development. The relative amount of variance in the data captured by each significant eigenvalue from KL expansion is 79.9, 12.1, 4.1 and 2.7 %.

The number of AINNs depends on the number of these significant eigenvalues. As a result, four individual AINNs are used to simulate the four eigenfunctions represented by the four eigenvectors. Figure 6 shows the values of the four eigenfunctions as well as their corresponding AINN approximations. In the figure, the red points represent eigenfunction values at discrete time points determined from the KL expansion while the black lines represent the approximated outputs from the networks. Each AINN has a high accuracy of learning their input-output mappings (the range of R^2 is between 0.96–0.97).

The next step is to incorporate stochastic measures into the neural network. These stochastic measures are representations of white noise with zero mean and variance equal to eigenvalues corresponding to the eigenfunctions, As a result, the proposed stochastic neural network can be considered as a linear combination of the product of eigenfunctions and their corresponding White noise. The stochastic neural network was assembled as in Eq. 18 and Fig. 3, and some realisations obtained from the developed stochastic neural network are shown in Fig. 7. They are remarkably similar to the realisations obtained from the original function shown in Fig. 1. In order to confirm the validity of the proposed stochastic neural network,

Fig. 6 The four
approximated eigenfunctions
from the AINN (*black line*)
superimposed on
eigenfunction values from the
KL expansion (*red dots*)

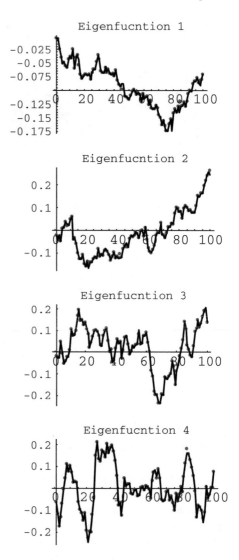

the covariance function of the developed stochastic neural network can be compared with that obtained from realistic realisations. Figure 8 displays the predicted
covariance function for 200 realisations extracted from the developed network and
it is very similar to the actual covariance matrix in Fig. 4.

In order to compare the actual and predicted covariance matrices, error distribution (difference between the predicted values and actual values in the COV
matrix) as well as R^2 were assessed as shown in Fig. 9. Figure 9a shows that error
ranges from -0.002 to 0.006 indicating that the approximated covariance matrix
closely follows the actual covariance matrix. According to Fig. 9b, there is a strong

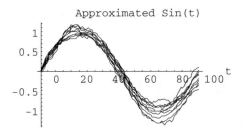

Fig. 7 Ten realisations obtained from the stochastic neural network

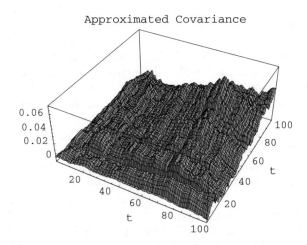

Fig. 8 Covariance matrix of the approximated Sine Function obtained from 200 realisations of the stochastic neural network

(a)   **(b)**

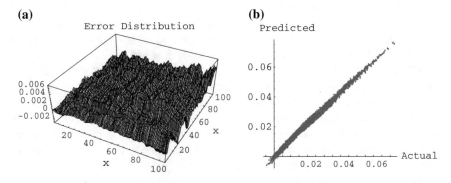

Fig. 9 **a** Distribution of error between the actual and predicted covariance matrices; **b** the linear relationship between the actual and predicted covariance

linear relationship between the predicted and the actual covariance matrices ($R^2 = 0.993$).

4.2 Displacement Fields of Wood in Compression

The previous example revealed that the proposed stochastic neural network has high accuracy in simulating the Sine function with noise. Here, this idea is applied to a much more complex realistic problem involving simulation of localised displacement fields of wood in compression. For comprehensive details, refer to [10].

Wood like most biological and cellular materials has a very complicated internal structure that leads to variability in properties. In a recent study, image processing methods have been used to obtain displacements at a large number of points (i.e., displacement fields) in a small area on the surface of a loaded wood specimen in compression parallel-to-grain [15, 16]. These provided a limited number of realisations of displacement and it is anticipated that stochastic memory of a stochastic neural network can help recall more realisations of the behaviour of this wood specimen based on the existing data and help understand subtle structural influence on mechanical behaviour of wood. The wood specimens were of the dimensions $41 \times 44 \times 136$ mm where the latter is the height which is parallel to the wood grain, and they were cut from kiln-dried structural grade New Zealand radiate pine (*Pinus radiata*) boards [15, 16]. These were tested on a computer controlled material testing facility that measured the applied load while a camera took images of the specimen at various load levels. By comparing the displaced images to the initial undisplaced images for an area around 20×20 mm on one surface of the specimen using a Digital Image Correlation (DIC) method, displacements of a large number of points were determined. Although, the overall area analysed is small, it is believed that this area contains all the microstructural effects that can be found elsewhere in the same specimen and in this particular wood. The data determined this way contains two different displacements for each point: vertical and horizontal displacements. The proposed stochastic neural network is applied here to analyse the structural influence on localised displacements of wood.

When a 20 kN (kilo Newtons) compression load is applied parallel-to-grain on a specimen, vertical displacement (u) measures the amount of contraction in the same direction of loading while horizontal displacement (v) measures the amount of expansion in the perpendicular direction to loading. Figure 10 shows both vertical and horizontal displacement realisations obtained from images as discussed before. Here, one vertical displacement (u) realisation corresponds to one column of 30 points in the image and one horizontal displacement (v) realisation to one row of 30 points in the image. There are 21 such realisations in each case. In Fig. 10, it can be seen that the influence of structure in loading parallel-to-grain on horizontal displacement (Fig. 10b) is more complex and fluctuating than on vertical displacement (Fig. 10a).

Fig. 10 Vertical and horizontal displacement profiles for a wood specimen loaded 20 kN in compression

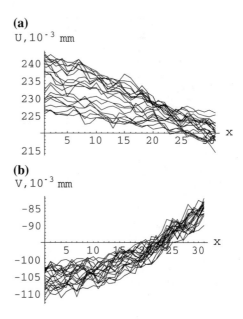

(a)

U, 10^{-3} mm

(b)

V, 10^{-3} mm

Figure 11 displays the covariance matrix of vertical displacement (Fig. 11a) as well as horizontal displacement (Fig. 11b) and reveals that the covariance matrix for horizontal displacement is much more complex.

Based on KL expansion, the covariance matrices of both vertical and horizontal displacements were decomposed in order to create deterministic input-output mappings. For the vertical displacement, there were only three significant eigenvalues from the distribution of all eigenvalues. However, eight significant eigen values were needed to capture the variance in the original data for the horizontal displacements. The higher number of significant eigenvalues for horizontal displacements also indicates that there is a lot more noise or complexity in this direction as also supported by Figs. 10b and 11b. Thus, three AINNs were needed for the vertical displacement and eight AINNs for the horizontal displacement to approximate their corresponding eigenfunctions. All these neural networks were developed and when the learning step is completed, most components of the proposed stochastic neural network have already been determined. The next step is to add the relevant White noise into their corresponding AINNs in order to achieve stochastic properties of the network as depicted in Fig. 3.

Figure 12a, b display some realisations obtained from the developed stochastic neural networks for the vertical displacement and the horizontal displacement, respectively. They are remarkably similar to the actual realisations shown in Fig. 10. Figure 13 shows the approximated covariance matrices from the stochastic neural networks. They closely resemble the respective actual covariance matrices depicted in Fig. 11 (R^2 values 0.971 and 0.969, respectively).

Fig. 11 a Covariance matrix
of the vertical displacement
(u); **b** covariance matrix of the
horizontal displacement
(v) from original experimental
realisations [10]

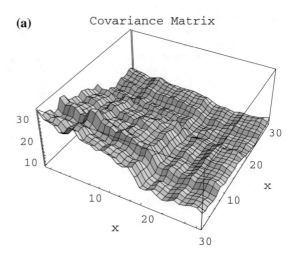

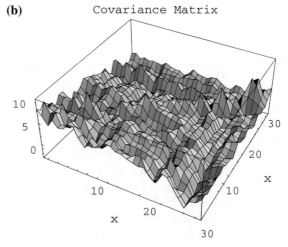

4.3 *Analysing Internal Workings of SNNs*

After confirming the validity of the proposed stochastic neural network, more
attention can be paid to analysing internal workings of the network. The purpose of
this is to completely understand why and how White noise can help a stochastic
neural network achieve its stochastic properties for simulating a stochastic process
or a stochastic system. In this section, the focus is on the following aspects: analyse
how all neurons produce the outcomes of the network in a particular realization and
compare any two different realisations obtained from the stochastic neural network.
The process of development of SNN in the examples presented in the previous
section is the same but different networks have different number of neurons as well

Fig. 12 Predicted realisations from the stochastic neural network: **a** vertical displacement; **b** horizontal displacement

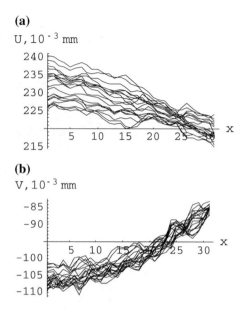

as different characters of White noise. The number of neurons in each stochastic neural network depends on the number of significant eigenfunctions indicated by the KL expansion of the covariance matrix. The characters of each White noise component depend on their corresponding eigenvalues. Therefore, one network is illustrated here as an example to analyse internal workings of a stochastic neural network. The SNN for simulating stochastic Sine Function was selected for this purpose (See Fig. 1).

In terms of this stochastic neural network, there are four significant eigenfunctions and therefore, four neurons in the stochastic neural network. The formulation involved in calculating the output of this stochastic neural network is shown in Eq. 18 and the structure of this stochastic neural network is displayed in Fig. 3. Figure 3 shows that the output of SNNs is a linear combination of the expected value at the each discrete position and the summation of the product of each eigenfunction and their corresponding White noise at the same discrete position. The expected value $E[\xi(t)]$ as well as eigenfunctions $\varphi_i(t)$ are fixed by the data and the deterministic network. The reason is that the expected value $E[\xi(t)]$ depends on the input space defined by the data collected from the real stochastic system and the eigenfunctions $\varphi_i(t)$ depend on the approximated AINN that have been already trained prior to developing the stochastic neural network. The distribution of the mean value $E[\xi(t)]$ is shown in Fig. 14 while the distribution of each eigenfunction has been shown in Fig. 6. Therefore, the product of each eigenfunction and their corresponding White noise mainly contributes to the difference between any two realisations obtained from SNNs. This means that the stochastic properties of SNNs arise from the summation of the product of each eigenfunction and their corresponding White noise.

(a)

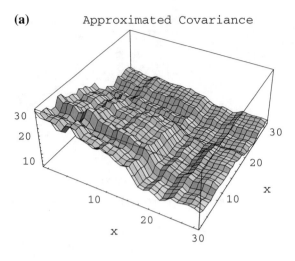

(b)

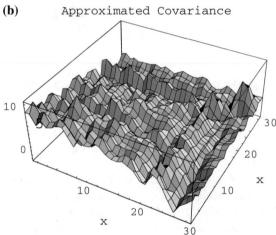

Fig. 13 Approximated covariance matrices: **a** vertical displacement (u); **b** horizontal displacement (v)

Let's look at a particular realisation from the stochastic neural network. Here, the four corresponding White noise components are constant for a particular realisation. According to Eq. 18, the output function of the network, which is shown in Fig. 15a (red line), can be written as $\mathrm{Sin}(t) = E[\mathrm{Sin}(t)] + (0.24\varphi_1(t) + 0.68\varphi_2(t) - 0.29\varphi_3(t) - 0.25\varphi_4(t))$, where constants are from White noise processes. In Fig. 15a, the black line is the expected value $E[\mathrm{Sin}(t)]$. Furthermore, the distance between the mean value and the output of the network is the summation of the product of each eigenfunction and its corresponding White noise process. This summation plays an important role in achieving stochastic properties of the network. Figure 15b shows the distribution of this product across the time domain and

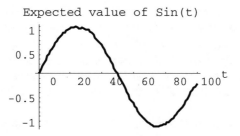

Fig. 14 The distribution of the mean value at each discrete position t

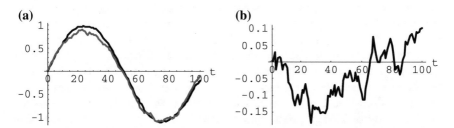

Fig. 15 a A realisation extracted from the stochastic neural network for simulating Sine function superimposed on expected values for discrete time points; **b** the distribution of the summation of the product of each eigenfunction and its corresponding White noise process

it indicates that this product also represents random fluctuations. Consequently, the difference in value between two discrete points mainly depends on the corresponding expected value and the summation of the product of each eigenfunction and its corresponding White noise.

However, the output of the stochastic neural network changes when another realisation is generated (see Fig. 16a). The reason for the difference is that the value for each White noise component changes when a new realisation is generated using

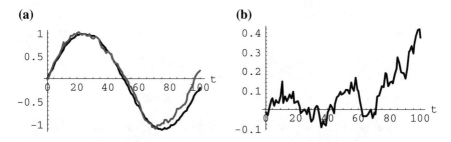

Fig. 16 a Another realisation extracted from the same stochastic neural network for simulating Sine function; **b** distribution of the summation of the product of each eigenfunction and its corresponding White noise process

the stochastic neural network. For example, the output function of the network in this case is $\text{Sin}(t) = E[\text{Sin}(t)] + (-0.86\varphi_1(t) + 0.9\varphi_2(t) + 0.65\varphi_3(t) - 0.16\varphi_4(t))$. The distribution of the summation of the product of each eigenfunction and their corresponding White noise for this case is displayed in Fig. 16b. Comparing Figs. 15 and 16, it is easy to see a significant difference between the output of the stochastic neural network as well as the summation of the product of eigenfunctions and White noise.

In the analysis of internal behaviour of the networks, another focus is how all neurons affect the final outcomes of the network. In this case, there are four stochastic neurons in this stochastic neural network. According to the previous discussion, these four stochastic neurons together make a contribution to produce the distance between the expected value and the final output of SNNs. In order to assess this, 50 realisations were generated and contribution of each neuron to the desired output was evaluated. Figure 17 shows how each individual stochastic neuron contributes to the distance in four selected realisations. In Fig. 17, the black line displays the distance between the expected values and the final output for each realisation and the colour lines display the contribution of each individual stochastic neuron of the network to this value. It was found that different realisations are dominated by different neurons. Furthermore, in a single realisation, individual neurons make varied contributions to the final output. However, after analysing 50 realisations, it was revealed that the proportion of each stochastic neuron's contribution to the final output is quite similar to the ratio of the corresponding eigenvalue to the sum of all eigenvalues (i.e., proportion of total variance captured by each eigenvalue). This means that in generating a large number of realisations,

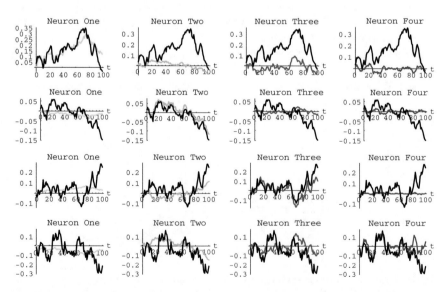

Fig. 17 Activation of the four stochastic neurons superimposed on the network output for four selected realisations

the contribution each stochastic neuron makes is equal to the portion of the total variance in data captured by it.

A similar analysis was made on the realisations obtained from the two SNNs developed for modelling wood displacements. The SNN representing the vertical displacement (u) had 3 neurons and that for horizontal displacement (v) had 8 neurons. Observations from both of these in terms of individual neuron contributions were conceptually similar to those from the SNN for Sine function with noise. For example, when 50 extracted realisations were analysed, neuron 1 became prominent 93.2 % of the time, neuron 2—1.47 % of the time, and neuron 3— 1.18 % of the time. These correspond to the contribution of the corresponding eigenvalues to the sum of eigenvalues (i.e, total variance in the data). Three selected examples for vertical displacement are shown in Fig. 18.

In summary, although the mean value and each eigenfunction do not change in different realisations, the changes of White noise components make the summation of the product of each eigenfunction and their corresponding White noise vary in these realisations. Furthermore, it is not possible to predict the exact realisation generated from the stochastic neural network due to the randomness of White noise processes. Thus SNN has stochastic properties and it can be used to recall more novel realisations that are characteristic of the system and are within the range of the statistical properties of the collected realisations.

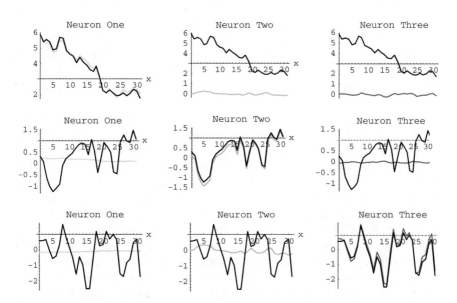

Fig. 18 Output of each of the three stochastic neurons in three selected realisations superimposed on the network output for modelling vertical displacements in wood

5   Conclusions

A major contribution of this chapter is the development of SNNs from the observations of stochastic processes. Two examples (stochastic Sine function and noisy wood displacements) provided evidence to confirm the validity of the theoretical results as well as give more confidence in developing a neural network to simulate real stochastic systems and stochastic processes. Taking the whole process of developing a stochastic neural network into account, the most important aspect is creating a possible and feasible mathematical model for neural networks to achieve stochastic properties by using a series of mathematical transformations. As a deterministic neural network can only approximate non-random mappings, creating a deterministic input-output mapping from stochastic processes or stochastic systems was a major part of the work. This is because the weights and other parameters are defined through learning of these deterministic input-output mappings. These then are used in conjunction with White noise to develop a suitable stochastic neural network for simulating stochastic processes. Furthermore, this chapter also provided an analytical view of the network outcomes and internal workings of the network. This information helps understand how a stochastic neural network simulates real stochastic processes and stochastic systems.

This kind of neural network is suitable to operate in a non-deterministic environment. The most significant advantage of developing SNNs is to successfully model an existing dataset and retrieve more information out of the collected dataset. After completing the training, a stochastic neural network can act as a stochastic process itself. This means that the stochastic neural network can generate more realisations of a stochastic process above and beyond the collected realisations. These realisations can be viewed as representing the behaviour of the stochastic process and they can help understand in-depth the properties of stochastic processes.

The proposed stochastic neural network however has some limitations. When a neural network is developed to learn a deterministic input-output mapping, it is very difficult to guess the initial values of weights, other parameters of the networks and the number of neurons in the networks. However, these factors can directly affect whether a suitable neural network is found or not. A precise standard to confirm whether the proposed neural network is the most optimum with the highest possible accuracy is currently lacking. Furthermore, neuron activation functions (AINNS) that are simpler to be modelled by the learning algorithm were used. Although this is good for the examples presented here, more powerful activation functions have the potential to perform more complex mapping suited to more complex stochastic processes in biological or environmental systems. Therefore, there is much scope for improving and optimizing the developed stochastic neural networks in future.

References

1. F.R. Araujo, A.S. Henriques, A Stochastic neural model for fast identification of spatiotemporal sequences. Neural Process. Lett. **16**, 277–292 (2002)
2. J. Bear, Hydrodynamic dispersion, in *Flow Through Porous Media* (Academic Press, New York, 1969)
3. M. Conti, C. Turchetti, Approximation of dynamical systems by continuous-time recurrent approximate identity neural network. Neural Parallel Sci. Comput. **2**, 299–322 (1994)
4. G. Cybenko, Approximation by superposition of sigmoidal function. Math. Control Syst. Signals **2**, 303–314 (1989)
5. K. Funahashi, On the approximate realization of continuous mappings by neural networks. Neural Netw. **2**, 183–192 (1989)
6. I.I. Gilman, A.V. Skorohod, *The Theory of Stochastic Processes* (Springer, Berlin, 1974)
7. K. Hornik, M. Stinchcombe, H. White, Multilayer feedforward networks are universal approximators. Neural Netw. **2**, 395–403(1989)
8. D. Kulasiri, W. Verwoerd, *Stochastic Dynamics Modeling of Solute Transport in Porous Media*. North Holland Applied Mathematics and Mechanics Series, vol. 44. (Elsevier Science, Amsterdam, 2002)
9. C.-C.J. Lin, J. Ghaboussi, Generating multiple spectrum compatible accelerograms using stochastic neural networks. Earthq. Eng. Struct. Dyn. **30**, 1021–1042 (2001)
10. L. Ling, S. Samarasinghe, D. Kulasiri, Modelling variability in full-field displacement profiles and Poisson ratio of wood in compression using stochastic neural networks. Silva Fennica **43** (5), 871–887 (2009)
11. J. Park, W. SandbergI, Universal approximation using radial-basis-function network. Neural Comput. **3**, 246–257 (1991)
12. P.E. Pfeiffer, *Probability for Application* (Springer, New York, 1990)
13. D. Rosaci, Stochastic neural networks for transportation systems modeling. Appl. Artif. Intell. **12**, 491–500 (1997)
14. S. Samarasinghe, *Neural Networks for Applied Sciences and Engineering: From Fundamentals to Complex Pattern Recognition* (Auerbach, Boca Raton, 2006)
15. S. Samarasinghe, G.D. Kulasiri, Displacement fields of wood in tension based on image processing: Part 1. Tension parallel- and perpendicular-to-grain and comparison with isotropic behaviour. Silva Fennica **34**(3), 51–259 (2000)
16. S. Samarasinghe, G.D. Kulasiri, Displacement fields of wood in tension based on image processing: Part 2. Crack-tip displacement in mode-I and mixed-mode fracture. Silva Fennica **34**(3), 261–274 (2000)
17. C. Turchetti, *Stochastic Models of Neural Networks* (IOS Press, Amsterdam, 2004)
18. R.J.M. Wiest, Fundamental Principles in groundwater flow, in *Flow Through Porous Media* (Academic Press, New York, 1969)

Curvelet Interaction with Artificial Neural Networks

Bharat Bhosale

Abstract Modeling helps simulate the behavior of a system for a variety of initial conditions, excitations and systems configurations, and that the quality and the degree of the approximation of the models are determined and validated against experimental measurements. Neural networks are very sophisticated techniques capable of modeling extremely complex systems employed in statistics, cognitive psychology and artificial intelligence. In particular, neural networks that emulate the central nervous system form an important part of theoretical and computational neuroscience. Further, since graphs are the abstract representation of the neural networks; graph analysis has been widely used in the study of neural networks. This approach has given rise to a new representation of neural networks, called Graph neural networks. In signal processing, wherein improving the quality of noisy signals and enhancing the performance of the captured signals are the main concerns, graph neural networks have been used quite effectively. Until recently, wavelet transform techniques had been used in signal processing problems. However, due to their limitations of orientation selectivity, wavelets fail to represent changing geometric features of the signal along edges effectively. A newly devised curvelet transform, on the contrary, exhibits good reconstruction of the edge data; it can be robustly used in signal processing involving higher dimensional signals. In this chapter, a generalized signal denoising technique is devised employing graph neural networks in combination with curvelet transform. The experimental results show that the proposed model produces better results adjudged in terms of performance indicators.

1 Introduction

The paradigm, methods and results of the ongoing research in Artificial Neural Networks (ANN) are quite fascinating. First inspired by the human brain and its natural structure, ANN combined with other Artificial Intelligence (AI) technologies

B. Bhosale (✉)
University of Mumbai, (M.S), Mumbai, India
e-mail: bnbhosale15@gmail.com

© Springer International Publishing Switzerland 2016
S. Shanmuganathan and S. Samarasinghe (eds.), *Artificial Neural Network Modelling*, Studies in Computational Intelligence 628,
DOI 10.1007/978-3-319-28495-8_6

109

may in return help create 1 day an artificial or hybrid life-form. With the multitude of commercial applications and research advancing on many fronts, ANN is attracting a lot of attention [1]. The recent advancement in the theory of neural networks has inspired new applications in the field of neuroscience such as study of models of neural networks, anatomical, and functional connectivity based upon functional magnetic resonance imaging, electroencephalography and magnetoencephalography [2]. Of late, wavelet based multi-resolution techniques have been widely used in image/signal processing, biological and computer vision, scientific computing, optical data analysis. For example, applications of wavelet transform have been explored in analyzing solitons [3], bio-informatics [4], neural networks [5] and many more. Since Olshausen and Field's work in *Nature* [6], researchers in biological vision have discussed the similarity between vision and multi-scale image processing. However, wavelets do not provide good direction selectivity, which is also an important response property of simple cells and neurons at stages of the visual pathway. To overcome this limitation, through several attempts, significant progress in the development of directional wavelets has been made in recent years. The complex wavelet transform was devised to improve directional selectivity to some extent; however, owing to the difficulties in its design with reconstruction properties and filter characteristics, it has not been widely used. In 1999, an anisotropic geometric wavelet transform, named ridgelet transform, was proposed by Candes and Donoho but it has also limited applicability to objects with global straight-line singularities. Although the ridgelet transform optimally represents straight-line singularities, the global straight-line singularities are rarely observed in real applications and hence limited its scope. To analyze local line or curve singularities, the obvious way is to consider partition of the signal/image, and then to apply the ridgelet transform to the resulting blocks/sub-images. This block ridgelet-based transform, named curvelet transform, was first proposed by Candès and Donoho [7]. Later, a considerably simpler second-generation curvelet transform based on a frequency partition technique was proposed by the same authors. This second-generation curvelet transform has been proved a versatile and efficient tool for wide range of applications from diverse fields such as signal/image processing, seismic data exploration, fluid mechanics, and solving partial differential equations encountered in non-linear physical phenomena. This newly devised curvelet transform is a special multi-scale pyramid with many directions and positions at each decomposition scale and therefore more suitable than all other multi-scale transforms used in signal and image processing applications including, filtering, enhancement, compression, de-noising, and watermarking. Hence, recently, image denoising using curvelet transform has been potentially used in many fields for its ability to obtain high quality images [8]. As most of the natural signals are assumed to have additive random noise, which is modeled as Gaussians, removing additive Gaussian noise by nonlinear methods such as curvelet denoising has better results than classical approaches [9]. As an example, from the analysis of CT scan images, denoising by curvelet transform recovers the original image from the noisy one using lesser coefficients than denoising using the wavelet transform [10]. Further, the curvelets resemble local plane waves; curvelet induced sparse representation of the local seismic events can be effectively used for preserving

wave fronts in seismic processing [11]. Another study presented an integrated classification machine (ICM), which is a hierarchy of artificial neural network, trained to
classify the seismic waveforms [12]. A velocity model inversion approach using
artificial neural networks for the study of aftershocks from the 2000 Tottori, Japan,
earthquake located around station SMNH01was proposed by Moya and Irikura [13].

Various techniques are reportedly used in signal extraction and related applications. Prominent amongst them are: one, separating noisy mixed signals using
fast Independent Component Analysis (ICA) algorithm and then applying curvelet
thresholding and the other one, using neural network thresholding to denoise the
mixed signals. The present work elucidates a systematic transition from classical
wavelets to curvelets; and proposes a model that unifies both these approaches.

2   Classical Wavelets

Wavelets are the mathematical functions which analyze data according to the scale
or resolution. Wavelets help in studying a signal in different windows or in different
resolutions. Practically wavelet transform is a convolution of the signal with a
family of functions obtained from a basic wavelet by shifts and dilations. In precise
terms and notations, the classical wavelet transform, also called Continuous
Wavelet transform (CWT), is a decomposition of a function, $f(x)$, with respect to a
basic wavelet, $\psi(x)$, given by the convolution of a function with a scaled and
translated version of $\psi(x)$

$$W_\psi(a,b)[f] = |a|^{-1/2} \int f(x)\psi^*\left(\frac{x-b}{a}\right)dx = \left\langle f, \frac{1}{\sqrt{|a|}}\psi\left(\frac{x-b}{a}\right)\right\rangle \quad (1)$$

The functions, f and ψ are square integrable functions and ψ satisfies the
admissibility condition, $C_\psi = \int \frac{|\hat{\psi}(\omega)|^2}{|\omega|}d\omega < \infty$, where C_ψ is the admissibility
constant.

With the substitution for $f(x)$ as the inverse Fourier transform,
$f(x) = \frac{1}{2\pi}\int_{-\infty}^{\infty}\exp(i\omega x)\hat{f}(\omega)d\omega$, wavelet transform (1) takes the form

$$W_\psi[f(x)](a,b) = \frac{1}{2\pi}|a|^{1/2}\int_{-\infty}^{\infty}\exp(i\omega b)\overline{\hat{\psi}(a\omega)}\hat{f}(\omega)d\omega \quad (2)$$

In discrete form, the wavelet transform (1) becomes,

$$W_\psi(m,n)[f] = \frac{1}{\sqrt{a_0^m}}\int_{-\infty}^{\infty}f(x)\psi\left(\frac{x-nb_0a_0^m}{a_0^m}\right)dx = \frac{1}{\sqrt{a_0^m}}\int_{-\infty}^{\infty}f(x)\psi\left(a_0^{-m}x - nb_0\right)dx$$

$$(3)$$

with the scale and translation parameters as $a = a_0^m$ and $b = n b_0 a_0^m$, where a_0 and b_0 are the discrete scale and translation step sizes, respectively.

The signal $f \in L^2(R)$ can be uniquely represented in wavelet expansion, $f = \sum_{j,k} c_{j,k}(f) \psi_{j,k}$, where $c_{j,k}(f) = \langle f, \psi_{j,k} \rangle$ are the wavelet coefficients and $\psi_{j,k}$ is a family of dilated and translated functions $\{ \psi_{j,k} = 2^{j/2} \psi(2^j, -k) : j, k \in Z \}$, generated by the mother wavelet $\psi \in L^2(R)$.

For a good frequency localization of the wavelet basis, the basic idea is to construct a wavelet basis that provides a partition of the frequency axis into (almost) disjoint frequency bands (or octaves). Such a partition can be ensured if the Fourier transform of the dyadic wavelet $\hat{\psi}$, $\hat{\psi}_{j,k}(\xi) = 2^{-j/2} e^{-i2^{-j}\xi k} \hat{\psi}(2^{-j}\xi)$, has a localized or even compact support and satisfies the admissibility condition, $\sum_j \left| \hat{\psi}(2^{-j}\xi) \right|^2 = 1, \xi \in R$. This admissibility condition also ensures the typical wavelet property, $\hat{\psi}(0) = \int_{-\infty}^{\infty} \psi(x) dx = 0$, which in turn ensures that the family of function $\{ \psi_{j,k} : j, k \in Z \}$ forms a tight frame of $L^2(R)$.

The same construction principle can be transferred to the 2D case for signal/image analysis by incorporating certain rotation invariance in proposed curvelet systems.

3   Classical Wavelets to Curvelets

The wavelet transform, especially, the discrete wavelet transform (DWT) has been an impressive tool for mathematical analysis and signal processing, but it suffers from the disadvantage of poor directionality, which has undermined its usage in many applications. To overcome this limitation, through several attempts, significant progress in the development of directional wavelets has been made in recent years. The modifications went through several versions: the complex wavelet transform to ridgelet transform, and then to block ridgelet-based transform, named curvelet transform, proposed by Candès and Donoho in 2000.

Curvelet transform is a new extension of wavelet transform which aims to deal with interesting phenomena occurring along curved edges in 2D images. In the 2D case, the curvelet transform allows optimal sparse representation of objects with singularities along smooth curves. Moreover, the curvelet methods preserve the edges and the structures better than wavelet transform. Owing to such advantages over the traditional multi-scale techniques, recently, the curvelets have been applied to study the non-local geometry of eddy structures and the extraction of the coherent vortex field in turbulent flows. In fluid mechanics, turbulence analysis- an efficient compression of a fluid flow with minimum loss of the geometric flow structures- is a crucial problem in the simulation of turbulence. Yet another novel application of the curvelet transform to the compressed sensing or compressive sampling (CS) is an inverse problem with highly incomplete measurements, which carries imaging and compression simultaneously. The CS based data acquisition depends on its sparsity rather than its bandwidth and it has an important impact for designing of measurement devices in various engineering fields such as medical magnetic resonance (MRI) imaging and remote sensing, especially for cases involving incomplete and inaccurate measurements limited by physical constraints, or very expensive data acquisition. In sum, curvelets outperform the wavelets in many ways.

4 Curvelet Transform

This formulation is due to Candes and Donoho [7], which involves two main components: one, considering polar coordinates in frequency domain and two, constructing curvelet elements being locally supported near wedges.

The Continuous Curvelet Transform (CCT), $f \rightarrow \Gamma_f(a, b, \theta)$, of functions $f(x_1, x_2)$ on R^2, into a transform domain with continuous scale $a > 0$, location $b \in R^2$, and orientation $\theta \in [0, 2\pi)$ is formulated as follows:

Consider the polar coordinates in frequency domain, $r = \sqrt{\xi_1^2 + \xi_2^2}$, $\omega = arctan \frac{\xi_1}{\xi_2}$ corresponding to $x = (x_1; x_2)^T$, the spatial variable, and $\xi = (\xi_1, \xi_2)^T$, the variable in frequency domain. Also consider the pair of windows $W(r)$ and $V(t)$, called radial window and angular window respectively

$$V(t) = \begin{cases} 1, |t| \leq 1/3 \\ \cos[\pi/2v(3|t| - 1)], 1/3 \leq |t| \leq 2/3 \\ 0, else \end{cases}$$

$$W(r) = \begin{cases} \cos[\pi/2v(5 - 6r)], 2/3 \leq r \leq 5/6 \\ 1, 5/6 \leq r \leq 4/3 \\ \cos[\pi/2v(3r - 4)], 4/3 \leq r \leq 5/3, 0, else \end{cases} \quad (4)$$

where v is a smooth function satisfying

$$v(x) = \begin{cases} 0, x \leq 0 \\ 1, x \geq 1 \end{cases}, \quad v(x) + v(1 - x) = 1, x \in R$$

The windows (4) should satisfy the admissibility conditions (5)

$$\int_0^\infty W(ar)^2 \frac{da}{a} = 1, \forall r > 0, \quad \int_{-1}^1 V(t)^2 dt = 1 \quad (5)$$

The window functions are used to construct a family of complex-valued waveforms, called curvelets, with three parameters, the *scale* $a \in (0; 1]$, the *location* $b \in R^2$, and the *orientation* $\theta \in [0; 2\pi)$. Using polar coordinates $(r; \omega)$ in frequency domain, a-scaled window, called polar wedge, is given as

$U_a = a^{3/4}W(ar)V\left(\frac{\omega}{\sqrt{a}}\right)$, for some a with $0 < a \le 1$, where the support of U_a depends on the supports of windows W and V.

Let the basic element $\gamma_{a,0,0} \in L^2(R^2)$ be given by its Fourier transform as $\hat{\gamma}_{a,0,0} = U_a(\xi)$.

By translation and rotation of basic element $\gamma_{a,0,0}$, the family of analyzing elements, called curvelets, is generated as

$$\gamma_{ab\theta}(x) = \gamma_{a00}(R_\theta(x - b)) \tag{6}$$

where R_θ is the 2×2 rotation matrix effecting planar rotation by θ radians

$$R_\theta = \begin{pmatrix} \cos\theta & -\sin\theta \\ \sin\theta & \cos\theta \end{pmatrix} \tag{7}$$

Applying this family of high frequency elements, $\{\gamma_{a,b,\theta} : (0,1],$ $b \in R^2, \theta \in [0, 2\pi\}$, the CCT, Γ_f of $f \in L^2(R^2)$, is given by

$$\Gamma_f(a, b, \theta) = \langle \gamma_{a,b,\theta}, f \rangle = \int_{R^2} \gamma_{a,b,\theta}(x)\overline{f(x)}dx \tag{8}$$

The corresponding inversion/reproducible formula is

$$f(x) = \int \Gamma_f(a, b, \theta)\gamma_{ab\theta}(x)\mu(da\,db\,d\theta) \tag{9}$$

$$\|f\|_{L^2}^2 = \int |\Gamma_f(a, b, \theta)|^2 \mu(da\,db\,d\theta)$$

where, μ denotes the reference measure, $d\mu = \frac{da}{a^3}db\,d\theta$

This formula is valid for $f \in L^2$ that has a Fourier transform vanishing for $|\xi| < 2/a_0$.

A discrete version of the continuous curvelet transform, called Discrete curvelet transform, is derived by a suitable sampling at the range of scales, orientations and locations [8].

Choosing the scales $a_j = 2^{-j}, j \ge 0$; the equidistant sequence of rotation angles $\theta_{j,l}, \theta_{j,l} = \frac{\pi l 2^{-\lfloor j/2 \rfloor}}{2}, l = 0, 1, \ldots, 4.2^{\lfloor j/2 \rfloor} - 1$; and the positions $b_k^{j,l} = b_{k_1,k_2}^{j,l} = R_{\theta_{j,l}}^{-1}\left(\frac{k_1}{2^j}, \frac{k_2}{2^{j/2}}\right), k_1, k_2 \in Z$.

This leads to a discrete curvelet system that forms a tight frame.

In this system, function, $f \in L^2(R^2)$, will be representable by a curvelet series, $f = \sum_{j,k,l} c_{j,k,l}(f)\gamma_{j,k,l} = \sum_{j,k,l} \langle f, \gamma_{j,k,l}\rangle\gamma_{j,k,l}$, for which the Parseval identity, $\sum_{j,k,l} |\langle f, \gamma_{j,k,l}\rangle|^2 = f_{L^2(R^2)}^2, \forall f \in L^2(R^2)$ holds.

The terms in the above series, $c_{j,k,l}(f) = \langle f, \gamma_{j,k,l} \rangle$, are the curvelet coefficients. The curvelet coefficients, $c_{j,k,l}$, are obtained by Plancherel's theorem for $j \geq 0$ as

$$c_{j,k,l}(f) = \int_{R^2} f(x)\overline{\gamma_{j,k,l}}(x)dx = \int_{R^2} \hat{f}(\xi)\overline{\hat{\gamma}_{j,k,l}}(\xi)d\xi = \int_{R^2} \hat{f}(\xi)U_j(R_{\theta_{j,l}}\xi)e^{i\langle b_k^{j,l}, \xi \rangle}d\xi$$

$$(10)$$

where $\hat{f}(\xi) = \frac{1}{2\pi}\int_{R^2} f(x)e^{-i\langle x, \xi \rangle}dx$.

The family of curvelet functions $\gamma_{j,k,l}(x)$, is generated from the basic curvelet,

$\hat{\gamma}_{j,0,0}(\xi) = U_j(\xi)$ as $\gamma_{j,k,l}(x) = \gamma_{j,0,0}\left(R_{\theta_{j,l}}\left(x - b_k^{j,l}\right)\right)$, where the scaled windows take the form

$$U_j(r, \omega) = 2^{-3j/4}W\left(2^{-j}r\right)V\left(\frac{2^{j/2}\omega}{2\pi}\right) = 2^{-3j/4}W\left(2^{-j}r\right)V\left(\frac{\omega}{\theta_{j,l}}\right) \qquad (11)$$

It is seen that the curvelet functions, $\gamma_{j,k,l}(x)$, in frequency domain are obtained as the inverse Fourier transform of a suitable product of the windows W and V:

$$\hat{\gamma}_{j,k,l}(\xi) = e^{-i\langle b_k^{j,l}, \xi \rangle}U_j\left(R_{\theta_{j,l}}, \xi\right) = e^{-i\langle b_k^{j,l}, \xi \rangle}2^{-3j/4}W\left(2^{-j}r\right)V\left(\frac{\omega + \theta_{j,l}}{\theta_{j,l}}\right) \qquad (12)$$

In practical implementations, Cartesian arrays instead of the polar tiling of the frequency plane are convenient to use. The Cartesian counterpart of curvelet-like functions is given by

$\gamma_{j,k,l}(x) = \gamma_{j,0,0}\left(S_{\theta_{j,l}}\left(x - b_k^{j,l}\right)\right)$, generated from the basic curvlets,

$\hat{\gamma}_{j,0,0}(x) = U_j(\xi)$, with shear matrix $S_\theta = \begin{pmatrix} 1 & 0 \\ -\tan\theta & 1 \end{pmatrix}$.

The curvelet coefficient in Cartesian arrays will be therefore obtained by

$$c_{j,k,l}(f) = \int_{R^2} \hat{f}(\xi)U_j\left(S_{\theta_{j,l}}^{-1}\xi\right)e^{i\langle b_k^{j,l}, \xi \rangle}d\xi = \int_{R^2} \hat{f}(S_{\theta_{j,l}}\xi)U_j(\xi)e^{i\langle k_j, \xi \rangle}d\xi, \qquad (13)$$

where $k_j = \left(k_1 2^{-j}, k_2 2^{-\lfloor -j/2 \rfloor}\right), (k_1, k_2)^T \in \mathbb{Z}^2$

Yet another version of curvelet transform, called Fast Digital Curvelet Transform (FDCT), is based on wrapping of Fourier samples. This takes 2D signal/image as an input in the form of a Cartesian array, $f[m, n]$, where $0 \leq m < M, 0 \leq n < N$ where M and N are the dimensions of the array; and the output will be a collection of curvelet coefficients, $c^D(j, l, k_1k_2)$, indexed by a scale j, an orientation l and spatial location parameters k_1 and k_2, given as

$$c^D(j, l, k_1k_2) = \sum_{0 \le n \le N}^{0 \le m \le M} f[m, n]\gamma_{j,k,k_1k_2}^D[m, n], \qquad (14)$$

for any bi-variate function, $f \in L^2(R^2)$, associated with signal vector.

Here, each $\gamma_{j,k,k_1k_2}^D$ is a digital curvelet waveform, where superscript D stands for "digital".

The wrapping based curvelet transform, FDCT, is a multi-scale pyramid which consists of several sub-bands at different scales consisting of different orientations and positions in the frequency domain. At a high frequency level, curvelets are so fine and looks like a needle shaped element whereas they are non-directional coarse elements at low frequency level.

Figure 1 demonstrates the image represented in spectral domain in the form of rectangular frequency tiling by combining all frequency responses of curvelets at different scales and orientations. Shaded region represents a typical wedge. It can be seen that curvelets are needle like elements at higher scale. Moreover, the curvelet becomes finer and smaller in the spatial domain and shows more sensitivity to curved edges as the resolution level is increased, thus allowing to effectively capturing the curves in an image, and curved singularities can be well-approximated with fewer coefficients as compared to any other multi-scale techniques [14].

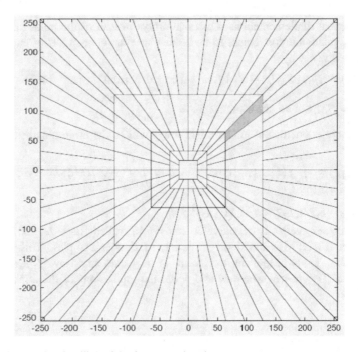

Fig. 1 The pseudopolar tiling of the frequency domain

The curvelet transform is usually implemented in the frequency domain to achieve a higher level of efficiency. First a 2D FFT is applied to the image and for each scale and orientation, a product of U_j^l "wedge" is obtained; the result is then wrapped around the origin. Finally, 2D IFFT is applied resulting in discrete curvelet coefficients. That is,

$$Curvelet\ transform = IFFT[FFT(Curvelet) \times FFT(Image)]$$

5 Neural Network Models

The simplest definition of a neural network, generally referred to as an 'artificial' neural network (ANN), due to the inventor of first neuro-computers, Dr. Robert Hecht-Nielsen, is, ...*a computing system made up of a number of simple, highly interconnected processing elements, which process information by their dynamic state response to external inputs.*

ANNs are processing devices (algorithms or actual hardware) that are loosely modeled after the neuronal structure of the mammalian cerebral cortex. Neural networks are typically organized in layers made up of a number of interconnected 'nodes' which contain an 'activation function'. It is a system composed of several *artificial* neurons and *weighted* links binding them. Every neuron in its layer, receives some type of stimuli as input, processes it and sends through its related links an output to neighboring neurons [1]. In machine learning and cognitive science, a class of artificial neural networks is a family of statistical learning algorithms inspired by biological neural networks (the central nervous systems of animals, in particular, the brain) and they are used to estimate or approximate functions that depend on a large number of unknown inputs.

A typical neural network is an adaptive system made up of four main components:

- A node as a unit that activates upon receiving incoming signals (inputs);
- Interconnections between nodes;
- An activation function (rule) which transforms inside a node, input into output;
- An optional learning function for managing weights of input-output pairs.

Most ANNs contain some form of 'learning rule' which modifies the weights of the connections according to the input patterns that it is presented with. Figure 2 illustrates the architecture of the basic ANN. This class of ANN is called *feed-forward* networks [15].

Figure 3 depicts the Feed-forward networks characterized by the layout and behaviour of their inner nodes is.

Essentially, ANNs are simple mathematical models defining a function $f : X \rightarrow Y$ or a distribution over X or both X and Y. Mathematically, a neuron's network function $f(x)$ is defined as a composition of other functions $g_i(x)$. A widely used type

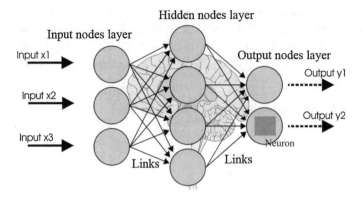

Fig. 2 Basic structure of an artificial neural network (ANN)

Fig. 3 Internal structure of a node with its inputs (x_i), weighted inputs (xw_i) and sigmoid function

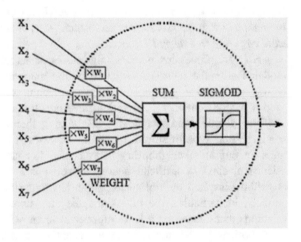

of composition is the *nonlinear weighted sum*, $f(x) = K\left(\sum_i w_i g_i(x)\right)$, where $K($commonly referred to as the activation function) is some predefined function, such as the hyperbolic tangent.

5.1 Applications of Artificial Neural Network

Nowadays, ANN structures have extended their domain of applicability from natural biological model to a model that relies more on statistical inference techniques and signal processing. This has led ANN to many applications in diverse fields such as medical sciences (cancer diagnosis, cardio-vascular system modeling), business (sales forecasting, customer research, target marketing, counterfeit prevention) and manufacturing (resource allocation and scheduling). Advances in speech recognition, pattern recognition and image analysis through the implementation of ANN techniques, have led to the significant advancement in the emerging areas like robotics and intelligent software. Moreover, by integrating self-learning neural network concepts into flight control software, recently NASA has launched the Intelligent Flight Control System (IFCS) research with the goal to develop adaptive and fault-tolerant flight control systems to improve aircraft performance in critical conditions [1].

Another useful ANN representation is a **Gaussian network model (GNM)** that depicts a biological macromolecule as an elastic mass-and-spring network to study, understand, and characterize mechanical aspects of its long-scale dynamics. GNM has also a wide range of applications from small proteins such as enzymes composed of a single domain, to large macromolecular assemblies such as a ribosome or a viral capsid. In the model, proteins are represented by nodes corresponding to alpha carbons of the amino acid residues. Similarly, DNA and RNA structures are represented with one to three nodes for each nucleotide.

5.2 Neural Network Models in Signal Processing

A pulse, or a signal or an image can be represented by a neural network. As discussed, neural networks are models which are structurally very simple because they consist of very simple building blocks, neurons that are organized into layers. A neuron is a basic element of a network that is mathematically presented as a point in space toward which signals are transmitted from surrounding neurons or inputs. Recently various approaches have been unified in neural network models under graph neural networks (GNN) which is used for processing the data represented in graph domains. Graph theoretic approach can be thus invoked to abstract representation of neural networks wherein the computational elements, neurons of the network, can be shown as nodes. GNN implements a function $\tau(G, n) \in R^m$ that maps a graph G and one of its nodes n onto an m-dimensional Euclidean space. The GNNs have been proved to be a sort of universal approximator for functions on graphs and have been applied to several problems, including spam detection, object localization in images, molecule classification [16].

The GNN is a larger class of neural networks which includes both Biological neural networks (BNN) and Artificial neural networks (ANN) [17]. BNNs are of objective existence, in which the neurons are linked as a network in a certain order, e.g. human neural network, whereas ANNs are aimed at modeling the organization principles of central neural system, with the hope that the biologically inspired computing capabilities of ANNs will allow the cognitive and sensory tasks to be performed more easily and satisfactorily [18]. ANNs are also finding of increasing use in noise reduction problems. The main design goal of these neural networks is to obtain a good approximation for some input output mapping [19].

6 Curvelet Interaction with Neural Networks

Curvelets in combination with neural networks have been successfully used in seismic data processing. Seismic data records the amplitudes of transient/reflecting waves during receiving time. The main problem in seismic data processing is to preserve the smoothness along the wave fronts when one aims to remove noise. As the curvelets resemble local plane waves, sparse representation of the local seismic events can be effectively used for preserving wave fronts in seismic processing. The study exploring the capability of the multilayer perceptron neural network to classify seismic signals recorded by the local seismic network of Agadir (Morocco) demonstrated that the classification results on a data set of 343 seismic signals have more than 94 % accuracy [11].

6.1 Signal/Image Denoising

Digital signals are invariably contaminated by noise. Noise is undesired information that degrades the signal/image. Noise arises due to imperfect instruments used in signal processing, problems with the data acquisition process, and interference which can degrade the data of interest.

In the signal/image de-noising process, information of the type of noise present in the original image plays a significant role. Images can be mostly corrupted with noise modeled with either a uniform, Gaussian, or salt and pepper distribution. Another type of noise is a speckle noise which is multiplicative in nature. Thus, the noise is present in image either in an additive or multiplicative form.

Mathematically, the image with additive noise is expressed as $w(x,y) = s(x,y) + n(x,y)$, whereas the image with multiplicative noise is expressed as $w(x,y) = s(x,y) \times n(x,y)$, where (x,y) is original signal, $n(x,y)$ is the noise introduced into the signal to produce a noisy image $w(x,y)$, and (x,y) is the pixel location. The above image algebra is done at pixel level. Image addition also has applications in image morphing. Image multiplication means the brightness of the image is varied.

7 Linear Gaussian Factor Analysis

The digital image acquisition process transforms an optical image into a continuous electrical signal that can be sampled. In every step of the process there are fluctuations caused by natural phenomena, adding random value to the exact brightness value for a given pixel. Independent component analysis (ICA) is widely used in statistical signal processing, medical image processing, and telecommunication applications [20]. The method of blind source separation (BSS) is also applied for extracting underlying source signals from a set of observed signal mixtures with little or no information of the nature of these source signals [21]. ICA is used for finding factors or components from multivariate statistical data and is one of the many solutions to the BSS problem [22].

The basic ICA model is expressed as,

$$\mathbf{x}(t) = \mathbf{A}\mathbf{s}(t) + \mathbf{\eta}(t) \tag{15}$$

where $x(t)$ is an N dimensional vector of the observed signals at the discrete time instant t, A is an unknown mixing matrix, $s(t)$ is original source signal of $M \times N (M \leq N)$ and $\eta(t)$ is the observed noise vector and M is number of sources.

Signal can be represented by GNN. A linear Gaussian factor analysis (GFA) model that represents the signal by GNN [23] can be formulated as,

$$x_i(t) = \sum_j A_{ij} s_j(t) + a_i + \eta_i(t) \tag{16}$$

where i indexes different components of the observation vector x_i, representing weighted sums of underlying latent variables, j indexes different factors and A_{ij} are the weights of the factors s, also known as factor loadings. The factors s and noise η are assumed to have zero mean. The bias in x is assumed to be caused by a. The effect of the inaccuracies and other causes is summarized by Gaussian noise η.

If the variances of the Gaussian noise terms, $\eta(t)$, are denoted by σ_i^2, the probability which the model gives for the observation $x_i(t)$ can be written as

$$P(x_i(t)|s(t), \mathbf{A}, a_i, \sigma_i^2) = \frac{1}{\sqrt{2\pi\sigma_i^2}} exp\left(-\frac{\left[x_i(t) - \sum_j A_{ij} s_j(t) - a_i\right]^2}{2\sigma_i^2}\right) \tag{17}$$

In vector form, the GFA can be expressed as

$$\mathbf{x}(t) = \mathbf{A}\mathbf{s}(t) + \mathbf{a} + \mathbf{\eta}(t) \tag{18}$$

where x, s, a and η are vectors and A is a matrix.

Equivalently, $\mathbf{x}(t) \sim N(\mathbf{A}\mathbf{s} + \mathbf{a}, \mathbf{\sigma}^2)$.

The purpose of the component analysis is to estimate original signal, $s(t)$, from the mixed signal, $x(t)$. That is equivalent to estimating the matrix A. Assuming that there is a matrix W, which is the de-mixing matrix or separation inverse matrix of A, then the original source signal is obtained by

$$s(t) = \mathbf{W}\mathbf{x}(t) \tag{19}$$

The ICA algorithm assumes that the mixing matrix A must be of full column rank and all the independent components $s(t)$, with the possible exception of one component, must be non-Gaussian [10].

8 Curvelet Induced Signal Denoising Model

In physical systems, transmitted signals are usually distributed partially, or some-times almost completely, by an additive noise from the transmitter, channel, and receiver. Noise arises due to imperfect instruments used in signal processing, problems with the data acquisition process, and interference which can degrade the data of interest. Also, noise can be introduced due to compression and transmission errors [24]. Denoising or the noise reduction is an essentially required process to enhance the estimation process of signal/image reconstruction of the captured signal. It is considered as a continuous mapping process of the noisy input data to a noise free output data. Improving quality of noisy signals has been an active area of research for many years.

Several signal denoising techniques are proposed in reported literature. Among them, curvelet thresholding technique has been prominently used in signal pro-cessing. Curvelet thresholding is a simple operation, which aims at reducing noise in a noisy signal. It is performed by selecting the FDCT coefficients below a certain threshold and setting them to zero

$$y_\lambda = \begin{cases} y_\lambda, |y_\lambda| \geq t_\lambda \\ 0, |y_\lambda| < t_\lambda \end{cases} \tag{20}$$

where t_λ is the threshold and λ is the index.

The threshold used in this technique is $t_\lambda = k\sigma_\lambda\sigma$, for some scale k, where σ is an estimation of the standard deviation of the noise estimated from the finest scale coefficients.

Another approach adopted recently in signal denoising problems is thresholding neural network (TNN), wherein thresholding function is used instead of activation functions in feed forward neural networks and threshold value is adjusted using [9],

$$t_{\lambda+1} = t_\lambda + \Delta t_\lambda, \text{ where } \Delta t_\lambda = -\propto \left. \frac{\partial MSE}{\partial t} \right|_{t=t_\lambda}, \propto \text{ is learning rate}, \tag{21}$$

t_λ is a threshold, $t_\lambda = \sigma\sqrt{2\log L}$, σ is a noise variance and $L = N^2$ is size of the signal.

Now, recalling the signal representation (16) and applying FDCT (14) to $x_i(t)$, to decompose the signal into a sequence of curvelet coefficients,

$$x^D(j, l, k_1 k_2) = \sum_{\substack{0 \leq m \leq M \\ 0 \leq n \leq N}} x_i[m, n] \gamma^D_{j,k,k_1 k_2}[m, n] \tag{22}$$

The curvelet coefficients of noisy signal so obtained are used as the input to the threshold function,

$$f(x, t_\lambda) = x - \frac{x}{exp\left[(x/t_\lambda)^2 - 1\right]} + \frac{1/8x}{exp(x/0.71t_\lambda)^2} \tag{23}$$

where x depicts the curvelet coefficients obtained from noisy signal $x_i(t)$.

Observe that the thresholding function $f(x, t_\lambda) \to 0$ in $[-t_\lambda, t_\lambda]$ and $f(x, t_\lambda) = x$ outside the threshold interval. That is, it converges to curvelet coefficients themselves. Thus, the noisy coefficients in the threshold interval $[-t_\lambda, t_\lambda]$ are shrieked. For chosen learning rate and convergence value, the universal threshold value (21) is obtained during learning process. In test phase, computed threshold value in learning phase is used by the thresholding function (23) to demise curvelet coefficients of test signals.

Finally, inverse curvelet transform (9) is applied to get the denoised signal $\hat{x}_i$.

The proposed curvelet induced GFA signal denoising model can be put to performance factor analysis. Among the various performance factors, the peak signal to noise ratio (PSNR) and root mean square error (RMSE) are the most commonly used measures of quality of reconstruction in signal denoising. Higher the value of PSNR with minimum value of RMSE, better the performance of the denoising model.

The PSNR and RMSE are calculated using the expressions

$$PSNR = 10 \log_{10} \left(\frac{\sum_{i=1}^{N} x^2(i)}{\sum_{i=1}^{N} [x(i) - \hat{x}(i)]^2} \right), \quad RMSE = \frac{1}{N} \sqrt{\sum_{i=1}^{N} [x(i) - \hat{x}(i)]^2} \tag{24}$$

where $x(i)$ is the original source signal, $\hat{x}(i)$ is the separated signal, i is the sample index and N is the number of samples of the signal.

The PSNR and the other statistical measures such as mean and standard deviation of the PSNR for different signal sub-bands/samples can be compared to determine the performance of the signal denoising model. Simulations can be performed on these noisy mixed signalson any Matlab® R 7.9 on a core i7 2.2 GHz PC.

To illustrate the performance of the proposed GFA model, simulations are performed on noisy mixed 256 × 256 'Leena' image and sample CT image on Mat lab® R 7.9 on a core i7 2.2 GHz PC, and curvelet transform via USFFT software package.

The denoising results (PSNR in dB) in both the experiments are presented in Table 1.

From the experimental results, it is seen that the curvelet induced GFA gives increased PNSR with minimum RMSE for the used images. Similar simulations can be performed for satellite images, seismic data, medical images etc.

Table 1 Experimental values of PSNR and RMSE

Image	Noisy image	Soft curvelet thresholding	TNN thresholding	Curvelet induced GFA	RMSE
Lena	$20.01(\sigma = 10)$	24.89	32.05	34.25	0.177533
CT (Random noise)	21.09 (m = 0, $\sigma^2 = 0.01$)	26.89	28.74	32.14	0.164434
CT (Gaussian noise)	19.77 (m = 0, $\sigma^2 = 0.01$)	27.70	29.64	31.17	0.14564

9 Conclusions

The present topic integrates the Independent component analysis, graph neural networks and curvelet thresholding adopted in signal denoising processes so as to harvest the advantages and to overcome the inherent limitations of each of them. Denoising using the curvelet transform recovers the original signal from the noisy one using lesser coefficients than the wavelet transform. Moreover, curvelet transform provides high PSNR- the most commonly used measure of quality of reconstructed signal- and removes the Random as well as Gaussian white noises. Further, the wrapping based Curvelet transform is conceptually simpler, faster and invertible with rapid inversion algorithm of the same complexity than the existing techniques. Thus curvelet outperforms the wavelet transform in several ways.

The proposed curvelet induced Gaussian factor analysis model in Graph neural network settings unifies earlier denoising models, consequently enhancing the performance in terms of the recovery and quality of the extracted signals of higher dimensions that arise in wide range of applications including medical image processing, seismic exploration, fluid mechanics, solutions of partial different equations, and compressed sensing.

References

1. M. Caundill, Neural networks primer part I. AI Expert **2**(12), 46–52 (1987)
2. C.J. Stam, J. Reijneveld, Graph theoretical analysis of complex networks in brain. Nonlinear Biomed. Phys. **1**(3), 215–223 (2007)
3. B. Bhosale, A. Biswas, Multi-resolution analysis of wavelet like soliton solution of KdV equation. Int. J. Appl. Phys. Math. **3**(4), 270–274 (2013)
4. B. Bhosale et al., Wavelet based analysis in bio- informatics. Life Sci. J. **10**(2), 853–859 (2013)
5. B. Bhosale et al., On wavelet based modeling of neural networks using graph theoretic approach. Life Sci. J. **10**(2), 1509–1515 (2013)
6. B. Olshausen, D. Field, Emergence of simple-cell receptive filed properties by learning a sparse code for natural images. Nature **381**, 607–609 (1996)

7. E. Candes, D. Donoho, Continuous curvelet transform: resolution of the wave front set. Appl. Comput. Anal. **19**(2), 162–197 (2005)

8. N. Kota, G. Reddy, Fusion based gaussian noise removal in the images using curvelets and wavelets with gaussian filter. Int. J. Image Proc. **5**(4), 230–238 (2011)

9. N. Yaser, J. Mahdi, A novel curvelet thresholding function for additive Gaussian noise removal. Int. J. Comput. Theory Eng. **3**(4), 169–178 (2011)

10. R. Sivakumar, Denoising of computer tomography images using curvelet transform. ARPN J. Eng. Appl. Sci. **2**(1), 26–34 (2007)

11. E. Hassan et al., Seismic signal classification using multi-layer perceptron neural network. Int. J. Comput. Appl. **79**(15), 35–43 (2013)

12. Y. Shimshoni, N. Intrator, Classification of seismic signals by integrating ensembles of neural networks. IEEE Trans. Sig. Process. **46**(5), 45–56 (1998)

13. A. Moya, K. Irikura, Inversion of a velocity model using artificial neural networks. Comput. Geosci. **36**, 1474–1483 (2010)

14. E. Candes et al., Fast discrete curvelet transforms, multiscale modeling and simulation. Appl. Comput. Anal. **5**(3), 861–899 (2006)

15. A. Weitzenfeld et al., The neural simulation language: a system for brain modeling (MIT Press, 2002)

16. F. Scarselli et al., The graph neural network model, neural networks. IEEE Trans. Syst. **20**(1), 61–80 (2008)

17. I. Podolak, Functional graph model of a neural network. IEEE Trans. Syst. **28**(6), 876–881 (1998)

18. X.U. Jim, B. Zheng, Neural networks and graph theory. Sci. China F **45**(1), 1–24 (2002)

19. L. Badri, Development of neural networks for noise reduction. Int. Arab J Inf. Technol. **7**(3), 156–165 (2010)

20. A. Hyvarinen, E. Oja, Independent component analysis: algorithms and applications. Neural Netw. **13**(4–5), 411–430 (2000)

21. S. Shehata et al., Analysis of blind signal separation of mixed signals based on fast discrete curvelet transform. Int. Electr. Eng. J. **4**(4), 1140–1146 (2013)

22. J.F. Cardoso, A. Souloumiac, Blind beam forming for non-gaussian signals. IEEE Proc. Part F **140**(6), 362–370 (1993)

23. H. Valpola, Bayesian ensembel learning for non-linear factor analysis. Acta Ploytechnica Scand. Math. Comput. Ser. **108**, 54–64 (2000)

24. M. Motwani et al., Survey of image denoising techniques, in *Proceedings 2004 Global Signal Processing Expo and Conference* (2004), pp. 27–30

Hybrid Wavelet Neural Network Approach

Muhammad Shoaib, Asaad Y. Shamseldin, Bruce W. Melville
and Mudasser Muneer Khan

Abstract Application of Wavelet transformation (WT) has been found effective in dealing with the issue of non-stationary data. WT is a mathematical tool that improves the performance of Artificial Neural Network (ANN) models by simultaneously considering both the spectral and the temporal information contained in the input data. WT decomposes the main time series data into its sub-components. ANN models developed using input data processed by the WT instead of using data in its raw form are known as hybrid wavelet models. The hybrid wavelet data driven models, using multi-scale input data, results in improved performance by capturing useful information concealed in the main time series data in its raw form. This chapter will cover theoretical as well as practical applications of hybrid wavelet neural network models in hydrology.

1 Introduction

Reliable simulation of runoff generated from a watershed in response to a rainfall event is an important area of research in hydrology and water resource engineering. It plays an important role in planning, design and sustainable management of water resources projects such as flood forecasting, urban sewer design, drainage system design and watershed management. The rainfall-runoff process is a complex

M. Shoaib (✉) · A.Y. Shamseldin · B.W. Melville
Department of Civil and Environmental Engineering,
The University of Auckland, Private Bag 92019, Auckland, New Zealand
e-mail: msho127@aucklanduni.ac.nz

A.Y. Shamseldin
e-mail: a.shamseldin@auckland.ac.nz

B.W. Melville
e-mail: b.melville@auckland.ac.nz

M.M. Khan
Department of Civil Engineering, Bahauddin Zakariya University, Multan, Pakistan
e-mail: mkha222@aucklanduni.ac.nz

© Springer International Publishing Switzerland 2016   127
S. Shanmuganathan and S. Samarasinghe (eds.), *Artificial Neural
Network Modelling*, Studies in Computational Intelligence 628,
DOI 10.1007/978-3-319-28495-8_7

non-linear outcome of numerous hydrological factors. These include precipitation intensity, evaporation, watershed geomorphology, infiltration rate of soil as well as interactions between groundwater and surface water flows, all of which cannot be modelled by simple linear models. Since the establishment of rational method in 1850 [1] for calculation of peak discharge, numerous hydrological models have been developed. Approaches used for simulating the rainfall-runoff transformation process include two main categories: physically-based and data driven approaches. Each approach has associated pros and cons. The physically based models include, for example, systeme hydrologique Européen with sediment and solute transport model (SHETRAN) [2] and Gridded Surface Subsurface Hydrologic Analysis model (GSSHA) [3]. These models usually involve the solution of a system of partial differential equations to model various components of hydrological cycle in the catchment. However, although, these physically-based models are helpful in understanding the underlying physics of hydrological processes, their practical application is limited because of the requirements of large input data and computational time. The black-box data-driven models, on the other hand, are based primarily on the measured data and map the input-output relationship without giving consideration to the complex nature of the underlying process. Among data-driven models, ANNs have appeared as powerful black-box models and received great attention during the last two decades. The ANN approach has been successfully applied for different modelling problems in various branches of science and engineering. In the field of hydrology, French et al. [4] were the first to use ANN for forecasting rainfall. Shamseldin [5] pioneered the use of ANN in modelling rainfall-runoff relationships. The ANN has been successfully applied in many hydrological studies (e.g., [6–14]). The merits and shortcomings of using ANNs in hydrology are discussed in Govindaraju [15], Govindaraju [16] and Abrahart et al. [17].

Despite the good performance of data driven models in simulating non-linear hydrological relationships, these models may not be able to cope with non-stationary data if pre-processing of input and/or output data is not performed [18]. Application of WT has been found to be effective in dealing with non-stationary data [19]. WT is a mathematical tool which simultaneously considers both the spectral and the temporal information contained in the data. Generally, WT can be carried out in a continuous (Continuous Wavelet Transformation (CWT)) form as well as in a discrete (Discrete Wavelet Transformation (DWT)) form. WT decomposes the main time series data into its sub-components and these sub-constituents are used as external inputs to the data driven models. The resulting model is known as the hybrid wavelet model. These hybrid wavelet data driven models, using multi scale-input data, result in improved performance by capturing useful information concealed in the main time series data in its raw form. Different studies used WT in order to increase accuracy of data driven hydrological models. A comprehensive review of applications of WT in hydrology can be found in Nourani et al. [20].

The performance of hybrid wavelet data driven models is very much dependent on the selection of numerous factors. These factors include the selection of WT type (continuous or discrete). Likewise, WT requires a mother wavelet function to perform transformation. The performance of hybrid models is also extremely sensitive to the selection of appropriate wavelet function from a number of available wavelet functions. The performing efficiency of hybrid wavelet models further relies heavily on the selection of suitable decomposition level. Furthermore, choice of data driven model is another vital factor as there are different types of data driven models available, including the Multilayer Perceptron Neural Network (MLPNN), the Generalized Feed Forward Neural Network (GFNN), Radial Basis Function Neural Network (RBFNN), the Modular Neural Network (MNN), Neuro-Fuzzy Neural Network (NFNN), the Time Delay Neural Network (TDLNN) and the Gene Expression Programming (GEP) model based on theory of evolution. The present chapter will therefore focus on the review of hybrid wavelet models developed and applied in hydrology in order to formulate general guidelines for the successful implementation of hybrid wavelet models.

2 Wavelet Transformation

Mathematical transformations are applied to time series data in order to retrieve additional information from the data that is not readily accessible in its raw form expressed in time domain. In most cases, vital information is obscure in the frequency domain of the time series data. Mathematical transformations are aimed to extract frequency information from the time series data.

There are a number of transformations available and probably the Fourier transformation (FT) was the first mathematical technique used for mining the frequency information contained in time series data/signal. FT transforms the signal from time domain to frequency domain. Mathematically, FT of a continuous function $f(t)$ is represented by the following equation;

$$F(\omega) = \int_{-\infty}^{+\infty} f(t)e^{-j\omega t}dt \tag{1}$$

which is an integral of the function $f(t)$ over time multiplied by a complex exponential. The FT maps a function of single variable, time t, into another function of a single variable, frequency ω. The FT uses a mathematical tool called inner product for measurement of similarity between the function $f(t)$ and an analysing function. Conceptually, multiplying each value of $f(t)$ by a complex exponential (sinusoid) of frequency ω gives the constituent sinusoidal components of the original signal. Graphically, it can be represented as shown in Fig. 1.

The FT converts the signal from the time domain to the frequency domain with the loss of time information. The FT provides the information of frequency in the signal without giving any information about time. This works well for stationary signals where all of the frequencies are present at all time but the FT does not suits non-stationary signals where different frequencies may be present at different times. In order to overcome this difficulty, Gabor [21] introduced a windowing technique called Short Term Fourier Transform (STFT) to analyse only a small section of the

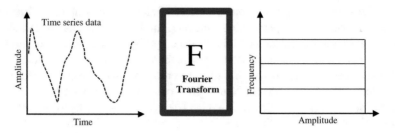

Fig. 1 Fourier transformation

signal where it is considered as stationary. Mathematically, it can be represented by the following equation;

$$F(\omega, \tau) = \int f(t)\omega(t - \tau)e^{-j\omega t}dt \qquad (2)$$

where $f(t)$ is the signal itself and $\omega(t - \tau)$ is the window function. As can be seen from Eq. (2), an STFT maps a function of one variable 't' into a function of two variables 'ω' and 'τ'. The STFT attempts to provide information on both time and frequency in contrast with the FT which only provides information about the frequency. Graphically, STFT of a signal can be represented as shown in Fig. 2. However, information on time and frequency can be provided by this method with limited precision depending on the size of the window selected. The major disadvantage associated with the STFT is that once a particular window size is selected it cannot be changed.

Grossmann and Morlet [22] presented wavelet transformation (WT) which is capable of providing time and frequency information simultaneously, hence giving time-frequency representation of the temporal data. The wavelet in WT refers to the window function that is of finite length (the sinusoids used in the FT are of unlimited duration) and is also oscillatory in nature. The width of the wavelet function varies as the transform is computed for every single spectral component and hence making WT suitable for analysis of the non-stationary data. The variable

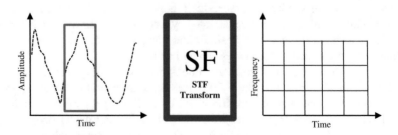

Fig. 2 Short term fourier transformation

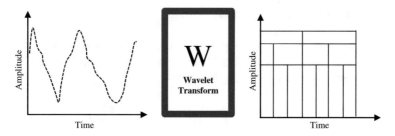

Fig. 3 Wavelet transformation

size of window in WT makes it possible to have more precise information of time and frequency information of a signal as shown in the Fig. 3.

In WT, wavelet function is used to decompose the time series data in different components at different resolution levels. Wavelet function $\psi(t)$ called the mother wavelet has finite energy and is mathematically defined as:

$$\int_{-\infty}^{\infty} \psi(t)dt = 0 \qquad (3)$$

where $\psi_{a,b}(t)$ is the wavelet function and can be obtained by the following equation as:

$$\psi_{a,b}(t) = |a|^{-\frac{1}{2}}\psi\left(\frac{t-b}{a}\right) \qquad (4)$$

where a and b are real numbers; a is the scale or frequency parameter; b is the translation parameter. Thus, the WT is a function of two parameters a and b. The parameter a is interpreted as dilation ($a > 1$) or contraction ($a < 1$) factor of the wavelet function $\psi(t)$ corresponding to different scales. The parameter b can be interpreted as a temporal translation or shift of the function $\psi(t)$. The temporal data can be analysed either by translation (moving the wavelet at different locations along the time axis of the data) or by squeezing or stretching of the wavelet (referred as scale or dilation) for transformation of the data under consideration.

Wavelet transformation is believed to be capable of revealing aspects of the original time series data such as trends, breakdown points, and discontinuities that other signal analysis techniques might miss [23, 24].

2.1 Types of Wavelet Transformation

There are two types of wavelet transformations, namely, CWT and DWT.

2.1.1 Continuous Wavelet Transformation

The Continuous Wavelet Transform (CWT) is a modified form of STFT and it is introduced to address the issue of fixed window size in the STFT. Mathematically, it can be represented by the following equation:

$$CWT_{a,b}(t) = \int_{-\infty}^{+\infty} f(t)\frac{1}{\sqrt{a}}\psi^*\left(\frac{t-b}{a}\right)dt \qquad (5)$$

where $CWT_{a,b}$ is the wavelet coefficient, * refers to the complex conjugate of the function while $\psi(t)$ is called the wavelet function or the mother wavelet. The entire range of the signal is analysed by the wavelet function by using two parameters, namely, a and b. The parameters a and b are known as the dilation (scale) parameter and translation (position) parameter, respectively, while $f(t)$ is the original temporal data required to be transformed. The product of wavelet $\Psi(t)$ and the function $f(t)$ in Eq. (5) are integrated over the entire range of the temporal data and mathematically, it is called convolution. In CWT, the analysing wavelet function $\Psi(t)$ compares the temporal data to be transformed at different scales and locations. The scale parameter a is associated with the frequency of the signal in CWT and it varies as $a \propto \frac{1}{\omega}$. This means that low scale corresponds to compress wavelet and it capture rapidly changing details of the signal (High frequency) while high scale corresponds to stretched wavelet and capture slowly changing features of the signal (low frequency). The CWT algorithm consists of selecting a particular type of mother wavelet or simply wavelet and then comparing it to the section of original signal at the start. The CWT coefficients '$CWT_{a,b}(t)$' are calculated using Eq. (5) for particular values of scale 'a' and the translation 'b'. The value of '$CWT_{a,b}(t)$' varies between 0 (no similarity) and 1 (complete similarity). Different values of '$CWT_{a,b}(t)$' are calculated by varying 'a' and 'b' in CWT using Eq. (5) and are plotted graphically as a contour map known as scalogram.

2.1.2 Discrete Wavelet Transformation

The calculation of CWT coefficients at each scale a and translation b results in a large amount of data. However, if the scale and translation are chosen on powers of two (dyadic scales and translation), then the amount of data can be reduced considerably resulting in more efficient data analysis. This transformation is called the discrete wavelet transformation (DWT) and can be defined as Mallat [25]:

$$\psi_{m,n}\left(\frac{t-b}{a}\right) = a_o^{-\frac{m}{2}}\left(\frac{t-nb_oa_o^m}{a_o^m}\right) \qquad (6)$$

where m and n are integers that govern the wavelet scale/dilation and translation, respectively; a_o is a specified fine scale step greater than 1; and b_o is the location parameter and must be greater than zero. The most common and simplest choice for parameters a_o and b_o are 2 and 1, respectively. This power of two logarithmic scaling of the dilations and translations is known as dyadic grid arrangement and is the simplest and most efficient case for practical purposes [25]. For a discrete time series $f(t)$, the DWT becomes:

$$DWT(m,n) = 2^{-\frac{m}{2}} \sum_{t=0}^{N-1} \psi^*(2^{-m}t - n)f(t) \tag{7}$$

where $DWT(m, n)$ is the wavelet coefficient for the discrete wavelet of scale $a = 2^m$ and location $b = 2^m n$. $f(t)$ is a finite time series ($t = 0, 1, 2,..., N - 1$), and N is an integer power of 2 ($N = 2^M$); this yields the ranges of m and n as, respectively, $0 < n < 2^{M-m} - 1$ and $1 < m < M$. At the largest wavelet scale (i.e., 2^m where $m = M$) only one wavelet is required to cover the time interval, and only one coefficient is produced. At the next scale (2^{m-1}), two wavelets cover the time interval, hence two coefficients are produced, and so on, down to $m = 1$. At $m = 1$, the a scale is 2^1, i.e., 2^{M-1} or $N/2$ coefficients are required to describe time series data at this scale. The total number of wavelet coefficients for a discrete time series of length $N = 2^M$ is then $1 + 2 + 4 + 8 + \cdots + 2^{M-1} = N - 1$.

In addition to this, a signal smoothed component, $\bar{T}$, is left, which is the signal/time series data mean. Thus, a time series of length N is broken into N components, i.e. with zero redundancy. The inverse discrete transformation is given by:

$$f(t) = \bar{T} + \sum_{m-1}^{M} \sum_{n=0}^{2^{M-m}-1} DWT_{m,n} 2^{-\frac{m}{2}} \psi*(2^{-m}t - n) \tag{8}$$

Or in a simple format as:

$$f(t) = \bar{T}(t) + \sum_{m=1}^{M} W_m(t) \tag{9}$$

where $\bar{T}(t)$ is called approximation sub-signal at level M and $W_m(t)$ are detail sub-signals at levels m = 1, 2, ..., M.

DWT operates on two sets of functions called scaling function (low pass filter) and the wavelet function (high pass filter). In DWT, the signal/original time series data is passed through the low-pass and high-pass filters and subsequently decomposes the signal into approximation (a_s) and detail (d_s) components respectively. This decomposition process is then iterated with successive approximations being decomposed in turn, so that the signal is broken down into many lower resolution components as shown in Fig. 4. At each decomposition level, the low pass and high pass filters produce signals spanning only half the frequency band. This makes the frequency resolution double as uncertainty in frequency is reduced by half. The high pass filters are used to analyse the high frequencies while the low pass filters, on the other hand, are used to analyse the low frequency content of the signal.

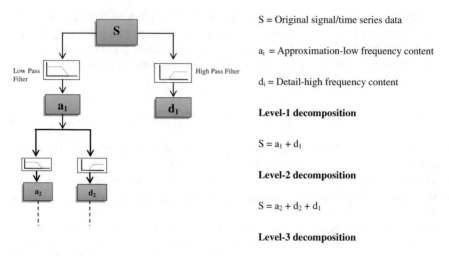

Fig. 4 n-level wavelet decomposition tree

The approximations are the high scale, the low frequency components of the signal while detail represents the low scale, high frequency components. The low frequency content of the signal is the most significant part providing the signal its identity while on the other hand, the high frequency content imparts flavour or nuance. The detail signals can catch trivial attributes of interpretational value in data while the approximation shows the background information of data [11, 26].

2.2 Wavelet Families

There are different wavelet families available and each family comprises of different wavelet functions. These wavelet functions are characterized by their distinctive features including their region of support and the corresponding number of vanishing moments. The wavelet support region is associated with the span length of the given wavelet function which affects its feature localization abilities. However, the vanishing moment limits the wavelet's ability to represent the polynomial behaviour or information of the data. For example, the db1, coif1, sym1 wavelet functions, with one moment, encrypts one coefficient polynomials, or constant signal components. The db2, coif2, sym2 functions encrypts two-coefficient polynomials, i.e., a process having one constant and one linear signal component. Likewise, the db3, coif3, sym3 functions represents three-coefficient polynomials which encode a process having constant, linear and quadratic signal components.

3   Critical Issues for the Hybrid Wavelet Models

3.1   Selection of Wavelet Transformation Type

As stated earlier, there are two types of wavelet transformations, namely, the CWT and the DWT. Different studies have used both. Nakken [27] used the CWT with the Morlet wavelet function to identify the temporal variability of rainfall-runoff data. Labat et al. [28, 29] applied the CWT with the Morlet wavelet and the DWT with the Haar wavelet to explain the non-stationarity of karstic watershed data. Wang and Ding [30] used the DWT with a trous algorithm for making predictions in hydrology. One day ahead stream flow forecasting models was developed using CWT by Anctil and Tape [31]. The CWT was used by Mwale and Gan [32] to identify homogenous zones of rainfall variability and predictability in East Africa. Likewise, Mwale et al. [33] applied CWT to analyse the nonstationary spatial, temporal and frequency regimes of regional variability in southern African summer. Adamowski [34, 35] developed wavelet coupled models using CWT for forecasting daily stream flows. Kuo et al. [36, 37] employed CWT for estimating seasonal streamflow and variability, tele-connectivity and predictability of the seasonal rainfall in Taiwan, respectively. Moreover, CWT was applied to develop a model for forecasting wave height by the Özger [38].

However, application of CWT is associated with inherit demerit of generation of large amount of data because of calculation of CWT coefficients at each scale and translation. Furthermore, it was also argued in favour of DWT that hydrologist does not have at his or her disposal a continuous-time signal process, but rather a discrete- time signal. Therefore most of the wavelet coupled models are limited to the use of DWT. Partal and Kişi [39], Partal and Cigizoglu [40], Nourani et al. [11], Kisi and Shiri [41], Ramana et al. [42] applied DWT for precipitation forecasting. Wang et al. [43] and Zhou et al. [44] applied DWT for discharge prediction. Kişi [45], Kişi [46], Partal [47], Adamowski [48], Pramanik et al. [49], Tiwari and Chatterjee [26], Shiri and Kisi [50], Wang et al. [51], Maheswaran and Khosa [52] applied DWT for stream flow forecasting. Cannas et al. [18] developed hybrid wavelet-ANN models using both CWT and DWT with the db4 wavelet function. Shoaib et al. [53] compared the performance of CWT and DWT for wavelet coupled rainfall-runoff models. The study advocated the use of DWT over CWT because of its superior performance.

3.2   Selection of Wavelet Function

The performance of hybrid wavelet models is also dependant on the selection of particular type of mother wavelet functions. As stated earlier, there are different wavelet families available and each containing a number of mother wavelet functions. In general, a limited number of studies evaluated the effects of the wavelet function types on the performance of the ANN models. Nourani et al. [11] found that the db4 and the Meyer wavelets provide performance superior to that of the Haar wavelet when developing a Wavelet-ANN model for prediction of monthly precipitation. Similarly, Nourani et al. [54] developed multivariate ANN-wavelet rainfall-runoff models and tested the performance of seven wavelet functions, namely Haar, db2, db3, db4, Sym2, Sym3 and Coif1 functions. The results show that the Haar and the db2 mother wavelet functions provide better results than the other five wavelet functions. In the context of Wavelet-ANN flood forecasting models, Singh [24] found the db2 wavelet function yields better performance than the db1 function. In a similar vein, Maheswaran and Khosa [55] noted that the db2 function has better performance than five wavelet functions, namely db1, db3, db4, Sym4 and B-spline functions when developing a Wavelet-Volterra hydrologic forecasting models. The performance of 23 mother wavelet functions were tested in a study conducted by. Shoaib et al. [53]. The db8 wavelet function of Daubechies wavelet family was found to outperform the other wavelet functions tested in the study.

3.3 Selection of Decomposition Level

The choice of suitable decomposition level in the development of wavelet coupled data driven models is very vital as it relates to the seasonal and periodic features embedded in the hydrological data. The maximum number of decomposition level depends on the length of data available. A DWT decomposition consists of Log_2N levels/stages at most. Aussem [56, 11] and Nourani et al. [57] used the following formula given in Eq. (10) to calculate the suitable decomposition level.

$$L = int[log(N)] \tag{10}$$

where L is the level and N is the total data points in data. This equation was derived for fully autoregressive data by only considering the length of data without giving attention to the seasonal signature of the hydrologic process [54]. Likewise, some other previous hydrological studies used different decomposition levels without giving any logical reasoning. Adamowski and Sun [23] used a level eight decomposition for developing wavelet coupled neural network models, Partal and Kişi [39] employed decomposition at level ten for developing wavelet coupled neuro-fuzzy model. However, Kisi and Shiri [41] used three level decomposition for developing wavelet based genetic programming and neuro-fuzzy models.

Shoaib et al. [53, 58, 59] suggested the use of level nine decomposition of data for the development of hybrid wavelet data driven models. Decomposition at level nine contains one large scale (lower frequency) sub-signal approximation (a_9) and nine small scale (higher frequency) sub-signals details (d_1, d_2, d_3, d_4, d_5, d_6, d_7, d_8 and d_9). The detail sub-series d_4 corresponds to time series 16-day mode, d_5 to 32-day mode (about monthly mode), d_6 to 64-day mode, d_7 to 128-day mode (about four months), d_8 to 256-day mode (about eight and half months) and d_9 to 512-day mode (about seventeen months). Furthermore, decomposition at level nine contains d_8 and d_9 sub-series which are responsible for detecting seasonal variations in the input rainfall data on almost annual basis. This annual periodicity is considered the extremely significant and leading seasonal cycle in the hydrological time series data.

3.4   Selection of Data Driven Model

There are various types of data driven models available and selection of a suitable type is one of the most important and difficult tasks in the development of hybrid wavelet models. The MLPNN, GFFNN, RBFNN, MNN, ANFIS and GEP models are the most commonly used data driven models in hydrology. Different studies employed different types of neural networks in order to map non-linear hydrological processes. The MLPNN type has been used in many rainfall-runoff studies (e.g., [7, 53, 60–67, 68]). Likewise, Waszczyszyn [69], Motter and Principe [70], Van Iddekinge and Ployhart [71], Anthony and Bartlett [72], Kişi [46] and Dibike et al. [61] employed RBFNN type for modelling rainfall-runoff process. Integration of neural network with fuzzy-rule based systems has introduced a new model type known as the adaptive neuro fuzzy inference system (ANFIS). The potential of ANIFS has been explored in many previous hydrological studies (e.g., [58, 73–81]). The modular neural network is another type of neural network which has been successfully used for rainfall-runoff modelling in some of previous studies including Zhang [82] and Rajurkar et al. [83]. In recent years, the data driven approach known as the genetic programming (GP) [84] based on evolutionary computing emerged as a powerful modelling tool for solving hydrological and water resource problems (e.g., [85–91]). Gene Expression Programming (GEP) is a variant of GP and has been found to show better performance than other data driven approaches such as the artificial neural network (ANN) and the adaptive neuro-fuzzy inference system (ANFIS) (e.g., [85, 92–96]). During the last decade, various studies successfully applied WT in order to increase forecasting efficiency of neural network models. However, the MLPNN type is found to be the most widely used neural network type for the purpose of developing wavelet coupled neural network models (e.g., [11, 18, 23, 24, 26, 30, 53, 57, 97]).

3.5   Identification of Dominant Wavelet Sub-series

A major shortcoming associated with the wavelet coupled model is the use of large numbers of inputs as WT decomposes the input data into several sub-data series. The use of large numbers of inputs results not only in increasing the simulation time, but also increasing the computational complexity and makes the network calibration more difficult. Furthermore, some of the data series obtained by the WT contains more information regarding hydrological signature of the catchment while others may be simply noisy or contain no/less significant information. Identification of wavelet sub-series of data containing significant information regarding the system is also another important issue in the successful implementation of wavelet coupled hydrological models.

Analysis of linear correlation coefficient between input sub-series/predictor and output has been widely used for determination of important wavelet sub-series to be used as input for the development of wavelet coupled ANN models. Additionally, there are two different approaches found in literature which employ linear correlation coefficients for this purpose. In the first approach, a correlation analysis is performed between each decomposed sub-series and the observed output. The wavelet sub-series having strong correlation with the observed output are only considered as the input for the wavelet coupled ANN models, while the ones having weak correlation are simply ignored. This approach has been successfully used in some of the hydrological studies. (e.g., [98, 91]). Likewise, another approach regarding selection of dominant wavelet sub-series based on the cross correlation analysis has also been employed in developing ANN and neuro-fuzzy hydrological models (e.g., [39, 41, 50, 58, 97]). In this approach, a correlation analysis is performed between each decomposed sub- series and the observed output. The wavelet sub-series having very weak correlation are ignored and the new input data series is obtained by adding the wavelet sub-series having good correlation with the observed output. This new data series is subsequently used as input for the development of wavelet coupled ANN models. Nevertheless, Nourani et al. [99, 100] criticize the selection of dominant wavelet sub-series on the basis of linear correlation, since a strong non-linear relationship may exist between input and target output despite the presence of weak linear correlation. Likewise, some other studies tested different mathematical methods to select dominant wavelet sub-series in the development of wavelet coupled models in hydrology. Vahid et al. [101] advocated the use of self-organizing feature maps (SOP) and entropy based approaches to select the dominant wavelet sub-series for developing a wavelet coupled rainfall-runoff model. It is, therefore, obvious from the above literature cited that some of the previous hydrological studies used all wavelet sub-series as input, while some others identified dominant wavelet sub-series using different methods and subsequently used them as input in the development of wavelet coupled models. It is therefore vital to compare these different strategies for the development of wavelet based models, so that an optimal strategy can be identified.

4 Conclusion

An extensive review of hybrid wavelet hydrological modes is carried out in this chapter in order to formulate general guidelines for successful implementation of hybrid wavelet models. The following conclusions are hereby be drawn.

- DWT should be employed compared to CWT because of its superior performance.
- The performance of the hybrid wavelet models is sensitive to the selection of the wavelet function. It is found that the db8 wavelet function is better in performance compared with other wavelet functions. This may be due to good time-frequency localization property of the db8 wavelet function in addition to its property of having reasonable support width.
- The level nine decomposition should be adopted as it contains most important annual and seasonal variations in it.
- The use of MLPNN type of neural network is advocated because of its better results.
- Further studies should be conducted in order to find the best technique for identification of dominant wavelet sub-series. This will not only reduce computational burden of the neural network but it will also decrease the computation time.

References

1. T.J. Mulvany, On the use of self-registering rain and flood gauges, in *Making Observations of the Relations of Rain Fall and Flood Discharges in a Given Catchment. Transactions and Minutes of the Proceedings of the Institute of Civil Engineers of Ireland*, vol. 1 (Dublin, Ireland, Session, 1850)
2. S.J. Birkinshaw, *SHETRAN Hydrological Model* (2013), http://research.ncl.ac.uk/shetran/
3. C. Downer, F.L. Ogden, GSSHA: A model for simulating diverse streamflow generation processes. J. Hydrol. Eng. **9**(3), 161–174 (2004)
4. M.N. French, W.F. Krajewski, R.R. Cuykendall, Rainfall forecasting in space and time using a neural network. J. Hydrol. **137**(1), 1–31 (1992)
5. A.Y. Shamseldin, Application of a neural network technique to rainfall-runoff modelling. J. Hydrol. **199**(3–4), 272–294 (1997)
6. M.A. Antar, I. Elassiouti, M.N. Allam, Rainfall-runoff modelling using artificial neural networks technique: a Blue Nile catchment case study. Hydrol. Process. **20**(5), 1201–1216 (2006)
7. K. Aziz et al., Application of artificial neural networks in regional flood frequency analysis: a case study for Australia. Stoch. Env. Res. Risk Assess. **28**(3), 541–554 (2014)
8. C.W. Dawson, R. Wilby, An artificial neural network approach to rainfall-runoff modelling. Hydrol. Sci. J. **43**(1), 47–66 (1998)
9. K.L. Hsu, H.V. Gupta, S. Sorooshian, Artificial neural network modeling of the rainfall-runoff process. Water Resour. Res. **31**(10), 2517–2530 (1995)
10. A. Jain, K.P. Sudheer, S. Srinivasulu, Identification of physical processes inherent in artificial neural network rainfall runoff models. Hydrol. Process. **18**(3), 571–581 (2004)

11. V. Nourani, M.T. Alami, M.H. Aminfar, A combined neural-wavelet model for prediction of Ligvanchai watershed precipitation. Eng. Appl. Artif. Intell. **22**(3), 466–472 (2009)
12. N. Sajikumar, B. Thandaveswara, A non-linear rainfall–runoff model using an artificial neural network. J. Hydrol. **216**(1), 32–55 (1999)
13. A.R. Senthil Kumar et al., Rainfall-runoff modelling using artificial neural networks: comparison of network types. Hydrol. Process. **19**(6), 1277–1291 (2005)
14. A.S. Tokar, P.A. Johnson, Rainfall-runoff modeling using artificial neural networks. J. Hydrol. Eng. **4**(3), 232–239 (1999)
15. R.S. Govindaraju, Artificial neural networks in hydrology. I: Preliminary concepts. J. Hydrol. Eng. **5**(2), 115–123 (2000)
16. R.S. Govindaraju, Artificial neural networks in hydrology. II: hydrologic applications. J. Hydrol. Eng. **5**(2), 124–137 (2000)
17. R.J. Abrahart et al., Two decades of anarchy? Emerging themes and outstanding challenges for neural network river forecasting. Prog. Phys. Geogr. **36**(4), 480–513 (2012)
18. B. Cannas et al., Data preprocessing for river flow forecasting using neural networks: Wavelet transforms and data partitioning. Phys. Chem. Earth, Parts A/B/C **31**(18), 1164–1171 (2006)
19. G.P. Nason, R.V. Sachs, *Wavelets in time-series analysis*, vol. 357 (1999), pp. 2511–2526
20. V. Nourani et al., Applications of hybrid wavelet–Artificial Intelligence models in hydrology: A review. J. Hydrol. **514**, 358–377 (2014)
21. D. Gabor, Theory of communications. Part 1:The analysis of information. J. Inst. Electr. Eng. **95**(38), 429–441 (1948)
22. A. Grossmann, J. Morlet, Decomposition of Hardy functions into square integrable wavelets of constant shape. SIAM J. Math. Anal. **15**(4), 723–736 (1984)
23. J. Adamowski, K. Sun, Development of a coupled wavelet transform and neural network method for flow forecasting of non-perennial rivers in semi-arid watersheds. J. Hydrol. **390**(1), 85–91 (2010)
24. R.M. Singh, Wavelet-ANN model for flood events, in *Proceedings of the International Conference on Soft Computing for Problem Solving (SocProS 2011)* December 20–22, 2011 (Springer, 2012)
25. G.S. Mallat, A theory for multiresolution signal decomposition: the wavelet representaiton. IEE Trans. Pattern Anal. Mach. Intell. **11**(7), 674–693 (1989)
26. M.K. Tiwari, C. Chatterjee, Development of an accurate and reliable hourly flood forecasting model using wavelet–bootstrap–ANN (WBANN) hybrid approach. J. Hydrol. **394**(3–4), 458–470 (2010)
27. M. Nakken, Wavelet analysis of rainfall–runoff variability isolating climatic from anthropogenic patterns. Environ. Model Softw. **14**(4), 283–295 (1999)
28. D. Labat, R. Ababou, A. Mangin, Rainfall–runoff relations for karstic springs. Part II: continuous wavelet and discrete orthogonal multiresolution analyses. J. Hydrol. **238**(3), 149–178 (2000)
29. D. Labat, R. Ababou, A. Mangin, Introduction of wavelet analyses to rainfall/runoffs relationship for a karstic basin: The case of Licq-Atherey karstic system (France). Groundwater **39**(4), 605–615 (2001)
30. W. Wang, J. Ding, Wavelet network model and its application to the prediction of hydrology. Nat. Sci **1**(1), 67–71 (2003)
31. F. Anctil, D.G. Tape, An exploration of artificial neural network rainfall-runoff forecasting combined with wavelet decomposition. J. Environ. Eng. Sci. **3**(S1), S121–S128 (2004)
32. D. Mwale, T.Y. Gan, Wavelet analysis of variability, teleconnectivity, and predictability of the september–november east african rainfall. J. Appl. Meteorol. **44**(2), 256–269 (2005)
33. D. Mwale et al., Wavelet empirical orthogonal functions of space-time-frequency regimes and predictability of southern Africa summer rainfall. J. Hydrol. Eng. **12**(5), 513–523 (2007)
34. J.F. Adamowski, Development of a short-term river flood forecasting method for snowmelt driven floods based on wavelet and cross-wavelet analysis. J. Hydrol. **353**(3), 247–266 (2008)

35. J.F. Adamowski, River flow forecasting using wavelet and cross-wavelet transform models. Hydrol. Process. **22**(25), 4877–4891 (2008)
36. C.-C. Kuo, T.Y. Gan, P.-S. Yu, Wavelet analysis on the variability, teleconnectivity, and predictability of the seasonal rainfall of Taiwan. Mon. Weather Rev. **138**(1), 162–175 (2010)
37. C.-C. Kuo, T.Y. Gan, P.-S. Yu, Seasonal streamflow prediction by a combined climate-hydrologic system for river basins of Taiwan. J. Hydrol. **387**(3), 292–303 (2010)
38. M. Özger, Significant wave height forecasting using wavelet fuzzy logic approach. Ocean Eng. **37**(16), 1443–1451 (2010)
39. T. Partal, Ö. Kişi, Wavelet and neuro-fuzzy conjunction model for precipitation forecasting. J. Hydrol. **342**(1–2), 199–212 (2007)
40. T. Partal, H.K. Cigizoglu, Prediction of daily precipitation using wavelet—neural networks. Hydrol. Sci. J. **54**(2), 234–246 (2009)
41. O. Kisi, J. Shiri, Precipitation forecasting using wavelet-genetic programming and wavelet-neuro-fuzzy conjunction models. Water Resour. Manage. **25**(13), 3135–3152 (2011)
42. R.V. Ramana et al., Monthly rainfall prediction using wavelet neural network analysis. Water Resour. Manage. **27**(10), 3697–3711 (2013)
43. W. Wang, J. Jin, Y. Li, Prediction of inflow at three gorges dam in Yangtze River with wavelet network model. Water Resour. Manage. **23**(13), 2791–2803 (2009)
44. H.-C. Zhou, Y. Peng, G.-H. Liang, The research of monthly discharge predictor-corrector model based on wavelet decomposition. Water Resour. Manage. **22**(2), 217–227 (2008)
45. Ö. Kişi, Stream flow forecasting using neuro-wavelet technique. Hydrol. Process. **22**(20), 4142–4152 (2008)
46. Ö. Kişi, Neural networks and wavelet conjunction model for intermittent streamflow forecasting. J. Hydrol. Eng. **14**(8), 773–782 (2009)
47. T. Partal, River flow forecasting using different artificial neural network algorithms and wavelet transform. Can. J. Civ. Eng. **36**(1), 26–38 (2008)
48. J. Adamowski, K. Sun, Development of a coupled wavelet transform and neural network method for flow forecasting of non-perennial rivers in semi-arid watersheds. J. Hydrol. **390** (1–2), 85–91 (2010)
49. N. Pramanik, R. Panda, A. Singh, Daily river flow forecasting using wavelet ANN hybrid models. J. Hydroinformatics **13**(1), 49–63 (2011)
50. J. Shiri, O. Kisi, Short-term and long-term streamflow forecasting using a wavelet and neuro-fuzzy conjunction model. J. Hydrol. **394**(3), 486–493 (2010)
51. Y. Wang et al., Flood simulation using parallel genetic algorithm integrated wavelet neural networks. Neurocomputing **74**(17), 2734–2744 (2011)
52. R. Maheswaran, R. Khosa, Wavelets-based non-linear model for real-time daily flow forecasting in Krishna River. J. Hydroinformatics **15**(3), 1022–1041 (2013)
53. M. Shoaib, A.Y. Shamseldin, B.W. Melville, Comparative study of different wavelet based neural network models for rainfall-runoff modeling. J. Hydrol. **515**, 47–58 (2014)
54. V. Nourani, Ö. Kisi, M. Komasi, Two hybrid Artificial Intelligence approaches for modeling rainfall–runoff process. J. Hydrol. **402**(1), 41–59 (2011)
55. R. Maheswaran, R. Khosa, Wavelet-Volterra coupled model for monthly stream flow forecasting. J. Hydrol. **450–451**, 320–335 (2012)
56. A. Aussem, J. Campbell, F. Murtagh, Wavelet-based feature extraction and decomposition strategies for financial forecasting. J. Comput. Intell. Finan. **6**(2), 5–12 (1998)
57. V. Nourani, M. Komasi, A. Mano, A multivariate ANN-wavelet approach for rainfall–runoff modeling. Water Resour. Manage. **23**(14), 2877–2894 (2009)
58. M. Shoaib et al., Hybrid wavelet neuro-fuzzy approach for rainfall-runoff modeling. J. Comput. Civil Eng. (2014)
59. M. Shoaib et al., Runoff forecasting using hybrid Wavelet Gene Expression Programming (WGEP) approach. J. Hydrol. **527**, 326–344 (2015)
60. C.W. Dawson et al., Evaluation of artificial neural network techniques for flow forecasting in the River Yangtze, China. Hydrol. Earth Syst. Sci. Dis. **6**(4), 619–626 (2002)

61. Y.B. Dibike, D. Solomatine, M.B. Abbott, On the encapsulation of numerical-hydraulic models in artificial neural network. J. Hydraul. Res. **37**(2), 147–161 (1999)
62. A. El-Shafie et al., Performance of artificial neural network and regression techniques for rainfall-runoff prediction. Int. J. Phys. Sci. **6**(8), 1997–2003 (2011)
63. A. Jain, A.M. Kumar, Hybrid neural network models for hydrologic time series forecasting. Appl. Soft Comput. **7**(2), 585–592 (2007)
64. A.W. Minns, M.J. Hall, Artificial neural networks as rainfall-runoff models. Hydrol. Sci. J. **41**(3), 399–417 (1996)
65. R. Modarres, Multi-criteria validation of artificial neural network rainfall-runoff modeling. Hydrol. Earth Syst. Sci. **13**(3), 411–421 (2009)
66. P. Phukoetphim, A.Y. Shamseldin, B.W. Melville, Knowledge extraction from artificial neural network for rainfall-runoff models combination system. J. Hydrol. Eng. (2013)
67. S. Riad et al., Predicting catchment flow in a semi-arid region via an artificial neural network technique. Hydrol. Process. **18**(13), 2387–2393 (2004)
68. S. Srinivasulu, A. Jain, A comparative analysis of training methods for artificial neural network rainfall–runoff models. Appl. Soft Comput. **6**(3), 295–306 (2006)
69. Z. Waszczyszyn, *Fundamentals of Artificial Neural Networks* (Springer, 1999)
70. M. Motter, J.C. Principe. *A gamma memory neural network for system identification*. in *Neural Networks, 1994. IEEE World Congress on Computational Intelligence., 1994 IEEE International Conference on*. 1994. IEEE
71. C.H. Van Iddekinge, R.E. Ployhart, Developments in the criterion-related validation of selection procedures: a critical review and recommendations for practice. Pers. Psychol. **61**(4), 871–925 (2008)
72. M. Anthony, P.L. Bartlett, *Neural Network Learning: Theoretical Foundations* (Cambridge University Press, 2009)
73. K. Aziza et al., *Co-Active Neuro Fuzzy Inference System for Regional Flood Estimation in Australia*. (Editorial Board, 2013), p. 11
74. C.-T. Cheng et al., Long-term prediction of discharges in Manwan Hydropower using adaptive-network-based fuzzy inference systems models, in *Advances in Natural Computation*. (Springer, 2005), pp. 1152–1161
75. A.P. Jacquin, A.Y. Shamseldin, Development of rainfall–runoff models using Takagi-Sugeno fuzzy inference systems. J. Hydrol. **329**(1), 154–173 (2006)
76. A. Lohani, R. Kumar, R. Singh, *Hydrological Time Series Modeling: A Comparison Between Adaptive Neuro Fuzzy, Neural Network And Auto Regressive Techniques* (Journal of Hydrology, 2012)
77. P.C. Nayak et al., A neuro-fuzzy computing technique for modeling hydrological time series. J. Hydrol. **291**(1–2), 52–66 (2004)
78. P.C. Nayak, K.P. Sudheer, K.S. Ramasastri, Fuzzy computing based rainfall–runoff model for real time flood forecasting. Hydrol. Process. **19**(4), 955–968 (2005)
79. Nayak, P.C., et al., Short-term flood forecasting with a Neurofuzzy model. Water Resour. Res. **41**(4) (2005)
80. P.C. Nayak, K.P. Sudheer, S.K. Jain, Rainfall-runoff modeling through hybrid intelligent system. Water Resour. Res. **43**(7) (2007)
81. A. Talei, L.H.C. Chua, C. Quek, A novel application of a neuro-fuzzy computational technique in event-based rainfall–runoff modeling. Expert Syst. Appl. **37**(12), 7456–7468 (2010)
82. B. Zhang, R.S. Govindaraju, Prediction of watershed runoff using Bayesian concepts and modular neural networks. Water Resour. Res. **36**(3), 753–762 (2000)
83. M.P. Rajurkar, U.C. Kothyari, U.C. Chaube, Modeling of the daily rainfall-runoff relationship with artificial neural network. J. Hydrol. **285**(1), 96–113 (2004)
84. Koza, J.R., *Genetic Programming: On the programming of Computers by Means of Natural Selection* (MIT Press, Cambridge, MA, 1992)
85. A. Aytek, M. Alp, An application of artificial intelligence for rainfall-runoff modeling. J. Earth Syst. Sci. **117**(2), 145–155 (2008)

86. V. Babovic, M.B. Abbott, The evolution of equations from hydraulic data Part I: Theory. J. Hydraul. Res. **35**(3), 397–410 (1997)
87. V. Babovic, M.B. Abbott, The evolution of equations from hydraulic data Part II: Applications. J. Hydraul. Res. **35**(3), 411–430 (1997)
88. Drecourt, J.-P., *Application of neural networks and genetic programming to rainfall-runoff modeling.* D2 K Technical Rep, 1999(0699-1): p. 1
89. L.-C. Chang, C.-C. Ho, Y.-W. Chen, Applying multiobjective genetic algorithm to analyze the conflict among different water use sectors during drought period. J. Water Resour. Plann. Manage. **136**(5), 539–546 (2009)
90. S.T. Khu et al., Genetic programming and its application in real-time runoff forecasting1. JAWRA J. Am. Water Resour. Assoc. **37**(2), 439–451 (2001)
91. T. Rajaee, Wavelet and ANN combination model for prediction of daily suspended sediment load in rivers. Sci. Total Environ. **409**(15), 2917–2928 (2011)
92. H.M. Azamathulla et al., Gene-expression programming for the development of a stage-discharge curve of the Pahang River. Water Resour. Manage **25**(11), 2901–2916 (2011)
93. A. Guven, A. Aytek, New approach for stage–discharge relationship: gene-expression programming. J. Hydrol. Eng. **14**(8), 812–820 (2009)
94. O. Kisi, J. Shiri, B. Nikoofar, Forecasting daily lake levels using artificial intelligence approaches. Comput. Geosci. **41**, 169–180 (2012)
95. O. Kisi, J. Shiri, M. Tombul, Modeling rainfall-runoff process using soft computing techniques. Comput. Geosci. **51**, 108–117 (2013)
96. J. Shiri et al., Daily reference evapotranspiration modeling by using genetic programming approach in the Basque Country (Northern Spain). J. Hydrol. **414**, 302–316 (2012)
97. O. Kisi, Wavelet regression model as an alternative to neural networks for river stage forecasting. Water Resour. Manage. **25**(2), 579–600 (2011)
98. R. Maheswaran, R. Khosa, Comparative study of different wavelets for hydrologic forecasting. Comput. Geosci. **46**, 284–295 (2012)
99. V. Nourani, M. Komasi, M.T. Alami, Hybrid wavelet–genetic programming approach to optimize ANN modeling of rainfall–runoff Process. J. Hydrol. Eng. **17**(6), 724–741 (2011)
100. V. Nourani et al., Using self-organizing maps and wavelet transforms for space–time pre-processing of satellite precipitation and runoff data in neural network based rainfall–runoff modeling. J. Hydrol. **476**, 228–243 (2013)
101. N. Vahid, K. TohidRezapour, B. AidaHosseini, *Implication of Feature Extraction Methods to Improve Performance of Hybrid Wavelet-ANN Rainfall?Runoff Model*, in *Case Studies in Intelligent Computing* (Auerbach Publications, 2014), pp. 457–498

Quantification of Prediction Uncertainty in Artificial Neural Network Models

K.S. Kasiviswanathan, K.P. Sudheer and Jianxun He

Abstract The research towards improving the prediction and forecasting of artificial neural network (ANN) based models has gained significant interest while solving various engineering problems. Consequently, different approaches for the development of ANN models have been proposed. However, the point estimation of ANN forecasts seldom explains the actual mechanism that brings the relationship among modeled variables. This raises the question on the model output while making decisions due to the inherent variability or uncertainty associated. The standard procedure though available for the quantification of uncertainty, their applications in ANN model are still limited. In this chapter, commonly employed uncertainty methods such as bootstrap and Bayesian are applied in ANN and demonstrated through a case example of flood forecasting models. It also discusses the merits and limitations of bootstrap ANN (BTANN) and Bayesian ANN (BANN) models in terms of convergence of parameter and quality of prediction interval evaluated using uncertainty indices.

Keywords Bayesian · Bootstrap · Point estimation · Uncertainty

1 Introduction

The applications of ANN model have been well acknowledged over the last two decades in water resources and environmental related studies. It is found that most of ANN applications mainly fall in the category of prediction, in which finding an

K.S. Kasiviswanathan · J. He
Department of Civil Engineering, Schulich School of Engineering,
University of Calgary, 2500 University Drive NW, Calgary T2N 1N4, Canada

K.P. Sudheer (✉)
Department of Civil Engineering, Indian Institute of Technology Madras,
Chennai 600036, India
e-mail: sudheer@iitm.ac.in

© Springer International Publishing Switzerland 2016   145
S. Shanmuganathan and S. Samarasinghe (eds.), *Artificial Neural
Network Modelling*, Studies in Computational Intelligence 628,
DOI 10.1007/978-3-319-28495-8_8

unknown relationship exists between a set of input factors and an output is of primary interest [1, 2]. The training process of ANN model leads to discover the underline relationship between the variables with available time series data. The major limitation is the ANN models do not explain any internal physical mechanism that controls the processes. However, the use of ANN model has been still encouraged in most of the occasions due to the accurate prediction or forecast compared to physics and conceptual models [2].

Despite the large number of applications, ANNs still remain something of a numerical enigma. Apart from the major criticism that ANNs lack transparency, many researchers mention that ANN development is stochastic in nature, and no identical results can be reproduced on different occasions unless carefully devised [3, 4]. This is a significant weakness, and it is hard to trust the reliability of networks while addressing real-world problems without the ability to produce comprehensible decisions. Therefore, a significant research effort is needed to address this deficiency of ANNs [2]. In fact, there is a belief that ANN point predictions are of limited value where there is uncertainty in the data or variability in the underlying system. These uncertainties mainly arise from input, parameter and model structure. The input (measured/forecasted precipitation in case of hydrologic models) uncertainty is mainly due to measurement and sampling error. The parameter uncertainty lies in inability to identify unique set of best parameters of the model. The simplification, inadequacy and ambiguity in description of real world process through mathematical equation leads a model structure uncertainty.

Statistically, the ANN output approximates the average of the underlying target conditioned on the neural network model input vector [5]. However, ANN predictions convey no information about the sampling errors and the prediction accuracy. The limited acceptance of the ANN model applications can be plausibly attributed to the difficulty observed in assigning confidence interval (or prediction interval) to the output [6], which might improve the reliability and credibility of the predictions. Therefore an investigation into quantifying the uncertainty associated with the ANN model predictions is essentially required.

Several uncertainty analysis techniques are available in literature, which quantifies the prediction uncertainty of hydrologic models with respect to its parameter, input and structure; however, each of these methods differs in their principle, assumption, level of complexity and computational requirement. In addition, there is no clear evidence that one method outperforms another in terms of accuracy in estimated prediction interval of model output [7]. It is to be also noted that application of standard procedures to carry out uncertainty analysis in theory driven models cannot be applied directly to ANN models due to their complex parallel computing architecture. The difficulty in performing an uncertainty analysis of ANN outputs lies in the fact that the ANNs have large degrees of freedom in their development.

According to Shrestha and Solomatine [6], the uncertainty methods have been classified into four different approaches such as (a) probabilistic based method (b) analyzing the statistical properties of the errors of the model in reproducing historically observed data (c) resampling techniques, generally known as ensemble methods, or the Monte Carlo method and (d) fuzzy based method. In probabilistic based method, the model variables associated with uncertainty (i.e. parameter or input) is defined as a probability distribution functions (PDFs). The number of random samples depending on the size of the problem is drawn from the distribution, so as to obtain the predictive distribution of model output [8, 9]. The Bayesian approach falls in this category and some of the studies reported the use of Bayesian approach in quantifying the uncertainty of neural network models [10, 11]. The second approach estimates the uncertainty of model output through statistical properties of the error while reproducing the observed data [6]. Numerous studies reported the third approach while estimating the uncertainty of ANN models [12, 13]. The major reason could be attributed to the meaningful estimation of uncertainty with limited assumptions. The bootstrapped is a sampling based approach which fall into the third category. The major assumptions of bootstrap method are: (a) the bootstrapped samples follow the statistical characteristics of population data and (b) the resamples mimic the random component of the process to be modeled. In the fuzzy method, the uncertainty of ANN model can be analyzed for the model parameters such as weights and biases or the model inputs. In such approach, the model inputs [14] and/or parameters [15] are represented as fuzzy number for defining the variability and to quantify the prediction uncertainty.

Different methods though available for quantifying the uncertainty, their comprehensive evaluation is limited. Hence, still there is a scope to compare various methods for evaluating uncertainty of ANN, so as to improve the confidence of model prediction. Hence, the objective of this Chapter is to illustrate the potential of different methods for the meaningful quantification of uncertainty. Since the Bayesian and bootstrap method has been reported to be a promising approach, it has been selected for the demonstration. The hourly rainfall and runoff data collected from Kolar river basin, India is used for developing the ANN based flood forecasting model.

2 Study Area and Data Description

In order to demonstrate the proposed method of uncertainty analysis of ANN models, a case study on an Indian River basin is presented herein. For the study, hourly data of rainfall and runoff from Kolar basin (Fig. 1) in India is used.

The Kolar River is a tributary of the river Narmada that drains an area about 1350 km^2 before its confluence with Narmada near Neelkant. In the present study the catchment area up to the Satrana gauging site is considered, which constitutes an area of 903.87 km^2. The 75.3 km long river course lies between latitude 21° 09' N to 23° 17' N and longitude 77° 01' E to 77° 29' E. An ANN model for forecasting the river flow has been developed in this study and analyzed for uncertainty. Hourly rainfall and runoff data is collected during monsoon season (July, August, and September) for three years (1987–1989). Note that areal average values of rainfall data for three rain gauge stations have been used in the study.

Fig. 1 Map of the study area
(Kolar Basin)

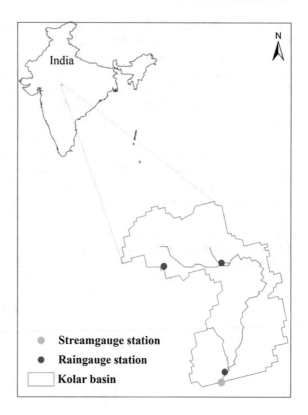

3   Methodology

The flowchart (Fig. 2) illustrates the overall methodology which includes the
standard methods used for ANN model training (i.e. input selection, data division
for calibration and validation, identification of model architecture, besides the
description of bootstrap and Bayesian methods. The subsequent section describes
each component with clear illustration.

3.1   Input Selection

The determination of significant input variable is the most important steps in discovering the functional form of any data driven models [16, 17]. The model prediction majorly depends on correlation between available input and output variables,
model complexity, and the learning difficulty of trained ANN. Generally some degree
of a priori knowledge is used to specify the initial set of candidate inputs [18, 19].
However, the relationship between the variables may not be clearly known a priori,
and hence often an analytical technique, such as cross-correlation, is employed

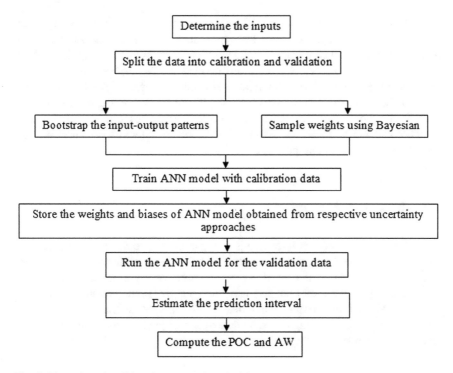

Fig. 2 Flow chart describing the proposed methodology

[20, 21]. The major disadvantage associated with using cross-correlation is that it is only able to detect linear dependence between two variables, while the modeled relationship may be highly nonlinear. Nonetheless, the cross-correlation methods represent the most popular analytical techniques for selecting appropriate inputs [16, 17]. The current study employed a statistical approach suggested by Sudheer et al. [21] to identify the appropriate input vector. The method is based on the heuristic that the potential influencing variables corresponding to different time lags can be identified through statistical analysis of the data series that uses cross-, auto-, and partial auto correlations between the variables in question.

3.2 Data Division

Before proceeding with data division, the data should be normalized. This is mainly for the ANN activation function that is generally bounded and sensitive to prede-fined ranges. The linear transformation is often reported while normalizing the data [21]. Hence, the actual dataset is changed into a normalized domain in order to contain the values within the interval [0, 1], or [−1, 1].

While developing an ANN model, generally the total available examples are divided into training and validation sets (split sample validation) prior to the model building, and in some cases a cross validation set is also used. The majority of ANN applications in hydrology, the data are divided on an arbitrary basis into the required subsets to develop ANN models. However, many studies have shown that the way the data are divided can have a significant impact on the results obtained [22]. It has been suggested that the statistical properties (e.g. mean, standard deviation) of the various data subsets need to be considered to ensure that each subset represents the same population [2].

3.3 Bootstrap Method

It has been emphasized that bootstrap is a simple method to quantify the parameter and prediction uncertainty of neural network [12]. The quantification of uncertainty using bootstrap method is carried out by sampling of input-output patterns with replacement [23]. Out of total available dataset, 'b' such random samples are bootstrapped each time to train neural network for different realization of parameter sets combinations. The simple arithmetic average of predictions would be considered as true regression of model.

$$\hat{Q}_i = \frac{1}{B} \sum_{b=1}^{B} f(x_i; p_b) \tag{1}$$

where $\hat{Q}_i$ represents the average forecast corresponding to the ith input data point, x_i; B denotes the total number of ensemble of ANN models; and p_b and f denotes the parameter and functional form of a ANN model.

It is to be noted that variability in predictions exists mainly due to the random sampling of input-output patterns. The network parameter is optimized by Levenberg-Marquardt algorithm [24].

3.4 Bayesian Method

The traditional learning of network uses error minimization function which attempts to find a set of deterministically optimized weights. In contrast, the Bayesian learning involves in training the neural network for the posterior probability distribution of weights from assumed prior probability distribution using Bayes's theorem. The posterior probability distribution is then used to evaluate the predictive probability distribution of network outputs.

According to Bayes' rule, the posterior probability distribution of parameters of ANN model 'M' given the input-output pattern (X, Y) is,

$$p(\theta|X, Y, M) = \frac{p(X, Y|\theta, M)\, p(\theta|, M)}{p(X, Y|M)} \qquad (2)$$

where, $p(D|M)$ is a normalization factor which ensures the total probability is one. M denotes the model with specified connection weights for selected network architecture; $p(X, Y|\theta, M)$ is the likelihood of the parameter θ. It is assumed that the model residuals follow Gaussian distribution and which can be written as,

$$p(X, Y|\theta, M) \propto \exp\left(-\frac{\beta|Y - f(X, \theta)|^2}{2}\right) \qquad (3)$$

$p(\theta|, M)$ is the prior probability distribution of parameter θ is assumed to follow Gaussian distribution and it is written as,

$$P(w) \propto \exp\left(-\frac{\alpha|w|^2}{2}\right) \qquad (4)$$

where, α, β are called hyper parameters of distribution which follows Inverse-gamma distribution. These values are updated using Bayes's theorem given the input-output patterns. The model prediction is integration of posterior distribution of weight vectors given the data and is represented as

$$E[Y_{n+1}] = \int f(X_{n+1}, \theta)\, p(\theta|(X, Y))d\theta \qquad (5)$$

Solving Eq. 5 analytically is computationally a non-trivial task. Therefore, it requires suitable sampling technique to numerically solve. This study used Marcov Chain Monte Carlo (MCMC) algorithm to sample the parameters with initial and actual sampling phases [9]. During initial sampling phase, only the parameters of ANN are updated, however the hyper parameters are fixed at certain values. This prevents taking biased values of hyper parameter before ANN parameter reaches reasonable values. Once these starting values are fixed, actual sampling phase is used to determine the values of hyper parameters. This progressively changes the shape of distribution for the effective sampling of ANN parameters. In such way, many combinations of finally converged parameters from posterior distribution are stored and that are used to predict the variable of interest for given input.

3.5 ANN Model Architecture

The selection of suitable ANN model architecture solely depends on the complexity of the problem. The model architecture is in general defined by the number of hidden layer and the number of hidden neurons that each layer contains. Despite, more number of hidden layer offer more flexibility, it might introduce the over training of network and thus result in poor generalization. It is reported that a single hidden layer is sufficient for the majority of ANN applications in water resources [25]. The number of hidden neurons is fixed by trial and error approach, although sophisticated optimization procedure available. The reason is that the trial and error method is simple to apply and has less computational burden [26, 27]. In such procedure, one starts with simplest model structure, and each time the model is calibrated by adding one more hidden neuron at a time. This process is repeated until there is no further significant improvement in the model performance in the model calibration.

3.6 Indices Used for Evaluating Model Performance and Prediction Uncertainty

The model performance is evaluated using the selected four statistical indices: the correlation coefficient (CC), Nash-Sutcliff efficiency (NSE) [28], root mean square error (RMSE), and mean biased error (MBE). These indices are calculated based on the observations and the final model outputs using the following equations.

$$CC = \frac{\sum\limits_{i=1}^{n} \left[(Q_i - \bar{Q}) \cdot \left(\hat{Q}_i - \bar{\hat{Q}} \right) \right]}{\sqrt{\sum\limits_{i=1}^{n} (Q_i - \bar{Q})^2 \cdot \sum\limits_{i=1}^{n} \left(\hat{Q}_i - \bar{\hat{Q}} \right)^2}} \tag{6}$$

$$NSE = \left\{ 1 - \frac{\sum\limits_{i=1}^{n} (Q_i - \hat{Q}_i)^2}{\sum\limits_{i=1}^{n} (Q_i - \bar{Q})^2} \right\} * 100 \tag{7}$$

$$RMSE = \sqrt{\frac{1}{n} \sum\limits_{i=1}^{n} (Q_i - \hat{Q}_i)^2} \tag{8}$$

$$MBE = \frac{1}{n} \sum\limits_{i=1}^{n} (Q_i - \hat{Q}_i) \tag{9}$$

where, Q_i and $\hat{Q}_i$ are the observed and forecasted value respectively; $\bar{Q}$ and $\bar{\hat{Q}}$ are the mean of the observed and forecasted values, respectively; and n is the total number of data points.

The prediction ranges from the minimum to the maximum modelled values from all models. The uncertainty is assessed using two indices, the percentage of coverage (POC) also known as coverage probability and the average width (AW). These two indices have been often used for evaluating the prediction uncertainty [11, 29].

The POC indicates the number of observations falling within the prediction interval; while the AW measures the average of the interval width, which is the difference between the upper and lower bounds of the prediction interval. The POC and the AW are calculated by

$$POC = \left(\frac{1}{n} \sum_{i=1}^{n} c_i \right) * 100 \tag{10}$$

$$AW = \frac{1}{n} \sum_{i=1}^{n} \left[\hat{Q}_i^U - \hat{Q}_i^L \right] \tag{11}$$

where, $\hat{Q}_i^U$ and $\hat{Q}_i^L$ are the upper and lower bounds of the prediction interval corresponding to the ith input data point; $c_i = 1$ if the observation falls within the prediction band $\left[\hat{Q}_i^U, \hat{Q}_i^L \right]$, otherwise $c_i = 0$.

4 Results and Discussion

4.1 Model Development

Based on the methodology suggested by Sudheer et al. (2002), the following inputs were identified for the ANN model: [R(t-9), R(t-8), R(t-7), Q(t-2), Q(t-1)], where R (t) represents the rainfall, Q(t) represents the runoff at time period 't'. The output of the network was considered as Q(t).

The data was normalized between the range of zero and one. From the total available data for three years 6525 patterns (input-output pairs) were identified for the study, which was split into training (5500 sets) and validation (1025 sets) data sets. A single hidden layer is considered in the study based on various research studies conducted on this basin [26, 30]. The optimal number of hidden neurons was found to be 3 after trial and error. The ANN model structure used in this study is shown in Fig. 3. Hence the model structure is assumed to be fixed, and only the parameter uncertainty is quantified.

4.2 Evaluation of Model Performance

In BTANN, out of 5500 pattern, 4500 patterns were sampled each time for developing a single ANN model. In BANN, the complete set (5500 pattern) of calibration data was used, however the parameters were sampled from respective PDFs. The number of ensemble is fixed as 100 in both model presented in this chapter, which has been normally adopted in the previous studies [27, 31].

The ANN model performance obtained through selected approaches is presented in Table 1, in terms of various statistical measures of performance. The indices such

Fig. 3 The final ANN
architecture identified for
Kolar Basin

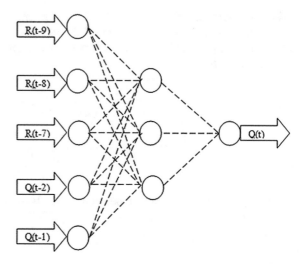

as Correlation Coefficient (CC), Nash Sutcliffe efficiency (NSE), Root mean square error (RMSE) and Mean biased error (MBE) are considered for performance evaluation of the developed model. It may be noted that listed performance indices values are estimated for the ensemble mean generated from 100 simulations of respective methods.

It is observed from Table 1 that the BTANN model produces high amount of RMSE and MBE when compared BANN, which indicates an inferior performance of the methods. The reason for such poor performance may be attributed to random sampling of input-output patterns for training without considering the preservation of statistical characteristics across bootstrapped samples. The BANN model showed a slight improved performance during both calibration (results are not presented here) and validation (Table 1). The values of CC and NSE do not show any significant change across the methods during calibration as well as validation. The Bayesian method has an NSE of 98.27 % in calibration and 97.04 % in validation. The major reason for superior performance of BANN model lies in reaching appropriate range of model parameters by keeping the same input data, whereas the BTANN searches the parameter based on the bootstrapped input samples during model training. The positive values of MBE of BTANN and BANN shows consistent underestimation of model during validation (Table 1). Overall, the ensemble mean obtained through BANN model closely matches the observed values with comparably better performance than BTANN.

Table 1 Model performance by BTANN and BANN

Model performance index	BTANN	BANN
Correlation coefficient	0.96	0.99
Nash-Sutcliff efficiency (%)	92.62	97.04
Root mean square error (m^3/s)	50.64	32.02
Mean biased error (m^3/s)	5.77	0.98

4.3 Parameter Uncertainty

The parameter uncertainty produced by BTANN and BANN is compared. Figure 4 illustrates the mean and minimum to maximum range of finally optimized parameters obtained from 100 number of ensemble of simulation.

In both methods, the mean value of most of the connection weight approaches zero. This is plausibly due to averaging the ensemble of parameters that has canceled the positive and negative values. Hence the comparison is mainly based on min-max range of respective methods. It is observed that the parameter variability of Bootstrap method is considerably higher than Bayesian method. In which the higher variability is found at the biases which connect hidden nodes and all other connection has almost equal variability around the mean value. This could be ascribed as the model bias towards the selection of patterns for training. In BANN, the weights connecting antecedent flow information has high parameter variability due to high autocorrelation to the output variable. However, the Bayesian approach produced consistent model parameters with narrow interval. The reason could be attributed to convergence of sampling algorithm towards appropriate range of parameter space in the form of PDF's [10]. This in turn resulted in the better predictive capability of Bayesian method with less parameter uncertainty.

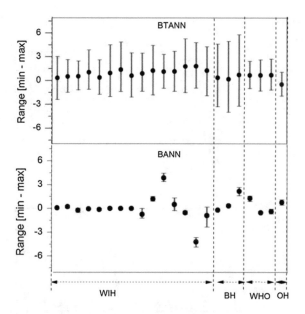

Fig. 4 Parameter uncertainty of BTANN and BANN (The *black dot point* and *red line* represent the mean and ranges of finally calibrated parameters) [The following nomenclature is used to represent the links: for the connection between input nodes and hidden nodes (WIH), weights connecting hidden nodes to the output node are designated as WHO and the bias connection to hidden and output nodes are represented as BH and BO respectively]

4.4 Prediction Uncertainty

The quality of prediction interval produced by BTANN and BANN is evaluated using uncertainty indices POC and AW (Table 2).

It may be noted that the presented results correspond to validation data only. In order to assess the uncertainty on different magnitude of flow, the flow values are statistically categorized as low, medium and high [26]. Out of 1025 patterns in validation, low flow values contain 843 patterns and 167, 15 patterns fall in medium, high flow respectively. In theory, a better model will have less AW with more number of observed values falls over the prediction band (i.e. maximum POC). In other words, quantitatively an ideal prediction band has POC of 100 % with AW approaches to lower values. The prediction interval obtained through different method has varying magnitude of uncertainty in terms of POC and AW estimate (Fig. 5). In general, the BTANN model has higher AW width values across different flow domains as illustrated in Fig. 5, which indicates larger biases in model performance across the pattern which is selected for model training. Consequently it leads higher variability from the mean prediction and higher values of POC. It is obvious that the higher the value of POC is a result of increased width. On the other hand, BANN model produces consistently less AW value across

Method	BTANN		BANN	
Flow series	AW (m^3/s)	POC (%)	AW (m^3/s)	POC (%)
Complete flow Series	75.95	98.14	3.42	46.58
Low flow (x < μ)	38.06	99.88	0.44	49.46
Medium flow (μ ≤ x ≤ + 2σ)	198.63	95.81	6.78	33.53
High flow (x > μ + 2σ)	839.56	20	133.49	26.67

Table 2 Comparison of predictive uncertainty by different method

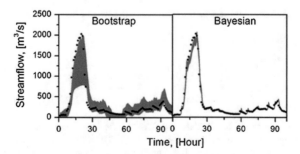

Fig. 5 Prediction interval obtained through different methods (The *dotted point* represents the measured streamflow value)

complete, low and medium flow series with 3.42, 0.44 and 6.78 m^3/s respectively. The forgoing discussion clearly indicates that the BANN model is good in terms of lesser AW and bootstrap method is good with better estimate of maximum POC. The choice of particular model is based on the application in which the model is going to be employed. However, it should be noted that any misleading information between these two conflict indices will lead to a biased model selection (Fig. 5). Overall, the BANN model produces ensemble of simulation which are very close to observed values though the POC is less.

5 Conclusion

In this chapter, the Bayesian and bootstrap based uncertainty evaluation method is applied in ANN models. The presented approach is demonstrated through case study of flood forecasting models for the Kolar basin, India. In both methods, the model structure is assumed to be deterministic and only a parameter uncertainty is evaluated. It is found that the BTANN and BANN model produce satisfactory results, while considering the ensemble mean. The comparison between these models suggests that the bootstrap method is simple and easy to apply than Bayesian method since it involves rigorous sampling of parameters through Monte Carlo simulation. However, the uncertainty produced by BANN model is better than BTANN in terms of narrow prediction interval and also with less variance in parameter convergence. The major reason for such reduction in parameter uncertainty of BANN model might be due to the potential of MCMC while converging towards optimal parameter space. It is clear from the illustrated example that meaningful quantification of uncertainty in ANN model would improve the reliability while making decisions. This suggests the choice of model should not only be based on the model performance, however with appropriate analysis of uncertainty. It should also be noted that each method differs by its own complexity, principle and computational efficiency.

References

1. K.P. Sudheer, Knowledge extraction from trained neural network river flow models. J. Hydrol. Eng. **10**(4), 264–269 (2005)
2. H.R. Maier, A. Jain, G.C. Dandy, K.P. Sudheer, Methods used for the development of neural networks for the prediction of water resource variables in river systems: current status and future directions. Environ. Model Softw. **25**, 891–909 (2010)
3. A. Elshorbagy, G. Corzo, S. Srinivasulu, D.P. Solomatine, Experimental investigation of the predictive capabilities of data driven modeling techniques in hydrology—Part 2: concepts and methodology. Hydrol. Earth Syst. Sci. **14**(10), 1931–1941 (2010)
4. A. Elshorbagy, G. Corzo, S. Srinivasulu, D.P. Solomatine, Experimental investigation of the predictive capabilities of data driven modeling techniques in hydrology—Part 2: application. Hydrol. Earth Syst. Sci. **14**(10), 1943–1961 (2010)

5. C.M. Bishop, Neural Networks for Pattern Recognition (Oxford University Press, New York, 1995)
6. D.L. Shrestha, D.P. Solomatine, Machine learning approaches for estimation of prediction interval for the model output. Neural Netw. **19**, 225–235 (2006)
7. T. Wagener, H.V. Gupta, Model identification for hydrological forecasting under uncertainty. Stoch. Environ. Res. Risk Assess. **19**(6), 378–387 (2005)
8. K.J.C. MacKay, A practical Bayesian framework for backpropagation. Netw. Neural Comput. **4**, 448–472 (1992)
9. N.M. Radford, *Bayesian Learning for Neural Networks*. Lecture Notes in Statistics, vol. 118 (Springer, NewYork, (1996)
10. M.S. Khan, P. Coulibaly, Y. Dibike, Uncertainty analysis of statistical downscaling methods. J. Hydrol. **319**, 357–382 (2006)
11. X. Zhang, F. Liang, R. Srinivasan, M. Van Liew, Estimating uncertainty of streamflow simulation using Bayesian neural networks. Water Resour. Res. **45**(2), 1–16 (2009)
12. R.K. Srivastav, K.P. Sudheer, I. Chaubey, A simplified approach to quantifying predictive and parametric uncertainty in artificial neural network hydrologic models. Water Resour. Res. **43** (10), W10407 (2007)
13. S.K. Sharma, K.N. Tiwari, Bootstrap based artificial neural network (BANN) analysis for hierarchical prediction of monthly runoff in Upper Damodar Valley Catchment. J. Hydrol. **374** (3–4), 209–222 (2009)
14. R.R. Shrestha, F. Nestmann, Physically Based and data-driven models and propagation of Input uncertainties in river flood prediction. J. Hydrol. Eng. **14**(December), 1309–1319 (2009)
15. S. Alvisi, M. Franchini, Fuzzy neural networks for water level and discharge forecasting with uncertainty. Environ. Model Softw. **26**(4), 523–537 (2011)
16. G.J. Bowden, H.R. Maier, G.C. Dandy, Input determination for neural network models in water resources applications. Part 1—background and methodology. J. Hydrol. **301**(1–4), 75–92 (2005)
17. G.J. Bowden, H.R. Maier, G.C. Dandy, Input determination for neural network models in water resources applications. Part 2—case study: forecasting salinity in a river. J. Hydrol. **301** (1–4), 93–107 (2005)
18. M. Campolo, A. Soldati, P. Andreussi, Forecasting river flow rate during low-flow periods using neural networks. Water Resour. Res. **35**(11), 3547–3552 (1999)
19. K. Thirumalaiah, M.C. Deo, Hydrological forecasting using neural networks. J. Hydrol. Eng. **5** (2), 180–189 (2000)
20. D. Silverman, J.A. Dracup, Artificial neural networks and long-range precipitation in California. J. Appl. Meteorol. **31**(1), 57–66 (2000)
21. K.P. Sudheer, A.K. Gosain, K.S. Ramasastri, A data-driven algorithm for constructing artificial neural network rainfall-runoff models. Hydrol. Process. **16**(6), 1325–1330 (2002)
22. A.S. Tokar, P.A. Johnson, Rainfall-runoff modeling using artificial neural networks. J. Hydrol. Eng. **4**(3), 232–239 (1999)
23. B. Efron, Bootstrap methods: another look at jackknife. Ann. Stat. **7**(1), 26 (1979)
24. A.P. Piotrowski, J.J. Napiorkowski, A comparison of methods to avoid overfitting in neural networks training in the case of catchment runoff modelling. J. Hydrol. **476**, 97–111 (2013)
25. R.J. Abrahart, L. See, Comparing neural network and autoregressive moving average techniques for the provision of continuous river flow forecasts in two contrasting catchments. Hydrol. Process. **14**(11–12), 2157–2172 (2000)
26. P.C. Nayak, K.P. Sudheer, D.M. Rangan, K.S. Ramasastri, Short-term flood forecasting with a neurofuzzy model. Water Resour. Res. **41**(4), 1–16 (2005). December 2004
27. K.S. Kasiviswanathan, R. Cibin, K.P. Sudheer, I. Chaubey, Constructing prediction interval for artificial neural network rainfall runoff models based on ensemble simulations. J. Hydrol. **499**, 275–288 (2013)
28. J.E. Nash, J.V. Sutcliffe, River flow forecasting through conceptual models: 1. A discussion of principles. J. Hydrol. **10**, 282–290 (1970)

29. K.S. Kasiviswanathan, K.P. Sudheer, Quantification of the predictive uncertainty of artificial neural network based river flow forecast models. Stoch. Environ. Res. Risk Assess. **27**(1), 137–146 (2013)
30. M. Chetan, K.P. Sudheer, A hybrid linear-neural model for river flow forecasting. Water Resour. Res. **42**, W04402 (2006). doi:10.1029/2005WR004072
31. M.K. Tiwari, J. Adamowski, Urban water demand forecasting and uncertainty assessment using ensemble wavelet-bootstrap-neural network models. Water Resour. Res. **49**(10), 6486–6507 (2013)

Classifying Calpain Inhibitors for the Treatment of Cataracts: A Self Organising Map (SOM) ANN/KM Approach in Drug Discovery

I.L. Hudson, S.Y. Leemaqz, A.T. Neffe and A.D. Abell

Abstract Calpain inhibitors are possible therapeutic agents in the treatment of cataracts. These covalent inhibitors contain an electrophilic anchor ("warhead"), an aldehyde that reacts with the active site cysteine. Whilst high throughput docking of such ligands into high resolution protein structures (e.g. calpain) is a standard computational approach in drug discovery, there is no docking program that consistently achieves low rates of both false positives (FPs) and negatives (FNs) for ligands that react covalently (via irreversible interactions) with the target protein. Schroedinger's GLIDE score, widely used to screen ligand libraries, is known to give high false classification, however a two-level Self Organizing Map (SOM) artificial neural network (ANN) algorithm, with KM clustering proved that the addition of two structural components of the calpain molecule, number hydrogen bonds and warhead distance, combined with GLIDE score (or its partial energy subcomponents) provide a superior predictor set for classification of true molecular binding strength (IC50). SOM ANN/KM significantly reduced the number of FNs by 64 % and FPs by 26 %, compared to the glide score alone. FPs were shown to be mostly esters and amides plus alcohols and non-classical, and FNs mainly aldehydes and ketones, masked aldehydes and ketones and Michael.

I.L. Hudson (✉)
School of Mathematical and Physical Sciences, The University of Newcastle,
NSW, Australia
e-mail: irene.hudson@newcastle.edu.au; irenelena.hudson@gmail.com

S.Y. Leemaqz
Robinson Research Institute, The University of Adelaide, Adelaide
SA 5005, Australia

A.T. Neffe
Institute of Biomaterial Science, Helmholtz-Zentrum Geesthacht,
Teltow, Germany

A.D. Abell
School of Physics and Chemistry, The University of Adelaide, Adelaide
SA 5005, Australia

© Springer International Publishing Switzerland 2016   161
S. Shanmuganathan and S. Samarasinghe (eds.), *Artificial Neural Network Modelling*, Studies in Computational Intelligence 628,
DOI 10.1007/978-3-319-28495-8_9

1   Introduction and Overview

Artificial neural networks (ANNs) have a proven ability to model complex relationships between pharmaceutically appropriate properties and chemical structures
of compounds, specifically for the analysis and modelling of nonlinear relationships
between molecular structures and pharmacological activity [1–4]. ANN applications in drug discovery include the prediction of biological activity, selection of
screening candidates (cherry picking), and the extraction of representative subsets
from large compound collections such as combinatorial libraries [5]. ANNs as such
have proven potential to improve diversity and the quality of virtual screening. Our
focus in this study is the Self Organising Map (SOM) [6] which is a nonlinear
multidimensional mapping tool. The SOM represents competitive learning based
clustering of artificial neural networks. The SOM is either an unsupervised or
supervised data method [7, 8] that can visualise high-dimensional data sets in lower
(typically 2) dimensional representations. Many practical applications of the SOM
exist in compound classification, drug design, and chemical similarity searching [9,
10].

   In this chapter we develop clustering and visualisation tools to verify a predicted
protein-ligand binding mode when structural information of protein-ligand complexes is not available, and to identify indicators for the prediction of binding
affinity in order to distinguish between high affinity binders and essential
non-binders. Protein–ligand binding is central to both biological function and
pharmaceutical activity. It is known that some ligands inhibit protein function,
while others induce protein conformational variations, thus modulating key
cell-signaling pathways. Whichever scenario pertains, achieving a desired therapeutic effect is dependent upon the magnitude of the binding affinity of ligand to the
target receptor. Designing tight-binding ligands while conserving the other ligand
properties crucial to safety and biological efficacy is a major objective of
small-molecule drug discovery. As such a primary goal of computational chemistry
and computer aided drug design is the accurate prediction of protein–ligand free
energies of binding (i.e., the binding affinities).

   Specifically in this chapter we investigate the SOM as a mathematical and
computational tool for the evaluation of docking experiments of calpain ligands
(small drug molecules), where calpain is the large protein to which the inhibitor
(drug) binds. Calpain inhibitors are possible therapeutic agents in the treatment of
cataracts caused by the over-activation of calpain. Discovery of new calpain inhibitors could clearly benefit cataract treatment [11] and this would have significant
impact in drug discovery and treatment [11–20].

   The aims of this study are to verify the predicted binding mode of the calpain
ligands, identify structural and docking parameters for improved prediction of
binding affinity, and to distinguish between high affinity binders and essential
non-binders of calpain. The methodology is validated for a set of "literature and in
house" potential binders to the protease *calpain*, for which modeling and analytical

data is available. These covalent inhibitors contain an electrophilic anchor ("warhead"), an aldehyde that reacts with the active site cysteine. Many therapeutics based on protease inhibitors are currently in late clinical trials, or are already available as drugs. However, inhibitors of the cysteine protease family are to date much under-represented, primarily because of flaws in their design: existing inhibitors are conformationally flexible and biologically unstable structures with a 'reactive warhead' that makes them un-drug like [21]. Covalent linkage formation as such is a very important mechanism for many covalent drugs to work. However, to date, partly due to the limitations of computational tools for covalent docking, covalent drugs have not been discovered systematically through structure based virtual screening (SBVS) techniques. Figure 7 gives a schematic of SBVS in drug discovery (http://www.click2drug.org/).

Whilst high throughput docking of such small molecules (ligands) into high resolution protein structures (e.g. calpain) is a standard computational approach in drug discovery [22], there is no docking program proved as yet to achieve low rates of both false positives and false negatives [23–27]. This is true particularly for ligands that react covalently with the target protein [28, 29]. For example it is known that Schroedinger's GLIDE score [30], widely used to screen libraries of ligands (of millions of compounds) for prediction of molecular binding affinities (http://www.schrodinger.com), can give a high false positive classification [31–37]. It has been suggested that GLIDE prediction does not correlate well with the true binding affinity especially for covalent binders (see the review by Mah et al. [38]). Generally in docking experiments binding of the ligand dock is assessed by the single score (e.g., GLIDE, such a score is an approximate predicted binding measure), with experimentally derived IC_{50} (gold standard for effective binding affinity and a value often reported in the chemistry and cheminformatics literature) —IC_{50} only being determined for those compounds that are synthesised.

Testing covalent binders requires a structured *feature extraction* system that allows for a meaningful analysis and for some pattern recognition if possible. In this spirit our study aims to highlight the need to improve on predictor sets beyond GLIDE, for example, by adding structural information to the exercise—specifically information on the number of hydrogen bonds and warhead distance. We propose the identification of molecules with an appropriate warhead group (e.g. the aldehyde functional group) as a possible pre-test or filter for further drug development. This has the potential to help reduce the cost and time required for the discovery of a potent drug which may be a false positive (or false negative). Advantages to rational drug design of structure-based and ligand-based drug design (SBVS) suggest the value of their complementary use, in addition to integrating them with experimental routines (see the review by Macalino et al. [39]). In our study each ligand's gold standard measure of activity, the half maximal inhibitory concentration, $IC_{50,}$ has been experimentally derived, adding strength to our classification problem.

2   The Self Organising Map—Clustering and Chemical Space

The Self Organising Map (SOM) is one of the best known artificial neural network (ANN) algorithms. Self Organising Maps also known as the Kohonen feature map, were introduced by Teuvo Kohonen in 1982 [40]. SOMs create an imagined geography in which data is presented, and are a useful tool to visualise high-dimensional data. SOM converts complex, nonlinear statistical relationships between high-dimensional data into simple geometric relationships on a low-dimensional display. The SOM, basically a vector quantization method, is a type of topological map, which organises itself based on the input patterns that it is trained with [41]. The output and data structure can be varied dynamically based on the data given, rather than statically defining the input structure. The map is thus self-organising in that when the contents change, its form also changes [42]. In the context of classification and data analysis, the SOM technique highlights the neighbourhood structure between clusters. The correspondence between this clustering and the input proximity is called the topology preservation. As such the SOM represents competitive learning based clustering of neural networks. In this study calpain inhibitors (ligands) were clustered into so-called classes of identifiable 'good' versus 'poor' binders.

SOMs are one of two common visualisation methods used to represent *chemical space* [43–45]. The other is principal component analysis (PCA) [46, 47]. Practical applications of the SOM involve compound classification, drug design, and chemical similarity searching. There is growing consensus that analysis of macromolecular (target) features should parallel feature extraction from sets of known ligands to obtain desired novel designs. The applicability of the SOM for mapping elements of protein structure-like secondary structure elements or surface pockets has also been demonstrated.

In drug discovery a common theme is *molecular feature extraction* which involves the transformation of raw data to a new co-ordinate system, where the axes of the new space represent "factors" or "latent variables"—features that might help to explain the shape of the original distribution. A popular statistical feature extraction method in drug design which belongs to the class of factorial methods is principal component analysis (PCA). PCA performs a linear projection of data points from the high-dimensional space to a low-dimensional space. In contrast to PCA, self-organising maps (SOMs) are non-linear projection methods, as are Sammon mapping and encoder networks, all of which can be employed in drug design. Since knowledge of target values (e.g., inhibition constants, properties) or class membership (e.g., active/inactive assignments) are not required by both PCA or SOM, they are characterised as "unsupervised". Unsupervised procedures are generally used to perform a first data analytic steps, complemented later by supervised methods [7, 8] through the adaptive molecular design process.

The relative distance between compounds in the PCA derived *chemical space* becomes a measure of their similarity with respect to the particular molecular representation used. In contrast to PCA, in SOMs, a set of objects is typically mapped into a rectangular array of nodes. Similar objects are mapped into the same or proximal nodes. In contrast, dissimilar objects map into distant nodes. Each neuron is assigned a number of weights that correspond to the number of input variables (i.e., descriptors). In the learning stage of a Kohonen network, the values of the weights in the nodes are first assigned as random numbers. Then, a molecule of the data set is projected into the neuron that has the closest weight values to the input variables of the molecule. Such a neuron is named the winning neuron. In the iterative steps the weight of the winning neuron and neighbouring neurons are updated. After the adjustments of weights, a second molecule from the data is taken and a single neuron is selected as a winner, the weights then adjusted and the process is repeated until all molecules have been assigned to a specific neuron. Feature analysis can be then performed by comparing adjacent neurons (Fig. 1).

Figure 2 shows the stages of SOM adaptation. Here a planar (10 × 10) SOM was trained to map a two-dimensional data distribution (small black spots). The receptive fields of the final map are indicated by Voronoi tessellation in the lower left projection. **A** and **B** denote two "empty" neurons, i.e., there are no data points captured by these neurons [43]. One limitation of the original SOM algorithm is that the dimension of the output space and the number of neurons must be pre-defined a priori to SOM training. It has been noted that self-organising networks with adapting network size and dimension can provide more advanced and at times more adequate solutions to data mining and feature extraction [41, 43].

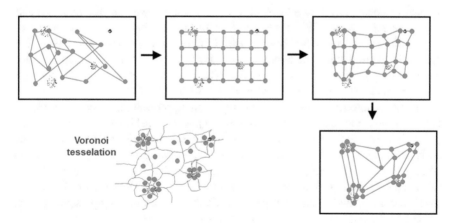

Voronoi
tesselation

Fig. 1 Stages of SOM adaptation (reproduced from [43])

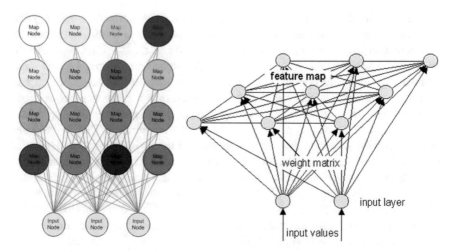

Fig. 2 SOM Lattice (http://www.generation5.org/content/2004/aiSomPic.asp and http://www. nnwj.de/kohonen-feature-map.html accessed October, 2008)

2.1   The SOM Algorithm

Kohonen's algorithm represents an efficient mapping of similar patterns, given as vectors close to each other in input space, onto contiguous locations in the output space. Each map node has a specific position (an x, y coordinate in the lattice) and contains a vector of weights that has the same dimension as the input vectors. As shown in Fig. 2 there are no lateral connections between nodes within the lattice. Instead of specifying a target output, SOM determines whether the node weights match the input vector. When the node weights and the input vector matches, that area of the lattice is selectively optimised to more closely resemble the data. The algorithm requires many iterations in order for SOM to determine a map of stable zones, which acts as a feature classifier. The unseen input vectors will simulate nodes in the zone which have similar weight vectors [48].

The SOM algorithm includes an iterative training procedure, where an elastic net that folds onto the 'data cloud' is formed [41]. This is achieved by mapping data points that are close together (in the Euclidean sense) onto adjacent map units. After an input vector is selected, SOM examines all the remaining nodes in the lattice to identify the Best Matching Unit (BMU) (Fig. 3), which is the node corresponding to the weight vector that is nearest to the input vector. The closeness or nearness is quantified using Euclidean distance. As such SOM is a topology-preserving map, given that the map preserves the neighbourhood relations [40].

Once the neighbourhood radius is idenstified, it is relatively simple to determine the nodes that lie within the BMU's neighbourhood. One only needs to iterate through the nodes and compare the neighbourhood radius and the distance between the node and the BMU. By repeating the process with a different random input vector for each iteration, a topology-ordered map will be produced. The black mesh

Fig. 3 BMU Neighborhood
Radius

Fig. 4 SOM process
(reproduced from [41])

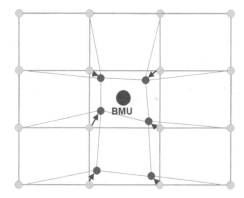

shown in Fig. 4 is the initial lattice. The BMUs are indicated by the grey dots. The grey lattice is the final trained map. The initial (black) lattice is altered after the SOM process to produce a trained (grey) map.

2.2 Visualisation of the SOM

Due to the characteristics of SOM, its visualisation can be used to inspect correlations even if they are in different parts of the data space. Several different methods of visualisation have been developed, these include 2D and 3D projection and colour coding methods [49]. In this chapter we apply one method of visualisation, namely Sammon Mapping [50], which is based on reference vectors. Each component of a node is represented by a plane using a colour representation, and partial correlations may be found within these planes. SOM attempts to seek an optimum non-linear projection for high-dimensional data, such that the resultant vectors can be represented in a 2D map, but which still retain their relative mutual Euclidean distances. The U-Matrix (see Fig. 5) is a popular method for displaying SOMs, as it gives an overall view of the shape of the map. Each element in the matrix represents

Fig. 5 Example of the
Unified Distance Matrix
(U-Matrix)

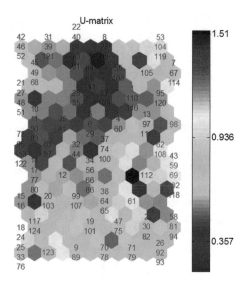

a neuron. The colours indicate the relative distance between neurons. Figure 5 is a
U-Matrix with a hexagonal topology. The colour scale at the right-hand side shows,
for example, blue colours to represent a smaller relative mutual Euclidean distance,
while a red colour represents a larger relative distance. This colour coding may be
used to link different types of visualisation, and colours can be assigned according
to the cluster structure of the map.

2.3   SOM in MATLAB

Matlab: 2015 [51] is a widely used programming environment for technical computing by MathWorks, Inc. Matlab and is available for Windows and various Unix
platforms [50, 52]. Specifically, the SOM Toolbox is applied in this chapter for
analyses using the Matlab: 2015 computing environment. Other packages that can
be used are the SOM_PAK, which is a program package implementing the SOM
algorithm in C-code. The SOM Toolbox contains functions for using SOM_PAK
programs from Matlab, and has functions for reading and writing data files in
SOM_PAK format. Another package that can be used in relation to SOM is what is
known as the LVQ_PAK for the learning vector quantisation (LVQ) algorithm. The
SOM_PAK and the LVQ_PAK are two public-domain software packages down-
loadable from Helsinki University of Technology's Computer and Information
Science   website: http://www.cis.hut.fi/research/som_lvq_pak.shtml   [40].
A graphical user interface (GUI) from the SOM Toolbox was used via command
line versions of the functions. Note that SOMs can also be performed in the R
statistical platform using Kohonen [53]. We refer the reader to the recent paper on
self organisation and missing values in SOM by Vatanen et al. [54].

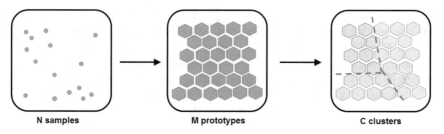

N samples   **M prototypes**   **C clusters**

Fig. 6 Two-level Clustering Approach (reproduced from [41])

2.4 Two-Level Clustering Approach

A two-level clustering approach is adopted in this chapter. This approach involves SOM followed by K-means clustering (KM) [55] according to the development of Vessanto and Alhoniemi [41]. In this SOM/KM hybrid cluster analysis method of relevant inputs (e.g. BMUs) are used to discover structures in data. Recent applications by Hudson of SOM/KM involve a study of sleep and work scheduling of railway drivers [56], and earlier of railway crossing accidents [57]. Visually the SOM/KM involves, as shown in Fig. 6, a large set of M prototypes formed by the SOM, which can then be interpreted as "proto-clusters". Each data belongs to its nearest prototype. In this case, the input data is "proto-clustered" into BMUs by SOM. Then K-means clustering is performed on SOM, i.e. the BMUs are clustered by K-means, to form.C clusters. The main benefit of using this two-level clustering approach is that it reduces the computational cost, by the use of "proto-clusters, and an additional benefit is noise reduction, in that the prototypes are less sensitive to random variations than are the input data, since they are local averages of the raw data. The impact of outliers on the vector quantization result is also limited.

3 Docking and Virtual High-Throughput Screening

Molecular docking [59, 60] remains a significant and useful tool for identifying potential ligands for binding to a particular enzyme or receptor in so-called molecular libraries of compounds. In particular, it allows the prediction of likely modes of binding (docking) and the estimation of relative free binding energies of the resulting protein-ligand complexes (scoring/ranking). Accurately predicting relative binding affinities and biological potencies for ligands that interact with proteins however remains a significant challenge for computational chemists. Although docking tools are generally reliable in predicting the binding mode, a detailed analysis of the protein-ligand component structure under test is not always possible. Hence, a computer based method (or model) to verify the binding assay data of the compound and binding models is required. Most evaluations of both docking and scoring algorithms have focused on enhancing ligand affinity for a protein by optimising docking poses and using enrichment factors during virtual screening. But to date there is still relatively limited information on the accuracy of commercially available docking and scoring software programs [24, 29, 59–62] for correctly predicting binding affinities and biological activities of structurally related inhibitors of different enzyme classes [27, 63].

3.1 Scoring and Comparison of Docking Programs

Drug discovery utilises chemical biology and computational drug design for the effective identification and optimisation of lead compounds. Chemical biology is primarily involved in the interpretation of the biological function of a target and the mechanism of action of a so-called chemical modulator. Computer-aided drug design (CADD) [58] also makes use of the structural knowledge of either the target (structure-based) or known ligands with assumed bioactivity (ligand-based) to enable determination of promising candidate drugs. Indeed knowledge of the 3D structure of a target protein is a major source of information for computer-aided drug design. Of particular interest are the size and form of the active site, and the distribution of functional groups and lipophilic areas. The number of solved X-ray structures of proteins is rapidly increasing, as is the volume of information available to address questions about conserved patterns of functional groups, or common ligand binding motifs or coverage of the protein structure universe. Automatic procedures are clearly needed for analysis, prediction, and comparison of macro-molecular structures, in particular potential binding sites in proteins, as visual inspection of structural models is inadequate. Virtual screening techniques thus aim to reduce the cost and time for the discovery of a potent drug.

Studies have suggested the complementary use of structure-based (SB) and ligand-based virtual screening (VS) drug design in integrated rational drug development (see the review by Macalino et al. [39] and by Kawatkar et al. [29]). The major steps of structure-based virtual screening (SBVS) are docking and scoring. Figure 7 gives a schematic of SBVS in drug discovery [64]. The processes involve input from databases of ligands, 3D structures or molecule fragments into VHTS, which may be either ligand based (pharmacophore) or structure based (docking). These steps then result in lead optimization and in databases to develop scoring functions or to develop ADME(T) or toxicity estimates; where in the latter case the primary measure of effectiveness is the ligand's binding activity (or affinity) with a protein, influencing a metabolic function, is the ligand's ADME(T) (absorption, distribution, metabolism, excretion, toxicity) properties.

Since Kuntz et al. [65] published the first docking algorithm DOCK in 1982, numerous docking programs have been developed during the past two decades [30, 66–75]. Several comprehensive reviews of the advances of docking algorithms and applications have been published [29, 59–62, 76, 77]. Scoring (ranking) the

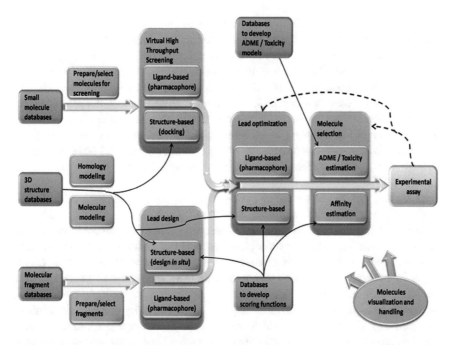

Fig. 7 Schematic of structure based virtual screening (SBVS) in drug discovery, incorporating Virtual high-throughput screening (VHTS) [26, 64]. (http://www.click2drug.org/ accessed August 30, 2015)

compounds retrieved from a database is performed simultaneously with the docking simulation. Molecular docking is a typical optimisation problem, for it is difficult to obtain the global optimum solution. The optimised process scoring function should be fast and accurate enough, allowing simultaneous ranking of the retrieved 3D structural poses in the optimisation process. Based on this idea, several scoring functions have been developed [24, 29, 33, 59–61, 78, 79]. Unfortunately, there is no scoring function developed so far that can reliably and consistently predict a ligand-protein binding mode and binding affinity at the same time [24, 29, 33, 59–62, 79]. Indeed the development of a fast and accurate scoring function in virtual screening remains an unresolved issue in current computer-aided drug research. Different scoring functions focus on diverse aspects of ligand binding, and no single scoring can satisfy the peculiarities of each target system [80]. Therefore, heuristic docking and consensus score strategies are often used in virtual screening [61, 81, 82].

3.2 Filters and False Positives and Negatives

Virtual high-throughput screening (HTS) remains one of the most powerful approaches available for identifying new lead compounds for the growing catalogue of validated drug targets. The principle that molecular structure determines meta-bolic and drug activity has meant that the preliminary High Throughput Screening (HTS) of molecules in search for new drugs has traditionally focused on identifying molecules with a similar structure to a known active molecule. High throughput docking of small molecules (ligands) into high resolution protein structures is now standard in *computational* approaches to drug discovery, where the receptor (pro-tein) structure is kept fixed, while the optimal location and conformation of the ligand is sought via sampling algorithms [36]. The cost of new drug discovery is expensive, with many cases recorded where candidate molecules have progressed through most of the development stages, to only fail in the final clinical trials assessment [83].

However, just as virtual and experimental HTS have accelerated lead identifi-cation of potential ligand candidates and changed drug discovery, they have also introduced a large number of "peculiar" molecules. Some of these are of interest for further optimisation, others are considered as dead ends after unsuccessful attempts to optimise their activity. Such false positive hits are still one of the key problems in the field of HTS and in the early stages of drug discovery in general. In the lead-discovery environment false positives can result from multiple mechanisms, including: nonspecific hydrophobic binding, poor solubility (protein or substrate precipitation), the tendency of small molecules to form aggregates, reactive func-tional groups, low purity, incorrect structural assignment or compound concentra-tion, interference with the assay or its readout, and experimental errors. False negatives can be the result of poor solubility, chemical instability, low purity, lower than expected compound concentration, interference with the assay or its readout, and experimental errors.

Compounds that do not adopt a sensible binding mode, and/or whose calculated free binding energy is lower than an arbitrarily defined cut-off value, are often excluded (filtered) by analysis and manipulation of the docking results. This sig-nificantly reduces the number of possible compounds to be synthesised and assayed for a drug discovery program, thus saving time and money. Many studies have been devoted to understanding the origins of false-positives, and the findings have been incorporated in filters and methods that can predict and eliminate problematic molecules from further consideration [25]. However, active compounds that possess a novel mode of binding, and/or an incorrectly calculated free binding energy, can inadvertently be excluded during such a process. This then removes potentially interesting candidates from the drug discovery process (false negatives). Furthermore, libraries of compounds based on existing molecules may emphasize molecules easy to synthesize rather than covering all active molecules [127].

3.3 Covalent Binding Modes

Docking is generally reliable at predicting *non-covalent* ligand binding modes, when a good protein model is available. This however can be a problem for proteins with a highly flexible binding site leading to an induced fit. Furthermore, e.g. water molecules participating in a trimolecular binding event have to be identified. By contrast, scoring and ranking are rarely exact [24, 31, 84]. In particular, the use of existing scoring functions results in inaccuracies which range from needing to compare structurally different compounds and to score multiple poses of the same ligand. Thus scoring functions are not always well suited to prioritising binding and often fail to accurately predict the binding affinity of all ligands (i.e. strong and weak binders). Strategies to overcome these deficiencies include the use of a consensus scoring of several scoring functions [85, 86] and the use of generous cut-off values so that many ligands and/or poses pass the scoring filter. However, this can result in the generation of a large number of compounds for synthesis and testing, many of which ultimately prove to be inactive (poor or non-binders), so-called false positives. The number of these false positives can however be reduced during the analysis, by tailoring the scoring/ranking process to the particular biological target under consideration. For example, a 'post-docking filter' can be used to exclude compounds that lack one or more of the identified key interactions [87, 88]. As such a simple filter is able to better distinguish between binders and non-binders of a particular target, thus significantly reducing the number of false positives. If generally applied, this might lead to scoring functions that compare the binding affinity of high affinity ligands without having to parameterise the function for a correct evaluation of low-affinity binders.

In general, docking programs have been developed to specifically describe *non-covalent* protein-ligand interactions, since these are associated with most of the known enzyme-inhibitor and receptor targeting based therapeutics. However, there are many *covalent* inhibitors that demand detailed molecular analysis if they are to be further developed [89]. The scoring of these types of binders is complex, since both the initial non-covalent complex and the bond forming process have to be correctly modelled. There have been some covalent docking studies reported for these purposes quite recently by Bianco et al. [90], by Kumalo et al. [91] and Mah et al. [91] but their use is not routine or general (see also the older investigations by Fradera [92] and by Katritch [89]). Indeed there is currently no available docking program that has achieved low rates of false positives and of false negatives, particularly for ligands that react covalently with the target protein [23, 29, 34–37, 93]. The challenge remains to accurately predict binding affinities of drug candidates, by accommodating protein flexibility, solvation and entropic effects [94]. Whether machine-learning approaches, generalised/universal scoring functions or new protein-specific/targeted scoring functions can achieve this is as yet unclear (see the review by Yuriev and Ramstad [27] and also by Yuriev et al. [95]).

4   Calpains and the Data Set

Calpain is a calcium activated cysteine protease with several known isoforms. Calpains were chosen as the target protein for this study since their over activity is crucial to medical conditions associated with cellular damage; including traumatic brain injury, stroke and cataract [12, 15, 18–20, 96]. Given the enzyme is an important modulator of these physiological and pathological conditions, specific calpain inhibitors are attractive therapeutic agents. Whilst a good deal of X-ray crystallographic and other data is available to aid in the design and synthesis of such inhibitors, methods for their analysis by molecular docking techniques are still lacking [12, 14, 15, 18, 20]. Whilst calpain inhibitors have demonstrated efficacy in animal models of calpain related diseases, the progression of the inhibitors into clinical trials has been fraught partly due to lack of calpain isoform selectivity and the general reactivity of the inhibitors [96]. Exploration of compounds that bind to allosteric sites of the enzyme may avoid these problems and afford new drug leads. Advances made in enhancing the cellular uptake of peptide calpain inhibitors, improving the pharmacokinetic properties of the inhibitors and site specific targeting of calpain inhibitors are still needed [15, 16, 18–20].

4.1   The Mechanism of Binding

In our study of the calpain inhibitor ligands in binding it is the electrophilic anchor ("warhead"), e.g. an aldehyde, that reacts with the active site cysteine, and a di- or tripeptide backbone (see Fig. 8) that adopts a *l*2 strand geometry on binding, as is

Fig. 8 *l*2 strand geometry of the backbone of calpain inhibitors (smaller grey molecule in the middle) when binding to the protein with the defining hydrogen bonds to Gly271 and Gly208 (3 interactions shown)

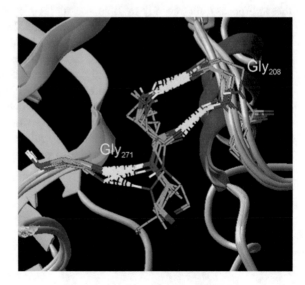

Fig. 9 Mechanism of ligand-protein binding

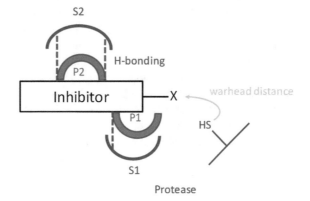

the case for almost all known binders to proteases. Figure 8 depicts binding of a ligand to 3 binding sites of the protein.

We describe a paradigm by which to explain the mechanism of the ligand-protein binding. A schematic for a ligand (inhibitor) binding to calpain (protein) is shown in Fig. 9, with both non covalent (reversible) and covalent (irreversible) interactions being important. Non covalent interactions are typified by hydrogen bonds (as depicted by the dashed lines in Fig. 9) and hydrophobic/ hydrophobic and hydrophilic/hydrophilic binding. With these non covalent inter-actions in place, the reactive warhead (typically an aldehyde functional group) of the ligand is correctly positioned to covalently react with an active site cysteine of the calpain (designated HS in Fig. 9). This results in long-lasting attachment of the ligand to calpain and hence inhibition of its activity. Thus both hydrogen bonding and warhead distance (separation of the warhead and cysteine) are critical param-eters in defining the binding affinity. The non covalent interactions between the ligand and calpain are depicted as P (1 to n) and P2 (1 to n) depending upon their proximity to the warhead as depicted in Fig. 9. S1 and S2 in Fig. 9 are the binding pockets of the enzyme that accommodate the P1 and P2 groups.

Proteases almost uniformly bind their substrates and inhibitors in a conformation whereby the peptide (or a peptide-like) backbone of the inhibitor adopts a β-strand geometry. This mode of binding is primarily dictated by the geometry of the active site subsites that accommodate the amino acid side chains of the substrate or inhibitor. An important approach to inhibitor design is to then introduce a carefully designed conformational constraint into the structure to pre-organise its backbone into a β-strand conformation, thereby reducing entropy loss associated with ligand-receptor binding, while also enhancing biostability. This typically involves chemically linking either the P1 and P3 or the P1' and P3' residues of a peptidomimetic-based protease inhibitor.

Chua et al. [21] have recently progressed this area and have computationally designed and prepared (using ring closing metathesis and click chemistry) potent cyclic inhibitors of cysteine proteases that overcome the basic problems of flaws in their design: existing inhibitors are conformationally flexible and biologically unstable structures with a 'reactive warhead' that makes them un-drug like. The constituent cycle constrains the inhibitor into a β-strand geometry that is known to favour binding to the enzyme, resulting in improved biostability, an entropic advantage to inhibitor binding, and increased potency without the need for a 're-active warhead'. These authors have shown that such inhibitors stop the progression of cataracts in lens culture and also in animal trials by inhibiting a cysteine protease —the study is entering a commercial phase. The research in this chapter involves the stereoselective synthesis of examples of these macrocyclic inhibitors and their assay against a range of proteases (including the proteasome) and an investigation into their potential to stop the growth of various cancer cell lines. Whilst a good deal of X-ray crystallographic and other data is available to aid in the design and synthesis of such inhibitors, it is known that methods for their analysis by molecular docking techniques are somewhat still lacking [12–16, 18–20].

5 The Data Set

A dataset of 124 molecules (ligands), which are considered as possible calpain inhibitors are analysed in this chapter. Whilst the binding affinity of a molecule (ligand) to a protein (calpain) is generally used as a measure of binding activity as discussed earlier, generally, however, strong binding affinity, alone, is not sufficient a criterion for further development of a molecule as a potential drug. This is because pharmacokinetics is also an important factor in ligand binding. Indeed, without acceptable pharmacokinetics, the molecule may fail further testing in the drug development process. Hence the three-dimensional structure of a molecule optimally needs to be examined, so as to understand its activity and underlying pharmacokinetics. Hence using the GLIDE computer program, the 124 molecules (ligands) were docked into a computerised model of calpain and the structures of each docked compound were analysed (see http://helixweb.nih.gov/schrodinger-2013.3-docs/glide/glide_user_manual.pdf). The poses of each ligand were then compared, and the pose with the lowest estimated free binding energy (i.e. glide score) were advanced for further analysis of its binding mode (see Neffe and Abell [11]). Most of the 124 ligands analysed had a highly congruent binding mode of up to three hydrogen bonds (e.g. Gly208, Gly271, see Fig. 8 above) [11].

The GLIDE algorithm also provides, apart from the global GLIDE score [97], the so-called Emodel score [97], where both the GLIDE and Emodel scores are linear combinations of the seven partial energies, g.lipo, g.hbond, g.eburp, g.evdw, g.ecoul, g.erotb, and g.esite, which are in the model notation [83, 98]. See Table 1 (shaded section) for definitions of these terms. The full set of seven partial energies is also denoted by g7. The data set contains a number of descriptors for the 124 ligand complexes, including the GLIDE (g.score), Emodel (g.emodel), the ligand's true binding strength, and the partial energies from GLIDE, namely the subcomponents that make up the one-off glide score (denoted by g.score in Table 1). The number of hydrogen bonds of the molecule's backbone in addition to the ligand's

Table 1 List of molecular descriptors and GLIDE: definition of the partial energies and full scores

Full scores				
GScore	g.score	Glide Score, used for comparing poses of different ligands		
Emodel	g.emodel	Combination of GScore, Coul, vdW, and Intern used for selecting poses of a given ligand		
Distance	Warhead distance	Distance		
Hbonds	Number of hydrogen bonds	Hbonds		
Partial scores				
Lipo	g.lipo	Lipophilic contact term	g7	g8
HBond	g.hbond	Hydrogen-bonding term		
BuryP	g.eburp	Penalty for buried polar groups		
vdW	g.evdw	Scaled van der Waals energy		
Coul	g.ecoul	Scaled Coulomb energy		
RotB	g.erotb	Penalty for freezing rotatable bonds		
Site	g.esite	Polar interactions in the active site		
Intern	g.einternal	Internal energy of the ligand, not included in the calculation of the Glide Score		

warhead distance and warhead class (Table 3), and the molecule's true binding strength classification (IC_{50} and BS) are also part of the dataset.

The following relationships hold for the scores given in Table 1: (i) Free binding energy = (energy of the protein-ligand complex)—(energy of the free protein)—(energy of the free ligand) [68, 73]; (ii) The energy of the protein-ligand complex is classically the sum of all interaction energies (e.g. Coulomb energy (Coul, electrostatic interactions) + van der Waals energy (vdW, lipophilic interactions) + energy of hydrogen bonds etc.) + internal energies coming from non-ideal angles, bond lengths etc.; and (iii) the emodel energy score is calculated from all the other energies.

The prediction of binding strength is based on the gold standard of molecules activity IC_{50} [83]. Molecules with high IC_{50} are poor binders, while molecules with low IC_{50} are good binders. Table 2 shows the expert-driven classification of binding strength (by organic chemists, Abell and Neffe) of ligands based on IC_{50}. These

Table 2 Classification of true Binding Strength (BS) based on IC_{50}

Binding Strength	IC_{50}	True BS category	Totals
0. Super Binders	<500 nM	Good	20
1. Good Binders	500 nM–5 μM	Good	23
2. Weak Binders	5–100 μM	Poor	22
3. Non-binders	>100 μM	Poor	59
Total			124

four categories of binding strength are dichotomised, so weak binders and non-binders are considered as "poor binders" and good binders and super binders are considered as "good binders".

5.1 Warhead Classification and Binding Affinity

Typical calpain inhibitors have an electrophilic warhead, e.g. aldehyde, which target the active site cysteine and a di- or tripeptide backbone conferring specificity. Four classifications of warhead are used in this study. Each molecule belongs to one of the four following warhead classification groupings: {Aldehydes and Ketones}, {Masked Aldehydes, Ketones, Michael Acceptors, Epoxides}, {Esters and Amides}and {Alcohols and non-classical}. Table 3 shows a cross tabulation of this 4 level warhead classification versus true binding strength (dichotomised), where the latter are the classification of true binding strength of ligands based on cut-points or ranges for IC_{50} in Table 2. The associated chi-square statistic is 42.20 and the Likelihood Ratio (LR) statistic is 53.78 with corresponding P < 0.00001, which indicates that there is a highly significant association between warhead classification and true binding strength. It can be seen that warhead classification 1 (Aldehydes + Ketones) and 2 (Masked Aldehydes, Ketones, Michael Acceptors, Epoxides) contain the majority of molecules with good binding strength; and warhead classification 3 (Esters + Amides) contains molecules with poor binding strength only.

In this study a major question is whether the false positive and false negatives are highly correlated to the warhead classification of the ligand. The false negatives (FNs) in this application are considered to be molecules (calpain-ligand) classified as 'poor binders' despite having high binding strength (low IC_{50}) and being plausible to the 'chemical eye' (Fig. 10a). False positives (FPs) are compounds classified as 'good binders', but which exhibit low binding strength (high IC_{50}) (Fig. 10b). Molecules that are found to be FNs or FPs may in fact need to be retested and their true binding strength recalculated.

6 Results

The 16 candidate models obtained via SOM/KM are discussed in this section Although it has been known that the GLIDE score [30] does not necessarily correlate well with binding affinity, as assessed by IC_{50}, it is still used in drug discovery [32, 99]. In this chapter our so-called GLIDE only model is used as a baseline to investigate a possible better suite of predictors of binding strength that classify a higher number of true positives and true negatives. For each model, classification tables, ROC analyses, sensitivity, specificity [100] and the results of a Cochran-Mantel-Haenszel (CMH) test [101, 102] are given, in addition to a

Table 3 Warhead classification by true binding strength

Warhead classification			True binding strength		Total
			Good	Poor	
1	Aldehydes +Ketones	(structures: R–CHO, R–CO–R, R=C....)	33	20	53
2	Masked Aldehydes, Ketones	(structures: R'–CH(OH)OH, R'–CH(OR)OH, R'–CH=NR–OH, R'–CH(NR)=R)	9	12	21
	Michael Acceptors				
	Epoxides				
3	Esters + Amides	(structures: R'–CO–OR, R'–CO–NR$_2$)	0	35	35
4	Alcohols + non-classical	(structures: R–CH$_2$OH, R–thiadiazole)	1	14	15

$\chi^2 = 42.20$, LR $= 53.78$ (P < 0.00001)

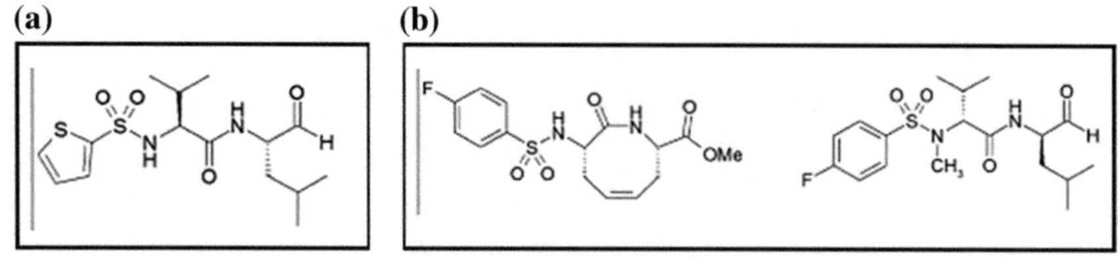

(a)   **(b)**

Fig. 10 Molecular structure of **a** false negative and **b** false positive

tabulation of the false positives and false negatives by warhead class and a plot of the mean of the predictors in the model per (good and poor) cluster/grouping. The results for the best model are discussed in detail, then, all models are compared on summary descriptives, on ROC diagnostics and by an examination of molecular warhead classification across the true (and predicted) good and poor binders.

6.1 Selection of the Best Predictor Sets (Models)

The best models are chosen based on ROC analyses [103, 104] which are based on criteria such as sensitivity (r) and specificity (s) values and area under the curve (AUC) indices [106], derived from pROC [107] and epi [112] in R. Table 4 gives the sensitivity and specificity with corresponding 95 % confidence intervals, sorted by decreasing $r + s$. The numbers of molecules as true and false positives/negatives are also shown. Models with a high value of $r + s$ exceeding 1.0 as the lowest threshold [105, 108] are considered as optimal models, as are models with high levels of AUC. From Table 4 the best candidate models as based on ($r + s$) are: the g7 models (g7 + hbonds + distance, g7, g7 + hbonds), followed by the simple 2 parameter glide + distance model, then the g8 variant models (g8 + hbonds, g8 + distance, and g8) and finally the (glide +hbonds) model. Table 5 provides a summary of the 16 models (sorted by decreasing AUC) with model specific values of the index of validity (I_v), $r + s$ and AUC, along with corresponding 95 % confidence intervals for the given AUC. The Z statistic (Zstat) and corresponding P value relates to tests comparing the AUC of each model with respect to the AUC of a given reference (ref) model. These reference models are the glide only model (P1) and the optimal g7 + hbonds + distance model (P2) (Table 5). The top models based on AUC are: g7 + hbonds + distance, g7 + distance, g7, g7 + hbonds, glide + distance, emodel + distance and the 3 parameter (glide +hbonds + distance) model. The glide only model was ranked 9th and 13th of the 16 models when compared to all models (Tables 4 and 5). Moreover, the glide + hbonds + distance, emodel + distance and g7 + distance models are equivalent (based on AUC) to the g7 + hbonds + distance model. These are also equivalent in terms of AUC to the g7, g7 + hbonds, glide + distance, glide + hbonds + distance, emodel + distance and the g7 + distance model. The top models will be further discussed in Sect. 6.1.1.

Table 4 The 16 models (sorted by decreasing r + s) with r, s and 95 % confidence intervals, numbers of false/true positives or negatives {FP, FN, TN, TP}

Model	r	95 % CI for r	s	95 % CI for s	FP	FN	TN	TP	r + s
g7 + hbonds + distance	0.651	(0.446,0.791)	0.654	(0.454,0.762)	28	15	53	28	1.315
g7	0.698	(0.513,0.829)	0.568	(0.366,0.688)	35	13	46	30	1.266
g7 + hbonds	0.628	(0.436,0.768)	0.630	(0.430,0.750)	30	16	51	27	1.258
glide + distance	0.674	(0.484,0.814)	0.568	(0.374,0.692)	35	14	46	29	1.242
g8 + hbonds	0.605	(0.424,0.746)	0.580	(0.394,0.712)	34	17	47	26	1.185
g8 + distance	0.535	(0.356,0.693)	0.580	(0.412,0.716)	34	20	47	23	1.115
glide + hbonds	0.558	(0.391,0.717)	0.543	(0.371,0.683)	37	19	44	24	1.101
g8	0.605	(0.433,0.754)	0.494	(0.324,0.638)	41	17	40	26	1.098
glide	0.209	(0.089,0.354)	0.827	(0.692,0.917)	14	34	67	9	1.036
emodel	0.349	(0.199,0.507)	0.654	(0.503,0.788)	28	28	53	15	1.003
emodel + hbonds	0.535	(0.372,0.697)	0.407	(0.267,0.564)	48	20	33	23	0.942
emodel + hbonds + distance	0.535	(0.373,0.711)	0.395	(0.257,0.568)	49	20	32	23	0.930
glide + hbonds + distance	0.488	(0.597,0.807)	0.370	(0.487,0.767)	51	22	30	21	0.859
g8 + hbonds + distance	0.395	(0.537,0.807)	0.444	(0.589,0.797)	45	26	36	17	0.840
emodel + distance	0.372	(0.535,0.811)	0.457	(0.622,0.802)	44	27	37	16	0.829
g7 + distance	0.372	(0.715,0.875)	0.346	(0.699,0.861)	53	27	28	16	0.718

Table 5 Summary of 16 models (sorted by decreasing AUC) with Index of Validity (Iv), $r + s$, area under the curve (AUC) and corresponding 95 % confidence interval

Model	I_v	$r + s$	AUC	95 % CI for AUC	Comparison of AUC with glide ref model Z1 stat	P1	Comparison of AUC with optimal model Z2 stat	P2
g7 + hbonds + distance	0.653	1.305	0.653	(0.564,0.742)	−2.278	0.023	Ref	ref
g7 + distance	0.355	0.718	0.641	(0.551,0.731)	−2.035	0.042	0.375	0.708
g7	0.613	1.266	0.633	(0.539,0.719)	−1.865	0.062	0.946	0.344
g7 + hbonds	0.629	1.258	0.629	(0.532,0.710)	−1.883	0.060	0.580	0.562
glide + distance	0.605	1.242	0.621	(0.501,0.684)	−1.241	0.214	1.849	0.065
g8 + hbonds	0.589	1.185	0.592	(0.465,0.650)	−0.654	0.513	2.348	0.019
emodel + distance	0.427	0.829	0.586	(0.494,0.677)	−1.150	0.250	1.062	0.288
g8 + hbonds + distance	0.427	0.840	0.580	(0.488,0.672)	−1.050	0.294	2.263	0.024
glide + hbonds + distance	0.411	0.859	0.571	(0.478,0.663)	−1.247	0.212	1.224	0.221
g8 + distance	0.565	1.115	0.558	(0.458,0.644)	−0.456	0.648	1.968	0.049
glide + hbonds	0.548	1.101	0.551	(0.457,0.641)	−0.511	0.610	2.258	0.024
g8	0.532	1.098	0.549	(0.445,0.630)	−0.314	0.754	2.268	0.023
glide	0.613	1.036	0.518	(0.444,0.592)	Ref	ref	2.278	0.023
emodel	0.548	1.003	0.502	(0.413,0.591)	0.260	0.795	2.932	0.003
emodel + hbonds	0.452	0.942	0.471	(0.378,0.564)	0.826	0.409	2.674	0.007
emodel + hbonds + distance	0.444	0.930	0.465	(0.372,0.557)	0.933	0.351	2.706	0.007

The 6 most optimal SOM/KM models according to **both** the AUC and (r + s) criteria, are: g7 + hbonds + distance, g7, g7 + hbonds, g7 + distance, glide + distance, glide +hbonds + distance (Tables 4 and 5). These 6 models best classify bindi strength of calpain, in that approximately 65 % of true "high binding strength" ligand complexes are correctly classified as a "good-binder" [95 % confidence interval (CI) (46–79 %)]; whilst 65 % of true "poor binder" ligand complexes are correctly classified as "poor-binders" [95 % CI (45–76 %)]. When using glide score alone, only a low 21 % of true "high binding strength" are correctly classified as a "good-binder" [95 % CI (9 % –35 %)], in contrast to a high 83 % of true "poor binders" correctly classified as "poor-binders" [95 % CI (69–92 %)]. The ROC curves in Fig. 11 allow us a visual comparison of the optimal g7 + hbonds + distance model against the other eleven, g7 and g8 based, candidate models (omitting the emodels). It is noteworthy that the g8 based models, containing the additional g.einternal score, the internal energy of the ligand (not used in the calculation of the glide score) are not in the top 6 models.

6.1.1   The Optimal g7 + distance + hbonds Model

This model contains the 7 partial energies from GLIDE, which include g.lipo, ghbond, g.burp, g.evdw, g.ecoul, g.erotb, and g.esite (Table 1), along with the number of hydrogen bonds of the molecule's backbone and warhead distance. Its corresponding SOM derived U-matrix in Fig. 12 contains all the components in this model. It self-organised the input data based on the 9 predictor variables given in the optimal g7 + hbonds + distance model. By observing the colours in the map, it can be seen that the map can be roughly separated into two parts, with a red-area on the bottom left corner and a blue-area in the rest of the map (Fig. 12). This means that molecules with similar characteristics are grouped together. For instance, molecules with ID, 7, 65, 67—in the top left corner of the map, are very different to molecules 16, 19,124 which are in the bottom right corner.

From the separate components graphic in Fig. 13 it can be seen that molecules with higher levels of g.lipo lie on the top of the map, while molecules with higher levels of g.hbond lie on the top right corner, higher levels of g.eburp are associated with a particular region of the U matrix, namely the top left corner. Higher levels of g.evdw are located at the top right corner, and higher levels of g.ecoul with the right side, higher levels of g.erotb with the top right corner (poor binders), higher levels of g.esite with the top, higher number of hydrogen bonds of the molecule's backbone with the left side, and also lower levels of warhead distance positioned on the left. Large warhead distance is associated with low numbers of hydrogen bonds (bottom right hand corner, poor binders) and with low levels of g.eburp, g.esite, and g.erotb.

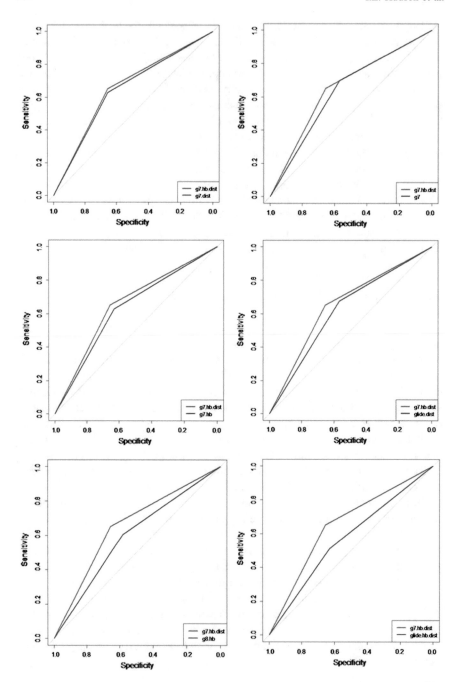

Fig. 11 ROC curves for the eleven g7, g8 and glide based models versus the optimal g7 + hbonds + distance (g7.hb.dist) model. The vertical (y) axis is sensitivity and the horizontal (x) axis is (1-specificity). ROC curves. The vertical (y) axis is sensitivity and the horizontal (x) axis is (1-specificity)

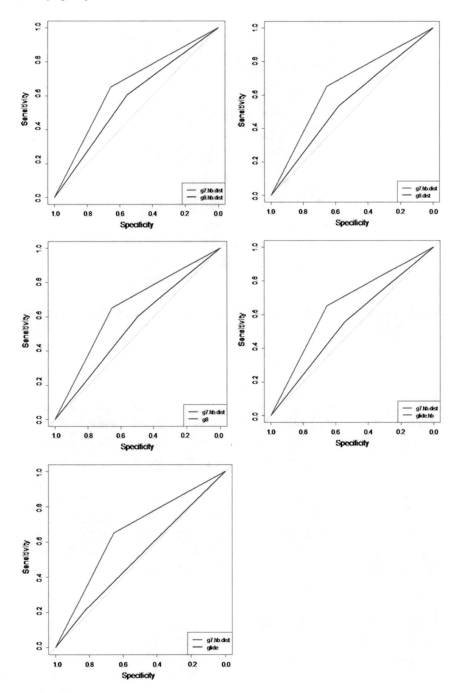

Fig. 11 (continued)

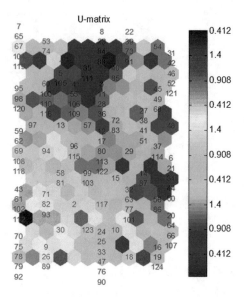

Fig. 12 U-matrix (k = 2) of the g7 + distance + hbonds model (predictor set of 7 glide subcomponents, number of hydrogen bonds and warhead distance), where k is the defined number of clusters

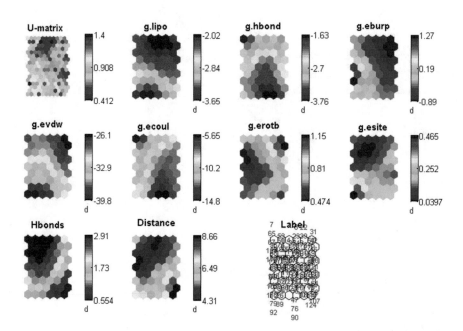

Fig. 13 SOM separate components graph (k = 2) of the g7 + distance + hbonds model

6.2   Summary of the 16 Candidate Models

To gain a more detailed view on the performance of the best model compared to the remaining models, we perform the following suite of tests and descriptors. The means of predictors by cluster membership are tabulated, the Cochran-Mantel-Haenszel (CMH) test for the 16 models are performed to investigate whether binding status is statistically significantly related to warhead class. A full list of false positives and negatives for the best models are given and discussed in this section.

6.2.1   Typical Profile of Predicted Good and Poor Binders

Table 6 provides a comparison of the means of the individual predictors in the model for cluster 1, which identifies molecules as "good binders" versus cluster 2, which classifies molecules as "poor binders" for the best SOM/KM model, g7 + distance +hbonds. Figure 14 is the corresponding barchart for the optimal g7 + distance +hbonds model. It can be seen that molecules in cluster 1 (classified as good binders) have lower mean values compared to cluster 2 on warhead distance, lower log IC_{50} and lower emodel score, also lower mean levels of g.lipo, g. hbond, g.evdw, g.ecoul and g.erotb. Whilst cluster 1 molecules exhibit a higher

Table 6 Descriptor means of the SOM clusters for the g7 + distance + hbonds model (k = 2)

SOM predicted binding affinity cluster/predictors	Cluster 1 (good)	Cluster 2 (poor)	P value
Distance	5.35	6.12	<0.015
hbonds	2.48	1.60	<0.05
g.lipo	−2.62	−2.42	<0.01
g.hbond	−2.61	−2.32	<0.01
g.eburp	0.17	−0.62	<0.005
g.evdw	−32.94	−30.16	<0.05
g.ecoul	−10.75	−7.60	<0.01
g.erotb	0.60	0.80	<0.05
g.esite	0.18	0.15	NS
g.einternal	8.23	6.41	<0.01
Glide score	−5.82	−6.23	<0.03
log IC_{50}	−5.16	−4.19	<0.05
emodel score	−57.23	−48.97	<0.005
# true positives	28	−	
# false positives	28	−	
# true negatives	−	53	
# false negatives	−	15	
Total no. molecules	56	68	

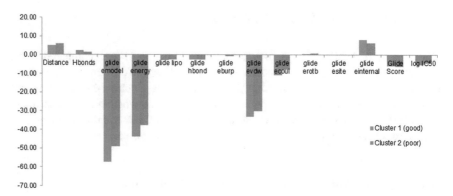

Fig. 14 g7 + distance + hbonds model: Plot of the predictor means for the classified good and poor binding clusters

mean number of hydrogen bonds of the molecule's backbone, and a higher mean glide score (g.score), along with higher means of g.eburp, g. esite, and g.einternal compared to cluster 2 (classified as poor binders).

6.2.2 False Positives and False Negatives of the SOM Models

The relationship between predicted binding affinity cluster and warhead class (WH) (at 4 levels or 2 groupings) is investigated in this section for the k = 2 SOM clusters. Predicted binding (good vs. poor) by WH class for the g7, g8 and glide variant candidate models are shown in Table 7, where these 12 models are given in decreasing order by (r + s). The percentage of predicted good binders for the 74 molecules with warhead classification 1 or 2 (aldehydes and ketones plus masked aldehydes, ketones, Michael Acceptors, epoxides) are given in column 8, and the percentage of predicted poor binders for the 50 molecules in warhead class 3 or 4 (mostly esters and amides plus alcohols and non-classical) are given in column 9 of Table 7 for each model.

Predicted binding affinity cluster membership is statistically significantly related to warhead class for all models, except 4 of the (predictor sets) models, according to the likelihood ratio (LR) P-values < 0.05 in the last column of Table 7. The models for which there is no statistically significant association between predicted binding class and warhead classification are: the glide + hbonds and glide model, along with the glide + hbonds + distance and the g7 model, where the latter 2 models, are however significant at the 10 % level of significance. From Table 5 all the optimal models with $(r + s) > 1.0$ and whose AUCs are equivalent to that of the best g7 + hbonds + distance model, with AUC = 0.653 (95 %CI (0.564, 0.742)), exhibit a statistically significant association between predicted binding status and warhead class. Note that of the molecules in warhead class 1 or 2, the percentage of predicted good binders is in the range 68–70 %; and in warhead class 3 or 4, the percentage of

Table 7 Predicted binding class by warhead (WH) class for the g7, g8 and glide models (in decreasing order of $(r + s)$

SOM/KM models	Predicted SOM group (k = 2)	Warhead (WH) class					predicted good in WH (1,2) (%)	predicted poor in WH (3,4) (%)	P-value
		1	2	3	4	Total			
g7 + hbonds + distance	Good	25	13	8	10	56	68		0.006
	Poor	28	8	27	5	68		47	
g7	Good	30	14	12	9	65	68		0.07
	Poor	23	7	23	6	59		49	(10 % sig)
g7 + hbonds	Good	28	12	8	9	57	70		0.01
	Poor	25	9	27	6	67		49	
glide + distance	Good	32	13	10	9	64	70		0.01
	Poor	21	8	25	6	60		52	
g8 + hbonds	Good	27	12	10	11	60	65		0.02
	Poor	26	9	25	4	64		45	
g8 + distance	good	24	11	10	12	57	61		0.007
	Poor	29	10	25	3	67		42	
glide + hbonds	Good	25	11	15	10	61	59		0.46
	Poor	28	10	20	5	63		40	
g8	Good	28	14	13	12	67	63		0.02
	Poor	25	7	22	3	57		44	
glide	Good	12↓	5↓	4	2	23	83		0.47
	Poor	41	15	31	13	101		44	
g8 + hbonds + distance	Good	27	8	23	4	62	52		0.04
	Poor	26	13	12↓	11↓	62		37	
glide + hbonds + distance	Good	27	10	27	8	72	51		0.051
	Poor	26	11	8↓	7↓	52		29	(10 % sig)
g7 + distance	Good	29	7	27	6	69	52		0.005
	Poor	24	14	8↓	9↓	55		31	
	Total	53	21	35	15	124	74	50	

predicted poor binders is in the range 47–52 %. The less optimal models are those for which there is a significant association between predicted binding status and warhead class, but which have an $(r + s) < 1.0$ and whose AUC is not significantly different to the AUC of the optimal g7 + hbonds + distance model (Table 5). For these less optimal models we have a reduced rate of prediction, in that of the 74 molecules belonging to warhead class 1 or 2, the percentage of predicted good binders is in the range 51–61 %, and for the 50 ligands in warhead class 3 or 4, the percentage of predicted poor binders is lower, between 31 and 44 %. Note that even for the 2 parameter glide + distance model, which is still considered a reliable model, 70 % of the molecules in warhead class 1 or 2 are predicted as good binders and the percentage of predicted poor binders from warhead class 3 or 4 is acceptable at 52 %. In contrast for the non-optimal glide (only) model of the ligands in warhead class 1 or 2, a very high 83 % are predicted to be good binders and for the molecules in warhead class 3 or 4, the percentage of predicted poor binders is as expected a rather low, 44 %.

Table 8 displays a three-way cross tabulation of the SOM (k = 2) predicted cluster (good vs. poor binder) by true binding strength (BS) by warhead class (1, 2) versus (3, 4) for the best g7 + distance + hbonds model. There is a clear difference in the relationship between predicted class and true binding affinity across warhead class (1, 2) versus (3, 4). From Table 8 we note that for molecules in warhead class 1 or 2 (aldehydes and ketones plus masked aldehydes, ketones, Michael Acceptors, epoxides);

- of the 38 predicted good binders, 74 % are true good binders, with 26 % being true poor binders
- of the 36 predicted poor binders, 39 % are true good binders, with greater than 1.5 times more, 61 %, being true poor binders
- of the 42 true good binders, 67 % are classified as good binders, with 33 % classified as poor binders
- of the 32 true poor binders, 31 % are classified as good binders, with more than twice, 69 % classified as poor binders.

For molecules in warhead class 3 or 4 (mostly esters and amides plus alcohols and non-classical);

- of the 18 predicted good binders, 0 % are true good binders, with all 100 % being true poor binders
- of the 32 predicted poor binders, 3 % are true good binders, with 97 % true poor binders
- the 1 true good binder is classified as a poor binder

Table 8 Cross tabulation of the SOM (k = 2) predicted clusters by true binding strength (BS) and by warhead classes for the g7 + distance + hbonds model

Warhead class 1 or 2		True binding strength		% per class of true good (of true poor, in brackets)	% predicted classes for true good and for true poor (in brackets)
SOM (k = 2)	Cluster	Good	Poor		
g7 + hb + distance model	Good	28	10	74 % (26 %)	67 % (33 %)
	Poor	14	22	39 % (61 %)	31 % (69 %)
	Total	42	32		
Warhead class 3 or 4		True Binding Strength		% per class of true good (of true poor, in brackets)	% predicted classes for true good and for true poor (in brackets)
SOM (k = 2)	Cluster	Good	Poor		
g7 + hb + distance model	Good	0	18	0 % (100 %)	0 % (100 %)
	Poor	1	31	3 % (97 %)	37 % (63 %)
	Total	1	49		

- of the 49 true poor binders ligand, 37 % are classified as good binders, with 1.7 times more, 63 %, being classified as poor binders.

For each model Table 9 gives the Cochran-Mantel-Haenszel (CMH) test statistics [101, 102] with corresponding P-values and the odds ratio (OR) of a molecule being a "good binder" for warhead classification 1 or 2, with corresponding 95 % confidence intervals. The odds ratios for warhead classes 3 and 4 cannot be obtained since there are no molecules with true good binding strength predicted as a good binder, thus the OR is mathematically undefined for in this case. For the top g7 + hbonds + distance model the odds ratio of "good binders" between each warhead class 1 and 2 is not homogeneous (CMH = 6.96, P = 0.008). Note that the OR estimate for "good binders" in warhead class 1 is 4.48 and the corresponding OR is 9.92 for molecules in warhead class 2 (for the top g7 + hbonds + distance model). This indicates that there is a higher likelihood, a doubling of odds, of being classified a "good binder" for molecules in warhead class 2 compared to warhead class 1 (Table 9), but still a significantly high odds > 1.0 for warhead class 1. Note that most true good binders (95 %) belonging to warhead class 1 and 2, whereas of the true poor binders 61 % belong to warhead classes 3 and 4 (Table 8). This agrees with the earlier observation that true binding strength is significantly associated with warhead classification (χ^2 = 42.20, LR statistic = 53.78, P < 0.0001) (see Table 3).

The results can be further verified using Table 10 which displays the list of each molecules' catalogue ID for both the false positive and false negative ligands for the g7 + hbonds + distance model. The FPs and FNs are also listed for three less optimal (predictor sets) models involving g8 or glide as descriptors, but without

Table 9 Summary Table of the CMH and Odds Ratio (OR) for all 16 candidate models

Model	CMH Statistic	Common Odds	95 % CI for common odds	P-value	Odds ratio (WH 1)	Odds ratio (WH 2)
g7 + hbonds + distance	6.9641	3.524	(1.388, 8.951)	0.008	4.476	9.918
g7	3.9087	2.745	(1.089, 6.922)	0.048	2.935	2.394
g7 + hbonds	2.3371	2.180	(0.906, 5.245)	0.126	2.270	4.521
Glide + distance	1.2702	1.874	(0.762, 4.608)	0.260	1.973	1.405
g8 + hbonds	1.6866	1.975	(0.804, 4.851)	0.194	2.008	1.934
g8 + distance	0.8967	1.673	(0.683, 4.102)	0.344	1.948	1.237
Glide + hbonds	1.3774	1.827	(0.749, 4.455)	0.241	1.580	2.661
g8	0.3544	1.410	(0.575, 3.458)	0.552	1.643	1.000
Glide	0.0185	0.972	(0.348, 2.716)	0.892	1.274	0.266
emodel + hbonds	1.9763	0.487	(0.199, 1.194)	0.160	0.664	0.109
emodel + hbonds + distance	2.3031	0.469	(0.189, 1.162)	0.129	0.459	0.184
Glide + hbonds + distance	0.0036	0.837	(0.353, 1.987)	0.952	1.061	0.376
g8 + hbonds + distance	1.9477	0.489	(0.198, 1.210)	0.163	0.404	0.712
emodel + distance	0.9189	0.593	(0.243, 1.447)	0.338	0.703	0.419
g7 + distance	5.3585	0.294	(0.112, 0.771)	0.021	0.364	0.138
emodel	0.8933	1.792	(0.686, 4.683)	0.345	1.722	3.502

Table 10 Listing of FP and FN molecules: Catalogue ID with their corresponding warhead classification (given in brackets) for a selection of 4 models

	g7 + hb + dist		g8 + dist		g8 + dist + hb		Glide + hb	
	FP (WH)	FN (WH)	FP (WH)	FN (WH)	FP (WH)	FN (WH)	FP (WH)	FN (WH)
Unique catalogue ID	7 (1)	3 (1)	7 (1)	2 (1)	10 (1)	1 (1)	7 (1)	6 (1)
	9 (1)	6 (1)	8 (1)	3 (1)	11 (1)	2 (1)	8 (1)	60 (1)
	14 (1)	60 (1)	14 (1)	6 (1)	13 (1)	4 (1)	9 (1)	61 (2)
	27 (3)	68 (1)	27 (3)	60 (1)	15 (3)	5 (1)	11 (1)	63 (2)
	31 (3)	71 (1)	29 (3)	61 (2)	16 (3)	59 (1)	13 (1)	66 (1)
	38 (3)	78 (1)	32 (4)	62 (2)	17 (2)	62 (2)	18 (2)	67 (1)
	47 (3)	83 (1)	34 (3)	68 (1)	18 (2)	63 (2)	23 (3)	68 (1)
	57 (2)	89 (1)	42 (3)	71 (1)	19 (3)	66 (1)	24 (3)	71 (1)
	65 (2)	90 (1)	43 (3)	78 (1)	21 (3)	67 (1)	27 (3)	78 (1)
	70 (2)	91 (1)	44 (3)	83 (1)	22 (3)	72 (1)	28 (1)	83 (1)
	75 (4)	94 (4)	47 (3)	89 (1)	23 (3)	84 (1)	29 (3)	84 (1)
	76 (4)	223 (1)	57 (2)	90 (1)	24 (3)	85 (1)	30 (3)	91 (1)
	77 (4)	321 (2)	58 (2)	91 (1)	25 (3)	87 (1)	31 (3)	103 (1)
	80 (3)	344 (1)	65 (2)	223 (1)	26 (3)	88 (1)	38 (3)	223 (1)
	81 (1)	809 (1)	69 (2)	317 (2)	28 (1)	94 (4)	39 (3)	321 (2)
	95 (3)		70 (2)	321 (2)	30 (3)	100 (1)	41 (3)	322 (1)
	98 (3)		75 (4)	335 (1)	31 (3)	103 (1)	42 (3)	342 (1)
	99 (4)		76 (4)	337 (1)	32 (4)	308 (2)	43 (3)	344 (1)
	101 (4)		77 (4)	344 (1)	34 (3)	316 (2)	44 (3)	809 (1)
	102 (4)		79 (4)	809 (1)	35 (3)	322 (1)	45 (3)	
	104 (4)		80 (3)		37 (3)	335 (1)	53 (1)	
	105 (3)		81 (1)		38 (3)	336 (1)	57 (2)	
	205 (4)		95 (3)		39 (3)	342 (1)	58 (2)	
	238 (2)		97 (1)		41 (3)	343 (1)	65 (2)	
	244 (2)		98 (3)		45 (3)	370 (2)	70 (2)	
	338 (1)		99 (4)		46 (3)	372 (2)	75 (4)	
	805 (4)		101 (4)		48 (3)		77 (4)	
	808 (4)		104 (4)		49 (3)		79 (4)	
			205 (4)		50 (3)		80 (3)	
			238 (2)		51 (2)		95 (3)	
			338 (1)		52 (1)		97 (1)	
			341 (1)		53 (1)		99 (4)	
			805 (4)		54 (1)		101 (4)	
			806 (4)		55 (1)		102 (4)	
					56 (1)		104 (4)	
					69 (2)		205 (4)	
					74 (2)		805 (4)	
					82 (1)			
					92 (4)			
					93 (1)			
					96 (3)			
					340 (1)			
					341 (1)			
					806 (4)			
					808 (4)			
N	28	15	34	20	45	26	37	19

warhead distance included in the predictor set. It can be seen that for the top model g7 + hbonds + distance (with 15/43 = 35 % false negatives (FNs) and 28/81 = 35 % false positives (FPs)) of the 15 FNs, 13 (87 %) are in warhead class 1 and 2 (aldehydes and ketones etc.) i.e. the chemical group that is suitable for covalent attachment. By contrast of the 28 FPs, 18 (64 %) are in warhead class 3 and 4 (esters and amides plus alcohols) i.e. the chemical groups that are not capable of covalently interacting with the calpain). This generalisation makes good chemical sense, with all active compounds requiring a suitable warhead to covalently react with the calpain and hence block its activity.

7   The Structure of the False Negatives and False Positives

Examination of the molecular structures showed that almost all the false positives, compounds which have a high IC_{50} but are classified as good inhibitors have a warhead distance < 10 angstroms, i.e. they are predicted to interact strongly with the calpain active site. This begs the question as to why are these compounds actually true poor binders as defined by a high IC_{50} value? An investigation of their structures confirms that such false positives lack a suitable warhead chemical group and/or a suitable backbone composition for binding (see Fig. 15), and as such these

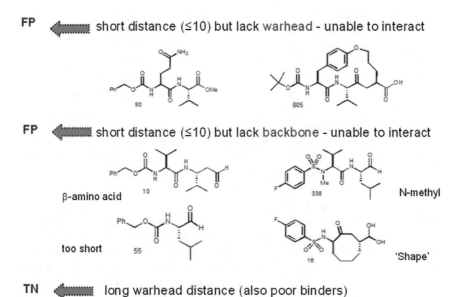

Fig. 15 Structural features of 2 false positive ligands (1st and 2nd structures) and one true negative calpain ligand (third chemical structure at the bottom of the figure)

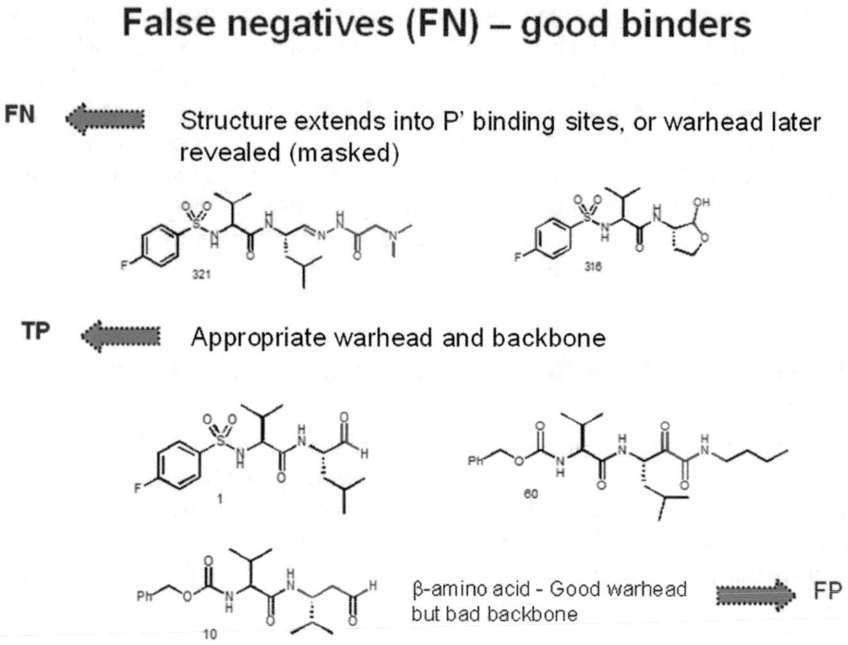

False negatives (FN) – good binders

FN ◄▪▪▪▪▪▪ Structure extends into P' binding sites, or warhead later
revealed (masked)

TP ◄▪▪▪▪▪▪ Appropriate warhead and backbone

β-amino acid - Good warhead ▪▪▪▪▪▪► FP
but bad backbone

Fig. 16 Structural features of false negative and true positive calpain ligands

ligands are unable to bind to calpain and thus show low activity. On the other hand, the false negatives (experimentally determined good binders as defined by low IC_{50}) either have the reactive warhead 'masked' or they contain suitable functionality in their backbone that is able to extend into additional P' binding sites of calpain (see Fig. 16). This second case allows tight binding of a nature not possible for other derivatives in the data set, i.e. these ligands are chemically unique.

Re-evaluation is advised for compounds which have shown higher activities in the literature or in the cases where a higher activity was expected after comparison with closely related compounds. Examples in this study are the biphenlyic compound, which show a higher activity in the original literature, and the Michael type acceptors which might need different assay conditions to be implemented due to their slow binding kinetics (Figs. 17, 18, 19 and 20).

As stated above most of the false negatives have a masked aldehyde warhead. It is likely that under physiological conditions, the aldehyde warhead is generated and therefore the docked species differs from the tested species (Fig. 20). Differences in the activities to the unmasked compounds generated in situ can thus be explained by the lower active concentration of inhibitor. The second group of false negatives have large substituents e.g. in the P3 position (or P prime position, see schema of the mechanism of binding, Fig. 9). Here, an induced fit mechanism is likely for the assay, however this has not been evaluated in the docking experiments.

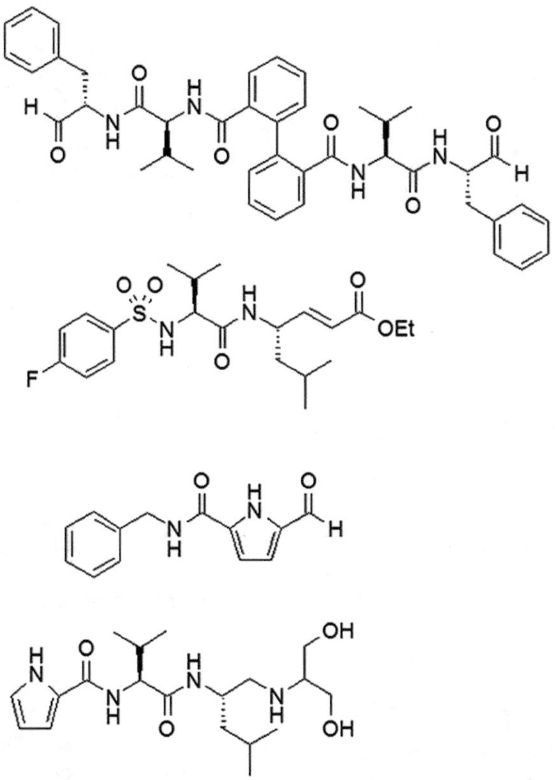

Fig. 17 False Negatives: biphenyls, Michael acceptor; 'short', interesting new compound

8 SOM and HUM Analysis of 3 Levels of Binding Affinity

In this section we briefly explore the applicability of extending our best SOM g7 + distance + hbonds (k = 2) model to 3 dimensions (k = 3). To this end we utilised HUM (Hypervolume Under the Manifold) [109] in the R statistical platform (http://www.R-project.org/) to perform a multi-category receiver operating characteristic (ROC) analysis. The aim of this section is to test the SOM classification of molecular binding strength at 3 levels. The 3 levels of *true* binding affinity are set at: best (true binding strength (BS) at levels 0 ($IC_{50} < 500$ nM) or BS at level 1 (500 nM $< IC_{50} < 5$ μM)); moderate (true BS at level 2 (5 μM $< IC_{50} < 100$ μM)) and worst binding affinity (true BS at level 3, i.e. $IC_{50} > 100$ μM)). HUM [109–111] is a recent tool for ROC curve analysis in 2D- and 3D-space, which allows for visualisation of the ROC curve for two or three class labels. HUM thus extends traditional ROC analyses of FPs and FNs to 3 dimensions (see Obuchowski et al. [104] who discuss conventional ROC Curves in Clinical Chemistry in terms of uses

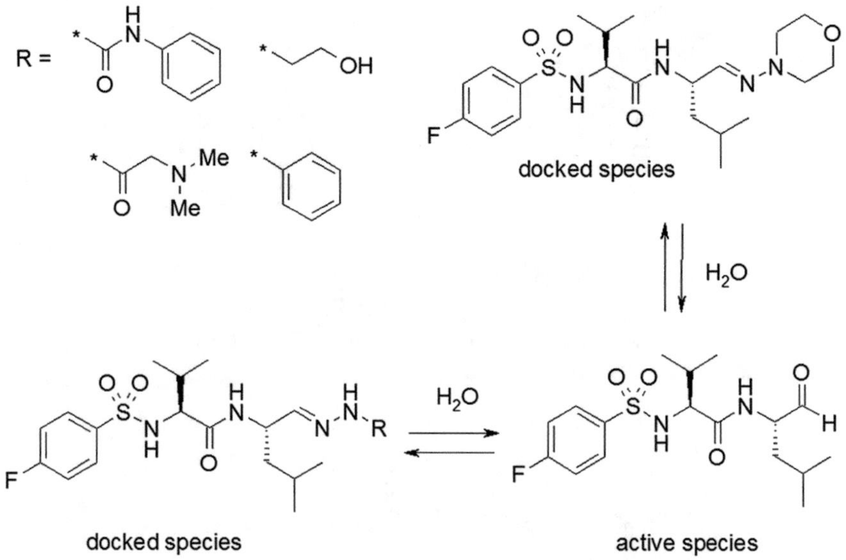

Fig. 18 Hydrolysis of hydrazones and semicarbazones to the corresponding aldehyde

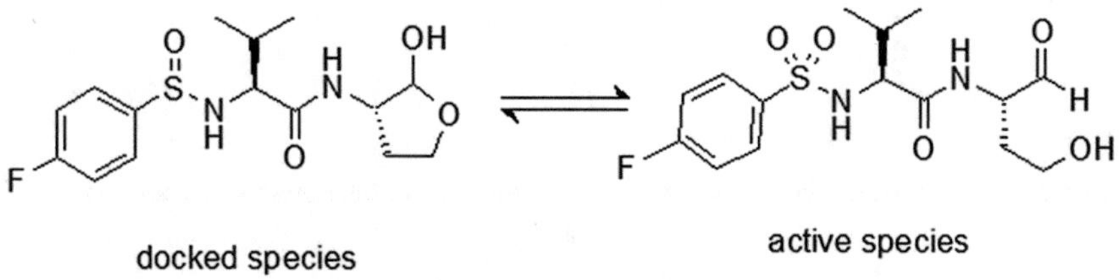

Fig. 19 Equilibrium between cyclic hemiacetal and $l3$-hydroxy aldehyde

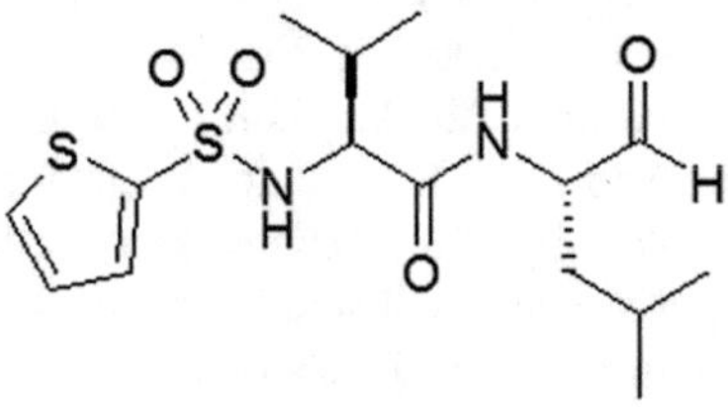

Fig. 20 The only 'real' false negative: compound shows high activity, and is promising to the 'chemical eye' but is classified as a poor binder

and misuse). HUM's main quantitative job is to compute the maximal HUM value between all possible permutations of class labels for all the individual features, selected for analysis (for the 3D problem).

8.1 SOM (K = 3)

The SOM (k = 3) U matrix and separate components SOM graph for the (k = 3)
g7 + distance + hbonds model are given in Figs. 21 and 22, respectively. The
resultant 3 SOM classes and their predictor means are given in Table 11, with
corresponding barchart also presented (Fig. 23). These show good visual and
quantitative separation between the predicted moderate, worst and best binding
strength classes of molecules (P < 0.003). Table 12 gives a breakdown of the 3
predicted SOM classes (moderate, worst and best binding affinity classes) each with
41, 34 and 49 molecules, in that order of predicted binding strength (BS). SOM
class membership is significantly associated with true BS (LR statistic = 14.1,
P < 0.025) (Table 12). Of the 81 true poor binders, 57 (70 %) are in SOM's
moderate and worst classes, with only 24 (30 %), in the best binders class. Of the 43
true good binders, 25 (58 %) are grouped in the best class, with only 8 (19 %)
molecules classified in the worst predictor class.

The relationship between predicted binding class (moderate, worst, best) by true
binding affinity (at 2 levels, good vs. poor) for the g7 + hbonds + distance model for
each of the 2 warhead class groupings (1, 2) versus (3, 4) is shown in Table 13.
There is a statistically significant difference in the association between predicted
binding class membership and true binding affinity across warhead class (1 or 2)
and warhead grouping (3 or 4) (LR statistic = 14.1, P < 0.025).

From Table 13 we have that of the 74 ligands in warhead class 1 or 2 (aldehydes
and ketones plus masked aldehydes, ketones, Michael Acceptors, epoxides);

- of the 34 predicted best binders, 59 % are true good binders, 28 % are true poor
  binders

Fig. 21 U-matrix (k = 3) for
the g7 + distance + hbonds
model

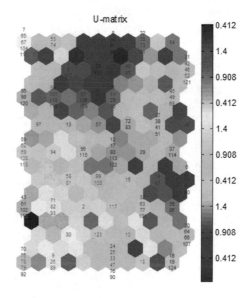

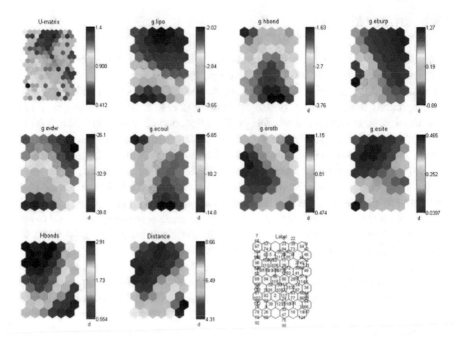

Fig. 22 SOM (k = 3) separate components graph for the g7 + distance + hbonds model

Table 11 Means of the molecular descriptors and glide subcomponents of the 3 SOM clusters: g7 + distance + hbonds model (k = 3)		Predicted BS (cluster)		
	Descriptor	best	moderate	worst
	distance	5.29	5.15	7.2
	hbonds	2.49	2.29	0.94
	g.emodel	−57.55	−51.25	−47.45
	g.lipo	−2.63	−2.49	−2.37
	ghbond	−2.72	−2.12	−2.46
	g.eburp	0.21	−0.51	−0.64
	g.evdw	−32.74	−31.74	−29.13
	g.ecoul	−11.11	−7.64	−7.68
	g.erotb	0.6	0.75	0.82
	g.esite	0.18	0.14	0.17
	g.einternal	8.44	6.58	6.29
	Glide Score	−5.91	−6.02	−6.26
	log IC_{50}	−5.14	−4.45	−4.09
	N	49	41	34

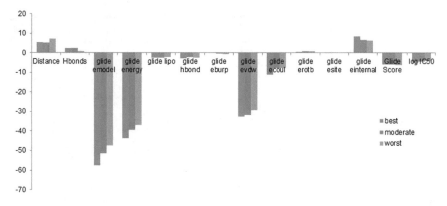

Fig. 23 Barchart of the means of the molecular descriptors and glide components for the 3 SOM classes: g7 + distance + hbonds model

Table 12 Cross tabulation of the 3 SOM predicted clusters/classes (moderate, worst, best binders) by true binding strength (BS) for the g7 + distance + hbonds model

SOM (k = 3) model	Class categories	True binding strength				
		0	1	2	3	Total
g7 + hb + distance	Moderate	3	7	23	8	41
	Worst	6	2	20	6	34
	Best	14	11	16	8	49
	Total	23	20	59	22	124
		0 or 1	2 or 3			
g7 + hb + distance	Moderate	10	31			41
	Worst	8	26			34
	Best	25	24			49
	Total	43	81			124

- of the 22 predicted moderate binders, 22 % are true good binders, 41 % are true poor binders
- of the 18 predicted worst binders, 19 % are true good binders, 31 %, are true poor binders
- of the 42 true good binders, 60 % are classified as best, 19 % as worst and 21 % as best binders
- of the 32 true poor binders, 28 % are classified as best, 31 % as worst and 41 % as moderate binders.

From Table 13 we have that molecules in warhead class 3 or 4 (mostly esters and amides plus alcohols and non-classical);

- of the 15 predicted best binders, 0 % are true good binders, 100 % true poor binders

Table 13 Cross tabulation of 3 SOM predicted cluster/class (worst, moderate, best) by true binding strength (BS) for warhead class (1, 2) and (3, 4) for the g7 + distance + hbonds model

Warhead class 1 or 2		True binding strength		% per class of true good (of true poor, in brackets)	% predicted classes for true good and for true poor (in brackets)
SOM (k = 3)	Cluster	Good	Poor		
g7 + hb + distance model	Moderate	9	13	21 % (41 %)	22 % (41 %)
	Worst	8	10	19 % (31 %)	19 % (31 %)
	Best	25	9	60 % (28 %)	59 % (28 %)
	Total	42	32		
Warhead class 3 or 4		True binding strength		% per class of true good (of true poor, in brackets)	% predicted classes for true good and for true poor (in brackets)
g7 + hb + distance model	Moderate	1	18	37 % (100 %)	6 % (94 %)
	Worst	0	16	32 % (0 %)	0 % (100 %)
	Best	0	15	31 % (0 %)	0 % (100 %)
	Total	1	49		

- of the 18 predicted moderate binders, 6 % are true good binders, 94 % true poor binders
- of the 16 predicted worst binders, 0 % are true good binders, with 100 % being true poor binders.

HUM allows visualisation of a 3D ROC curve for the three class label problem (moderate, worst, best predicted binding classes). HUM computes point coordinates for obtaining and plotting 3D-points for a 3D-ROC curve, shown in Fig. 24 for the g7 + distance + hbonds (k = 3) model; and in Fig. 25 for the simple parameter predictor set, i.e. the glide +distance (k = 3) model. The optimal threshold values

Fig. 24 3D HUM ROC graph for the g7 + distance + hbonds SOM model (k = 3)

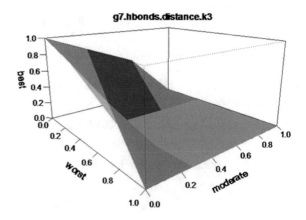

g7.hbonds.distance.k3

Fig. 25 3D HUM ROC
graph for the glide +distance
SOM model (k = 3)

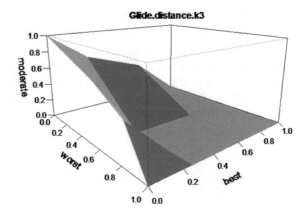

and the accuracy of the classifier are given in Table 14 for the 16 candidate models. Whereas the optimal threshold for the traditional two-class problem is the pair of sensitivity and specificity values for the selected feature classifier (e.g. class), the optimal threshold for our three-class problem is the 3D-point with the coordinates presenting the fraction of the correctly classified data objects for each class. These are the *r* (optimal), *s* (optimal) and threshold (optimal values) in Table 14. These are

Table 14 HUM analysis of the 16 SOM (k = 3) models

SOM Models (k = 3)	r (optimal)	s (optimal)	Threshold (optimal)	P value Predicted class by warhead
g7	0.455	0.000	0.837	0.11
g7.distance	0.500	0.407	0.465	0.12
g7.hbonds	0.458	0.000	0.860	0.07
g7.hbonds.distance	**0.729**	**0.000**	**0.581**	NS[a]
g8	0.424	0.791	0.000	0.015
g8.distance	0.000	0.000	1.000	0.017
g8.hbonds	0.458	0.860	0.000	0.045
g8.hbonds.distance	0.458	0.884	0.000	0.01
glide	0.488	0.424	0.364	NS
glide.distance	**0.559**	**0.419**	**0.455**	0.007
glide.hbonds	0.000	0.000	1.000	NS
glide.hbonds.distance	0.465	0.458	0.318	NS
emodel	0.682	0.000	0.581	NS
emodel.distance	0.000	0.814	0.409	NS
emodel.hbonds	0.407	0.364	0.349	NS
emodel.hbonds.distance	0.455	0.419	0.373	NS

[a]NS denotes not significant at the 5 % level of significance

(0.73, 0.00, 0.58) and (0.56, 0.42, 0.46) for the g7 + distance + hbonds (k = 3) model and for the glide + distance (k = 3) model, respectively. The r (optimal), s (optimal) and threshold (optimal) values for the g7 + distance (k = 3) model is (0.50, 0.417, 0.47), which is similar to the glide + distance model. Note that for the glide + distance model the 3 class prediction is significantly different across the warhead classes (P = 0.007 for the 3 level predicted classes by warhead interaction) (last column of Table 14).

8.2   Logistic and Multinomial Models for True Binding Affinity BS

In this section we test whether the following predictors, the molecule's predicted SOM cluster membership obtained from the optimal g7 + distance + hbonds model, (poor vs. good binder), along with the molecule's warhead class and their inter-action (cluster by warhead class) are significant predictors of the ligand's true binding strength, BS. Likewise we test prediction models for true BS using the ligand's glide score, number of hydrogen bonds, warhead distance and its warhead class as possible predictors, along with interactive effects of glide score and these structural parameters by warhead class (SOM cluster is not included in this prediction).

Such modelling of true binding strength (BS) at either 2 or 3 levels is performed using logistic and multinomial models, respectively, as given in the R statistical platform [112–114]. The aims of these analyses are to establish:

- whether structural parameters such as warhead distance, the number of hydrogen bonds (hbonds) plus predicted SOM cluster (poor vs. good binder), along with the molecule's warhead class are significant predictors of true binding strength (BS), and whether warhead class interacts with SOM cluster in this prediction (warhead class is considered at k = 2 or 4 levels)
- whether the glide score alone, along with both or either warhead distance and hbonds, plus warhead class are significant predictors of true BS. Cluster is not a predictor in these models. Again we also test whether warhead class interacts with any of the glide, distance or hbonds parameters in this prediction, with warhead class at k = 2 or 4 levels.

8.2.1   Logit and Multinomial Models of True Binding Affinity

The results of the logistic (logit) and multinomial models are reported in brief in this sub-section.

SOM predicted cluster status (poor vs. good binder) is a highly significant predictor (P < 0.005) of true BS, as is the molecule's warhead class, whether at 2 levels WH (1, 2) versus WH (3, 4) (P < 0.00007) or at 4 levels (P < 0.002). When SOM predicted cluster is in the model along with warhead class, then warhead distance and the number of hydrogen bonds (hbonds) are no longer significant predictors of true BS.

From the logistic regression the predicted probability of a ligand from the good SOM cluster, being a true poor binder (BS = 2, 3) is 0.28 and 0.96 for WH class (1, 2) and WH class (3, 4), respectively. Whereas the predicted probability of a ligand from the poor binder SOM cluster, being a true poor binder is higher at 0.59 and 0.99 for WH class (1, 2) and WH class (3, 4), respectively. The impact of warhead classification is similar across the SOM predicted binding statuses.

By contrast when using warhead at 4 levels in the logistic regression model, the predicted probability of a ligand classified as a good binder via SOM clustering, being a true poor binder is 0.20, 0.44, 1.0 and 0.91 for ligands in WH classes 1 to 4, in that order. The corresponding predicted probabilities of true good binding for this stratum of molecules is (0.80, 0.56, 0.00 and 0.09). Whereas the predicted probability of a ligand classified as a *poor* binder via SOM clusters, being a true poor binder is 0.53, 0.78, 1.0 and 0.99 for ligands in WH classes 1 to 4, in that order. The corresponding predicted probabilities of true good binding is (0.47, 0.22, 0.0 and 0.01).

For the logistic models using glide, distance, hbonds and warhead class as predictors of true BS (but not including SOMs cluster), the probability of a ligand being a poor binder is significantly impacted by the number of hbonds (P < 0.03), with less hbonds related to poorer BS. After adjusting for the ligand's number of hbonds, then poor binding is also significantly related to WH class at 2 levels (P < 0.0006) and at 4 levels WH (P < 0.005), with ligands in warhead class 4 having a significantly increased likelihood of poor binding compared to warhead class 1. The impact of the effect of increasing number of hbonds is similar across the warhead classes.

Multinomial models, with BS at 3 levels as the response variable, show similar trends to the logit models, with SOM predicted class (good vs. poor) impacting significantly on the likelihood of worst binding status *versus* the true moderate and the best binders (P < 0.03). True BS at 3 levels is also significantly impacted by warhead class (P < 0.00005). There is no interaction between SOM cluster status and warhead class. Multinomial models which include glide score, distance, hbonds and warhead class as predictors of BS (SOM cluster is not in the model) show that warhead classification alone is a significant predictor of true BS at 3 levels.

9 Conclusion

In the course of developing calpain inhibitors new tools for the evaluation of docking experiments were required and developed by Abell and his colleagues [11–20]. A combination of docking experiments, of analysis of the docked protein-ligand-complexes, and assay results to obtain IC_{50} has been applied to predict the binding mode of a data set of calpain ligands by our ANN methods. SOM/KM is shown to be valid for classification in our molecular docking experiments. Moreover, the SOM/KM prediction of binding affinity with glide score (or its subcomponents) along with distance and/or the number of hydrogen bonds (hbonds) enables reliable classification of compounds as (good) binders and (poor) non-binders of calpain—leading to a significantly lower number of false positives and also of false negatives than using the glide score alone.

Furthermore, the SOM analysis provides optimal sets of parameters for predicting true binding strength of calpain inhibitors with good specificity, sensitivity and AUC accuracy. We have established that the following models are optimal according to both the AUC and $(r + s)$ criteria (given our SOM method). These 6 models delineate specific subsets of predictors that best identify true binding affinity. The models are: g7 + distance + hbonds, g7, g7 + hbonds, g7 + distance and the glide + distance and the glide + hbonds + distance model. The aforementioned four g7 based models are shown to correctly identify 6-7 out of 10 true good binders and 6 out of 10 true poor binders. The two parameter glide score with distance model, also does well, correctly classifying 7 out of 10 true good binders and 6 out of 10 true poor binders. These rates are significantly superior to the glide score only model, which correctly identifies merely 2 out of 10 true good binders, in contrast to a high rate of correct classification for the true poor binders (8 out of 10). All 6 candidate models are equivalent in terms of prediction accuracy to the best g7 + distance + hbonds model.

We have demonstrated that the predictor sets containing; (i) warhead distance, and the number of hydrogen bonds, added to the seven partial energies of glide, or (ii) the simpler 3 parameter predictor set (model) containing warhead distance and hbonds added to just the glide score—give optimal classification of molecular binding affinity. These results verify that structure adds information to the exercise of virtual screening [115, 116]. All our models perform better than classification based on glide score only. Moreover as binding affinity in this data set is optimally described by just adding warhead distance to glide score alone (with or without adding the number of hydrogen bonds), we have a simple model by which to classify molecules, with no need to extract the 7 subcomponents (g7) of the glide score.

Molecules that are false positives are shown to be mostly esters and amides plus alcohols and non-classical, while the majority of false negatives are aldehydes and ketones, masked aldehydes and ketones and Michael acceptors and epoxides. There is a clear difference in the relationship between predicted class (good vs. poor) and true binding affinity across warhead class (1, 2) and warhead class (3, 4) for our best g7 + distance + hbonds model. This is also true for another 11 SOM/KM models. Specifically for our best g7 + distance + hbonds model an examination of the molecules in warhead class 1 or 2 shows that of the *predicted* good binders, 74 % are true good binders, with 26 % true poor binders; correspondingly of the predicted poor binders 39 % are true good binders with more than 1.5 times more, 61 %, being true poor binders. Likewise of the *true* good binders in warhead class (1, 2), 67 % are classified as good binders, with 33 % predicted as poor binders. Of the true poor binders 31 % are classified as good binders, with more than twice, 69 %, predicted as poor binders. By contrast for the ligands in warhead class 3 or 4 (esters and amides plus alcohols and non-classical), of the *predicted* good binders, 0 % are true good binders, 100 % being true poor binders, and of the predicted poor binders 3 % are true good binders, with 97 % being true poor binders. The one *true* good binder in warhead class 3 or 4, is classified as a poor binder, and of the true poor binders, 37 % are classified as good binders, with 1.7 times more, 63 %, classified as poor binders.

There is a clear statistically significant relationship between the 3 predicted classes (moderate, worst, best) and true binding affinity across warhead class (1, 2) and warhead grouping (3, 4) (for the optimal g7 + hbonds + distance model). An examination of the ligands in warhead class 1 or 2 shows that of the *predicted* moderate binders 21 % are true good binders, with 41 % true poor binders; of the predicted worst binders, 19 % are true good binders and 31 % are true poor binders, and of the predicted best binders 60 % are true good binders and 28 % are true poor binders,. Notably for the *true* good binders, 59 % are classified as the best binders, 22 % as moderate and 19 % predicted as worst binders. In contrast of the *true* poor binders, 28 % are classified as best, 41 % as moderate and 31 % as the worst binders. By contrast in warhead class 3 or 4 (esters and amides plus alcohols and non-classical) 100 % of the *predicted* worst and best binders are true poor binders.

There are still a number of false positives that our SOM/KM methods obtain. Re-evaluation is advised for such compounds which have shown higher activities in the literature or in the cases where a higher activity was expected after comparison with closely related compounds. Examples in this study are the biphenlyic compound, which showed a higher activity in the original literature, and the Michael type acceptors which might need different assay conditions to be implemented due to their slow binding kinetics. An investigation of the structures of all the false positives showed the majority of them to have a warhead distance <10 angstroms i.e. they are predicted to interact strongly with the calpain active site. An investigation of the structures confirms that, however, they lack a suitable warhead chemical group and/or a suitable backbone composition for binding—so confirming that these ligands are unable to bind to calpain and consequently demonstrate low activity. If several closely related compounds show the same classification and comparable activities re-modelling may also be necessary. Compared with other compounds, these inhibitors are much smaller. Therefore, the molecular weight and/or the length of the inhibitors might be an additional parameter to be evaluated. Some 'false positives' are actually much more active than was to be expected from their structure and have an IC_{50} just below the set threshold. Therefore, their clustering as a good binder may help reveal interesting new structures for further development.

On the other hand, compounds with a high assay activity but which are classified as poor binders ('false negatives') are rarer than false positives, generally half as many FNs to FPs. The false negatives (experimentally determined good binders as defined by low IC_{50}) either have the reactive warhead 'masked' or they contain suitable functionality in their backbone that is able to extend into additional P' binding sites of calpain. The second scenario allows tight binding of a nature not possible for other derivatives in the data set, i.e. these ligands are chemically unique and have large substituents e.g. in the P3 position (or P prime position). Here, an induced fit mechanism is likely for the assay, however this has not been evaluated in the docking experiments.

The false positives and false negatives give important feedback to docking experimentalists, as to where to change docking procedures where graphical location by visualisation is required to decide optimal molecular docking positions of the ligand to the protein molecule. Some such misclassified molecules need re-testing/re-modelling to assess their true binding affinity (IC_{50}), and importantly some false positive molecules have as shown, a N-Methyl group that decreases binding in assays. The methods described in this chapter should find broader application in docking experiments aimed at verifying the binding mode of a whole class of compounds in complex cases where no crystal structure nor NMR structure is available. Thereby, we should be able to discriminate between binders and non-binders in those very scenarios where traditionally used scoring functions may fail to give accurate results. Our new indicator model(s) (predictor sets) for binding affinity may assist the identification of poor binders that should be excluded from further FDA drug development—thus enabling developers of scoring functions to improve on parameterisation of scoring functions, by concentrating their efforts and focus on the likely binders, eliminating ahead of time the likely non binders.

Future research will entail performing hybrid SOM approaches with popular mixture type cluster methods (according to Mengersen et al. [117]) to investigate whether these models perform better than the SOM/KM methods [118]. Such approaches were recently applied in a research study of sleep, wake and duty profiles of Australian railway drivers [56], to research of personality indicators of depression and psychological distress [119] and the study of drug data bases [120, 129]. Specifically recent EMMIX skew type models developed by Lee and McLachlan [121–124] will be tested on this data set of 124 ligands. These models will be compared to Bayesian mixture approaches developed by Kim [128].

Additionally we will explore the full set of 16 candidate SOM models via the use of HUM in the R statistical platform (http://www.R-project.org/) for multi-category receiver operating characteristic (ROC) analysis to test ANN-SOM classification of molecular binding strength at 3 levels [109–111]. Various 3 levels of binding affinity that will be tested and compared to our current 3 categories of true binding affinity, namely, best (true binding strength (BS) at level 0 ($IC_{50} < 500$ nM) and BS at level 1 (500 nM $< IC_{50} < 5$ μM)), moderate (true BS at level 2 (5 μM $< IC_{50} < 100$ μM)) and worst binding affinity (true BS at level 3, i.e. $IC_{50} > 100$ μM). HUM will extend the traditional ROC analyses of false positives and false negatives to 3 dimensions [104].

Another important future direction is to re-analyse the study data set of calpains using recently developed software programs for covalent inhibitors (consider the work of London et al. [125]) and by using autodock (http://autodock.scripps.edu). Their implementation is described recently by Bianco et al. [90]. Of late, the merits and pitfalls of the theory and applications used in the development of covalent inhibitors and covalent docking in drug discovery was further discussed by Kumalo et al. [91] and by Mah et al. [38]. These covalent docking platforms and approaches were not available at the time our SOM/KM and glide analyses were performed. Use of CovalentDock is another candidate program recently developed for covalent docking, as described by Ouyang et al. [126]. These authors investigated automated covalent docking via parameterised covalent linkage energy estimation and specific molecular geometry constraints. These studies and that of Zhu et al. [28], on a parameter free approach to pose prediction and scoring in the docking of covalent inhibitors, underscores the recent momentum for covalent inhibitor research via docking.

References

1. V.G. Maltarollo, et al., *Applications of Artificial Neural Networks in Chemical Problems, Artificial Neural Networks—Architectures and Applications* (InTech, 2013)
2. V.G. Maltarollo et al., Applying machine learning techniques for ADME-Tox prediction: a review. Exp. Opin. Drug Metab. Toxicol. **11**, 259–271 (2015)
3. F. Marini et al., Artificial neural networks in chemometrics: History, examples and perspectives. Microchem. J. **88**, 178–185 (2008)
4. L. Wang, et al., Self-organizing map clustering analysis for molecular data, ed. by J. Wang, et al., in *Advances in Neural Networks,* ISNN 2006, vol. 3971 (Springer, Berlin, 2006), pp. 1250–1255
5. J.C. Gertrudes et al., Machine learning techniques and drug design. Curr. Med. Chem. **19**, 4289–4297 (2012)
6. R.G. Brereton, Self organising maps for visualising and modelling. Chem. Cent. J. **6**(Suppl 2), S1 (2012)
7. Y.D. Xiao, et al., Supervised self-organizing maps in drug discovery. 1. Robust behavior with overdetermined data sets. J. Chem. Inf. Model **45**, 1749–1758 (2005)
8. Y.D. Xiao, et al., Supervised self-organizing maps in drug discovery. 2. Improvements in descriptor selection and model validation. J. Chem. Inf. Model **46**, 137–144 (2006)
9. F. Marini et al., Class-modeling using Kohonen artificial neural networks. Anal. Chim. Acta **544**, 306–314 (2005)
10. M. Stahl et al., Mapping of protein surface cavities and prediction of enzyme class by a self-organizing neural network. Protein Eng. **13**, 83–88 (2000)
11. A.T. Neffe, A.D. Abell, Developments in the design and synthesis of calpain inhibitors. Curr. Opin. Drug Discov. Dev. **8**, 684–700 (2005)
12. A.D. Abell et al., Molecular modeling, synthesis, and biological evaluation of macrocyclic calpain inhibitors. Angew. Chem. Int. Ed. Engl. **48**, 1455–1458 (2009)
13. A.D. Abell et al., Investigation into the P3 binding domain of m-calpain using photoswitchable diazo- and triazene-dipeptide aldehydes: new anticataract agents. J. Med. Chem. **50**, 2916–2920 (2007)
14. M.A. Jones et al., Synthesis, biological evaluation and molecular modelling of N-heterocyclic dipeptide aldehydes as selective calpain inhibitors. Bioorg. Med. Chem. **16**, 6911–6923 (2008)
15. S.A. Jones et al., The preparation of macrocyclic calpain inhibitors by ring closing metathesis and cross metathesis. Aust. J. Chem. **67**, 1257–1263 (2014)
16. S.A. Jones et al., A template-based approach to inhibitors of calpain 2, 20S proteasome, and HIV-1 protease. ChemMedChem **8**, 1918–1921 (2013)

17. J.D. Morton et al., A macrocyclic calpain inhibitor slows the development of inherited cortical cataracts in a sheep model. Invest. Ophthalmol. Vis. Sci. **54**, 389–395 (2013)
18. A.D. Pehere et al., Synthesis and extended activity of triazole-containing macrocyclic protease inhibitors. Chemistry **19**, 7975–7981 (2013)
19. M. Pietsch et al., Calpains: attractive targets for the development of synthetic inhibitors. Curr. Top. Med. Chem. **10**, 270–293 (2010)
20. B.G. Stuart et al., Molecular modeling: a search for a calpain inhibitor as a new treatment for cataractogenesis. J. Med. Chem. **54**, 7503–7522 (2011)
21. K.C.H. Chua et al., Macrocyclic Protease Inhibitors with reduced peptide character. Angew. Chem. Int. Ed. **53**, 7828–7831 (2014)
22. J.P. Hughes et al., Principles of early drug discovery. Br. J. Pharmacol. **162**, 1239–1249 (2011)
23. A.D. Bochevarov et al., Jaguar: a high-performance quantum chemistry software program with strengths in life and materials sciences. Int. J. Quantum Chem. **113**, 2110–2142 (2013)
24. E. Kellenberger et al., Comparative evaluation of eight docking tools for docking and virtual screening accuracy. Proteins **57**, 225–242 (2004)
25. R. Sink et al., False positives in the early stages of drug discovery. Curr. Med. Chem. **17**, 4231–4255 (2010)
26. R. Macarron, Critical review of the role of HTS in drug discovery. Drug Discov. Today **11**, 277–279 (2006)
27. E. Yuriev, P.A. Ramsland, Latest developments in molecular docking: 2010–2011 in review. J. Mol. Recognit. **26**, 215–239 (2013)
28. K. Zhu et al., Docking covalent inhibitors: a parameter free approach to pose prediction and scoring. J. Chem. Inf. Model. **54**, 1932–1940 (2014)
29. S. Kawatkar et al., Virtual fragment screening: an exploration of various docking and scoring protocols for fragments using Glide. J. Comput. Aided Mol. Des. **23**, 527–539 (2009)
30. T.A. Halgren et al., Glide: a new approach for rapid, accurate docking and scoring. 2. Enrichment factors in database screening. J. Med. Chem. **47**, 1750–1759 (2004)
31. A.R. Leach et al., Prediction of protein-ligand interactions. Docking and scoring: successes and gaps. J. Med. Chem. **49**, 5851–5855 (2006)
32. T. Schulz-Gasch, M. Stahl, Scoring functions for protein-ligand interactions: a critical perspective. Drug Discov. Today Technol. **1**, 231–239 (2004)
33. P. Ferrara et al., Assessing scoring functions for protein-ligand interactions. J. Med. Chem. **47**, 3032–3047 (2004)
34. G.D. Geromichalos, Importance of molecular computer modeling in anticancer drug development. J. Buon. **12**(Suppl 1), S101–118 (2007)
35. A.J. Knox, et al., Considerations in compound database preparation—"hidden" impact on virtual screening results. J. Chem. Inf. Model. **45**, 1908–1919 (2005)
36. N. Moitessier et al., Towards the development of universal, fast and highly accurate docking/scoring methods: a long way to go. Br. J. Pharmacol. **153**(Suppl 1), S7–26 (2008)
37. T. Tuccinardi, Docking-based virtual screening: recent developments. Comb. Chem. High Throughput Screen. **12**, 303–314 (2009)
38. R. Mah et al., Drug discovery considerations in the development of covalent inhibitors. Bioorg. Med. Chem. Lett. **24**, 33–39 (2014)
39. S.J. Macalino, et al., Role of computer-aided drug design in modern drug discovery. Arch. Pharm. Res. (2015)
40. T. Kohonen, Essentials of the self-organizing map. Neural Netw. **37**, 52–65 (2013)
41. J. Vesanto, E. Alhoniemi, Clustering of the self-organizing map. IEEE Trans. Neural Netw. **11**, 586–600 (2000)
42. R. Rojas, *Neural Networks: A Systematic Introduction* (Springer, New York, Inc., 1996)
43. G. Schneider, Analysis of chemical space, in *Madame Curie Bioscience Database [Internet]* (Landes Bioscience, 2000). http://www.ncbi.nlm.nih.gov/books/NBK6062/
44. J. Sadowski, H. Kubinyi, A scoring scheme for discriminating between drugs and nondrugs. J. Med. Chem. **41**, 3325–3329 (1998)

45. V.N. Viswanadhan, et al., Atomic physicochemical parameters for three dimensional structure directed quantitative structure-activity relationships. 4. Additional parameters for hydrophobic and dispersive interactions and their application for an automated superposition of certain naturally occurring nucleoside antibiotics. J. Chem. Inf. Comput. Sci. **29**, 163–172 (1989)

46. M. Otto, *Chemometrics: Statistics and Computer Application in Analytical Chemistry Weinheim* (Wiley-VCH, New York, 1999)

47. D.W. Wichern, R.A. Johnson, *Applied Multivariate Statistical Analysis* (Prentice Hall, Upper Saddle River, 2007)

48. J.C. Fort, SOM's mathematics. Neural Netw. **19**, 812–816 (2006)

49. J. Vesanto, SOM-based data visualization methods. Intell. Data Anal. **3**, 111–126 (1999)

50. J. Himberg, et al., The Self-organizing map as a tool in knowledge engineering, in *Pattern Recognition in Soft Computing Paradigm*, ed. (World Scientific Publishing Co., Inc., 2001), pp. 38–65

51. MATLAB:2015, *version R2015a* (The MathWorks Inc., Natick, 2015)

52. J. Vesanto, et al., Self-organizing map in Matlab: the SOM Toolbox, in *Matlab DSP Conference*, 1999, pp. 35–40

53. R. Wehrens, L.M.C. Buydens, Self- and Super-organizing maps in R: The kohonen package. J. Stat. Softw. **21**, 19 (2007)

54. T. Vatanen et al., Self-organization and missing values in SOM and GTM. Neurocomputing **147**, 60–70 (2015)

55. B. Everitt et al., *Cluster Analysis* (Wiley, New York, 2011)

56. I.L. Hudson, et al., SOM clustering and modelling of Australian railway drivers' sleep, wake, duty profiles, in *28th International Workshop on Statistical Modelling*, Palermo, Italy, 2013, pp. 177–182

57. I.L. Hudson, J.A. Sleep, Comparison of self-organising maps, mixture, K-means and hybrid approaches to risk classification of passive railway crossings, in *23rd International Workshop on Statistical Modelling (IWSM)*, Utrecht, The Netherlands, 2008, pp. 396–401

58. F. Lopez-Vallejo et al., Integrating virtual screening and combinatorial chemistry for accelerated drug discovery. Comb. Chem. High Throughput Screen. **14**, 475–487 (2011)

59. S.Y. Huang, X. Zou, Advances and challenges in protein-ligand docking. Int. J. Mol. Sci. **11**, 3016–3034 (2010)

60. X. Li et al., Evaluation of the performance of four molecular docking programs on a diverse set of protein-ligand complexes. J. Comput. Chem. **31**, 2109–2125 (2010)

61. C. Bissantz et al., Protein-based virtual screening of chemical databases. 1. Evaluation of different docking/scoring combinations. J. Med. Chem. **43**, 4759–4767 (2000)

62. D.B. Kitchen et al., Docking and scoring in virtual screening for drug discovery: methods and applications. Nat. Rev. Drug Discov. **3**, 935–949 (2004)

63. W. Xu et al., Comparing sixteen scoring functions for predicting biological activities of ligands for protein targets. J. Mol. Graph. Model. **57**, 76–88 (2015)

64. Swiss Institute of Bioinformatics, *Click2Drug: Directory of Computer-Aided Drug Design Tools* (2013)

65. I.D. Kuntz et al., A geometric approach to macromolecule-ligand interactions. J. Mol. Biol. **161**, 269–288 (1982)

66. C.A. Baxter et al., Flexible docking using Tabu search and an empirical estimate of binding affinity. Proteins **33**, 367–382 (1998)

67. J.S. Dixon, Evaluation of the CASP2 docking section. Proteins **1**(Suppl), 198–204 (1997)

68. D.K. Jones-Hertzog, W.L. Jorgensen, Binding affinities for sulfonamide inhibitors with human thrombin using Monte Carlo simulations with a linear response method. J. Med. Chem. **40**, 1539–1549 (1997)

69. H. Li et al., GAsDock: a new approach for rapid flexible docking based on an improved multi-population genetic algorithm. Bioorg. Med. Chem. Lett. **14**, 4671–4676 (2004)

70. M.D. Miller et al., FLOG: a system to select 'quasi-flexible' ligands complementary to a receptor of known three-dimensional structure. J. Comput. Aided Mol. Des. **8**, 153–174 (1994)

71. G.M. Morris et al., Automated docking using a Lamarckian genetic algorithm and an empirical binding free energy function. J. Comput. Chem. **19**, 1639–1662 (1998)

72. E. Perola et al., Successful virtual screening of a chemical database for farnesyltransferase inhibitor leads. J. Med. Chem. **43**, 401–408 (2000)

73. M. Rarey et al., A fast flexible docking method using an incremental construction algorithm. J. Mol. Biol. **261**, 470–489 (1996)

74. C.M. Venkatachalam et al., LigandFit: a novel method for the shape-directed rapid docking of ligands to protein active sites. J. Mol. Graph. Model. **21**, 289–307 (2003)

75. W. Welch et al., Hammerhead: fast, fully automated docking of flexible ligands to protein binding sites. Chem. Biol. **3**, 449–462 (1996)

76. P.A. Buckley et al., Protein-protein recognition, hydride transfer and proton pumping in the transhydrogenase complex. Structure **8**, 809–815 (2000)

77. B.K. Shoichet et al., Lead discovery using molecular docking. Curr. Opin. Chem. Biol. **6**, 439–446 (2002)

78. H.-J. Böhm, M. Stahl, The use of scoring functions in drug discovery applications, in *Reviews in Computational Chemistry*, ed (Wiley, Inc., New York, 2003), pp. 41–87

79. H. Gohlke, G. Klebe, Approaches to the description and prediction of the binding affinity of small-molecule ligands to macromolecular receptors. Angew. Chem. Int. Ed. Engl. **41**, 2644–2676 (2002)

80. H. Li et al., An effective docking strategy for virtual screening based on multi-objective optimization algorithm. BMC Bioinformatics **10**, 58 (2009)

81. P.S. Charifson et al., Consensus scoring: A method for obtaining improved hit rates from docking databases of three-dimensional structures into proteins. J. Med. Chem. **42**, 5100–5109 (1999)

82. R.D. Clark et al., Consensus scoring for ligand/protein interactions. J. Mol. Graph. Model. **20**, 281–295 (2002)

83. I.J. Enyedy, W.J. Egan, Can we use docking and scoring for hit-to-lead optimization? J. Comput. Aided Mol. Des. **22**, 161–168 (2008)

84. T. Oprea, G. Marshall, Receptor-based prediction of binding affinities. Persp. Drug Discov. Des. **9–11**, 35–61 (1998)

85. S. Betzi, et al., GFscore: a general nonlinear consensus scoring function for high-throughput docking. J. Chem. Inf. Model **46**, 1704–1712 (2006)

86. M. Feher, Consensus scoring for protein-ligand interactions. Drug Discov. Today **11**, 421–428 (2006)

87. E. Perola, Minimizing false positives in kinase virtual screens. Proteins **64**, 422–435 (2006)

88. T.V. Pyrkov et al., Complementarity of hydrophobic properties in ATP-protein binding: a new criterion to rank docking solutions. Proteins **66**, 388–398 (2007)

89. V. Katritch, et al., Discovery of small molecule inhibitors of ubiquitin-like poxvirus proteinase I7L using homology modeling and covalent docking approaches. J. Comput. Aided Mol. Des. **21**, 549–558 (2007)

90. G. Bianco, et al., Covalent docking using autodock: two-point attractor and flexible side chain methods. Protein Sci. (2015)

91. H.M. Kumalo et al., Theory and applications of covalent docking in drug discovery: merits and pitfalls. Molecules **20**, 1984–2000 (2015)

92. X. Fradera et al., Unsupervised guided docking of covalently bound ligands. J. Comput. Aided Mol. Des. **18**, 635–650 (2004)

93. L. Wang et al., Accurate and reliable prediction of relative ligand binding potency in prospective drug discovery by way of a modern free-energy calculation protocol and force field. J. Am. Chem. Soc. **137**, 2695–2703 (2015)

94. G.A. Ross, et al., One size does not fit all: the limits of structure-based models in drug discovery. J. Chem. Theory Comput. **9**, 4266–4274 (2013)

95. E. Yuriev, et al., Challenges and advances in computational docking: 2009 in review. J. Mol. Recognit. **24**, 149–164 (2011)
96. I.O. Donkor, Calpain inhibitors: a survey of compounds reported in the patent and scientific literature. Expert Opin. Ther. Pat. **21**, 601–636 (2011)
97. E. Perola et al., A detailed comparison of current docking and scoring methods on systems of pharmaceutical relevance. Proteins **56**, 235–249 (2004)
98. R.A. Friesner et al., Glide: a new approach for rapid, accurate docking and scoring. 1. Method and assessment of docking accuracy. J. Med. Chem. **47**, 1739–1749 (2004)
99. T. Schulz-Gasch, M. Stahl, Binding site characteristics in structure-based virtual screening: evaluation of current docking tools. J. Mol. Model. **9**, 47–57 (2003)
100. A. Taube, Sensitivity, specificity and predictive values: a graphical approach. Stat. Med. **5**, 585–591 (1986)
101. A. Agresti, *Categorical data analysis*, 2nd edn. (Wiley, Hoboken, 2002)
102. N. Mantel, W. Haenszel, Statistical aspects of the analysis of data from retrospective studies of disease. J. Natl. Cancer Inst. **22**, 719–748 (1959)
103. T. Fawcett, An introduction to ROC analysis. Pattern Recogn. Lett. **27**, 861–874 (2006)
104. N.A. Obuchowski et al., ROC curves in clinical chemistry: uses, misuses, and possible solutions. Clin. Chem. **50**, 1118–1125 (2004)
105. W.J. Youden, Index for rating diagnostic tests. Cancer **3**, 32–35 (1950)
106. D. Hand, R. Till, A simple generalisation of the area under the ROC curve for multiple class classification problems. Mach. Learn. **45**, 171–186 (2001)
107. X. Robin et al., pROC: an open-source package for R and S + to analyze and compare ROC curves. BMC Bioinformatics **12**, 77 (2011)
108. D. Bohning et al., Revisiting Youden's index as a useful measure of the misclassification error in meta-analysis of diagnostic studies. Stat. Methods Med. Res. **17**, 543–554 (2008)
109. N. Novoselova et al., HUM calculator and HUM package for R: easy-to-use software tools for multicategory receiver operating characteristic analysis. Bioinformatics **30**, 1635–1636 (2014)
110. Z. Cai et al., Classification of lung cancer using ensemble-based feature selection and machine learning methods. Mol. BioSyst. **11**, 791–800 (2015)
111. J. Hu, et al., Comparison of three-dimensional ROC surfaces for clustered and correlated markers, with a proteomics application. Stat. Neerlandica, Wiley Online Library (2015)
112. B. Carstensen, et al. (2015). Epi: A Package for Statistical Analysis in Epidemiology. R package version 1.1.71. http://CRAN.R-project.org/package=Epi
113. W. Venables, B. Ripley. (2015). nnet: Feed-forward neural networks and multinomial log-linear models. R package version 7.3-11. http://CRAN.R-project.org/package=nnet
114. R Core Team. (2015). R: A language and environment for statistical computing. R Foundation for Statistical Computing. http://www.R-project.org/
115. C.N. Cavasotto, A.J. Orry, Ligand docking and structure-based virtual screening in drug discovery. Curr. Top. Med. Chem. **7**, 1006–1014 (2007)
116. C. McInnes, Virtual screening strategies in drug discovery. Curr. Opin. Chem. Biol. **11**, 494–502 (2007)
117. K.L. Mengersen, et al., *Mixtures: Estimation and Applications*, vol. 896 (Wiley, New York, 2011)
118. I.L. Hudson, et al., EMMIX skew classification of molecular ligand binding potency of calpain inhibitors. Mol. Inf. (in prep)
119. S. Lee, et al., Visualizing improved prognosis in psychiatric treatment via mixtures, SOMs and Chernoff faces, in *Australian Statistical Conference*, Adelaide, Australia, 2012, p. 131
120. I.L. Hudson, et al., Druggability in drug discovery: Self organising maps with a mixture discriminant approach, presented at the *Austraian Statistical Conference*, Adelaide, South Australia, 2012, p. 108
121. S.X. Lee, G.J. McLachlan, Model-based clustering and classification with non-normal mixture distributions. Stat. Methods Appl. **22**, 427–454 (2013)

122. S.X. Lee, G.J. McLachlan, On mixtures of skew-normal and skew t-distributions. Adv. Data Anal. Classif. **7**, 241–266 (2013)
123. S.X. Lee, G.J. McLachlan, EMMIX-uskew: An R package for fitting mixtures of multivariate skew t-distributions via the EM algorithm. J. Stat. Softw. **55**, 1–22 (2013)
124. S. Lee, G.J. McLachlan, Finite mixtures of multivariate skew t-distributions: some recent and new results. Stat. Comput. **24**, 181–202 (2014)
125. N. London et al., Covalent docking of large libraries for the discovery of chemical probes. Nat. Chem. Biol. **10**, 1066–1072 (2014)
126. X. Ouyang et al., CovalentDock: automated covalent docking with parameterized covalent linkage energy estimation and molecular geometry constraints. J. Comput. Chem. **34**, 326–336 (2013)
127. J. Polanski et al., Priveleged structures-dream or reality: preferential organization of azanaphthalene scaffold. Curr. Med. Chem. **19**(13), 1921–1945 (2012)
128. S.W. Kim, Bayesian and non-Bayesian mixture paradigms for clustering multivariate data: time series synchrony tests. PhD, University of South Australia, Adelaide, Australia (2011)
129. S. Zafar, et al., Linking ordinal log-linear models with correspondence analysis: an application to estimating drug-likeness in the drug discovery process, ed. by J. Piantadosi, R. S. Anderssen, J. Boland, MODSIM2013, in *20th International Congress on Modelling and Simulation* (Modelling and Simulation Society of Australia and NZ, 2013), pp. 1945–1951. ISBN: 978-0-9872143-3-1. http://www.mssanz.org.au./modsim2013/I1/zafar.pdf

Improved Ultrasound Based Computer Aided Diagnosis System for Breast Cancer Incorporating a New Feature of Mass Central Regularity Degree (CRD)

Ali Al-Yousef and Sandhya Samarasinghe

Abstract Ultrasound is one of the most frequently used methods for early detection of breast cancer. Currently, the accuracy of Computer Aided Diagnosis (CAD) systems based on ultrasound images is about 90 % and needs further enhancement in order to save lives of the undetected. A meaningful approach to do this is to explore new and meaningful features with effective discriminating ability and incorporate them into CAD systems. Some of the most powerful features used in cancer detection are based on the gross features of mass (e.g., shape and margin) that are subjectively evaluated. Recently, from an extensive investigation of ultrasound images, we extracted an easily quantifiable and easily measurable new geometric feature related to the mass shape in ultrasound images and called it Central Regularity Degree (CRD) as an effective discriminator of breast cancer. This feature takes into account a consistent pattern of regularity of the central region of the malignant mass. To demonstrate the effect of CRD on differentiating malignant from benign masses and the potential improvement to the diagnostic accuracy of breast cancer using ultrasound, this chapter evaluates the diagnostic accuracy of different classifiers when the CRD was added to five powerful mass features obtained from previous studies including one geometric feature: Depth-Width ratio (DW); two morphological features: shape and margin; blood flow and age. Feed forward Artificial Neural Networks (ANN) with structure optimized by SOM/Ward clustering of correlated weighted hidden neuron activation, K-Nearest Neighbour (KNN), Nearest Centroid and Linear Discriminant Analysis (LDA) were employed for classification and evaluation. Ninety nine breast sonograms—46 malignant and 53 benign- were evaluated. The results reveal that CRD is an effective feature discriminating between malignant and benign cases leading to improved accuracy of diagnosis of breast cancer. The best results were obtained by ANN where accuracy for training and testing using all features except CRD was 100 and 81.8 %, respectively, and 100 and 95.45 % using all features. Therefore, the overall improvement by adding CRD was about 14 %, a significant improvement.

A. Al-Yousef · S. Samarasinghe (✉)
Integrated Systems Modelling Group, Lincoln University, Christchurch, New Zealand
e-mail: sandhya.samarasinghe@lincoln.ac.nz

© Springer International Publishing Switzerland 2016   213
S. Shanmuganathan and S. Samarasinghe (eds.), *Artificial Neural Network Modelling*, Studies in Computational Intelligence 628,
DOI 10.1007/978-3-319-28495-8_10

Keywords Computer aided diagnostic system · Breast Cancer · Ultrasound · New mass feature · MLP-SOM model · Statistical Classifiers

1 Introduction

Breast tumors including benign and malignant solid masses start with few cells and grow faster than the normal cells and form a mass inside the breast tissues. A typical growing solid benign mass pushes and compresses the surrounding tissues to create a space which makes a well-defined boundary for the mass. Benign masses in most cases take an oval shape where the large axis of the oval is parallel to the skin line. Usually, the benign masses block the blood vessels which reduce the blood flow into the mass. Most of the diagnosed benign masses are fibroadinoma. Another example of benign tumors is cysts which are a fluid-filled mass with a clear margin. On the other hand, malignant masses grow and invade the surrounding tissues and not restricted to a limited area resulting in an irregular shape and ill-defined margin [1]. They also have an increased level of blood flow.

There are several changes in the breast anatomy during a female's life. In some cases, calcium salt gets deposited in the soft tissues that makes them harder; this is called calcification. Several factors contribute to this including age, past trauma to breast and breast inflammation. Also, calcification in a few cases is a sign of breast carcinoma. There are two types of calcification: the first is macro-calcification, where the size of calcium mass is large and probably benign. The second one is micro-calcification, where the size of calcium deposit is very small, less than 0.5 mm, and this type is probably malignant and requires further checking for micro-calcification morphology and distribution [2].

Ultrasound is one of the most frequently used methods for early detection of breast cancer. This method uses inaudible sound waves. The tumours are divided into four types according to their ability to return the sound waves (echogenicity). The first is anechoic lesions, and has no internal echo. This type of lesion is a benign cyst. The second one is hypoechoic lesions and has low level echoes throughout the lesion and is probably malignant. The third type is hyperechoic lesions with increased echogenicity compared to their surrounding fat, and are probably benign. The last type is isoechoic lesions with similar echogenicity compared to the fat echogenicity, and are probably benign [3, 4].

An ultrasound image is processed carefully to differentiate malignant from benign masses. Radiologists extract a number of mass features from an ultrasound image such as, shape, margin, orientation, echogenic pattern, posterior acoustic (shadow) features, effect on surrounding parenchyma and Calcifications (Table 1) [3, 5]. All these features are currently used to differentiate benign from malignant masses. Table 1 describes characteristics of each feature that help achieve this aim.

There are two main types of mass features in the image; combined morphological and geometric features and texture features. Morphological features are related to the structure of the mass such as, shape, margin, blood flow and the

Table 1 Ultrasound mass features and a brief description of each feature [19]

Feature	Description
Shape	Mass takes two main shapes, regular or irregular; the regular masses (or oval) are probably benign and irregular masses (speculated) are probably malignant
Margin	The border that separates the mass from the neighbouring normal tissue; it can be clear or well defined, which is suggestive of benign or it can be blurry or ill-defined which is probably malignant.
Orientation	The long axis of the mass can be parallel to the skin line, which is suggestive of benign
Posterior acoustic	The shadow behind the mass usually caused by malignant lesion
Echogenic pattern	This feature reflects the internal mass echogenicity (indication of density of the mass); hypoechoic lesions with low echogenicity are suggestive of malignant
Surrounding tissue	The effect of the mass on the surrounding tissues; depends on the mass type; for example, solid benign or malignant mass may compact the neighbouring tissues
Calcification	The presence of calcium inside the breast tissues; microcalcification is suggestive of malignant and macrocalcification is suggestive of benign
Blood flow	The speed of blood flow inside the mass tissues; high speed is suggestive of malignant
Envelop	Most benign masses are enveloped or partially enveloped

pattern of mass [6]. Geometric features, such as, volumes, spaces and measurements, can also be used to assess morphological features of the mass to extract new mass features. On the other hand, texture features are related to the arrangement of colours and other intrinsic pattern regularities, smoothness, etc., in the ultrasound image. This type of features are harder to extract as they require image processing algorithms to analyse the image.

Several studies have used texture features of ultrasound images for building Breast Cancer-Computer Aided Diagnosis Systems (BC-CAD). For example, 28 texture descriptors related to the intensities of the pixels in the Region Of Interest (ROI) including: the Mean of the Sum of the Average of intensities, Range of the Sum of the Entropy of pixels, pixel correlation,...,etc., and two Posterior Acoustic Attenuation Descriptors (PAAD) including—the difference between the average gray level within (ROI) and within regions of 32×32 pixels inside the (ROI), and the difference between the gray level in a region of 32×32 pixels inside the (ROI) and the average of the average of gray levels in the two adjacent 32×32 pixel regions to the left and right of ROI—were extracted and used for differentiating a cyst from solid mass [7]. Stepwise logistic regression was used for feature selection and the study obtained 95.4 % classification accuracy using Mean of the Sum of Average of intensities, Range of the Sum of Entropy of pixels and the second PAAD descriptor. In another study, the co-variance of pixel intensity in the Region Of Interest (ROI) and the pixel similarity were used to differentiate benign from malignant masses in ultrasound image with 95.6 % classification accuracy [8].

Another study [9] used the shape and texture features of the mass including: eccentricity (the eccentricity of the ellipse that has the same second moment as the mass region); solidity of the mass, ratio of the number pixels in the mass to that of the convex including the mass; convex hull's area minus convex rectangular area that contain the convex; Cross correlation left and cross correlation right (i.e., cross correlation value between the convex rectangular area that contain the convex and the left side and right side areas, respectively) to classify the masses into benign and malignant. These features were used as inputs to a Support Vector Machine (SVM) to obtain 95 % classification accuracy. Furthermore, texture features including auto-covariance coefficients of pixel intensities, smoothness of the mass and block difference of inverse probabilities were used with Support Vector Machines (SVM) in another study [6] that produced 94.4 % classification accuracy. Another study [10] extracted eight texture features, three geometric features and two pixel histogram features, and used them as input to a Fuzzy Support Vector Machine (FSVM), Artificial Neural Networks (ANN) and SVM to classify 87 lesions into benign and malignant. The FSVM obtained the best results where the accuracy was 94.2 % followed by the ANN with 88.51 % diagnostic accuracy.

Compared with texture features, there are few studies that used the morphological features for building BC-CADs. For example, shape, orientation, margin, lesion boundary, echo pattern, and posterior acoustic features were used as inputs for a BC-CAD with 91.7 % classification accuracy (88.89 % sensitivity and 92.5 % specificity) [11]. Also, another study [12] extracted 6 morphological features of the mass shape including: Form Factor, Roundness, Aspect Ratio, Convexity, Solidity and Extent; all of these features were based on a maximum and minimum diameters of the mass and area, convex and perimeter of the mass. These features were used with a Support Vector Machine and obtained 91.7 % classification accuracy (90.59 % sensitivity and 92.22 % specificity). Another study [13] extracted seven morphological features including the number of substantial protuberances and depressions, lobulation index, elliptic-normalized circumference, elliptic-normalized skeleton, long axis to short axis ratio, depth-to-width ratio, and size of the lesion. Multilayer Perceptron MLP neural networks was used for classification. The accuracy, sensitivity, specificity measures were 92.95 ± 2 %, 93.6 ± 3 % and 91.9 ± 5.3 %, respectively.

Although texture features are good in differentiating benign from malignant, their values depend on the settings of the ultrasound machine and require extensive processing of ultrasound images. On the other hand, the morphological features are system independent and therefore, they are easy to incorporate into CAD systems and more commonly used in diagnosis systems [14].

However, the accuracy of current CAD systems based on the morphological features of ultrasound images is around 90 % and needs further enhancement to save the lives of the undetected. A meaningful approach to do this is to explore new and meaningful features with enhanced discriminating power and incorporate them into CAD systems.

Mass shape is one of the most useful features in breast cancer CAD systems and has been used as an input for classifiers in several studies [11, 15–17]. This feature takes two main forms; regular or irregular shape. Irregular mass is indicative of malignancy; however, there are several types of benign mass that also take an irregular shape. Greater clarity on this issue may provide better ways to improve breast cancer diagnostic accuracy.

The aim of this chapter is to enhance the diagnostic accuracy of ultrasound based CAD systems based on morphological features by a thorough investigation of the shape of the mass in ultrasound images in order to extract a new mass feature with greater discriminating power and employ this feature in a CAD system along with few carefully selected morphological features. In doing so, we expand our work reported in [19].

2 Materials and Methods

This study evaluates 99 cases; 46 malignant and 53 benign. All cases were obtained from The Digital Database for Breast Ultrasound Images (DDBUI) [18]. The Second Affiliated Hospital of Harbin Medical University collected all images from 2002 to 2007. Each case in the database contains 1–6 images and a text file (Table 2) that lists important information of the patient and the lesions, such as, age, family history, shape, margin, size, blood flow, echo, micro-calcification number and shape. All these features were obtained by five experts.

Table 2 The contents of the text file in the database

Assessment: Malignant
Ultrasound Doctor's Depiction:
Assessment Sex: Female
Age: 32
Family History Criterion: none
Shape: Not Regular
Border: Blur
Echo: Unequable
Envelope: None
Side Echo: None
Micro-calcification Shape: None,
Micro-calcification Number: 0–0
Reduction: Has
Lymph Transformation: None
Blood: I or III Level
Mass Maximal Diameter: More than 2
Mammogram Total Number: 0
Radiologist Assessment: None
Biopsy Result: None
Biopsy Description:
Operation Result: Benign
Operation Description: breast fibroadenoma

2.1 Methods

The methodology is primarily aimed at extracting a new feature and assessing its
effectiveness following four main steps shown in Fig. 1. The first step is feature
extraction where all features indicative of breast cancer is extracted. Importantly,
this step also contains the extraction of the proposed new feature related to the mass
shape and called Central Regularity Degree (CRD) (see Sect. 2.1.1). All features are
normalised and then, the features that are strongly related to breast cancer are
selected for classification using hierarchical clustering and self organizing maps
(SOM). Next, the selected features are used as inputs to four classifier systems
based on neural networks and statistical methods (MFFNN—multi-layer feed forward neural network; KNN—K-nearest neighbour; LDA—Linear Discriminant
Analysis; NC—Nearest Centroid). Here, two sets of inputs are used: a set without
the new feature and a set including the new feature. Finally, individual classifiers
are evaluated using test datasets and the results of different CAD systems are
compared to assess the effectiveness of the new feature and the classifiers.

2.1.1 Feature Extraction

As mentioned, each case in the dataset has been described using a text file (Table 2)
and 1–6 images. The text file contains 12 features and some of these features are
described using linguistic variables such as shape: regular or irregular; envelop:

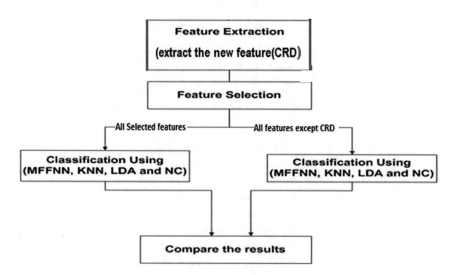

Fig. 1 The components of the proposed ultrasound CAD system (CRD—Central Regularity
Degree (proposed new feature); MFFNN—multi-layer feed forward neural network; KNN—
K-nearest neighbor; LDA—Linear Discriminant Analysis; NC—Nearest Centroid)

enveloped, not filled or none. To use these features in classifications, we converted the linguistic variables subjectively into numeric values as shown in Table 3.

Also we used the case images to extract two mass geometric features; Width to Depth ratio (WD) and the proposed new feature Central Regularity Degree (CRD). To compute the WD feature we applied the following steps:

1 For each case edit the ultrasound image of the mass using an image editor.
2 Draw the smallest rectangle that contains the mass as shown in Fig. 2. The X-axis of the rectangle parallels the skin line and represents the width of the mass, whereas the Y-axis represents the depth of the mass.

Table 3 Description of the mass features and the numeric value assigned to each description

Feature	Description	Coding
Shape	Regular	1
	Irregular	2
Border	Clear	1
	Blur	2
Echo	Equable	1
	Unequable	2
Envelope	None	1
	Not filled	2
	Yes	3
Sides Echo (sides shadow)	None	1
	Yes	2
Micro-calcification shape	None	1
	Needle	2
	Cluster	3
	Large	4
Micro-calcification number	0	0
	1	1
	1–2	1.5
	2–3	2.5
	3–4	3.5
	4–5	4
	5–6	5.5
Reduction (back shadow)	None	1
	Yes	2
Lymph Transformation	None	1
	Yes	2
Blood	Level 1	1
	Level 2 or 3	2
Mass Maximal Diameter	<1	1
	>1 and <2	2
	>2	3

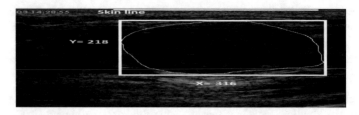

Fig. 2 The smallest rectangle containing the mass. *Thin line* represents the mass boundary and the *thick line* represents the boundary of the smallest rectangle that contain the mass. The *X* represents the width of the mass along horizontal axis and *Y* represents the depth of the mass along vertical axis

3 Compute the length of mass along X (W) and Y axes (D) (in pixels).
4 Compute WD = W/D = X/Y.

A new feature—mass Central Regularity Degree (CRD).

As mentioned previously, mass shape is either regular or irregular. The irregular shape of mass is an indication of malignancy but there are several benign masses that take an irregular shape too. Similarly, regular mass is suggestive of a benign case but there are few malignant masses that take a regular shape. Thus there is a need for further investigation into the mass shape to differentiate irregular benign mass from irregular malignant mass. In an in-depth investigation and closer examination of mass shape to discover potential new features, we were able to define a new geometric feature related to the mass shape called Central Regularity Degree (CRD). The CRD reflects the degree of regularity of the middle part of the mass. This was inspired by the fact that malignant masses typically have irregular shape and CRD is designed to specifically capture this irregularity in the more central part of the mass. As illustrated in Fig. 3, the mass boundary in this image was defined previously by experts as the white outline [18] and CRD involves defining the smallest width of the mass (red line).

To find CRD, we drew on ultrasound images (Fig. 2), a small rectangle that contains the complete mass using image editor software. The rectangle lines X and Y represent the mass width and the mass depth, respectively. Then we divided the rectangle horizontally into three equal parts; upper, middle and lower (Fig. 3). Next, for the middle part of the mass we found the length (Z) of the shortest line that is parallel to the horizontal axis (X). Finally, we found the ratio of Z to the rectangle width (X) (Eq. 1). The output value represents the Central Regularity Degree of the middle part of the mass.

$$CRD = Z \div X \tag{1}$$

After extracting and coding the mass features, there were a total of 14 features (age, 11 features listed in Table 3 and the two geometric features (WD, CRD)) as inputs for the next step. But still there is a need for data normalization to speed up

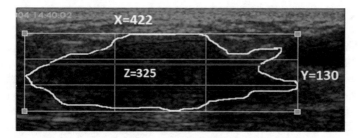

Fig. 3 New mass feature Central Regularity Degree (CRD). X is the rectangle width parallel to skin line, Y is the rectangle depth and Z is the shortest line in the middle part of the mass [19]

the classification process and to remove the effect of large spread of some features. For this, Min-Max normalization method was applied (Eq. 2).

$$X' = \frac{X - min_A}{max_A - min_A}(max_n - min_n) + min_n \qquad (2)$$

where X', A, n are; normalised values of X, old range and new range, respectively.

2.1.2 Feature Selection

In this step, we used a wrapper method to select a subset of ultrasound features that are strongly related to breast cancer. Now each case in the ultrasound dataset is represented as a vector of case features $S = \{S_1, \ldots, S_n\}$. The set of features that is strongly related to breast cancer is selected by using Hierarchal clustering (Johnson 1967) [19] and Self-Organizing Map (SOM) as follows:

1 Build a state space starting from empty set in the root and add features one by one until the set of all features is reached.
2 Use sequential search starting from the root to find the node that separates benign clusters from malignant clusters by applying the following steps:

   a. Apply Hierarchal clustering.
   b. Find the best cut off point that differentiates benign clusters from malignant clusters.
   c. Compute and save the accuracy and the node index.
   d. Repeat a-c until all nodes in the state space are visited

3 Select the node with the highest accuracy.
4 Validate the results using Self Organizing Map (SOM) combined with WARD clustering.

2.1.3 Classification

To evaluate the effect of the new ultrasound feature CRD on classification accuracy we applied four supervised classifiers: Multilayer Feed Forward Neural Network (MFFNN), Nearest Centroid (NC), K nearest neighbour (KNN) and Linear Discriminant Analysis (LDA) [20]. We divided the dataset randomly into two groups: training and testing, where, 77 cases were used for training and 22 for testing. Each classifier was trained and tested four times using different randomly extracted subsets. Each time the results were stored for the final evaluation.

Multilayer Feed Forward Neural Network (MFFNN)

Here, we used a Multilayer Feed Forward Neural Network with an input layer, one hidden layer and one neuron in the output layer. The network was trained using Scaled Conjugate Gradient Back Propagation Algorithm (SCGBP) and Logistic function as the neuron activation function.

Optimizing the number of hidden neurons in the network is still a challenge. Insufficient number of hidden neurons results in two problems: the first is under fitting that results from selecting a small number of hidden neurons giving rise to bias in generalisation. The second problem is over fitting that results from selecting a large number of hidden neurons giving rise to variance in generalisation. To overcome these problems, this study used a Self Organizing Map (SOM) based approach to optimize the number of hidden neurons in the network as described by Samarasinghe [21]. The algorithm starts with training a feed forward neural network with a relatively large number of neurons in the hidden layer. Then it reduces the number of hidden neurons by removing the redundant neurons that form correlated associations with other neurons in terms of the weighted hidden neuron activation that feeds the output neuron.

A summary of the MFFNN and its optimisation is given below. The net or weighted input u_i to neuron i and the output y_i of each neuron in the hidden layer is given by the following equations:

$$u_i = \sum_{m=1}^{r} x_m w_{im}; \qquad y_i = \frac{1}{1 + e^{-u_i}} \qquad (3)$$

where x is an input vector, r is the number of inputs including bias inputs and w_i is the weight vector between input vector x and neuron i. From Eq. 3, the weighted input to the output neuron v and the final output of the neural network are:

$$v = b_0 + \sum_{i=1}^{n} y_i b_i; \qquad z = \frac{1}{1 + e^{-v}} \qquad (4)$$

where, b_0 is the bias input to the output neuron, b is the weight vector between hidden neurons and the output neuron.

Form Samarasinghe [21], the weighted activation of redundant hidden neurons across the input vectors correlate. The weighted activation for neuron j is as shown in Eq. 5 where y_j is hidden neuron output and b_j is its weight to the output neuron.

$$N_j = b_j y_j \tag{5}$$

Now, we have a hidden neuron weighted activation matrix N where each row in the matrix represents one neuron j. To reduce the complexity of the neural network, we removed the redundant neurons. For this, we applied Self Organising Map (SOM) with correlation distance measure where the activation vector of each neuron was an input vector. The SOM brings similar input vectors closer together by distributing input vectors according to similarity. Similar SOM neurons was clustered using WARD clustering. The number of different clusters on the SOM indicates the optimum number of hidden neurons.

K-Nearest Neighbour (KNN)

In KNN, data vectors or the experimental samples are represented as marked points in the space where each point belongs to a known class (benign or malignant, for example). For a new instance, the classifier represents the instance in the same space and calculates the distance between it and the experimental samples. The label of the new instance depends on the labels of the K closest points to the new instance. The instance is labelled with the class label that has the largest number of points within K closest points [22]. To select the value of K we applied the following steps:

1   Set F = 0.
2   For I = 2 to N, where N is a large number:

   a. Set the value of K = I
   b. Classify the samples using KNN.
   c. Compute the overall *accuracy*.
   d. If *accuracy* ≥ F then set F = *accuracy* and Best = I.

3   Set K = Best

Nearest Centroid (NC)

In this classifier, the classification is done by calculating the mean (centroid) of both classes, malignant and benign, from the available data. For the new object x, the algorithm calculates the distance between the new object and the means of the classes; the object is then labelled with the label of the closest class centroid [23].

Linear Discriminant Analysis (LDA)

This classifier uses a covariance matrix to build a hyperplane between the benign and malignant classes by maximizing between to within variance ratio for the classes (Eq. 6) [24] such that

$$P(i|x) > p(j|x) \quad \forall i \neq j \tag{6}$$

The probability that x belongs to class i is not easy to compute; therefore, the simplest mathematical formula of LDA is:

$$f_i(x_k) = \mu_i C^{-1} x_k^T - \frac{1}{2}\mu_i C^{-1} \mu_i^T + \ln(p_i) \tag{7}$$

where μ_i is the mean vector of class i, C^{-1} is the inverse of covariance matrix of the dataset, p_i is the probability of class i and T is transpose.

The x_k belongs to class i if and only if:

$$f_i(x_k) \geq f_j(x_k) \quad \forall \, i \neq j \tag{8}$$

2.1.4 Evaluation

The aim of this step is to evaluate the performance of the two CAD systems (with and without the new feature CRD) using sensitivity, specificity, False Positive (FP) rate, False Negative (FN) rate and overall accuracy [25].

$$\text{Sensitivity} = \frac{\textit{Number of malignant samples correctly classified}}{\textit{Total number of malignant samples}} \tag{9}$$

$$\text{Specificity} = \frac{\textit{Number of benign samples correctly classified}}{\textit{Total number of benign samples}} \tag{10}$$

$$\text{Accuracy} = \frac{\textit{Number of samples correctly classified}}{\textit{Total number of samples}} \tag{11}$$

$$\text{FP} = \frac{\textit{Number of benign samples falsly classified}}{\textit{Total number of samples in malignant class}} \tag{12}$$

$$\text{FN} = \frac{\textit{Number of malignant samples falsly classified}}{\textit{Total number of samples in benign class}} \tag{13}$$

3 Results and Discussion

3.1 Feature Selection

Frequencies of specific ultrasound features in both malignant and benign cases are shown in Table 4. A feature would be considered effective or good if it clearly separated benign from malignant cases; for example, the margin was considered an

Table 4 Frequency of ultrasound features in the 99 cases (46 malignant (M) and 53 benign (B)) and their effectiveness as a feature that discriminates between malignant and benign cases [19]

Feature	M	B	Effectiveness
Age (mean)	46.4	38	Good
Shape			
Regular	2	29	Good
Irregular	44	24	
Margin			
Clear	8	38	Good
Blur	38	15	
Echo			
Equable	4	19	Not good
Not equable	42	34	
Envelope			
Enveloped	8	9	
Partially	5	9	Not good
No	33	35	
Microcalcification			
Big	2	0	Good
Cluster	2	1	
Needle	21	6	
None	21	46	
Blood level			
Level 1	14	41	Good
>1	32	12	
WD ratio			
> = 1.34	17	36	Good
<1.34	29	17	
CRD			
> = 0.7	10	39	Good
<0.7	36	14	
Diameter			
<= 1	2	8	Not good
>1 and <=2	24	26	
>2	20	19	

effective feature because most malignant cases (38 out of 46) were blurry and most benign cases (38 out of 53) were clear. On the other hand, the mass echo was not considered effective because most benign (34 out of 53) and malignant (42 out of 46) cases were not equable.

Table 4 shows seven effective features and these are: Age, Shape, Margin, Micro-calcification, Blood level, WD ratio and CRD. In particular, it shows that the new feature CRD is on a par with the most effective feature- margin. According to these effective features, a malignant mass has irregular shape, blurry margin, potential micro-calcification, higher speed of blood flow, lower WD ratio (below 1.34) indicative of less elongated shape and smaller CRD (below 0.7) indicative of more irregular shape.

Hierarchical clustering (HC) and self-organizing maps were used for feature selection. We started with hierarchical clustering to find a set of features that separates benign cases and malignant cases into different clusters. Hierarchical Clustering (HC) found age, shape, margin, level of blood flow, DW and our new feature CRD as the features that strongly demarcate breast cancer cases. HC divided the dataset into 9 different clusters. The distribution of malignant samples was: 39 out of 46 cases were distributed over 2 different clusters (clusters X and Y) with 0.84 sensitivity (ratio of malignant cases in the 2 clusters to total malignant cases) (Fig. 4). On the other hand, 42 out of 53 benign cases were distributed over 7 clusters with 0.793 specificity (ratio of benign cases in the 7 clusters to total benign cases) (Fig. 4). Hierarchical clustering produced 81.8 % accuracy.

To confirm the above results, we used SOM to find the distribution of the 99 ultrasound samples over SOM map using the same features (Fig. 5). The dataset was distributed over different regions on the SOM map where most of the malignant cases (41 out of 46) were distributed in the upper part of SOM and most benign cases (37 out of 53) were distributed in the lower part of SOM (Fig. 5c). The SOM U-matrix clearly divided the upper part of SOM into three clusters that appear in the U-matrix as dark blue regions (Fig. 5b). To clarify the boundary of each cluster in SOM map, we used K-mean clustering (k = 9), same number of clusters in hierarchal clustering, to cluster the neurons of SOM (Fig. 5a). By analysing the 9

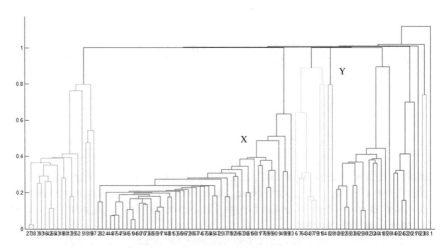

Fig. 4 Hierarchal clustering of the 99 cases using the 6 selected features. Each *color* represents a cluster. Hierarchal clustering has spread the 99 cases over 9 different clusters. (Clusters X and Y are the two clusters containing most of the malignant cases. Other clusters contain benign cases)

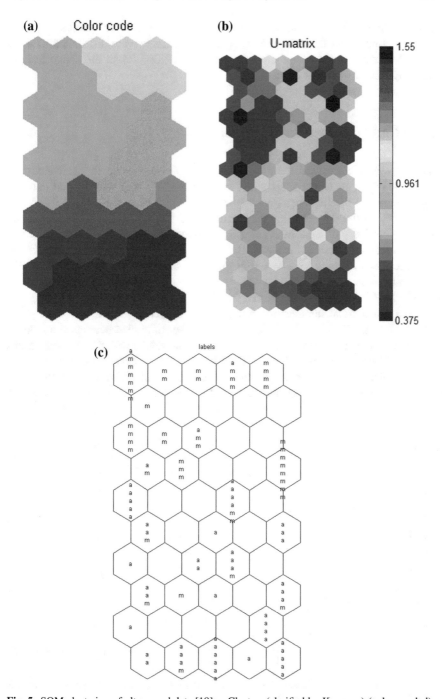

Fig. 5 SOM clustering of ultrasound data [19]: **a** Clusters (clarified by K-means) (colour coded). **b** SOM U-matrix—the distance of a neuron to the neighboring nodes in the SOM lattice is represented by a color according to the bar that appears on the right side of the figure; the distances range from *dark blue* (small distance) to *dark red* (large distance). Large distances indicate potential cluster boundaries. **c** The distribution of the benign (a) and malignant (m) cases over the SOM lattice which indicates that most malignant cases are in the upper part of SOM whereas most benign cases are in the lower part of SOM

clusters we found 89 % of malignant cases were distributed over 3 clusters (1, 2 and 3) in the upper part of SOM and 70 % of benign cases were distributed over the other 6 clusters. Both Hierarchical clustering and SOM found the above features strongly related to breast cancer.

3.2 Classifications and Evaluation

The selected features and the new feature (CRD) were tested for their ability to differentiate between malignant and benign cases. This step was divided into two main stages: In the first stage, we applied the four classifiers; KNN, MFFNN, NC and LDA, on the dataset using all features including CRD. For KNN, firstly, we must determine the value of K, which represents the number of neighbours that controls the class label of the new instance. For this, we started with a large K = 30 and decreased it down to K = 1 (Fig. 6). The best result was obtained when the value of k = 3.

The MFFNN is more complicated than KNN. In the MFFNN, we must take into account the optimal number of neurons in the hidden layer. For this, firstly, we trained and tested an MFFNN using a large number of hidden neurons and reduced the number gradually. Every time, we compared the results with the previous results until the best results were achieved. The best results were obtained using 15 hidden neurons. Secondly, despite the goodness of the results obtained with the 15 hidden neurons, we applied the previously described network pruning method based on

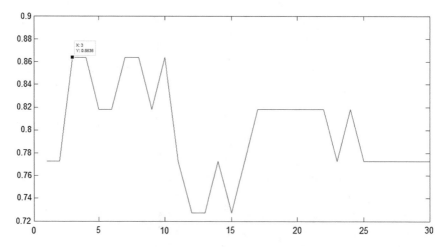

Fig. 6 The accuracies for different k values in KNN clustering. (*X-axis* represents the value of K and *Y-axis* represents the accuracy)

SOM clustering of correlated weighted hidden neuron activation patterns [21] to the MFFNN network just developed. The purpose was to determine whether there is a simpler network than the 15 hidden neuron network that would still give good results. Weighted activation of each hidden neuron for the original inputs constituted an input vector, each representing an individual hidden neuron, into SOM.

To build the SOM, we started with selecting the SOM topology. As the maximum possible clusters is 15 which is the number of hidden neurons, a 4×5 rectangular map was built and trained. Then we divided the map neurons into clusters using WARD clustering which resulted in 9 different clusters (Fig. 7). According to Samarasinghe [21], the number of different clusters represents the number of optimum hidden neurons. To verify the performance of the 9 hidden neurons against 15 neurons, we trained and tested a new MFFNN with 9 neurons using the same dataset. The output results obtained from 9 hidden neuron MFFNN were compared with the results from 15 hidden neuron MFFNN and the values of accuracy, sensitivity and specificity were found equal in both networks. The classification accuracy was 95.4 % with 100 % sensitivity and 90.9 % specificity. The results show that the SOM reduced the number of hidden neurons without any effect on the classification performance and reduced the complexity of the neural networks.

The outputs of different classifiers obtained from the first stage are shown in Table 5 (refer to numbers without brackets). The MFFNN was the superior

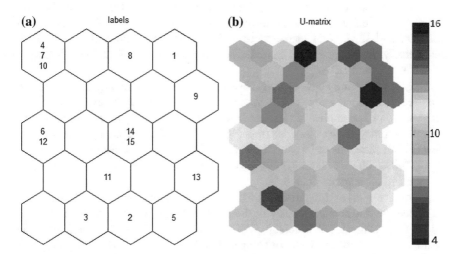

Fig. 7 SOM representing 15 hidden neurons. **a** Distribution of 15 neurons over the map; groups of hidden neurons (4, 7, 10), (6, 12) and (14, 15) each shared an SOM neuron (and formed three Ward clusters). **b** U-matrix for the 15 hidden neurons; neurons (9, 1, 8) were found close to each other (the blue color on the top right of the matrix) and considered to be one cluster by WARD clustering that divided the SOM neurons into 9 clusters

Table 5 Performance of different classifiers (SN is sensitivity, SP specificity and Ac accuracy)

Classifier	Training			Testing		
	SN (%)	SP (%)	Ac (%)	SN (%)	SP (%)	Ac (%)
KNN	85.7 (80)	92.8 (92.8)	89.6 (85.7)	81.9 (90.9)	90.9 (90.0)	86.4 (90.9)
NC	82.8 (82.8)	75.6 (75.6)	80.5 (80.5)	100 (90.9)	63.6 (63.6)	81.8 (77.3)
MFFNN	100 (100)	100 (100)	100 (100)	100 (81.8)	90.9 (81.8)	95.4 (81.8)
LDA	82.8 (80)	85.7 (83.3)	84.4 (81.8)	90.9 (90.9)	81.8 (72.7)	86.4 (81.4)

Numbers without brackets are for all features (including CRD) and those within brackets are for all features excluding CRD

classifier with 100 % sensitivity, 90.9 % specificity, 8.3 % False Positive (FP) rate, 100 % True Negative (TN) rate and 95.4 % classification accuracy. MFFNN has 100 % results in the training stage. NC is the worst classifier. KNN and LDA had similar accuracies but LDA was a better discriminator of malignant cases.

In the second stage, we applied the same classifiers (KNN, NC, MFFNN and LDA) on the same dataset using all features except CRD. The output results of different classifiers are shown in Table 5 (numbers within brackets). By comparing the results of different classifiers obtained from the first and second stages we found that, the sensitivities of KNN and LDA in the training phase were improved by 5.7 and 2.8 %, respectively, by adding CRD. Also, the specificity of LDA was increased from 83.3 to 85.7 %; furthermore the overall accuracies of LDA and KNN have increased. In the testing phase, the sensitivity of MFFNN and NC were improved by 18.2 and 9.1 %, respectively, by adding CRD. Also the specificity of MFFNN and LDA were increased to 90.9 and 81.8 % from 81.8 and 72.7 %, respectively. Adding the CRD increased the overall accuracy of the three classifiers, MFFNN, NC and LDA.

Only in the testing phase of KNN classifier that we found adding CRD decreased the classification accuracy. This is because of the disadvantages KNN that leads to an increase in the number of misclassified cases in the testing phase. However, the overall accuracy of the classifier (training and testing) is increased by adding the new CRD feature where the overall accuracy of KNN using all features except CRD is ((22*90.9/100) + (77*85.7))/99) % = 86.9 %; whereas, the overall accuracy of KNN using all features is 88.8 %.

From the above comparison, we found that adding the new feature CRD enhanced the diagnostic accuracy of the BC-CAD. The best results were obtained by MFFNN with 95.4 % accuracy, 90.9 % specificity and 100 % sensitivity in the testing phase. These results were compared with Wei et al. (2007) [11] study that used the same morphological features of masses except our new feature CRD as inputs to the CAD. The comparison found the accuracy of the current system 3.7 % better than their system. Furthermore, a significant improvement of the proposed BC-CAD over the system in [11] was in the 100 % sensitivity achieved by it compared to only 88.89 % achieved by the system in [11] which means that 11.11 % of cancer cases misdiagnosed in [11] could be diagnosed correctly by the current system and hence more lives could be saved.

Also, we compared the results of the current system with Chang et al. (2005) [12] results and we found that the current diagnostic system has 3.7 % higher accuracy and 9 % higher sensitivity, a significant improvement. Furthermore, the results of the current system was slightly better than the results obtained by Chen et al. [13] where the best accuracy for this system was 94.8 % compared with 95.4 % classification accuracy of the current system. But Chen et al. [13] system was more complicated than the current system because all the features used in their system were based on image processing algorithms that are computationally exhaustive and time consuming. Also, it is difficult for a radiologist to extract such features which make these features unusable in a direct diagnostic assessment made by the radiologist. In contrast, all the features used in the current system are extracted from the radiologist report, except DW ratio and our new feature CRD that are easy to extract and be used by a radiologist, without requiring any image processing. By these comparisons we found that the current system has superior diagnostic accuracy to previous ultrasound based BC-CAD systems.

4 Conclusions

We have presented in this chapter a new ultrasound based CAD system for early detection of breast cancer, incorporating a novel effective geometric feature—Central Regularity Degree (CRD). From a thorough investigation of the ultrasound images, we extracted this new geometric feature related to the shape of the mass in images, and it was inspired by the fact that most malignant masses are irregular. The CRD reflects the degree of regularity of the middle part of the mass. To demonstrate the effect of CRD in differentiating malignant from benign masses and the resulting improvement in the diagnostic accuracy of breast cancer based on ultrasound images, this study evaluated the diagnostic accuracy of four different classifiers when CRD was added to five known effective mass features: one geometric feature which is Width-Depth ratio (WD); two morphological features: shape and margin; level of blood flow and age.

Multilayer Feed Forward Neural Networks (MFFNN), K Nearest Neighbour (KNN), Nearest Centroid (NC) and Linear Discriminant Analysis (LDA) were used for classification and accuracy, sensitivity and specificity measures were used for evaluation. Ninety-nine breast sonograms—46 of which were malignant and 53 benign were evaluated. The results revealed that CRD is an effective feature discriminating between malignant and benign cases leading to improved accuracy of diagnosis of breast cancer. MFFNN obtained the best results, where the accuracies in the training and testing phases using all features except CRD were 100 and 81.8 %, respectively. On the other hand, adding CRD improves the accuracy of training and testing phases to 100 and 95.5 %, respectively. Therefore, the overall improvement by adding CRD was about 14 %, a significant improvement. Also, the new CRD feature makes the current system better than the previous CADs that used the morphological features of ultrasound images by a considerable margin.

References

1. R.J. Mason, C.V. Broaddus, T. Martin, T. King, D. Schraufnagel, J.F. Murray, J.A. Nadel, Benign tumors, in *Murray & Nadel's Textbook of Respiratory Medicine* (Saunders Elsevier, Philadelphia, 2005)
2. R.N. Panda, D.B.K. Panigrahi, D.M.R. Patro, Feature extraction for classification of microcalcifications and mass lesions in mammograms. Int. J. Comput. Sci. Netw. Secur. **9**, 255–265 (2009)
3. M.B. Popli, Pictorial essay: sonographic differentiation of solid breast lesions, Indian J. Radiol. Imaging **12**, 275–279 (2002)
4. G. Rahbar, A.C. Sie, G.C. Hansen, J.S. Prince, M.L. Melany, H.E. Reynolds, V.P. Jackson, J. W. Sayre, L.W. Bassett, Benign versus malignant solid breast masses: US differentiation 1. Radiology **213**, 889–894 (1999)
5. H. Madjar, E.B. Mendelson, *The Practice of Breast Ultrasound Technigues: Findings. differential Diagnosis*, 2 edn. (Thieme Verlag, New York, 2008)
6. Y.L. Huang, D.R. Chen, Y.R. Jiang, S.J. Kuo, H.K. Wu, W.K. Moon, Computer-aided diagnosis using morphological features for classifying breast lesions on ultrasound. Ultrasound Obstet. Gynecol. **32**, 565–572 (2008)
7. R. Sivaramakrishna, K.A. Powell, M.L. Lieber, W.A. Chilcote, R. Shekhar, Texture analysis of lesions in breast ultrasound images. Comput. Med. Imaging Graph.: Off. J. Comput. Med. Imaging Soc. **26**, 303–307 (2002)
8. W.J. Kuo, R.F. Chang, D.R. Chen, C.C. Lee, Data mining with decision trees for diagnosis of breast tumor in medical ultrasonic images. Breast Cancer Res. Treat. **66**, 51–57 (2001).
9. F. Zakeri, H. Behnam, N. Ahmadinejad, Classification of benign and malignant breast masses based on shape and texture features in sonography images. J. Med. Syst. **36**, 1621–1627 (2012). 2012/06/01
10. X. Shi, H.D. Cheng, L. Hu, Mass detection and classification in breast ultrasound images using fuzzy SVM. Proceedings of the Joint Conf. on Information Sciences, Taiwan (2006)
11. W.C. Shen, R.F. Chang, W.K. Moon, Y.H. Chou, C.S. Huang, Breast ultrasound computer-aided diagnosis using BI-RADS features. Acad. Radiol **14**(8), 928–939 (2007)
12. R.F. Chang, W.J. Wu, W.K. Moon, D.R. Chen, Automatic ultrasound segmentation and morphology based diagnosis of solid breast tumors. Breast Cancer Res. Treat. **89**, 179–185 (2005). 2005/01/01
13. C.M. Chen, Y.H. Chou, K.C. Han, G.S. Hung, C.M. Tiu, H.J. Chiou, S.Y. Chiou, Breast lesions on sonograms: computer-aided diagnosis with nearly setting-independent features and artificial neural networks 1. Radiology **226**, 504–514 (2003). February 1
14. J. Heinig, R. Witteler, R. Schmitz, L. Kiesel, J. Steinhard, Accuracy of classification of breast ultrasound findings based on criteria used for BI-RADS. Ultrasound Obstet. Gynecol. **32**, 573–578 (2008)
15. A.S. Hong, E.L. Rosen, M.S. Soo, J.A. Baker, BI-RADS for sonography: positive and negative predictive values of sonographic features. Am. J. Roentgenol. **184**, 1260–1265 (2005). April 1
16. G.N. Lee, D. Fukuoka, Y. Ikedo, T. Hara, H. Fujita, E. Takada, T. Endo, T. Morita, Classification of benign and malignant masses in ultrasound breast image based on geometric and echo features, in *Digital Mammography, Proceedings*, vol. 5116, ed. by E.A. Krupinski (Springer, Berlin, 2008), pp. 433–439
17. J.W. Tian, Y. Wang, J.H. Huang, C.P. Ning, H.M. Wang, Y. Liu, X.L. Tang, The digital database for breast ultrasound image. Presented at the proceedings of the 11th joint conference on information sciences, Paris (2008)
18. S. Johnson, Hierarchical clustering schemes. Psychometrika **32**, 241–254 (1967)
19. A. Al-yousef and S. Samarasinghe, Ultrasound based computer aided diagnosis of breast cancer: evaluation of a new feature of mass central regularity degree, in *Proceedings of the*

19th International Congress on Modelling and Simulation (MODSIM'11). Perth, Australia, 12–16 December 2011 (2011), pp. 1063–1069

20. S. Samarasinghe, Neural networks for water system analysis: from fundamentals to complex pattern recognition. Presented at the hydrocomplexity: new tools for solving wicked water problems, Paris (2010)

21. X. Wu, V. Kumar, J.R. Quinlan, J. Ghosh, Q. Yang, H. Motoda, G.J. McLachlan, A. Ng, B. Liu, P.S. Yu, Z.-H. Zhou, M. Steinbach, D.J. Hand, D. Steinberg, Top 10 algorithms in data mining. Knowl. Inf. Syst. **14**, 1–37 (2007)

22. G.A. Marcoulides, The elements of statistical learning: data mining, inference and prediction. Struct. Equ. Model. Multi. J. **11**, 150–151 (2004)

23. S. Balakrishnama, A. Ganapathiraju, Linear Discriminant Analysis—A brief Tutorial. Institute forSignal and information processing (1998). http://www.music.mcgill.ca/~ich/classes/mumt611/classifiers/lda_theory.pdf

24. D.G. Altman, J.M. Bland, Statistics notes: diagnostic tests 1: sensitivity and specificity. BMJ **308**, 1552 (1994). 1994-06-11 00:00:00

25. D.G. Altman, J.M. Bland, Statistics notes: diagnostic tests 1: sensitivity and specificity. BMJ **308**(6943), 1552 (1994)

SOM Clustering and Modelling of Australian Railway Drivers' Sleep, Wake, Duty Profiles

Irene L. Hudson, Shalem Y. Leemaqz, Susan W. Kim,
David Darwent, Greg Roach and Drew Dawson

Abstract Two SOM ANN approaches were used in a study of Australian railway drivers (RDs) to classify RDs' sleep/wake states and their sleep duration time series profiles over 14 days follow-up. The first approach was a feature-based SOM approach that clustered the most frequently occurring patterns of sleep. The second created RD networks of sleep/wake/duty/break feature parameter vectors of between-states transition probabilities via a multivariate extension of the mixture transition distribution (MTD) model, accommodating covariate interactions. SOM/ANN found 4 clusters of RDs whose sleep profiles differed significantly. Generalised Additive Models for Location, Scale and Shape of the 2 sleep outcomes confirmed that break and sleep onset times, break duration and hours to next duty are significant effects which operate differentially across the groups. Generally sleep increases for next duty onset between 10 am and 4 pm, and when hours since break onset exceeds 1 day. These 2 factors were significant factors determining current sleep, which have differential impacts across the clusters. Some drivers groups catch up sleep after the night shift, while others do so before the night shift. Sleep is governed by the RD's anticipatory behaviour of next scheduled duty onset and hours since break onset, and driver experience, age and domestic scenario. This has clear health and safety implications for the rail industry.

I.L. Hudson (✉)
School of Mathematical and Physical Sciences, The University of Newcastle,
NSW, Australia
e-mail: irene.hudson@newcastle.edu.au; irenelena.hudson@gmail.com

S.Y. Leemaqz
Robinson Research Institute, The University of Adelaide, Adelaide, SA 5005, Australia

S.W. Kim
Centre Epidemiology and Biostatistics, Flinders University, Adelaide, SA, Australia

D. Darwent · G. Roach · D. Dawson
Appleton Institute for Behavioural Science, Human Factors and Safety,
Central Queensland University, Adelaide, SA, Australia

© Springer International Publishing Switzerland 2016   235
S. Shanmuganathan and S. Samarasinghe (eds.), *Artificial Neural
Network Modelling*, Studies in Computational Intelligence 628,
DOI 10.1007/978-3-319-28495-8_11

1 Introduction

Fatigue in the rail industry is an important health and safety issue in Australia [1–4]. The Australian rail industry is large and varied, combining passenger and freight operations, and employing over 40,000 people, involved in 24-h operations, with clear impacts on sleep, fatigue and subsequently waking function [5–7]. From an organisational viewpoint, the fatigue associated with working time is managed primarily by regulating work hours [1, 3, 8]. Inherent to this strategy is the assumption that all employees use non-work time to obtain recovery sleep and that sleep periods are of equal value to all employees [9–11].

Fatigue is affected by many factors with sleep and circadian rhythms, two of the fundamental physiological factors. For railway drivers many factors such as environmental, physical conditions, and type of work also impact on fatigue. Domestic factors, marital status, presence of dependents and health status may also contribute to sleep behaviour [9, 11], along with shiftwork experience and age. Darwent et al. [12] demonstrated the extent to which social factors modify the timing and duration of sleep–wake behaviour over and above the purely physiological (e.g. sleep and circadian factors). These authors extended their initial work on using the timing and duration of work schedules to predict sleep–wake behaviour in long-haul airline pilots and to present a more general model that can be applied to a broader set of work places and occupations. This work provides significant support for the development of second-generation biomathematical models to deliver improved capacity to design schedules for adequate sleep opportunity (see Dawson [3]). Some recent studies of Australian railways shift workers [1, 11–13] have suggested that drivers tend to have similar sleep patterns despite differing work schedules and personal attributes—sleeping during the night and awake during the day. There are clear policy implications from these findings [10, 14, 15].

The focus of this chapter is the classification and modelling of the full data set or a subset of 190 railway drivers (RDs) sleep/wake states and also of their multivariate sleep duration time series profiles with respect to ANN-derived groups. Modelling these two types of sleep outcomes (sleep state or duration) was performed with respect to drivers' predictors of sleep onset time, current and next duty onset times, work and break onset times and break duration. RDs were asked to wear an activity monitor 24 h a day for 14 days and record details of sleep in a diary. Socio-demographics—marital status, number of dependents, RD age and experience were also collected. Generalised Additive Models for Location, Scale and Shape (GAMLSS) [16] were then used to model the two sleep outcomes with respect to the resultant ANN clusters and the predictors of sleep/duty/break and next duty onset and interactions with cluster.

1.1   A Brief Overview of Biomathematical Models of Sleep and Waking Alertness

Under normal environmental conditions, the sleep/wake cycle is synchronised with the exogenous day/night cycle. Sleep onset recurs several hours after dusk of each day and persists in a single consolidated bout for 6–8 h until sleep offset in the morning. The modern era of sleep research was foreshadowed by the rise in prominence of the biological sciences in the 19th and 20th centuries. Among the landmark events were the technological development of the electroencephalogram (EEG) and the discovery of rapid eye movement (REM) sleep. These formative events elevated the historic view of sleep as a simple and passive state of rest to its contemporary status as an essential and highly complex state of functional neurological activity.

Biomathematical models of sleep and waking alertness were originally developed to describe and evaluate competing hypotheses about the nature of the biological processes that regulate the sleep/wake cycle. The current generation of models have their origins in the two-process model of sleep-wake regulation [17, 18]. According to this model, the sleep/wake cycle is generated by two basic biological processes, including: (1) *a circadian process*—that determines 24-h oscillations in sleepiness and alertness; and (2) *a sleep homeostatic* process—that minimizes deviations from an optimal sleep amount by inducing sleepiness in response to wake and alertness in response to sleep. Sleep restriction below the optimum, i.e. <6–8 h/day, yields predictable dose-dependent deficits in waking function that can be masked by alertness-enhancing activities or drugs but which ultimately can only be recovered by sleep [19, 20].

Synchrony between the sleep/wake and exogenous day/night cycle is established via a master anatomical pacemaker located in the suprachiasmatic nucleus (SCN) of the anterior hypothalamus. The SCN is innervated by a rhetinohypothalamic tract that receives light input from the exogenous environment via photosensitive receptors in the retina. The phase and period of the pacemaker is dependent on the input of appropriately-timed light stimuli—which under normal conditions is largely determined by the daily light/dark cycle. Encoding of exogenous light into a biologic signal entrains the pacemaker to the day/night cycle, but this mechanism is reciprocally mediated by the sleep/wake cycle;—which alternately retards (i.e. closed eyes) and facilitates (i.e. open closed) passage of light into the retina.

The two-process model of sleep/wake regulation stimulated the development of augmented biomathematical models of sleep and alertness. These sought to explain nonlinear interactions between the circadian and homeostatic processes [21], the distribution of REM/Non-REM cycles within sleep [22, 23], the dynamics of photic entrainment [24, 25], and the alertness consequences of restricted and/or mistimed sleep [26]. The predictions made by these models demonstrated robust agreement with observed data collected in laboratory-based, experimental protocols. These successes encouraged researchers to generalize their models to predict the performance impairment associated with the disturbed sleep/wake cycles exhibited by shiftworkers in occupational settings [27–29]. The basic premise was to estimate sleep times on the basis of biological principles and then estimate any consequent performance impairment.

Models based on biological principles often yield poor agreement with the sleep/wake times observed in real-world settings, particularly in shiftwork operations that involve transmeridian travel [30]. The consequence of this failure is that the secondary algorithms used to estimate performance impairment based on sleep/wake times are similarly poor. Failure to accurately predict sleep/wake times in real-world settings occurs for three main reasons. First, light exposure in the technological age is no longer dictated by the geophysical day/night cycle and is easily altered with artificial light sources. Second, individuals have the capacity to self-select the timing of sleep/wake cycles by the use of artificial countermeasures (e.g. lights, noise, and drugs) to override endogenous sleep promoting signals. Third, in the laboratory, there is a relative absence of competing cultural/social/family priorities that in real-world situations place downward competitive pressure on the time that individuals allocate to sleep.

In response to these limitations, a novel class of models emerged that instead focused on predicting statistical distributions of sleep. These latter models were parameterised using data collected from shiftworkers e.g. train drivers [12] and aviation pilots [31], during their normal day-to-day lives. Calculations of performance, which unlike sleep times are primarily determined by circadian and homeostatic processes irrespective of cultural/social/familial factors, are then based on the estimated sleep times. This involves intermediate algorithms to estimate the state of the circadian and homeostatic processes given the likely distribution of sleep times. The implication is that improved accuracy of sleep estimates leads to improved performance predictions.

The remainder of this section gives a brief overview and motivation of the railway driver application of this chapter which aims to understand the impacts of duty and sleep time and break duration on sleep. Sleep enables people to recover from the tiredness of one wakeful period in preparation for the ensuing wake period [32–35]. Generally, the duration and quality of sleep determines the person's alertness and level of performance during the following period of wakefulness [36]. Researchers have shown that this relationship is positive, that is, performance and alertness increases as sleep duration increases [37, 38] although not linear [39].

As stated earlier a two-process model was proposed by Borbely [17] based on two basic physiological processes, a homeostatic process (S) and a circadian process (C) [33, 40–43] which allows for the regulation of (i) the minimum quantity of sleep for optimal daytime functioning (homeostatic), then (ii) the timing and structure (propensity) of sleep throughout the day (circadian rhythm) [32, 33, 35]. Indeed the effects of cumulative sleep loss become manifest in sleepiness and lack of focus during performance, with total sleep deprivation worsening a person's actual measures of subjective sleepiness [44], sleep latency [45], neurobehavioural performance [46–50], and complex cognitive functions [51–53]. Performance quality and alertness are also reduced with restricted sleep duration over one or more consecutive nights [54–57]. Irregular shift work schedules increase sleepiness at work [58–60] and also increase the risk of accidents [61].

Disturbed sleep is understood to have major consequences for shift workers [62–65], but the adverse effects may be diminished for those who effectively optimise sleep and socialisation opportunities during their scheduled rest periods. Darwent et al. [12] suggested that the primary factors governing optimal sleep are the timing and duration of rest periods, which was also suggested by Kurumatani et al. [66], as were the significant influences of local day-night cycles. Various studies [62, 67–69] have suggested that sub-optimal sleep is a deleterious consequence of working night shifts. Day time sleep tends to be shorter and of poorer quality due to higher levels of ambient noise and light encroaching on the sleeping environment [70, 71], and the desire for family and social interaction impacting further on the sleeper [72–82]. Studies suggest that nocturnal people usually take a daytime nap and this behaviour coincides with low alertness after lunch time [83–86]. Roach et al. [11] investigated the effects of break duration and time of break onset on the amount of sleep (sleep duration) obtained between consecutive work periods in a real work setting of RD's. Subsequently, Darwent et al. [12] proposed that when the opportunity for sleep coincides with natural circadian rhythms (C) and homeostatic processes (S) (outlined in the two-process model above), less sleep is achieved when one's sleep schedules conflict with or disrupt natural cycles [33]. The methodologies used in the one-step and two-step biomathematical models of human fatigue and performance is given schematically in Fig. 1 (from [33]). Unbroken lines represent the transfer of respective inputted values. Dashed lines represent the transfer of predicted values.

It has been suggested that railway drivers tend to have similar sleep patterns regardless of their differing work schedules and personal attributes—that is, they generally sleep during the night and remain awake during the day (see Kim [13], Darwent [1, 12]). There have been suggestions that rail drivers do not tend to adapt physiologically to irregular work schedules, with alertness and performance lowest at 2.15 am during the early morning shift. Adequate amounts of sleep (5–6 h) have been reported only by those drivers whose breaks began between 6 pm and 4 am, suggestive that drivers whose breaks commence outside of these hours need a longer break duration [1, 12, 13, 33].

1.1.1 The Data: Sleep, Wake and Duty Profiles of Australian Railway Drivers

A series of fourteen field-based studies (two depots chosen by each of seven members of Rail consortium) are used in this chapter. The Rail consortium comprised seven national rail organisations and the Rail, Tram and Bus Union. Railway drivers were observed for a period of 14 days between June 1996 and June 1997. For each study, drivers who participated in the study on a voluntary basis were asked to wear an activity monitor (actigraph) 24 h a day for 14 days and to record details of their sleep in a diary. The established reliability of self-reported subjective

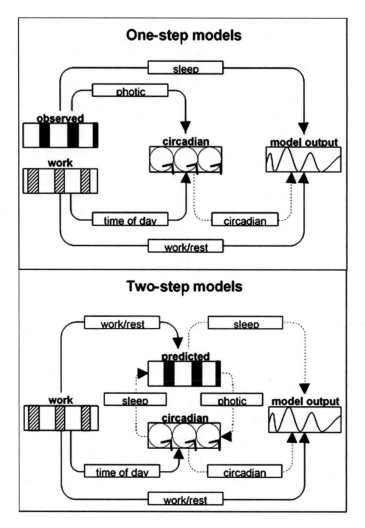

Fig. 1 One-step (*top*) and two-step (*bottom*) biomathematical models

sleepiness [87] makes it a valuable feedback procedure for safety management assessment of drivers [88, 89]. Work records were taken from the depot rosters for the first two field studies, but because rostered work times were sometimes found to deviate widely from realised work time, drivers who participated in the study were also asked to keep a record of their actual work schedule, and complete the standard shiftwork survey which covered questions on domestic and social situations, health and wellbeing along with sleep, wake, duty and break details (see Roach et al. [11]).

Australia wide, 253 drivers of an average age of 39.7 years (range 25–59) and an average length of shiftwork experience of 19.8 years (range 3–41 years), participated in the fourteen studies. Overall, 95 % of the actigraphy data was successfully collected [33]. Two hundred and fifty one drivers returned sleep diaries, 249 returned work diaries and 233 drivers returned their completed Standard Shiftwork Survey [13, 33]. Of the 253 drivers, there were 63 drivers who had missing sleep/wake or work behaviour records. Only sleep/wake and work behaviour data of 190 drivers (i.e. those without missing data) were analysed [13]. This data set of 190 time series constitutes 2814 sleep episodes (or transitions) (between 4 and 23 sleep periods for each driver) and with sleep duration ranging from 1 to 35 h. Six variables were calculated from the diaries: Break duration (total hours off-duty); Hours Since Break Onset; Hours To Next Duty; Next Duty Onset, Break and Sleep onset times (on a 24-h clock). Sleep duration (hrs) was calculated per sleep episode and socio-demographics (marital status, number of dependents, RD age, status of young kids and driver experience) collected.

1.1.2 Model and Methods

Two ANN approaches were performed. The first approach (see Sect. 3 and Part A of the Sect. 4 results) aimed to create RD sleep networks based on their sleep/wake/duty feature parameter vectors, comprising weights and transition probabilities between states (λ, Q) obtained via a transitional state process approach of Kim [13]. See also Kim et al. [90] and the book chapter of Hudson et al. [91]. Recently Kim's approach was generalised to a multivariate system applied by Hudson et al. [92, 93]. This multivariate extension of the univariate mixture transition distribution (MTDg) accommodated covariate interactions [13], thus generalising the MTD model of Berchtold and Raftery [94] which had no interaction terms. The SOM/MTD analysis in this chapter was performed on the full set of 190 RDs and allows RDs to have differing lengths of sleep time series (either of 14 days or shorter than 14 days). The mathematical details of the multivariate M-MTDg are given in Appendix A.

The second approach (see Part B results, Sect. 5) adopts a feature-based approach which uses ANN/SOMs to cluster the most frequently occurring patterns of sleep duration time series profiles per driver as based on a subset of 69 of the original 190 drivers, with a full record of 14 days sleep. ANN clustering of RDs' sleep series was performed via VANTED [95], which was used by Hudson et al. [92, 93] for climate change research and subsequently on this set of RDs data by Hudson et al. [96]. The latter approach involved modelling so-called RD-specific time series profiles, and created RD networks based on the RD's sleep/wake/duty feature parameter vectors.

Generalised Additive Models for Location, Scale and Shape (GAMLSS) [16] were then used to model the two sleep outcomes (namely, the states (λ, Q) and the sleep duration (hrs) per RD) with respect to the resultant ANN clusters and the predictors of sleep/duty/break and next duty onset and interaction terms with cluster. GAMLSS modelling of the VANTED SOMs of sleep duration also tested four additional predictors—hours to next duty and hours since break onset (following Hudson et al. [96]) along with sleep hours attained during 1 or 2 episodes prior to the current sleep period. GAMLSS models were recently used to investigate climatic effects and thresholds on the intensity of Eucalypt flowering [97, 98, 168]. Similarly we aim in this chapter to establish which levels/thresholds of work/break and timing of sleep/break/next duty trigger fatigue. GAMLSS model optimality is based on the AIC criterion, RD is treated as a random effect and cubic splines used. Circularity of time is accommodated for in Part B for all onset times on a 24 h clock (e,g, sleep onset, break onset, next duty onset). Two models- 1 and 2- are fitted using both stepwise and non-stepwise GAMLSS procedures. Model 1 contains the predictors; sleep at lag 1–2, SOM group factor, sleep, break and next duty onset times, break duration, hours to next duty (and their interactions with group). Model 2 is Model 1 plus the additional covariate, hours since break onset.

2 The Self Organising Maps

The Self Organising Map [99], known as the Kohonen feature map, converts complex, nonlinear statistical relationships between high-dimensional data into simple geometric relationships on a low-dimensional display, usually a 2D map. The SOM as such is a topological map which organises itself based on the input patterns that it is trained on. A non-time series SOM approach was used recently to map living standards in VietNam [100] and for accident risk classification of Australian railway crossings [101], the latter study developed a SOM with mixtures approach where the SOM best mapping units were clustered using model based clustering (MCLUST) of Fraley et al. [102]. The SOM algorithm converts complex, nonlinear relationships between high-dimensional data into simple networks and a map based on the most likely patterns in the multiplicity of time series that it trains. The aim of this chapter is to cluster and model the profiles using a SOM approach adapted recently for multivariate time series data to analyse flowering series and to derive a new metric for species synchronisation of flowering [92]. We adopt a feature-based approach which uses SOMs to cluster the most frequently occurring patterns of profiles of episodic sleep events. Generalised Additive Models for Location, Scale and Shape (GAMLSS) [16] of attained sleep are then used to model RDs' sleep with respect to resultant SOM cluster membership and the following predictors: hours to next duty and hours since break onset, sleep onset time, next duty onset time, break onset time and break duration; and their interactions with cluster, along with sleep hours attained 1 or 2 sleep episodes prior.

2.1 Time Series Clustering: Why SOMs?

Given a set of (possibly unlabeled) time series, one often wants to determine groups of time series that are similar [103]. This process is called time series clustering, a method organised into 3 groups depending upon whether they are applied directly to the raw data either in the time or frequency domain, or work indirectly with (i) features extracted from the raw time series data, or indirectly with (ii) models built from the raw data. The former raw data approach relies on a major modification of replacing the distance/similarity measure for static (non-time series) data with an appropriate one for time series. The latter approaches (i) and (ii) first convert raw time series data either into a feature vector of lower dimension or a number of model parameters, both result in forms of static data, so that existing algorithms for clustering static data can be used directly. Conventional clustering algorithms can then be applied to the extracted feature vectors or model parameters. These approaches are thus called a feature-based and model-based approach, respectively. There are two major approaches of feature and model-based methods: a statistical approach (e.g. AutoClass [104], which uses Bayesian statistical analysis to estimate the number of clusters), and a neural network approach. A well-known method of the neural network approach to clustering used in this chapter is Kohonen's self-organising feature map [99]. The 3 major procedural steps for time series clustering are: clustering algorithms; data similarity/distance measurement; and performance evaluation criterion of clusters. Various iterative algorithms have been developed to cluster different types of time series. Generally these modify existing algorithms for clustering of non-time-series (static) data—our focus is self-organising maps (SOMs).

Raw-data-based approaches usually imply working in a high dimensional space and apply either in the time or frequency domain. Any two time series being compared are conventionally sampled at the same interval, but their length (number of time points) may vary. Particular raw data-based applications are studies to find similar regions of activation of the brain using functional MRI (fMRI) data; DNA microarray studies and use of clustering and discriminant analysis of non-stationary time series. Feature-based clustering methods have been proposed to reduce dimensionality and to deal with time series that are highly variable in time (and possibly space), attributes which phenological time series often possess (see Hudson et al. 2011 [92]). Feature extraction methods tend to be generic in that the extracted features are usually application dependent, applications can involve clustering algorithms of spectra constructed from the original time series or clustering using the Haar wavelet transform, e.g. using fMRI time series to identify brain regions with similar activation patterns, whereby features extracted comprise cross-correlation measures.

The performance of a time series clustering method relies on the function used to measure the similarity between two data sets being compared to determine the appropriate number of clusters, M. Data may be of various forms—as raw values of equal or unequal length, vectors of feature-value pairs, and also transition matrices (as in Sects. 3 and 5, Part B). Whilst the following clustering algorithms, k-means, and fuzzy c-means algorithms, require the number of clusters M to be specified a priori, this is not necessarily the case for SOMs as used in batch mode. For example, M was determined for the SOM application in this chapter as that number of clusters for which the correlation between the RD sleep hour pairs (network arms) remain stable (for changing values of cluster number, as an iterative process). Information criteria (IC) such as the Akaike (AIC), Schwarz (BIC) and the integrated complete likelihood (ICL) [105] can be used if the data arise from an underlying mixture of Gaussian distributions with equal isotropic covariance matrices. The optimal number of clusters is then the one that yields the highest value of the information criterion. Under the framework of time series clustering, as described above, we investigate in this study a feature-based time series clustering approach which uses SOMs to investigate the most frequently occurring patterns (features) of a set of RD time series profiles of attained sleep (duration).

2.2   Clustering Based on the SOM Algorithm

Self-organising maps (SOMs) [99] are a popular tool for analysing variability in large, complex, multidimensional, multivariate datasets. Among the various existing neural network architectures and learning algorithms, the SOM is one of most popular neural network models. It is an unsupervised learning algorithm, fundamentally a pattern recognition process, in which intrinsic inter- and intra-pattern relationships within the data set are learnt without the presence of a potentially prejudiced or outside influence. SOMs provide a powerful, nonlinear technique to optimally summarise and visualise complex data using a preselected number of "icons" or SOM states, allowing rapid identification of preferred patterns and numerous aspects of data variability [106]. SOMs produce an expedient, concise organisation (to identify natural groupings) of multivariate data based on similarities among the patterns. This is achieved without imposing a structure on the data (as done by K means) [101, 107]. The SOM consists essentially of two layers of neurons (or nodes): the so-called input-layer and the output layer (Fig. 2). The SOM algorithm presents a regular, traditionally two-dimensional (2-D), grid of map nodes, where each node is represented by a prototype vector, which is of the input vector-dimension. The nodes are connected to adjacent units by a neighbourhood relation (see Fig. 2 and steps 2 and 3 in Sect. 2.2.1).

In general terms the SOM algorithm involves an iterative training procedure, where an elastic net that folds onto the 'data cloud' is formed [108]. This is achieved by mapping data points that are close together (in the Euclidean sense) onto adjacent map units. Learning in SOM is founded on competitive learning where the output nodes of the network compete amongst themselves to be activated or fired (Fig. 2). Only one output node, or one node per group, is considered to be on at any one time. The output nodes that win the competition are traditionally termed the winner-take-all nodes. During the self-organisation process, the cluster unit whose weight most closely matches the input data (typically, in terms of the minimum squared Euclidean distance) is chosen as the winner. The weights of the winning node and its neighbours are then updated proportionally in the iteration process.

Fig. 2 The architecture of the
Self Organising Map
(SOM) with M cluster nodes

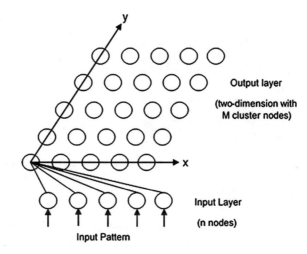

2.2.1 Initialize Weights, BMUs and Adjusting the Weights

SOM as applied in this chapter is a tool for extraction of common measurement patterns over time. The mathematics described in this follows the framework of Hudson et al. [92]. Consider a time series $Y(t) = S = (y_1, \ldots \ldots, y_n)$ and fix the sliding window's width at l. A set of time series segments $X(s) = \{s_i = (s_i, \ldots, s_{i+l-1}) | i = 1, \ldots \ldots, n-l+1\}$ can be formed, and will be presented to the SOM initial map as input patterns for training of the network. In contrast to the traditional SOM algorithm, where the input vector is the actual data vector and the final trained map groups data nodes with similar characteristics, the trained map in this study inputs *patterns of time series* and is expected to group a set of patterns $X_1, \ldots, X_k$, where k is the size of output map (or output layer plane). The SOM process can be split into two phases, the training phase and the assigning phase. During the training process, the temporal patterns represented by each node of the output layer change at each iteration according to the pattern inputted (Fig. 2). A pattern structure that represents the majority of the winning candidates and is different from its neighbours is formed in the final process. This final set of patterns, obtained from the output layer, then represents the most frequently appearing patterns according to the given time series inputted into SOM. Figure 3 exemplifies such a formation of frequently appearing patterns from two output nodes. Different iterations of the whole SOM training process lead to such an evolving process.

Firstly the SOM algorithm is described in more general terms of the best mapping unit (BMU) and centroids, and secondly by a more mathematical description of the algorithmic steps of the SOM (steps 1–3 below). The SOM approach constitutes firstly a training phase in which clusters of common input patterns in the data are identified. Secondly, a lookup phase which assigns each input vector to the best fitting cluster (centroid). At the beginning of each iteration, an input pattern vector is chosen randomly from the set of input patterns, denoted

(a)

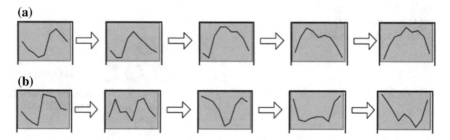

(b)

Fig. 3 Two examples of patterns gleaned during the pattern discovery process (http://www. researchgate.net/publication/228771755_Pattern_discovery_from_stock_time_series_using_self-organizing_maps accessed September, 2015)

by the subscript k. The chosen input pattern is then given by $X_i, i \in \{1, 2, \ldots, n\}$. After an input pattern vector X_i is selected, the SOM examines all the remaining nodes in the lattice (network or grid) to identify the so-called Best Matching Unit (BMU) (steps 1–5 below), considered to be the node corresponding to the weight vector that is nearest to the current input pattern vector.

For each input pattern X_i and each node j, the closeness between each input node and the initial weights is given by:

$$D_j = \sum_i (X_i - w_{ij})^2 \tag{1}$$

where w is the weight vector, and X are the input patterns.

Equation (1) is equivalent to the more general formulation, given in Eq. (2) of step 1 below. The BMUs are thus obtained by selecting input patterns that have the minimum Euclidean distance. These BMUs can be considered as the most frequently appearing patterns (these are shown in the RD-specific coloured boxes of Fig. 9 according to the input time series). After the BMUs are obtained, in each iteration, the neighbouring nodes within the BMU radius (neighbourhood) are found using Pythagoras theorem (step 2).

The final step in each training iteration is the adjustment of the weights for the nodes in the BMU's neighbourhood set (step 3). During this step, the BMU and its neighbours are moved closer to the input vector X_i in the input space. Nodes that are nearer to the BMU have a larger weight alteration than the ones further away. The weight adjustment is different for each neighbourhood node, and is dependent on the distance between the node and the BMU and the *learning rate* (step 2). By repeating this process using a different random input vector for each iteration, a topology-ordered map is produced by SOM [109]. Thus during the training phase, a SOM weight matrix is trained based on the mapping-data analysed. SOM adapts itself to common input patterns. The outputs of this phase are centroids, or BMUs.

An overview of the mathematical approach and algorithmic steps of the SOM follow. The SOM algorithm begins from an initial map containing random weights, which are standardised random values between 0 and 1, i.e. $0 < w < 1$. After the weights are initialised, the input pattern X_i is presented to the initial map or lattice.

Essentially each training-iteration consists of the following iterative steps: presentation of a randomly chosen input pattern vector from the input space, the evaluation of the network, and updating of the weight vectors (steps 1–3 below).

Step 1. Initialization of the training process: Initialize by assigning small random values to the weight vectors w of the neurons in the network. Each of the neurons i in the 2-D map is assigned a weight vector. At each training step t, a so-called training data $X(t) = x(t)$ (time series at training step t of length n), or input pattern from the input space, is chosen randomly. The Euclidean distances between the input pattern $x(t)$ and weight vector is computed for all M neurons in the network. A winner neuron (cluster) w_v is that which achieves the smallest Euclidean distance to $x(t)$:

$$v = \arg\min_i \|x(t) - w_i t\|, \quad i \in \{1, \ldots, M\} \tag{2}$$

for M the total number of neurons in the 2D map (or grid).

Step 2. SOM adjusts the weight of the winner neuron (cluster) and all the neighborhood neurons as follows. The weight of neuron i is updated depending upon whether it lies within a certain neighbourhood kernel $h_{vi}(t)$ around the winner neuron or not. The updating rule for the weight of the winner neuron is:

$$w_i(t+1) = w_i(t) + \alpha(t) \times h_{vi}(t) \times [x(t) - w_i(t)] \tag{3}$$

where $\alpha(t)$ is the learning rate and $h_{vi}(t)$ is the neighborhood kernel at training step (or time) t, respectively. The neighbourhood kernel $h_{vi}(t)$ is a function which is defined over the lattice points (Fig. 2). Both the size of the neighbourhood and the learning rate (or step size of weight adaptation) shrink monotonically with the iterations, i.e. decrease monotonically with time within 0 and 1.

Step 3. Update the weights of the neighbourhood neurons such that:

$$w_k(t+1) = w_k(t) + \alpha(t) \times h_{vk}(t)$$
$$\times \begin{cases} \left(\left[x(t) - w_v(t) + [w_v(t) - w_k(t)]\left(\frac{d_{vk}}{\Delta_{vk}\lambda} - 1\right)\right]\right), & \text{if } w_v(t) \text{ between } x(t) \text{ and } w_k(t) \\ \left(\left[x(t) - w_v(t) - [w_v(t) - w_k(t)]\left(\frac{d_{vk}}{\Delta_{vk}\lambda} - 1\right)\right]\right), & \text{if } w_k(t) \text{ between } x(t) \text{ and } w_v(t) \\ \left([x(t) - p] + [p - w_k(t)]\left(\frac{d_{vk}}{\Delta_{vk}\lambda} - 1\right)\right), & \text{otherwise} \end{cases}$$

where d_{vk} and Δ_{vk} are the distances between neurons v and k in the data space on the map, respectively. Here λ is a positive and pre-specified resolution parameter. It represents the desired inter-neuron distance reflected in the input space and depends on the size of the map, the variability in the data, and the required resolution of the map.

Step 4. Refresh the map randomly and choose a neuron weight.

Step 5. Repeat steps 1–4 until the map converges.

In each iteration, the winner neuron (cluster) in the output space is found and its center (centroid) is updated according to step 3. The number of centroids M needs to be determined before the training phase, but M can remain unspecified and vary in the batch mode of SOMs. Since the neighbouring neurons are updated at each step, there is a tendency that neighbouring neurons in the network represent neighbouring locations in the feature space. Hence the topology of the data in the input space is preserved during mapping via the SOM algorithm. The algorithm requires many iterations in order for SOM to determine a map of stable zones, which subsequently acts as a feature classifier. The unseen input vectors simulate nodes in the zone which have similar weight vectors [110].

2.2.2 Visualisation Software: VANTED

Clustering of the RDs' sleep duration time series of this chapter was performed using the VANTED (visualization and analysis of networks containing experimental data) software platform [111] (see Part B results). The VANTED system offers a variety of new functionalities for visual exploration, statistical calculations (t-test, outlier identification, correlation analysis), and data clustering with self-organising maps, available free at http://vanted.ipk-gatersleben.de. Applications of VANTED including a user guide and example data are available at the web site. To date, no other data visualisation tool apart from VANTED permits: (i) calculation of correlations within a complex data set; (ii) automatic generation of correlation networks from such correlations; and (iii) clustering of data via neural (or neuronal) network algorithms, i.e. SOM developed by Klukas et al. [111] and Junker et al. [95]). SOM is useful when correlations between vector components in the input data exist. SOM visualisation can also be used to inspect correlations based on clustering the *patterns* underlying the time series records, even if they are in different parts of the data space. Such correlations, for example, were given by Hudson et al. [92] for each of the phenological Eucalypt species pairs analysed in those climate studies (see also Kim [13]).

2.3   Verifying the SOM/KM Groups via GAMLSS Modelling

GAMLSS [16] methods were used to model the RDs' dynamic profiles of attained sleep (sleep duration) over time, in part to verify the multivariate MTD-SOM/KM groupings or SOM VANTED groups and also importantly to test for important differential effects of various predictors of sleep (break onset time, break duration, sleep onset and next duty onset time) across the MTD-SOM/KM groupings. Significant predictor by group interactions thus add insight as to what drives specific groups of RDs to sleep and wake as they do. The benefits of GAMLSS for time series analysis are that they can: identify the main drivers of the event of interest from a multiplicity of predictors, allow for possible non-linear impacts of the explanatory variables/predictors and detect statistically for change points and account for the auto-correlated nature of time series, by incorporating lag effects.

The GAMLSS framework of statistical modelling is implemented in a series of packages in R, which can be downloaded from the R library, CRAN, or from http://www.gamlss.com. The GAMLSS procedure used here [112] involved cubic spline smoothing functions, where 'spline' refers to a wide class of functions used in applications requiring data interpolation and/or smoothing. For each model tested it was assumed that the sleep duration series is represented by a Gaussian (normal) distribution. The RS algorithm, a generalisation of the algorithm of Rigby and Stasinopoulos [113, 114], was used to obtain the estimates of covariate (predictors of sleep hours or duration of sleep) as non-linear cubic spline terms.

3   MTDg Procedural Methods: Part A

The aim of Part A of this chapter is to study the multivariate relationship between the probability of sleep states (on/off), in relation to discrete states of the 6 predictors of shift, duty, next duty break and break duration along with characteristics of the driver's next scheduled duty. This is performed via a generalised multivariate generalisation of the mixture transition distribution (MTD) analysis, which allows for a different transition matrix for each lag (up to 2 sleep episodes backwards in time) to the present sleep episode, the so-called MTDg analysis (of Berchtold [115–117]). Following Hudson et al. [91, 93] and Kim [13] in this chapter we use our extended MTDg model to allow for interactions (between covariates) to account for changes in the transition matrices amongst the differing sleep lags. This work extends both the MARCH MTD software [116, 117] and generalises also the previous work of Kim et al. [90, 118] used to model historical flowering records of four eucalyptus species with respect to climate indicator time series. An early review of the original MTD is given by Berchtold and Raftery [94].

Our extended multivariate version of the MTD_g model [91, 93] as employed in this chapter accommodates interactions via the AD Model Builder^{TM} (ADMB) of Fournier [119] (see the recent review of Fournier et al. [120]). Our extended model is different to MARCH [116] given it is multivariate in nature, but also in the way it incorporates interactions between covariates and in its minimisation process, namely use of ADMB [119, 120]. ADMB employs auto-differentiation as a minimisation tool, which we showed to be computationally less intensive than MARCH [13, 90, 118]. Details of the mathematical formulation of our M-MTD_g are given in Appendix A.

The methodologies used in Part A, the multivariate-MTD_g (M-MTD_g) and the SOM analysis involve the following steps. Firstly an MTD_g analysis is performed to obtain the reparameterised time series profiles, thus creating new equal length time series per driver, which are of significantly reduced dimensions. Secondly a SOM/KM classification is adopted to cluster the M-MTD_g reparameterised time series profiles into groups with similar sleep duration profiles. Thirdly the groupings identified by the SOM/KM classification were used as a categorical predictor ('SOM/KM group' factor) in a subsequent analysis of the 190 time series, a record of RDs' daily sleep hours per sleep episode. The analytic method for modelling sleep duration was the generalised additive model for location scale and shape (GAMLSS) developed by Rigby and Stasinopoulos [16] and adapted to include interactions (see similar applications by Hudson et al. [97, 98, 121, 122]).

These GAMLSS results are given in Sect. 4. In part they aim to verify the significance of the four SOM/KM groupings established by our two-step M-MTD$_g$ then SOM/KM approach. More importantly the GAMLSS analysis is also able to identify significant differential effects of the following 6 predictors on attained sleep across groups—namely, break onset time, sleep onset and next duty onset time, break duration (hrs), hours since break onset and hours to next duty. Lagged differences of the current number of hours of sleep were added into GAMLSS as autoregressive (AR) lags. Cubic spline effects are denoted as cs($\cdot$). Railway driver effects were modelled as a random 'driver' effect in GAMLSS [13, 16].

3.1   The MTD$_g$ Data

3.1.1   The Data: Sleep, Wake and Duty Profiles of Australian Railway Drivers

Sleep/wake and duty and off-duty records are shown visually in Figs. 4 and 5 (see also Fig. 6). Using these raw records, the following six variables were derived:

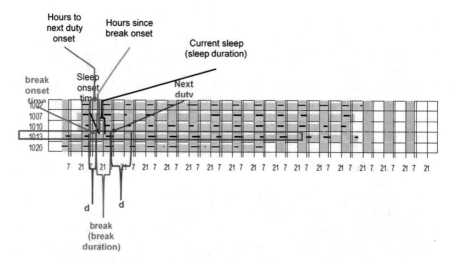

Fig. 4 Schematic of sleep/wake and break/duty data for drivers with ID 1002, 1007, 1010, 1013 and 1020. Horizontal axis indicates time on the 24 h clock (7: 7 am, 21: 9 pm). Each driver's duty schedule is colour coded: pink for duty; blue for off duty; grey for night (9 pm–7 am), white for no data. '—' indicates that the Railway driver is in the sleep state (see Kim [13])

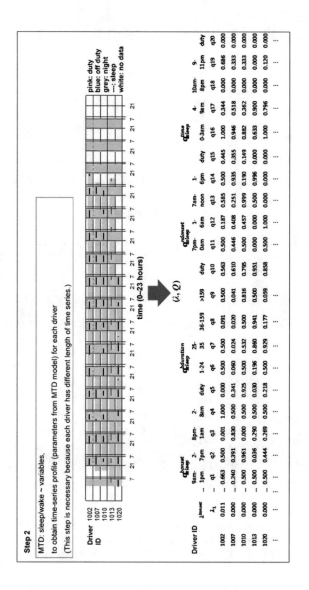

Fig. 5 Diagram of the M- MTD$_g$ (λ, Q) derived from the data (see Kim [13])

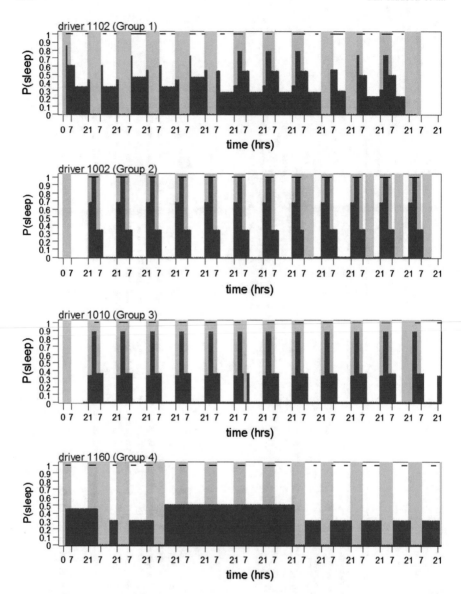

Fig. 6 Estimated *P* (*sleep*) obtained via Model C (*blue bars*) with observed sleep/wake states (*black lines*). *Pink shades* indicate duty and *grey shades* indicate night. A 7 on the x-axis indicates 7 am, 21 indicates 9 pm, and 0 indicates midnight (refer to Kim [13])

- Break onset time: the time (on a 24 h clock) when a driver starts his or her break (i.e. time at the first hour off-duty). If a driver is off-duty, the time when the off-duty episode began is their break onset time.
- Break duration: total number of hours of an off-duty period if a driver is off-duty.

- HrsSinceBreakOnset: number of hours since a break started, i.e. the number of hours since break onset when a driver goes to sleep
- HrsToNextDuty: number of hours left till the next duty begins, i.e. the number of hours left till the next duty.
- Sleep onset time: the time (on a 24-h clock) when a driver goes to sleep.
- Sleep duration: the duration (in hours) of a sleep period or episode.

The six variables are illustrated for five RDs, with unique ID's as given in Table 1 (see also Figs. 5 and 6). Note that driver 1002 (driver ID = 1002) had 14 sleep episodes recorded, the first two sleeps were of duration 8 and 9 h, which occurred at 10 pm ('sleephr' = 22 h) and 9 pm (21 h), respectively. These two sleep durations happened during driver 1002's break or off-duty period, both of which lasted 177 h and commenced at 10 pm (22 h). These sleep episodes occurred 8 and 32 h after driver 1002 commenced their break (hrs since break onset). This driver anticipated the next duty to start at 7 am, with 169 and 145 h to the next anticipated duty (at each of the first two sleep episodes). Note that all drivers can have a variable number of follow up days. For example, driver 1007 had 10 sleep episodes recorded, and their last two sleep episodes (9 and 10) were relatively short in duration (6 and 7 h). Driver 1007 commenced duty in the early morning, for break periods of a relatively short duration (9 and 22 h), and with sleep periods that occurred quite close to the next duty onset (15 and 4 h ahead). In total we analysed 190 railway drivers' sleep/wake/duty records which comprised 2,814 sleeps (4–23 sleeps for each driver) with a follow-up duration ranging from 1 to 35 h.

Table 1 Format of the data per railway driver (RD)

Driver	Sleep number (ID)[a]	Duration of sleep[b]	Sleep onset time[c]	Break onset time	Break duration	Hrs since break onset	Next duty onset time	Hrs to next duty onset
1002	1	8	22	22	177	8	7	169
1002	2	9	21	22	177	32	7	145
...	...	...	...	...	...	...	...	...
1002	13	7	23	16	31	14	23	17
1002	14	2	19	16	31	29	23	2
1007	1	10	23	16	57	17	1	40
1007	2	8	23	16	57	39	1	18
...	...	...	...	...	...	...	...	...
1007	9	6	1	22	24	9	22	15
1007	10	7	0	9	26	22	11	4
...	...	...	...	...	...	...	...	...
1020	1	7	0	0	64	7	16	57
...	...	...	...	...	...	...	...	...

[a]Sleep number denotes the number of sleep episodes for each driver
[b]Duration of sleep (hrs) per sleep episode for a given sleep period
[c]Sleep onset time is on the 24-h clock and defines the start of each sleep episode

3.1.2 Cut-Points for Categorisation of the Predictor Variables

The variables in the sleep series such as sleep onset time, break onset time, break duration, next duty onset time, number of hours since break onset and number of hours to next duty onset time, and time were categorised for the implementation of the MTD analysis. The categories for each predictor variable, as summarised and interpreted in terms of sleep research 'speak' are given in Table 2. Such

Table 2 Interpretation of the four categories of the 6 predictors of sleep behaviour gleaned by the GAMLSS analysis

Variable name	Category	Category description
sleepOnset	9 pm–11 pm	Sleep opportunity at night—morning types
	0 am–3 am	Sleep opportunity at night—evening types
	4 am–9 am	Sleep opportunity in the early morning
	10 am–8 pm	Sleep opportunity in the afternoon and daytime nappers
breakOnset	9 am–1 pm	Break after the night shift
	2 pm–7 pm	Break after the morning shift
	8 pm–1 am	Break after the afternoon shift
	2 am–8 am	Break after the evening shift
breakDuration	1–24 h	Break between consecutive shifts of the same type
	25–35 h	Break between consecutive shifts of a different type
	36–159 h	Break between non-consecutive shifts
	>159 h	Leave (not a normal rest period)
NextDutyOnset	7 pm–midnight	Start night shift (daytime sleep beforehand)
	1 am–6 am	Start morning shift (short night sleep before)
	7 am–noon	Start morning shift (normal night sleep before)
	1 pm–6 pm	Start afternoon shift (late night sleep before)
HrsSinceBreakOnset	1–15 h	Sleep opportunity within a day since break onset
	16–35 h	Sleep opportunity in 1-1.5 days since break onset
	36–71 h	Sleep opportunity during a longer break
	>71 h	Sleep opportunity during leave from work
HrsToNextDuty	1–8 h	Sleep opportunity of a sleep less than 8 h
	9–16 h	Sleep opportunity of a sleep longer than 8 h
	17–85 h	Sleep opportunity during a break longer than half a day
	>85 h	Sleep opportunity during leave

interpretations (in the category description column of Table 2) were obtained after consultation with sleep research experts (authors Dawson, Darwent and Roach).

3.2   MTDg Models: Analysis of the Dynamic Sleep and Wake States

In Kim [13] 10 MTD models were presented whereby each driver's dynamic discretised sleep and wake (sleep = 'off') states were analysed with respect to subsets of six predictors, break onset time ('bonset'), break duration ('bduration'), next duty onset time ('ndonset'), hours since break onset ('HSBO'), hours to next duty onset ('HTND') and time of the day ('time'), which were categorised into four levels (Table 2). Note that an extra 5th category, 'duty', was required for each predictor, so that a wake episode, when a driver is 'on duty', could be distinguished from waking episodes which interrupt a period when a driver is off-duty. Also 'sleephr' was categorised as a sleep/wake indicator (0 for wake, 1 for sleep). The notion of 'time' needed also to be investigated as a possible candidate predictor of sleep propensity or opportunity (Table 2).

The ten MTD models of Kim [13] were run and tested on each of the RDs' time series, e.g. 190 runs for each model, i.e. 1900 runs in all for 10 models. Firstly four main effects models were analysed, these are denoted by Model A, B, C and D as follows;

- Model A: 'bonset' + 'bduration' + 'ndonset' + 'HSBO'
- Model B: 'bonset' + 'bduration' + 'ndonset' + 'HTND'
- Model C: 'bonset' + 'bduration' + 'ndonset' + 'time'
- Model D: 'bonset' + 'bduration' + 'ndonset'.

Model C was found to be optimal for 173 of the 190 drivers [13], so subsequently a two-way interaction between specific predictor variables was added to Model C to test a further six interaction models.

Ten MTD models in total were compared, as in Kim [13], and their model specification is given below. These six interaction models based on Model C denoted by Model CIj ($j = 1, ..., 6$) were:

- Model CI1: Break onset time ('bonset') and break duration ('bduration')
- Model CI2: Break onset time ('bonset') and next duty onset time ('ndonset')
- Model CI3: Break onset time ('bonset') and time of the day ('time')
- Model CI4: Break duration ('bduration') and next duty onset time ('ndonset')
- Model CI5: Break duration ('bduration') and time of the day ('time')
- Model CI6: Next duty onset time ('ndonset') and time of the day ('time').

Table 4 shows the estimated values of the (λ, Q) parameter vectors, for a selection of five drivers (with IDs 1002, 1007, 1010, 1013 and 1020). See also Fig. 6. Each RD's sleep/wake time series is analysed by MTD model C, CI3 and CI6. Figure 6 shows the MTD_g predicted sleep probabilities, $P(sleep)$, as blue bars, with the driver's observed sleep/wake states shown as black lines, for four drivers (each from group j, $j = 1, \ldots, 4$), found subsequently by the cluster analysis of each RD's MTD_g (λ, Q) vector profiles (see Fig. 6, Tables 3 and 4).

Table 3 MTD_g parameters (λ, Q)

Model	Model specification
Main effects models	
C	$P(sleep) = \lambda^{bonset} q_{sleep}^{bonset} + \lambda^{bduration} q_{sleep}^{bduration} + \lambda^{ndonset} q_{sleep}^{ndonset} + \lambda^{time} q_{sleep}^{time}$
Interaction models (CIj, $j = 1, 2, \ldots, 6$)	
CI3	$P(sleep) = \lambda^{bonset} q_{sleep}^{bonset} + \lambda^{bduration} q_{sleep}^{bduration} + \lambda^{ndonset} q_{sleep}^{ndonset} + \lambda^{time} q_{sleep}^{time}$ $+ \lambda^{bonset*time} q_{sleep}^{bonset*time}$
CI6	$P(sleep) = \lambda^{bonset} q_{sleep}^{bonset} + \lambda^{bduration} q_{sleep}^{bduration} + \lambda^{ndonset} q_{sleep}^{ndonset} + \lambda^{time} q_{sleep}^{time}$ $+ \lambda^{ndonset*time} q_{sleep}^{ndonset*time}$
where the MTD_g weights are related to the following main and interaction effects,	
λ^{bonset} break onset time variable	
$\lambda^{bduration}$ break duration variable	
$\lambda^{ndonset}$ next duty onset time variable	
λ^{HSBO} hours break onset variable	
λ^{HTND} hours to next duty onset variable	
λ^{time} time of day variable	
$\lambda^{bonset*time}$ interaction between break onset time and time of day	
$\lambda^{ndonset*time}$ interaction between next duty onset time and time of day	
and the relevant vector of MTD_g transition probabilities (TPs) are:	
q_{sleep}^{bonset} transition probabilities from break onset time categories to sleep	
$q_{sleep}^{bduration}$ transition probabilities from the break duration categories to sleep	
$q_{sleep}^{ndonset}$ transition probabilities from the next duty onset time categories to sleep	
q_{sleep}^{HSBO} transition probabilities from the hours since the break onset categories to sleep	
q_{sleep}^{HTND} transition probabilities from the hours to the next duty onset categories to sleep	
q_{sleep}^{time} transition probabilities from the time of a day categories to sleep	
$q_{sleep}^{bonset*time}$ transition probabilities from the interaction between break onset time and time of a day	
$q_{sleep}^{ndonset*time}$ transition probabilities from the interaction between next duty onset and time of a day	

Table 4 Estimated (λ, Q) parameters from the MTD_g for a selection of RD's: Model C [13]

Driver ID	λ^{bonset}	$\lambda^{bduration}$	$\lambda^{ndonset}$	λ^{time}	q_1 break onset	...	q_6 break duration	...	q_{11} next duty onset	...	q_{16}	...
1002	0.011	0.000	0.000	0.989	0.663	...	0.500	...	0.500	...	1.000	...
1007	0.000	0.000	0.000	1.000	0.240	...	0.060	...	0.446	...	0.946	...
1010	0.000	0.000	0.000	1.000	0.500	...	0.500	...	0.500	...	0.882	...
1013	0.000	0.000	0.039	0.961	0.500	...	0.196	...	0.000	...	0.633	...
1020	0.000	0.000	0.051	0.949	0.500	...	0.500	...	0.500	...	1.000	...
1022	0.000	0.000	0.000	1.000	0.500	...	0.500	...	0.500	...	1.000	...
1029	0.000	0.000	0.091	0.909	0.991	...	0.071	...	1.000	...	1.000	...
1033	0.000	0.143	0.000	0.857	0.721	...	0.000	...	0.586	...	0.657	...
...	...	...	...	...	...	...	...	...	...	...	...	...

4 Results and Interpretation: Part A

4.1 SOM/KM of M-MTD and the GAMLSS Models

In this section we focus on two variants of MTD Model C, namely Model CI3 and Model CI6 (see Sects. 4.1.1–4.1.5) in our reporting of the results of the analysis of SOM/KM clustering of the multivariate RD-specific MTD parameters (λ, Q).

4.1.1 Summary of the CI6 Model

Figure 7 shows the U matrices of the SOM/KM clustering of the CI6 model (λ, Q) data, where model CI6 is {bonset + bduration + ndonset + time + ndonset*time}. SOM/KM found 4 groups of size 29, 59, 45 and 47 which have significantly different sleep duration, where the group effect is denoted by SOMKMj (j = 1, 2, 4) in the GAMLSS model (Table 5). Group 3 is the baseline contrast. Table 5 and Fig. 8 show that group 2 drivers have the least sleep and group 3–4 the most sleep. Table 5 shows that sleep patterns differ significantly between group 1 and 3 (p < 0.003) and between group 2 and 3 (p < 0.02). Current sleep is also highly positively related to attained sleep one episode prior (p < 0.00005) and at lag 2 (p < 0.05). The other significant main effects are break onset (p < 0.0003) and next duty onset time (p < 0.005). Sleep onset and break duration are not significant main

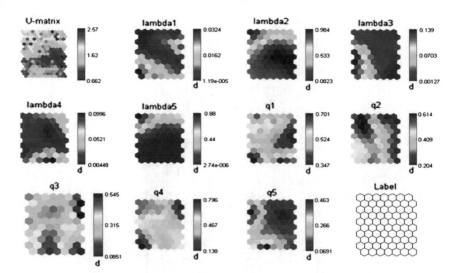

Fig. 7 SOM/KM clustering and U matrices for Model CI6 (λ, Q) data. From *top left* hand corner: U-matrix, $\lambda_1, \lambda_2, \lambda_3$; middle row λ_4, λ_5 and q_1, q_2 and bottom row q_3, q_4 and q_5. The lambda weights are for break onset time, break duration (hrs), next duty onset and sleep onset time and time of duty/shift

Table 5 GAMLSS model for Model CI6

	Estimate	SE	t value	Pr(> |t|)
(Intercept)	4.385	0.34	12.777	0.0000
Group differences				
SOMKM1	−1.51	0.5	−3.0541	0.002
SOMKM2	−1.32	0.52	−2.5263	0.012
SOMKM4	−0.767	0.48	−1.5944	NS†
Lag effects				
cs(lag1)	0.138	0.033	4.2174	0.00003
cs(lag2)	0.066	0.033	2.0069	0.045
Main effects				
cs(sleepOnset)	−8.30E-04	0.009	−0.0899	NS
cs(breakOnset)	4.85E-02	0.011	4.2214	0.00025
cs(breakDuration)	3.07E-03	0.003	1.2259	NS
cs(NextDutyOnset)	3.35E-02	0.012	2.8289	0.0047
Group by sleep onset interaction				
SOMKM1:cs(sleepOnset)	3.94E-02	0.013	3.0042	0.0027
SOMKM2:cs(sleepOnset)	4.23E-02	0.012	3.5060	0.00046
SOMKM4:cs(sleepOnset)	2.91E-02	0.013	2.2886	0.02
Group by break onset interaction				
SOMKM1:cs(breakOnset)	−1.12E-02	0.017	−0.6374	NS
SOMKM2:cs(breakOnset)	−1.47E-02	0.016	−0.8967	NS
SOMKM4:cs(breakOnset)	−8.67E-03	0.017	−0.5217	NS
Group by break duration interaction				
SOMKM1:cs(breakDuration)	−1.28E-03	0.003	−0.426	NS
SOMKM2:cs(breakDuration)	5.75E-05	0.003	0.0169	NS
SOMKM4:cs(breakDuration)	4.24E-03	0.003	1.5421	NS
Group by next duty onset interaction				
SOMKM1:cs(NextDutyOnset)	1.99E-02	0.019	1.0618	NS
SOMKM2:cs(NextDutyOnset)	3.80E-02	0.02	1.937	0.05
SOMKM4:cs(NextDutyOnset)	1.52E-02	0.018	0.8364	NS

[a]NS denotes not significant at the 5 % level of significance

effects. All significant factors are shown to have *non-linear* effects on current sleep. GAMLSS found significant differential effects on attained sleep of the following factors by group: sleep onset time by group (group 1 vs 3, $p < 0.003$; group 2 vs 3, $p < 0.0005$; group 4 vs 3; $p < 0.03$) and of next duty onset time by group (group 2 vs 3, $p < 0.005$). The impact of break onset time and break duration was similar across the groups. Interactions can be interpreted from Fig. 8 (LHS). Figure 8 shows the predicted sleep hours using the GAMLSS model for CI6 (left) and model CI3 (right) allowing interaction effects of group by sleep onset time, by break onset time, by break duration and by next duty onset time (group is colour-coded).

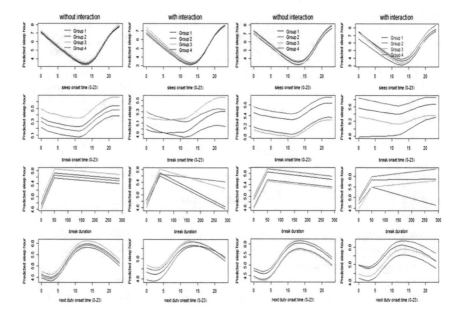

Fig. 8 Predicted sleep *hours* using the GAMLSS model for CI6 (*left*) and model CI3 (*right*) allowing interaction effects of group by sleep onset time, by break onset time, break duration and by next duty onset time (group is colour-coded)

4.1.2 Summary of the CI3 Model

For model CI3, namely {bonset + bduration + ndonset + time + bonset*time}, the SOM/KM clustering of the (λ, Q) data of the CI3 model also found 4 groups of size 43, 50, 49 and 48 which have significantly different sleep durations. Group 3 is the baseline contrast in GAMLSS modelling. Group 2 drivers have the least sleep and group 4 the most sleep. Sleep patterns differ significantly between group 1 and 3 ($p < 0.005$) only. Current sleep is also highly positively related to attained sleep one episode prior ($p < 0.00001$) and at lag 2 ($p < 0.00001$). The other significant main effects are break onset ($p < 0.000001$), break duration ($p < 0.000001$) and next duty onset time ($p < 0.000001$). All significant factors are shown to have *non-linear* effects on current sleep. Sleep onset time is not a significant main effect. GAMLSS found significant differential effects on attained sleep of the following factors by group: sleep onset time by group (group 2 vs 3, $p < 0.02$); and of break duration by group (group 2 vs 3, $p < 0.03$). The impact of break onset and next duty onset time was similar across the groups. Figure 8 shows the predicted sleep hours using the GAMLSS model for CI3 (right). Note that RD membership in CI3 and CI6 groups are not significantly related (LR = 6.45, $p > 0.6$).

4.1.3   Interpreting the Interactions in the GAMLSS Models

Adequate sleep (5–6 h) is reported only by RDs with breaks between 6 pm and 4 am except for group 2 with minimum sleep (CI3 model) and for group 1 the cluster with the next lowest attained sleep hours. Next duty onset time and break onset time are significant factors in determining current sleep, for both the CI3 and CI6 models. When break onset is between 7 am and 8 pm sleep increases similarly across groups for both models. Sleep onset time has a different effect across groups for both models, with differences particularly evident between the groups with maximum versus minimum sleep.

Next duty onset time and break duration also operate differently on current sleep across the RD groups, for models CI6 and CI3, respectively. Specifically for model CI6, group 3 RDs, with maximum sleep, rapidly increase sleep hours (duration) as their next duty onset occurs later than 12 noon; whereas group 2 RDs, with minimum sleep, have reduced sleep hours when their next duty onset occurs between midnight and 5 am, or from 7 pm to midnight.

Generally RDs increase sleep for break duration of 1-2 days. For model CI3 break onset time and next duty onset time have a similar effect across the RD groups. Most sleep occurs when the diver's next duty onset occurs between midday and 4 pm and least sleep duration, when next duty onset is between midnight and 3 am. For model CI6 generally sleep increases for next duty onset (NDO) between 10 am and 4 pm for all groups, except for group 2 (RDs with minimum sleep) for whom most sleep occurs when the RD's NDO occurs around midday to 6 pm. Least sleep is obtained for group 2 when their NDO is between midnight to 3 am. Note in contrast for group 2 when NDO is between 3 am and 10 am sleep increases with increasing lateness of the driver's NDO between 10 am and 12 noon. Generally groups with maximum sleep, rapidly increase sleep as their next duty onset is later than 12 noon, and the group with minimum sleep, has reduced sleep hours when next duty onset occurs between midnight and 6 am, or from 4 pm to midnight.

Generally RDs increase sleep when their break duration lasts 1–2 days. For the CI6 model as break duration increases beyond 2 days RDs in group 2, with minimum sleep, and in group 1, with next lowest attained sleep, reduce their sleep hours, with increasing length of break duration. These RDs seem unable to maintain increased sleep for longer break duration periods. However sleep hours are maintained for groups 3 and 4 with the highest sleep durations (model CI6). Similarly for the CI3 model, group 2 RDs, with least sleep, steadily decrease sleep as their break duration increases past 2 days. In contrast RDs in groups 4 and 3 for the CI3 model increase sleep as break duration exceeds 2 days, and notably sleep in group 1 plateaus.

4.1.4   Conditions for Most and Least Sleep Based on (λ, Q)

Table 6 gives a detailed description of conditions for most and least sleep for the RDs in each group for the CI6 MTD$_g$ model, according to the statistically

Table 6 Most sleep and least sleep categories of the 4 MTD-SOM groups for Model CI6's (λ, Q) MTDg parameterisation

	Group 1	Group 2	Group 3	Group 4
Next duty onset				
λ	**0.046**	0.007	**0.096**	**0.15**
Most sleep	7 pm–0 am	1 pm–6 pm	7 pm–0 am	7 pm–0 am
Least sleep	7 am–12 pm	7 pm–0 am	7 am–12 pm	1 am–6 am
Sleep onset time				
λ	**0.037**	**0.027**	0.009	**0.039**
Most sleep	0–3 am	0–3 am	4 am–9 am	4 am–9 am
Least sleep	10 am–8 pm	9 pm–11 pm	0 am–3 am	9 pm–11 pm
Break onset				
λ	0.006	0.007	**0.015**	0.008
Most sleep	2 am–8 am	9 am–1 pm	2 am–8 am	9 am–1 pm
Least sleep	3 pm–1 am	2 pm–7 pm	8 pm–1 am	8 pm–1 am
Break duration				
λ	**0.757**	**0.873**	**0.543**	**0.872**
Most sleep	1–24 h	1–24 h	1–24 h	1–24 h
Least sleep	36–159 h	36–159 h	36–159 h	36–159 h
*Next duty * sleep onset*				
λ	**0.046**	0.007	**0.096**	**0.015**
Most sleep	1–6 am * 9–11 am	7 pm–0 am * 9–11 am	1 am–6 am * 0–3 am	1 am–6 am * 0 am–3 am
Least sleep	7 am–12 pm * 10 am–8 pm	1 am–6 am * 10 am–8 pm	7 am–12 pm * 10 am–8 pm	9 am–12 pm * 10 am–8 pm

significant main and interaction effects based purely on the estimated (λ, Q) MTD$_g$ parameters. This differs to the GAMLSS models interpreted in Sect. 4.1.3.

Break onset time significantly impacts group 1 RDs' sleep duration ($\lambda = 0.012$). Both group 1 and 3 RDs sleep the most when break onset is between 2 am and 8 am i.e. the drivers break after the evening shift. In contrast groups 2 and 4 sleep the most when their break onset is between 9 am and 1 pm, i.e. the drivers break after the night shift. Groups 1, 3 and 4 sleep the least when break onset is between 8 pm and 1 am i.e. break after the afternoon shift. In contrast group 2 drivers with minimum sleep have least sleep when break onset is between 2 pm and 7 pm, i.e. break after the morning shift (Table 6).

Sleep is significantly impacted by **break duration** for all four groups of drivers. All RDs sleep most for break durations of 1–24 h i.e. break between consecutive shifts. Least sleep is obtained for all groups when their break occurs between non-consecutive shifts (i.e. break duration is between 36 and 159 h) (Table 6). Sleep attainment is significantly impacted by **sleep onset time**, particularly for group 4 and 1, with maximum and minimum sleep, respectively. Group 4 and 3 sleep most between 4 am and 9 am, in contrast group 1 and 2 RDs, who sleep most between 0 am and 3 am. Least sleep is attained between 9 pm and 11 pm for groups 2 and 4, and between 10 am and 8 pm for group 1 (with minimal sleep) and between 0 am and 3 am in group 3 (the same period 0 am–3 am, when group 1 and 2 sleep the most) (Table 6).

Sleep is significantly impacted by **next duty onset (NDO)** time for all groups, except for group 2. Most sleep is gained by RDs in groups 1, 3 and 4 when NDO is between 7 pm–midnight (i.e. drivers start night shift after a daytime sleep beforehand). In contrast group 2 (the group with least sleep) gains most sleep when NDO is between 1 pm and 6 pm (i.e. RDs next duty is the afternoon shift with a late night sleep beforehand) (Table 6). Groups 1 and 3 sleep the least for NDO between 7 am–noon (start morning shift, with a normal night sleep prior). Whereas group 4 is sleep deficit when NDO is between 1 am and 6 am (i.e. start morning shift with a short night sleep before). In contrast RDs in group 2, with least sleep across groups, are sleep deficit when their NDO is between 7 pm–midnight (i.e. drivers start night shift with daytime sleep beforehand) (Table 6).

NDO interacts with sleep onset time for all groups, especially for groups 3 and 1. For groups 3 and 4 (with higher sleep duration) most sleep is gained when RDs sleep between 0 and 3 am (sleep opportunity at night) prior to their next duty being an early morning shift (1 am–6 am, with a short night sleep prior). For this case NDO* time in Table 6 is given by (1 am–6 am)*(0 am–3 am). Group 1 RDs (with low sleep levels) gain the most sleep between 9 am and 11 am when their next duty onset time is the early morning shift (after a short sleep the night before). For group 2 (with the lowest sleep duration across groups) most sleep occurs also between 9 am and 11 am and when the RDs next duty is scheduled between 7 pm to midnight (i.e. the RD is about to start night shift, with a daytime sleep beforehand) (Table 6).

Least sleep is gained for all groups for daytime sleep between 10 am and 8 pm, but sleep is impacted significantly by the onset time of the RD's next duty. Specifically groups 1 and 3 are sleep deficit when they have daytime sleep (10 am–8 pm) (i.e. the daytime nappers) and their next anticipated duty is the morning shift (7 am–noon). Like groups 1 and 3, groups 2 and 4 have least sleep from 10 am–8 pm (daytime napping), when their next duty is the very early morning shift (1 am–6 am) for group 2; or is the later start morning shift between (7 am–noon) for group 4 (Table 6).

4.1.5 Socio-Demographics of the 4 Groups of Railway Drivers

Table 7 gives a summary of the domestic scenario and RD-specific experience for the groups based on the CI6 model data. Model CI6 groups are shown to differ according to age of the RD, years of shiftwork and the average number of dependents in the domestic situation (P = 0.05, MANOVA test). From Table 6 we note that group 2 (G2) RDs, with minimum sleep duration, had the maximum number of dependents (mean = 2.46) and were drivers of lowest mean age (39 years). Group 3 (G3) RDs with a high sleep duration had the greatest years of experience (mean = 21.9 years) and were the oldest (mean age = 40.7 years).

Table 7 Socio-demographic data per group: Model CI6 and CI3 (missing data not shown)

Model	Model CI6				Model CI3				Total
Groups	G1	G2	G3	G4	G1	G2	G3	G4	Total
Domestic situation									
Married/partner	32	41	33	31	35	36	31	35	137
Separated/divorced	0	3	5	1	1	1	5	2	9
Single	4	8	3	7	3	9	5	5	22
Widowed	0	0	0	1	1	0	0	0	1
Partner									
Yes	32 (82 %)	41 (70 %)	33 (73 %)	31 (66 %)	35 (81 %)	36 (72 %)	31 (63 %)	35 (73 %)	137
Young kids									
Yes	11 (28 %)	21 (36 %)	20 (44 %)	16 (34 %)	25 (28 %)	15 (36 %)	12 (44 %)	16 (34 %)	68
Means									Overall mean
Age (yrs)	39.2	39.0	40.7	40.4	39.0	40.7	38.5	41.6	40.0
No. of dependents	2.08	2.46	2.13	2.35	2.3	2.4	2.2	2.2	2.3
Shiftwork (yrs)	19.0	18.8	21.9	20.6	20.0	19.7	18.8	20.8	19.8
N	39	59	45	47	43	50	49	48	190

Model CI3 demographics given in Table 6 show that the groups differ according to the number of RDs with young kids in their domestic situation (LR statistic = 11.4, p < 0.02); with group 2 and 3 (with lowest and second highest mean attained sleep) having low numbers of young kids. However, from Table 6 group 2 (G2) RDs, with minimum sleep duration, had the maximum number of dependents (mean = 2.4) and the lowest years of driver experience (19.7 years). Group 4 (G4) RDs with maximum sleep duration had the highest years of shiftwork (mean = 20.8 years) and were the oldest RDs (mean age = 41.6 years). It seems that higher mean age and driver experience tends to be associated with more sleep and accommodation of shifts and breaks; and lowest sleep is related with increased number of dependents, lower years of experience and younger age.

5 Results and Interpretation: Part B

5.1 SOM/VANTED of the Multivariate Sleep Duration Series

SOM VANTED clustering was based on a subset of 69 of the 190 RDs used for the earlier M-MTD analysis (Part A, Sect. 4). These 69 RDs had a full record of 14 consecutive day work/shift/break activity. GAMLSS modelling of the sleep duration in this section follows some results recently presented by Hudson et al. [96], wherein two additional predictors for the GAMLSS modelling are included, compared to the analysis in Part A. These additional predictors are termed HrsSinceBreakOnset, which denotes the number of hours since a break started, i.e. the number of hours since break onset when a driver goes to sleep, and HrsToNextDuty, which denotes the number of hours left till the next duty begins. These 2 predictors are modelled along with break onset time, break duration and sleep onset and next duty onset time (as in Part A, see Table 5).

5.2 SOM/VANTED Clusters and GAMLSS Modelling Effects

SOM VANTED clustering found 4 clusters/groups 1–4 of size 18, 13, 12 and 26 (in that order) across which sleep patterns were significantly different (Fig. 9). RDs in cluster 1 (n = 18) had minimum sleep hours per episode (average = 6.96 h), cluster 2 RDs (n = 13) gained maximum sleep (average = 7.71 h), cluster 3 (n = 12) and 4 (n = 26) RDs average hours attained sleep is 7.44 and 7.35 h, respectively. Cluster 4 was the baseline contrast in the GAMLSS (Table 5).

For model M1 current sleep was highly positively related to attained sleep one episode prior (P < 0.001), but not at lag 2. The highly significant main effects of sleep onset time (P < 0.000003), break onset time (P < 0.03) and hours to next duty

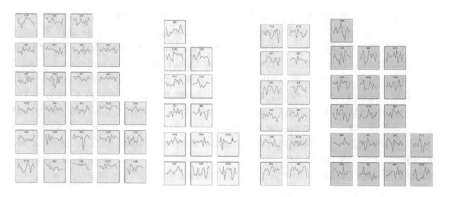

Fig. 9 SOM VANTED clusters of the sleep time series: clusters 4, 2, 3, and 1 from *left* to *right*, of sizes 26, 13, 12 and 18 (in that order)

onset (P < 0.0005) were shown to be the same across groups. Sleep was highly significantly related to next duty onset time (P < 0.000006), but hours since break onset was not a significant main effect (Fig. 10). GAMLSS found significant differential effects on attained sleep of the following factors by group: next duty onset time (group 4 vs 1; P < 0.005) and hours since break onset (group 4 vs 1; P < 0.08, significant at 10 %) (Fig. 11).

When hours since break onset was included (as in Model M2), it was shown to be an additional significant main effect (P < 0.05) and break duration also had a significant impact on attained sleep. This was not the case for model M1, where break onset time was a significant main effect (not break duration), as was hours to next duty (model M1). Stepwise variants of M1 and M2 found cluster/group as a significant main effect, with group 1 RDs attaining the least sleep across groups (detailed results not shown here).

From the SOM VANTED and GAMLSS modelling break and sleep onset times, break duration and hours to next duty are significant effects which operate similarly

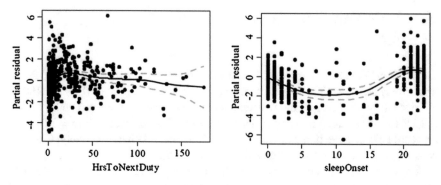

Fig. 10 GAMLSS term plots of effect of hours to next duty (P = 0.005) and sleep onset time (P = 0.000003)

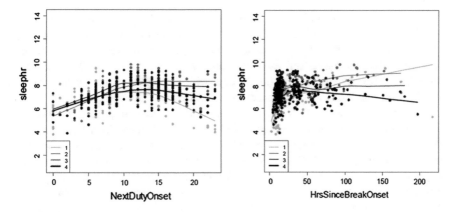

Fig. 11 Interaction plots for next duty onset time (P = 0.000006) and hours since break onset (P = 0.08)

across the groups. Next duty onset time and hours since break onset were found to be significant factors in determining current sleep, which have differential impacts on current sleep across the 4 groups. Group 2 RDs, with maximum sleep, rapidly increase sleep as their next duty onset is later than 12 noon, group 1, with minimum sleep, have reduced sleep hours when next duty onset occurs between midnight and 6 am, or from 4 pm to midnight. Group 2 and 1 RDs increase sleep as their hours since work/duty break exceeds 1 and 2 days, respectively. Group 3 and 4 RDs' sleep duration decreased when hours since break onset exceeds 1 day. Generally RDs increase sleep for break duration from 1–2 days.

6 Conclusion

The ANNs found 4 clusters of railway drivers. GAMLSS confirmed that both the timing of sleep, break and next duty, duration of break, and hours to next duty can significantly influence sleep. Break and sleep onset times, break duration and hours to next duty are significant factors which can operate differentially across the groups. Although RDs have different sleep patterns, the amount of sleep is generally higher at night and when break duration is 1–2 days or more. Sleep increases for next duty onset between 10 am and 4 pm, and when hours since break onset exceeds 1 day. These 2 factors were significant factors determining current sleep, which also had differential impact across the clusters. Some drivers catch up sleep after the night shift, while others do so before the night shift. Sleep is governed by the RD's anticipatory behaviour of next scheduled duty onset and hours since break onset.

Sleep onset times, break duration and hours to next duty are significant predictors which operate similarly across the groups. Adequate sleep (5–6 h) is reported only by RDs with breaks between 6 pm and 4 am. Next duty onset time and hours since break onset are found to be significant factors in determining current sleep, which have differential impacts on current sleep across groups. Generally RDs increase sleep for break duration from 1 to 2 days. Most drivers' sleep patterns were affected by time of day, with sleep patterns influenced also by break onset time and next duty onset time and by break duration.

The extent to which factors other than working time might affect the sleep behaviour of employees in the large and diverse Australian rail industry is not well documented. Sleep duration was greatest prior to night shifts, followed by afternoon shifts and morning shifts. Overall RDs with dependents got significantly less sleep than participants without dependents. Consistent with previous research, participants with dependents were found to obtain significantly less sleep than participants without dependents [123, 124]. It also seems that higher mean age and driver experience tends to be associated with more sleep and accommodation of shifts and breaks; and lowest sleep was related to increased number of dependents, lower years of experience and younger RD age. We have demonstrated that some predictors such as age of RD, years of shiftwork and number of dependents are significantly different between the identified clusters.

As far as the authors are aware Hudson et al. [96] was the first study to find that sleep patterns are governed by anticipatory behaviour in relation to *hours since break onset.* Earlier Kim [13] was the first study using M-MTD$_g$ to establish a significant anticipatory effect of *next duty onset time* on RD sleep patterns. Our results support this via SOM/KM clustering. We have some evidence from the SOM on the M-MTD$_g$ data and GAMLSS that social predictors such as age of RD, years of shiftwork and number of dependents can act differently between our identified RD clusters/groups. All significant factors are shown to have *non-linear* impacts on current sleep. A preliminary analysis of the dataset of 69 RDs, using multivariate Gaussian Hidden Markov Model (HMM) analysis [125], which likewise accommodates covariates and interactions supports the results of this chapter. Further investigations will involve adapting HMMs to semi-Markov HMMs allowing RD profiles with disparate lengths (<14 days and 14 days monitoring). All future study will also aim to establish the effects of consecutive shifts and to investigate the effects of two additional predictors, the number of hours to the RDs *next break* and the *duration of the next break.*

Appendix A: Mathematics of M-MTD$_g$

A.1 The Mixture Transition Distribution (MTD) Model

The classic Markov chain is a probabilistic model that represents dependences between successive observations of a random variable (usually over time). In this chapter a discrete state random variable (or multivariate analogues of) taking values in the finite set $\{1,\ldots, m\}$ is considered in the MTD formulation, which allows for a covariate interaction and modelling of high-order Markov chains (from a time series viewpoint). Markov chains are traditionally used to predict the current value as a function of the previous observations of this same variable (the so-called lagged dependency). The Markov chain was introduced by Andrej A. Markov [126] at the beginning of the twentieth century and has wide applicability in many areas such as: mathematical biology [127], internet applications [128], economics [129], meteorology [130], geography [131], biology [132], chemistry [133], physics [134], behavioural science [135], social sciences [136] and music [137]. For a comprehensive treatment of Markov chains and early applications see Bremaud [138]. Seneta [139] provides an account of Markov's motivations including an excellent discussion of the early development of the theory.

Raftery [140] introduced the mixture transition distribution (MTD) model to model high-order Markov chains. Berchtold [135, 141–149] subsequently developed software (called Markovian Models Computation and Analysis, MARCH) to model Markov chains using a suite of methods including the MTD and the double chain Markov model. The MTD model has been applied to genomic sequence and time series data [147, 150–153].

The aim of part A of this chapter is to study the multivariate relationship between the probability of sleep with 4-states each of sleep onset times, break onset times, next duty onset times, break duration, and time of sleep via a multivariate mixture transition distribution (M-MTD) which accommodates a different transition matrix from each lag to the present (MTD_g) analysis [154]. The issue of accommodating for interaction terms between covariates itself had not, till the work of Kim [155] and of Hudson et al. [156], been addressed in the MTD [140], MTD_g nor MARCH [149] literature. Kim et al. [154, 157] and Hudson et al. [158] first introduced the concept of interactions based on work in this current chapter, along with GAMLSS [159].

The idea of the original mixture transition distribution model was to consider independently the effect of each lag to the present instead of considering the effect of the combination of lags as in pure Markov chain processes. The assumption behind the MTD model, namely the assumed equality of the transition matrices among different lags, is a strong assumption. We further extend the MTD_g model to allow for interactions (between break, duration and next onset times with sleep times) to account for changes in the transition matrices amongst the differing covariates.

This work extends both the MARCH MTD software of Berchtold [149] and the previous work Hudson et al. [156] and of Kim et al. [160–162]. Our model is different to MARCH in terms of incorporating interactions between the covariates and also in its minimisation process [155]. It uses the AD Model Builder^{TM} [163], [164]. This M-MTD adaptation also utilises auto-differentiation as a minimisation tool, and was shown to be computationally less intensive than MARCH (see the papers of Kim et al. [161, 162] and Kim [155]). The AD Model Builder^{TM} platform (see Fournier et al. [164]) has great application in fisheries research and recently in computational mathematics and operations management (e.g. electronic systems models [165, 166]).

A.1.1 The MTD Model

Let $\{Y_i\}$ be a sequence of random variables taking values in the finite set $N = \{1,\ldots, m\}$. In a lth-order Markov chain, the probability that $i_l, \ldots, i_o \in N$ depends on the combination of values taken by $X_{t-l}, \ldots, X_{t-1}$. In the MTD model, the contributions of the different lags are combined additively, as follows:

$$
P(X_t = i_0 | X_0 = i_t, \ldots, X_{t-1} = i_1)
$$
$$
= \sum_{g=1}^{l} \lambda_g P(X_t = i_0 | X_{t-g} = i_g) = \sum_{g=1}^{l} \lambda_g q_{i_g i_0} \tag{A.1}
$$

where $i_l, \ldots, i_o \in N$, and where the probabilities $q_{i_g i_0}$ are elements of a $m \times m$ transition matrix $Q = [q_{i_g i_0}]$, each row of which is a probability distribution (i.e., each row sums to 1 and the elements are nonnegative) and $\lambda = (\lambda_l, \ldots, \lambda_1)'$ is a vector of lag parameters, such that

$$
0 \le \sum_{g=1}^{l} \lambda_g q_{i_g i_0} \le 1
$$

The vector λ is made subject to the following constraints, $\sum_{g=1}^{l} \lambda_g = 1$, and $\lambda_g \ge 0$. Equation (A.1) gives the probability for each individual combination of $i_l, \ldots, i_0$. The model can also be written in matrix form, giving the whole distribution of X_t [19].

Each row of the transition matrix Q is a probability distribution and as such sums to 1, where the matrix has $m(m-1)$ independent parameters. In addition, a lth-order model has l lag parameters $\lambda_1, \ldots, \lambda_l$, but only $(l-1)$ of them are independent. Thus a lth order MTD model has $m(m-1) + (l-1)$ independent parameters, which is far more parsimonious than the corresponding fully parameterised Markov chain which has $m^l(m-1)$ parameters. Moreover, each additional lag in a MTD model adds only one extra parameter. For the basic MTD model of Raftery [15], the same transition matrix Q is used to model the relationship between any of the lags and the present state.

A.1.2 The MTD$_g$ Model

Let $\{Y_i\}$ be a sequence of random variables taking values in the finite set $N = \{1,\ldots, m\}$. In an lth-order Markov chain, the probability that $X_t = i_0$, $i_0 \in N$, depends on the combination of values taken by $Y_{t-l}, \ldots, Y_{t-1}$. In the basic MTD model, the same transition matrix Q is used to model the relation between any of the lags and the present. The idea of the mixture transition distribution (MTD) model is to consider independently the effect of each lag to the present instead of considering the effect of

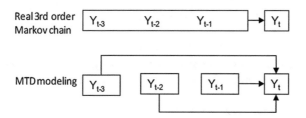

Fig. A.1 Comparison between a 3rd order Markov chain and its MTD model analogue. In a real high-order Markov chain, the combination of all lags influences the probability of the present

the combination of lags (Fig. A.1), as in the case of the more traditional pure Markov chain process. The constraints imposed by the use of only one transition matrix to represent the relation between each lag and the present is sometimes too strong to allow good modeling of the real high-order transition matrix. In this case, it is possible to replace the basic MTD model by an MTD_g model. The principle of the MTD_g model is to use a different transition matrix of size ($k \times k$) to represent the relationship between each lag and the present. The high-order transition probabilities are then written as follows,

$$P\left(Y_t = i_0 | Y_{t-1} = i_1, \ldots, Y_{t-f} = i_f\right) = \sum_{g=1}^{f} \lambda_g q_{g i_g i_0}$$

where $q_{g i_g i_0}$ is the transition probability from modality i_g observed at time $t\text{-}g$ and modality i_0 observed at time t in the transition matrix Q_g associated with the gth lag. In addition to the lag weight vector $[\lambda_1, \ldots, \lambda_f]$, the MTD_g model implies the estimation of f transition matrices $Q_1, \ldots, Q_f$, for a total of $f_k(k - 1) + (f - 1)$ independent parameters. This is much more than was involved in the basic MTD model, but this number of parameters remains small compared to the number of independent parameters of a real fully parameterised fth order Markov chain; thus the MTD_g and its extensions model prove useful in many situations, as shown in this chapter. Here, the contribution of each lag upon the present is considered independently. The MTD model thus approximates high-order Markov chains with far fewer parameters than the fully parameterised model. Though Markov chains are well suited to represent high-order dependencies between successive observations (of a random variable), as the order l of the chain and the number of possible values m increase, the number of independent parameters increases exponentially. The problem then becomes too large to be estimated efficiently, as is often the case for data sets of the size typically encountered in practice [145].

A.1.3 The MTD$_g$ Model with Interactions

The MTD$_g$ model with interactions can also have a different transition matrix of size $(k \times k)$ to represent the relationship between each lag and the present [155]. The high-order transition probabilities are then computed as follows

$$P\big(Y_t = i_0 | Y_{t-1} = i_1, \dots, Y_{t-f} = i_f, C_1 = c_1, \dots, C_e = c_e, M_1 = m_1, \dots, M_l = m_l\big)$$

$$= \sum_{g=1}^{f} \lambda_g q_{g i_g i_0} + \sum_{h=1}^{e} \lambda_{f+h} d_{h j_h i_0} + \sum_{u=1}^{l} \lambda_{f+e+u} s_{u v_u i_0}$$

where λ_{f+e+u} is the weight for the interaction term, $q_{g i_g i_0}$ is the transition probability from modality i_g observed at time $t-g$ and modality i_0 observed at time t in the transition matrix Q_g associated with the gth lag, $s_{u v_u i_0}$ is transition probability between covariate h_1 and covariate h_2 interaction term $(v_u = d_{h_1 j_{h_1}} \times d_{h_2 j_{h_2}})$ and Y_t, and where $\sum\limits_{g=1}^{f+e+l} \lambda_g = 1$ and $\lambda_g \geq 0$.

A.1.4 Parameter Estimation

The parameters λ and q of the MTD$_g$ model can be estimated by minimising the negative the log-likelihood (NLL) of the model:

$$NLL = - \sum_{i_1, \dots, i_0 = 1}^{m} n_{i_l, \dots, i_0} \log \left(\sum_{g=1}^{f} \lambda_g q_{g i_g i_0} + \sum_{h=1}^{e} \lambda_{f+h} d_{h j_h i_0} + \sum_{u=1}^{l} \lambda_{f+e+u} s_{u v_u i_0} \right)$$

where $n_{i_l, \dots, i_0}$ is the number of sequences of the form

$$Y_{t-1} = i_1, \dots, Y_{t-f} = i_f, C_1 = c_1, \dots, C_e = c_e, M_1 = m_1, \dots, M_l = m_l.$$

To ensure that the model defines a high order Markov chain, the negative log-likelihood is minimised with respect to the constraints delineated above. ADMB^{TM} was used to minimise the negative the log-likelihood (NLL). This uses auto-differentiation (AUTODIFF) [164] as the minimisation tool, shown to be computationally less intensive than MARCH [155]. Estimation algorithms relevant to this procedure can be found in Fournier [163, 167]. A major advantage of the new model is that its run-time is considerably shorter (less than one minute as compared with two days for the original MTD model [157]) and it can be run from a batch file in DOS. Hence, multiple models can be tested consecutively in remote mode. Outputs can also be appended into one file to allow easy access by any graphical software package [155].

References

1. D. Darwent et al., Managing fatigue: it really is about sleep. Accid. Anal. Prev. **82**, 20–26 (2015)
2. C. Bearman et al., *Evaluation of Rail Technology: A Practical Human Factors Guide* (Ashgate, United Kingdom, 2013)
3. D. Dawson, Fatigue research in 2011: from the bench to practice. Accid. Anal. Prev. **45** (Suppl), 1–5 (2012)
4. J. Dorrian et al., Work hours, workload, sleep and fatigue in Australian Rail Industry employees. Appl Ergon **42**, 202–209 (2011)
5. D.F. Dinges, An overview of sleepiness and accidents. J. Sleep Res. **4**, 4–14 (1995)
6. S. Folkard, P. Tucker, Shift work, safety and productivity. Occup Med (Lond) **53**, 95–101 (2003)
7. S. Folkard et al., Shiftwork: safety, sleepiness and sleep. Ind. Health **43**, 20–23 (2005)
8. A. Kosmadopoulos et al., The effects of a split sleep-wake schedule on neurobehavioural performance and predictions of performance under conditions of forced desynchrony. Chronobiol. Int. **31**, 1209–1217 (2014)
9. J.L. Paterson et al., Beyond working time: factors affecting sleep behaviour in rail safety workers. Accid. Anal. Prev. **45**(Suppl), 32–35 (2012)
10. M. A. Short, et al., A systematic review of the sleep, sleepiness, and performance implications of limited wake shift work schedules. Scandinavian J. Work Environ. Health (2015)
11. G.D. Roach et al., The amount of sleep obtained by locomotive engineers: effects of break duration and time of break onset. Occup. Environ. Med. **60**, e17 (2003)
12. D. Darwent et al., A model of shiftworker sleep/wake behaviour. Accid. Anal. Prev. **45** (Suppl), 6–10 (2012)
13. S.W. Kim, Bayesian and non-Bayesian mixture paradigms for clustering multivariate data: time series synchrony tests. PhD, University of South Australia, Adelaide, Australia (2011)
14. P. Gander et al., Fatigue risk management: organizational factors at the regulatory and industry/company level. Accid. Anal. Prev. **43**, 573–590 (2011)
15. F.M. Fischer et al., 21st International symposium on shiftwork and working time: the 24/7 society–from chronobiology to practical life. Chronobiol. Int. **31**, 1093–1099 (2014)
16. R.A. Rigby, D.M. Stasinopoulos, Generalized additive models for location, scale and shape (with discussion). Appl. Stat. **54**, 507–554 (2005)
17. A.A. Borbely, A two process model of sleep regulation. Hum Neurobiol **1**, 195–204 (1982)
18. S. Daan et al., Timing of human sleep: recovery process gated by a circadian pacemaker. Am. J. Physiol. **246**, R161–183 (1984)
19. G. Belenky et al., Patterns of performance degradation and restoration during sleep restriction and subsequent recovery: a sleep dose-response study. J. Sleep Res. **12**, 1–12 (2003)
20. H.P. Van Dongen et al., The cumulative cost of additional wakefulness: dose-response effects on neurobehavioral functions and sleep physiology from chronic sleep restriction and total sleep deprivation. Sleep **26**, 117–126 (2003)
21. A. A. Borbely, et al., Sleep initiation and initial sleep intensity: interactions of homeostatic and circadian mechanisms. J. Biol. Rhythms **4**, 149–160 (1989)
22. P. Achermann et al., A model of human sleep homeostasis based on EEG slow-wave activity: quantitative comparison of data and simulations. Brain Res. Bull. **31**, 97–113 (1993)
23. D.G. Beersma, P. Achermann, Changes of sleep EEG slow-wave activity in response to sleep manipulations: to what extent are they related to changes in REM sleep latency? J. Sleep Res. **4**, 23–29 (1995)
24. M.E. Jewett, R.E. Kronauer, Refinement of a limit cycle oscillator model of the effects of light on the human circadian pacemaker. J. Theor. Biol. **192**, 455–465 (1998)
25. R.E. Kronauer et al., Quantifying human circadian pacemaker response to brief, extended, and repeated light stimuli over the phototopic range. J. Biol. Rhythms **14**, 500–515 (1999)

26. P. McCauley et al., A new mathematical model for the homeostatic effects of sleep loss on neurobehavioral performance. J. Theor. Biol. **256**, 227–239 (2009)
27. T. Akerstedt, S. Folkard, Predicting duration of sleep from the three process model of regulation of alertness. Occup. Environ. Med. **53**, 136–141 (1996)
28. S.R. Hursh, et al., Fatigue models for applied research in warfighting. Aviat. Space Environ. Med. **75**, A44-53; discussion A54-60 (2004)
29. M. Moore-Ede et al., Circadian alertness simulator for fatigue risk assessment in transportation: application to reduce frequency and severity of truck accidents. Aviat. Space Environ. Med. **75**, A107–118 (2004)
30. K.J. Kandelaars, et al., A review of bio-mathematical fatigue models: where to from here?, in *2005 International Conference on Fatigue Management in Transport Operations* (2005), pp. 11–15
31. D. Darwent et al., Prediction of probabilistic sleep distributions following travel across multiple time zones. Sleep **33**, 185–195 (2010)
32. A.A. Borbely, A two process model of sleep regulation. Hum. Neurobiol. **1**, 195–204 (1982)
33. D. Darwent, The sleep of transportation workers in Australian rail and aviation operations. PhD, School of Education Arts and Social Sciences, University of South Australia, Adelaide (2006)
34. A.A. Borbely, et al., Sleep initiation and initial sleep intensity: interactions of homeostatic and circadian mechanisms. J. Biol. Rhythms. **4**, 37–48 (1989)
35. S. Daan et al., Timing of human sleep: recovery process gated by a circadian pacemaker. Am. J. Physiol.: Regul. Integr. Comp Physiol. **246**, 161–183 (1984)
36. N.J. Wesensten et al., Does sleep fragmentation impact recuperation?A review and reanalysis. J. Sleep Res. **8**, 237–245 (1999)
37. D.F. Dinges et al., Cumulative sleepiness, mood disturbance, and psychomotor vigilance performance decrements during a week of sleep restricted to 4–5 h per night. Sleep **20**, 267–277 (1997)
38. R.T. Wilkinson et al., Performance following a night of reduced sleep. Psychonomis Sci. **5**, 471–472 (1966)
39. M. Lumley et al., The alerting effects of naps in sleep-deprived subjects. Psychophysiology **23**, 403–408 (1986)
40. P. Achermann, A.A. Borbély, Mathematical models of sleep regulation. Front. Biosci.: J. Virtual Libr. **8**, s683–s693 (2003)
41. D.G. Beersma, Models of human sleep regulation. Sleep Med. Rev. **2**, 31–43 (1998)
42. D.J. Dijk, R.E. Kronauer, Commentary: models of sleep regulation: successes and continuing challenges. J. Biol. Rhythms **14**, 569–573 (1999)
43. M.M. Mallis et al., Summary of the key features of seven biomathematical models of human fatigue and performance. Aviat. Space Environ. Med. **75**, A4–A14 (2004)
44. M. Koslowsky, H. Babkoff, Meta-analysis of the relationship between total sleep deprivation and performance. Chronobiol. Int. **9**, 132–136 (1992)
45. M.H. Bonnet, D.L. Arand, Level of arousal and the ability to maintain wakefulness. J. Sleep Res. **8**, 247–254 (1999)
46. S.M. Doran et al., Sustained attention performance during sleep deprivation, evidence of state instability. Arch. Ital. Biol. **139**, 253–267 (2001)
47. D.A. Drummond et al., Why highly expressed proteins evolve slowly. Proc. Natl. Acad. Sci. USA **102**, 14338–14343 (2005)
48. X. Zhou et al., Sleep, wake and phase dependent changes in neurobehavioral function under forced desynchrony. Sleep **34**, 931–941 (2011)
49. D.A. Cohen, et al., Uncovering residual effects of chronic sleep loss on human performance. Sci. Transl. Med. **2**, 1413 (2010)
50. E.J. Silva et al., Circadian and wake-dependent influences on subjective sleepiness, cognitive throughput, and reaction time performance in older and young adults. Sleep **33**, 481–490 (2010)

51. I. Philibert, Sleep loss and performance in residents and nonphysicians; a meta-analytic examination. Sleep **28**, 1393–1402 (2005)
52. D.J. Kim et al., The effect of total sleep deprivation on cognitive functions in normal adult male subjects. Int. J. Neurosci. **109**, 127–137 (2001)
53. Y. Harrison, J.A. Horne, The impact of sleep deprivation on decision making: a review. J. Exp. Psychol. Appl. **6**, 236–249 (2000)
54. G. Belenky et al., Patterns of performance degradation and restoration during sleep restriction and subsequent recovery: a sleep dose-response study. J. Sleep Res. **12**, 1–12 (2003)
55. M. Gillberg, T. Akerstedt, Sleep restriction and SWS-suppression: effects on daytime alertness and night-time recovery. J. Sleep Res. **3**, 144–151 (1994)
56. L. Rosenthal et al., Level of sleepiness and total sleep time following various time in bed conditions. Sleep **16**, 226–232 (1993)
57. H.P.A. Van Dongen et al., The cumulative cost of additional wakefulness: dose-response effects on neurobehavioral functions and sleep physiology from chronic sleep restriction and total sleep deprivation. Sleep **26**, 117–126 (2003)
58. A. Hak, R. Kampmann, Working irregular hours: complaints and slate of fitness of railway personnel, ed. by A. Reinberg, et al. *Night and Shiftwork: Biological and Social Aspects* (Pergamon Press, Oxford, 1981), pp. 229–236
59. M. Härmä et al., The effect of an irregular shift system on sleepiness at work in train drivers and railway traffic controllers. J. Sleep Res. **11**, 141–151 (2002)
60. L. Torsvall, T. Akerstedt, Sleepiness on the job: continuously measured EEG changes in train drivers. Electroencephalogr. Clin. Neurophysiol. **66**, 502–511 (1987)
61. G.D. Edkins, C.M. Pollock, The influence of sustained attention on railway accidents. Accid. Anal. Prev.: Spec. Issue, Fatigue Transp. **29**, 533–539 (1997)
62. K. Kogi, Sleep problems in night and shift work. J. Hum. Ergol. (Tokyo) **11**(suppl), 217–231 (1982)
63. P. Naitoh et al., Health effects of sleep deprivation. Occup. Med. **5**, 209–237 (1990)
64. J. Rutenfranz, Occupational health measures for night- and shiftworkers. J. Hum. Ergol. (Tokyo) **11**(suppl), 67–86 (1982)
65. A.J. Scott, J. LaDou, Shiftwork: effects on sleep and health with recommendations for medical surveillance and screening. Occup. Med. **5**, 273–299 (1990)
66. N. Kurumatani et al., The effects of frequently rotating shiftwork on sleep and the family life of hospital nurses. Ergonomics **37**, 995–1007 (1994)
67. P. Knauth, J. Rutenfranz, in Duration of sleep related to the type of shiftwork, ed. by A. Reinberg, et al. *Advances in the Biosciences, Vol 30. Night and shiftwork: Biological and Social Aspects* (Pergamo Press, New York, 1980), pp. 161–168
68. R.R. Mackie, J.C. Miller, Effects of hours of service regularity of schedules and cargo loading on truck and bus driver fatigue, *US Department of Transport,* vol. HS-803 (1978), p. 799
69. C.D. Wylie, et al., *Commercial Motor Vehicle Driver Rest Periods and Recovery of Performance* (Transportation Development Centre, Safety and Security, Transport Canada, Montreal, Quebec, Canada, 1997)
70. P. Knauth, J. Rutenfranz, The effects of noise on the sleep of night-workers, ed. by P. Colquhoun, et al. *Experimental Studies of Shift Work* (Westedeutscher Verlag, Opladen, 1975), pp. 57–65
71. K.R. Parkes, Sleep patterns, shiftwork, and individual differences: a comparison of onshore and offshore control-room operators. Ergonomics **37**, 827–844 (1994)
72. T. Akerstedt, Shift work and disturbed sleep/wakefulness. Soc. Occup. Med. **53**, 89–94 (2003)
73. T. Akerstedt et al., Spectral analysis of sleep electroencephalography in rotating three-shift work. Scand. J. Work Environ. Health **17**, 330–336 (1991)
74. G. Costa, The impact of shift and night work on health. Appl. Ergon. **27**, 9–16 (1996)
75. G. Costa, The problem: shiftwork. Chronobiol. Int. **14**, 89–98 (1997)

76. C. Cruz et al., Clockwise and counterclockwise rotating shifts: effects on vigilance and performance. Aviat. Space Environ. Med. **74**, 606–614 (2003)

77. C. Cruz et al., Clockwise and counterclockwise rotating shifts: effects on sleep duration, timing, and quality. Aviat. Space Environ. Med. **74**, 597–605 (2003)

78. F.M. Fischer et al., Day- and shiftworkers' leisure time. Ergonomics **36**, 43–49 (1993)

79. M. Frese, C. Harwich, Shiftwork and the length and quality of sleep. J. Occup. Environ. Med. **26**, 561–566 (1984)

80. V.H. Goh et al., Circadian disturbances after night-shift work onboard a naval ship. Mil. Med. **165**, 101–105 (2000)

81. R. Loudoun, P. Bohle, Work/non-work conflict and health in shiftwork: relationships with family status and social support. Int. J. Occup. Environ. Health 3(suppl 2), 71–77 (1997)

82. J. Walker, Social problems of shift work, ed. by S. Folkard and T. Monk, in *Hours of Work—Temporal Factors in Work Scheduling* (Wiley, New York, 1985), pp. 211–225

83. M.J. Blake, Relationship between circadian rhythm of body temperature and introversion-extraversion. Nature **215**, 896–897 (1967)

84. W.P. Colquhoun, *Biological Rhythms and Human Performance* (Academic, London, 1971)

85. G.S. Richardson et al., Circadian variation of sleep tendency in elderly and young adult subjects. Sleep 5(suppl 2), S82–S94 (1982)

86. W.B. Webb, D.F. Dinges, Cultural perspective on napping and the siesta, ed. by D.F. Dinges, R.J. Broughton, in *Sleep and Alertness: Chronobiological, Behavioral, and Medical Aspects of Napping* (Raven Press, New York, 1989), pp. 247–265

87. E. Hoddes et al., The development and use of the stanford sleepiness scale. Psychophysiology **9**, 150 (1972)

88. H. Nabi, et al., Awareness of driving while sleepy and road traffic accidents: prospective study in GAZEL cohort. BMJ **333**, 75 (2006)

89. L.A. Reyner, J.A. Horne, Falling asleep whilst driving: are drivers aware of prior sleepiness? Int. J. Legal Med. **111**, 120–123 (1998)

90. S.W. Kim, et al., Modelling the flowering of four Eucalypts species via MTDg with interactions, in *18th World IMACS Congress and International Congress on Modelling and Simulation*, Cairns, Australia, 2009, pp. 2625–2631

91. I. Hudson, et al., Modelling the flowering of four eucalypt species using new mixture transition distribution models, ed. by I.L. Hudson, M.R. Keatley in *Phenological Research* (Springer, Netherlands, 2010), pp. 299–320

92. I.L. Hudson et al., Using self-organising maps (SOMs) to assess synchronies: an application to historical eucalypt flowering records. Int. J. Bio-meteorol. **55**, 879–904 (2011)

93. I.L. Hudson, et al., Modelling lagged dependency of flowering on current and past climate on Eucalypt flowering: a mixture transition state approach, in *International Congress of Biometeorology*, Auckland, New Zealand, 2011, pp. 239–244

94. A. Berchtold, A.E. Raftery, The mixture transition distribution model for high- order Markov chains and non-Gaussian time series. Stat. Sci. **17**, 328–356 (2002)

95. B.H. Junker et al., VANTED: a system for advanced data analysis and visualization in the context of biological networks. BMC Bioinformatics **7**, 109 (2006)

96. I.L. Hudson, et al., SOM clustering and modelling of Australian railway drivers' sleep, wake, duty profiles, in *28th International Workshop on Statistical Modelling*, Palermo, Italy, 2013, pp. 177–182

97. I.L. Hudson, et al., Climate effects and thresholds for flowering of eight Eucalypts: a GAMLSS ZIP approach, in *19th International Congress on Modelling and Simulation*, Perth, Australia, 2011, pp. 2647–2653

98. I. Hudson, et al., Climatic Influences on the flowering phenology of Four Eucalypts: a GAMLSS approach, in ed. by I.L. Hudson, M.R. Keatley, in *Phenological Research* (Springer Netherlands, 2010), pp. 209–228

99. T. Kohonen, *Self-organizing maps*, 3rd edn. (Springer, Berlin, 2001)

100. P.N. Nguyen et al., Living standards of Vietnamese provinces: a Kohonen map. Case Stud. Bus. Ind. Govern. Stat. **22**, 109–113 (2009)

101. J.A. Sleep, I.L. Hudson, Comparison of self-organising maps, mixture, K-means, and hybrid approaches to risk classification of passive railway crossings, in *23rd International Workshop on Statistical Modelling*, Utrecht, Netherlands, 2008

102. C. Fraley, et al., MCLUST Version 4 for R: normal mixture modeling and model-based clustering, classification, and density estimation, Department of Statistics, University of Washington Technical Report No. 504 (2013)

103. J.F. Roddick, M. Spiliopoulou, A survey of temporal knowledge discovery paradigms and methods. IEEE Trans. Knowl. Data Eng. **14**, 750–767 (2002)

104. P. Cheeseman and J. Stutz, Bayesian classification (AutoClass): theory and results, U.M. Fayyard, et al., in *Advances in Knowledge Discovery and Data Mining* (AAAI/MIT Press, Cambridge, 1996)

105. C. Biernacki et al., Assessing a mixture model for clustering with the integrated completed likelihood. IEEE Trans. Pattern Anal. Mach. Intell. **22**, 719–725 (2000)

106. D.B. Reusch, et al., North Atlantic climate variability from a self-organizing map perspective. J. Geophys. Res. **112** (2007)

107. A.F. Costa, Clustering and visualizing SOM results, in *11th International Conference on Intelligent Data Engineering and Automated Learning—IDEAL 2010*. Lecture Notes in Computer Science. vol. 6283 (Springer, Berlin, 2010), pp. 334–343

108. J. Vesanto, E. Alhoniemi, Clustering of the self-organizing map. IEEE Trans. Neural Netw. **11**, 586–600 (2000)

109. H. Yin, Learning nonlinear principal manifolds by self-organising maps, ed. by A. Gorban, et al. in *Principal Manifolds for Data Visualization and Dimension Reduction*, vol. 58 (Springer, Berlin, 2008), pp. 68–95

110. J.C. Fort, SOM's mathematics. Neural Netw. **19**, 812–816 (2006)

111. C. Klukas, The VANTED software system for transcriptomics, proteomics and metabolomics analysis. J. Pestic Sci. **31**, 289–292 (2006)

112. D.M. Stasinopoulos, R.A. Rigby, Generalized additive models for location scale and shape (GAMLSS) in R. J. Stat. Softw. **23**, 1–46 (2007)

113. R.A. Rigby, D.M. Stasinopoulos, MADAM macros to fit mean and dispersion additive models, ed. by A. Scallan, G. Morgan, in A. Scallan, G. Morgan, *GLIM4 Macro Library Manual, Release 2.0* (Numerical Algorithms Group, Oxford, 1996), pp. 68–84

114. R.A. Rigby, D.M. Stasinopoulos, Mean and dispersion additive models, ed. by W. Hardle, M.G. Schimek, in *Statistical Theory and Computational Aspects of Smoothing* (Physica-Verlag, Heidelberg, 1996), pp. 215–230

115. A. Berchtold, Markov chain computation for homogeneous and non-homogeneous data: MARCH 1.1 Users Guide. J. Stat. Softw. (2001)

116. A. Berchtold, *March v.3.00. Markovian models Computation and Analysis Users guide* (2006). http://www.andreberchtold.com/march.html

117. A. Berchtold, *March v.2.01. Markovian models Computation and Analysis Users guide* (2004). http://www.andreberchtold.com/march.html

118. S.W. Kim, et al., Modelling and synchronization of four Eucalypt species via Mixed Transition Distribution (MTD) and Extended Kalman Filter (EKF), in *23rd International Workshop on Statistical Modelling*, Utrecht, Netherlands, 2008, pp. 287–292

119. D.A. Fournier, *AD Model Builder, Version 5.0.1.* (Otter Research Ltd., Canada, 2000)

120. D.A. Fournier, et al., AD Model Builder: using automatic differentiation for statistical inference of highly parameterized complex nonlinear models. Optim. Methods Softw. **27**, 233–249 (2012)

121. I.L. Hudson, et al., Climatic influences on the flowering phenology of four Eucalypts: a GAMLSS approach, in *18th World IMACS Congress and International Congress on Modelling and Simulation. Modelling and Simulation Society of Australia and New Zealand*, Cairns, Australia, 2009, pp. 2611–2617

122. I.L. Hudson, et al., Climate impacts on sudden infant death syndrome: a GAMLSS approach, in *23rd International Workshop on Statistical Modelling (IWSM)*, Utrecht, Netherlands, 2008, pp. 277–280

123. T. Akerstedt et al., Accounting for partial sleep deprivation and cumulative sleepiness in the Three-Process Model of alertness regulation. Chronobiol. Int. **25**, 309–319 (2008)
124. P.M. Krueger, E.M. Friedman, Sleep duration in the United States: a cross-sectional population-based study. Am. J. Epidemiol. **169**, 1052–1063 (2009)
125. D. Nur, et al., Multivariate Gaussian hidden Markov models for sleep profiles of railway drivers, in *Australian Statistical Conference*, Adelaide, Australia, July 2012, p. 159
126. A.A. Markov, Extension of the limit theorems of probability theory to a sum of variables connected in a chain (reprinted in Appendix B), ed. by R. Howard, in *Dynamic Probabilistic Systems volume 1: Markov Chains* (Wiley, New York, 1971)
127. N. Goldman, Z. Yang, A codon-based model of nucleotide substitution for protein-coding DNA sequences. Mol. Biol. Evol. **11**, 725–736 (1994)
128. L. Muscariello et al., Markov models of internet traffic and a new hierarchical MMPP model. Comput. Commun. **28**, 1835–1851 (2005)
129. S.F. Gray, Modeling the conditional distribution of interest rates as a regime-switching process. J. Finan. Econ. **42**, 27–62 (1996)
130. C.J. Spanos, et al., The economic impact of choosing off-line, inline or in situ metrology deployment in semiconductor manufacturing, in *Semiconductor Manufacturing Symposium, 2001 IEEE International*, 2001, pp. 37–40
131. W.A.V. Clark, Markov chain analysis in geography: an application to the movement of rental housing areas. Ann. Assoc. Am. Geogr. **55**, 351–359 (1965)
132. B. Mau et al., Bayesian phylogenetic inference via Markov Chain Monte Carlo methods. Biometrics **55**, 1–12 (1999)
133. N.G. van Kampen, *Stochastic Processes in Physics and Chemistry*, 3rd edn. (Elsevier, Amsterdam, 2007)
134. J. Claerbout, *Fundamentals of Geophysical Data Processing: with Applications to Petroleum Prospecting* (McGraw-Hill Inc., New York, 1976)
135. A. Berchtold, G. Sackett, Markovian models for the developmental study of social behavior. Am. J. Primatol. **58**, 149–167 (2002)
136. D. Draper, Inference and hierarchical modeling in the social sciences. J. Educ. Behav. Stat. **20**, 115–147 (1995)
137. K. Verbeurgt, et al., Extracting patterns in music for composition via Markov chains, in *17th International Conference On Innovations in Applied Artificial Intelligence*, Ottawa, Canada, 2004, pp. 1123–1132
138. P. Bremaud, *Markov Chains: Gibbs Fields, Monte Carlo Simulation, and Queues* (Springer, New York, 1999)
139. E. Seneta, Markov and the birth of chain dependence theory. Int. Stat. Rev. **64**, 255–263 (1996)
140. A.E. Raftery, A model for high-order Markov chains. J. R. Stat. Soc. Ser. B (Stat. Methodol.) **47**, 528–539 (1985)
141. A. Berchtold, Autoregressive modelling of Markov chains, in *10th International Workshop on Statistical Modelling (IWSM)* (Springer, New York, 1995), pp. 19–26
142. A. Berchtold, Swiss health insurance system: mobility and costs. Health Syst. Sci. **1**, 291–306 (1997)
143. A. Berchtold, The double chain Markov model. Commun. Stat.: Theory Methods **28**, 2569–2589 (1999)
144. A. Berchtold, Estimation in the mixture transition distribution model. J. Time Ser. Anal. **22**, 379–397 (2001)
145. A. Berchtold, A.E. Raftery, The mixture transition distribution model for high- order markov chains and non-gaussian time series. Stat. Sci. **17**, 328–356 (2002)
146. A. Berchtold, High-order extensions of the double chain Markov model. Stoch. Models **18**, 193–227 (2002)
147. A. Berchtold, Mixture transition distribution (MTD) modeling of heteroscedastic time series. Comput. Stat. Data Anal. **41**, 399–411 (2003)

148. A. Berchtold. (2004, March v.2.01. Markovian models computation and analysis users guide. http://www.andreberchtold.com/march.html
149. A. Berchtold. (2006, March v.3.00. Markovian models computation and analysis users guide. http://www.andreberchtold.com/march.html
150. K. Fokianos, B. Kedem, Regression theory for categorical time series. Stat. Sci. **18**, 357–376 (2003)
151. S. Lèbre, P.-Y. Bourguignon, An EM algorithm for estimation in the mixture transition distribution model. J. Stat. Comput. Simul. **78**, 713–729 (2008)
152. J. Luo, H.-B. Qiu, Parameter estimation of the WMTD model. Appl. Math.: J. Chin. Univ. **24**, 379–388 (2009)
153. C.S. Wong, W.K. Li, On a mixture autoregressive model. J. R. Stat. Soc. Ser. B (Stat. Methodol.) **62**, 95–115 (2000)
154. S.W. Kim, et al., Modelling and synchronization of four Eucalypt species via Mixed Transition Distribution (MTD) and Extended Kalman Filter (EKF), in *23rd International Workshop on Statistical Modelling (IWSM)*, Utrecht, Netherlands, 2008, pp. 287–292
155. S.W. Kim, Bayesian and non-Bayesian mixture paradigms for clustering multivariate data: time series synchrony tests. PhD, University of South Australia, Adelaide, Australia (2011)
156. I. Hudson, et al., Modelling the flowering of four Eucalypt species using new mixture transition distribution models, ed. by I.L. Hudson, M.R. Keatley, in *Phenological Research* (Springer, Netherlands, 2010), pp. 299–320
157. S.W. Kim, et al., MTD analysis of flowering and climatic states, in *20th International Workshop on Statistical Modelling*, Sydney, Australia, 2005, pp. 305–312
158. I.L. Hudson, et al., Modelling the flowering of four Eucalypt eucalypts species using new mixture transition distribution models, ed. by I.L. Hudson, M.R. Keatley, in *Phenological Research: Methods for Environmental and Climate Change Analysis* (Springer, Dordrecht, 2010), pp. 315–340
159. S.W. Kim, et al., Analysis of sleep/wake and duty profiles of railway drivers using GAMLSS with interactions, in *Australian Statistical Conference*, Adelaide, South Australia, July 2012, p. 121
160. S.W. Kim, et al., Mixture transition distribution analysis of flowering and climatic states, in *20th International Workshop on Statistical Modelling*, Sydney, Australia, 2005, pp. 305–312
161. S.W. Kim, et al., Modelling the flowering of four eucalypts species via MTDg with interactions, in *18th World International Association for Mathematics and Computers in Simulation (IMACS) Congress and International Congress on Modelling and Simulation. Modelling and Simulation, Society of Australia and New Zealand and*, Cairns, Australia, 2009, pp. 2625–2631
162. S.W. Kim, et al., Modelling and synchronization of four Eucalypt species via Mixed Transition Distribution (MTD) and Extended Kalman Filter (EKF), in *23rd International Workshop on Statistical Modelling*, Utrecht, Netherlands, 2008, pp. 287–292
163. D.A. Fournier, *AD Model Builder, Version 5.0.1.* (Otter Research Ltd., Canada, 2000)
164. D.A. Fournier, et al., AD Model Builder: using automatic differentiation for statistical inference of highly parameterized complex nonlinear models. Optim. Methods Softw. **27**, 233–249 (2012)
165. N. Kilari et al., Block replacement modeling for electronic systems with higher order Markov chains. IUP J. Comput. Math. **4**, 49–63 (2011)
166. N. Kilari et al., Reengineering treatment in block replacement decisions using higher order Markov chains. IUP J. Oper. Manag. **10**, 22 (2011)
167. D.A. Fournier, *AUTODIFF, A C++ Array Language Extension with Automatic Differentiation for use in Nonlinear Modeling and Statistics* (Otter Research Ltd., Canada, 1996)
168. I.L. Hudson, et al., Climatic influences on the flowering phenology of four Eucalypts: a GAMLSS approach, ed. I.L. Hudson, M.R. Keatley, in *Phenological Research: Methods for Environmental and Climate Change Analysis* (Springer, Dordrecht, 2010), pp. 213–237

A Neural Approach to Electricity Demand Forecasting

Omid Motlagh, George Grozev and Elpiniki I. Papageorgiou

Abstract Electricity demand forecasting is significant in supply-demand management, service provisioning, and quality. This chapter introduces a short-term load forecasting model using Fuzzy Cognitive Map, a popular neural computation technique. The historic data of intraday load levels are mapped to network nodes while a differential Hebbian technique is used to train the network's adjacency matrix. The inferred knowledge over weekly training window is then used for demand projection with Mean Absolute Percentage Error (MAPE) of 5.87 % for 12 h lead time, and 8.32 % for 24 h lead time. A Principal Component Analysis is also discussed to extend the model for training using big data, and to facilitate long-term load forecasting.

Keywords Energy demand forecasting · Neural networks · Time series

1 Introduction

A significant issue for successful smart grid implementation is about management of supply and demand. Electricity is by its nature difficult to store and has to be available on demand. Hence, demand variability influences everything from quality and stability of electricity supply to long-term investment decisions at base and peak load plants, physical transmission network, power system protection, and various financial aspects. The classic protection strategy suggests periodic isolation of over-demanding nodes whilst maintaining rest of network in operation. However, this involves sophisticated and highly reliable protection hardware,

O. Motlagh (✉) · G. Grozev
CSIRO Land and Water, Bayview Ave., Clayton, Victoria 3168, Australia
e-mail: omid.motlagh@csiro.au

E.I. Papageorgiou
Department of Computer Engineering, Technological Educational Institute of Central Greece,
Old National Road Lamia-Athens, 35100 Lamia, Greece

© Springer International Publishing Switzerland 2016   281
S. Shanmuganathan and S. Samarasinghe (eds.), *Artificial Neural
Network Modelling*, Studies in Computational Intelligence 628,
DOI 10.1007/978-3-319-28495-8_12

which themselves need additional cost and maintenance. The risk of blackout still remains as there is no insight into unexpected demand behaviors. Dynamic demand can remedy the situation by delaying appliance operating cycles by a few seconds to increase the diversity factor of the set of loads. The government of the state of Queensland, Australia, plans to have devices fitted onto certain household appliances such as air conditioners, pool pumps, and hot water systems.

These devices would allow energy companies to remotely cycle the use of these items during peak hours. Demand response is another approach at user end that includes any reactive or preventative method (mainly in response to time-of-use tariff) which automatically can reduce, flatten or shift peak demand. However, both techniques require reliable forecasting at their fundamentals, first on short-term forecasting of load levels, and second on forecasting patterns of usage behaviors. In Victoria, more than 80 % of power stations burn brown coal making large on-off time lags [1]. Metrological and urban factors change dynamically which lead to unstable demand patterns. On the other hand, the extent and schedule of dynamic demand control are based on the difference between forecasted demand level and scheduled generation capacity in a particular region. Hence, forecasting plays a key role in all aspects of modern energy management systems. The modern approach suggests utilization of new algorithms and software solutions alongside ever advancing hardware.

Methods of artificial intelligence enable for reliable projection of demand, at regional substations down to neighborhoods and blocks. Long-term models of future electricity demand generally use global climate model (GCM) datasets that predict decadal trends of changing temperatures, humidity, rainfall, etc. Accordingly, they are appropriate for determining the future investments, such as building new peaking plants. On the other hand, short-term models are more crucial when it comes to power system protection. Network stability and consistent supply could be improved by prediction of intraday demand curves from patterns of other data such as climate data, demand profiles and usage behavior. Most short-term techniques attempt at drawing linear regression models of regional electricity demand with variables such as temperature, wind speed, humidity, etc. The state of the art suggests that a large proportion of the variability in electricity demand is dependent on the weather. Accordingly, it incorporates weather forecasting with focus on relationship between temperature and electricity demand. Despite some reliable methods in the literature, regression models are subject to criticism, unless all of the multi-variables are taken into account. Electricity demand is a function of tens of multivariate that have to be concurrently analyzed in a single model, a task far beyond current achievements.

This article first argues about the existing approaches towards short-term prediction of electricity demand in the literature (in Sects. 2.1 and 2.2). Section 2.3 discusses the disadvantages which briefly include the facts that (1) climate data are not always available, (2) climate forecast is subject to uncertainty which leads to augmented error in electricity demand projection, and (3) electricity demand is a function of many heterogeneously independent variables requiring multivariate regression models which are mathematically complex, inefficient, and incomplete due to variables which are not even known to experts. In Sect. 2.4, we outline an alternative strategy to model future demand as a function of time, i.e., focusing only on intraday demand curves themselves without need for any other data.

As the intended contribution, this strategy is implemented through the development of an unsupervised neural model first introduced in this article (Sect. 3). We understand the patterns of changes along intraday demand curves on weekly basis that is proven to be most effective. The model uses past and current data, to forecast future demand, and since it does not need other data (metrological, etc.) it can be regarded as a benchmark for more sophisticated techniques. Section 4 presents results and validation of methodology using actual dataset, followed by conclusion, and recommendations for future works (in Sect. 5).

2 Review of the Literature

Residential electricity demand is most volatile as compared against other components, namely commercial and industrial demand. It is also significant as most distribution and retailer companies make higher profit out of residential services. There are techniques for separating residential demand from other components, such as independent component analysis (ICA), and down to individual components of load profiles [2]. Another robust approach is based on decomposition by separating daily, weekly, monthly, and seasonal patterns [3] so that more volatile behaviours (associated with residential demand) could be spotted. However, having a different scope, and with the available mixed dataset from Australian Energy Market Operator [4], in this article we present the mixed mode demand curves at state level.

2.1 Factors Influencing Electricity Demand Behaviour

There are tens of various variables influencing patterns of electricity consumption at individual households or a power network at broader scope. Weather variables, temperature, precipitation, wind speed, wind direction, humidity, pressure, and solar radiation, are only one type of variables influencing electricity demand [5, 6] while such variables themselves may depend on one another. The complexity of relationships among climate variables and electricity demand even increases when the time dimension involves, i.e., issues such as global warming and climate changes. There has been abundance of research to project gradual impact of global warming and climate change [7, 8] mainly using multivariate regression techniques to draw relationships among one or more climate variables and energy demand.

Apart from climate factors, number of occupants, type of occupancy, occupants daily routines, patterns of behaviours and time-dependent activities, are just another group of variables [9, 10]. The probability of occupants be at home, the conditional probability to start an activity, and the probability distribution function for the duration of that activity, are examples of factors to be analysed in the first place [10] while understanding their relationship with electricity demand, and other classes of variables is another challenge. Another type of information is related to types and estimated numbers of electrical appliances used at individual households [10], which itself relates to many other financial and socio-cultural factors. It is for example understood that women indirectly contribute to increasing consumption as home appliances are purchased to alleviate time pressure [11]. In a broader scope where demand behaviour of a larger district is concerned, other variables come to play role such as residential building stock, geographical factors, demographics of gender, age, income and many others. Economy influences electricity demand [3, 12, 13] or vice versa [14]. In addition, demand behaviour also varies due to other external factors such as regulations set by government and other authorities. Dispatch and settlement routines, quota, time of use tariff, and billing policies [15–17], external dynamic demand control by suppliers, and demand response from user side, all vary in relation to one another.

$$E = f(v_1 \ldots v_m) \tag{1}$$

It is therefore concluded in Eq. 1 that electricity demand (E) is a non-linear function (f) of many heterogeneously inter-dependent variables $v_1 \ldots v_m$, e.g., related to demographics, geo-climate, social and urban development, as well as other unknown variables and factors yet to be discovered. And despite the fact that variables could be grouped into different types, sadly the non-linear function (f) could not be broken down leaving a complex multi-dimensional problem. Accordingly, exhaustive multivariate regression models are needed to obtain f for a given residential unit or district, or to understand their behaviours.

2.2 Soft Models Versus Conventional Models

An alternative to reduce the problem size as seen in most of the literature is to focus on more dominant variables (or variables of interest) and ignore the others under certain circumstances. For example, temperature is proven to have a major impact on electricity usage [18, 19]. A traditional way to describe the impact of temperature is in terms of heating degree days (HDDs) and cooling degree days (CDDs). HDDs and CDDs measure the difference between the daily mean temperatures above or below a regional base temperature [20, 21]. The aim is to draw a linear regression model between regional electricity demand and temperature [18] to forecast intraday demand behaviour. A justification for temperature-related demand forecasting is that apparent temperature [22] encompasses other climate variables, e.g., humidity, and therefore reduces the problem size. From this perspective, regression techniques such as smooth transition regression (STR), threshold regression (TR), and switching regression (SR) have been examined [19], as well as other linear [13], non-linear spline method [23], least square, and maximum likelihood [17] methods. Probabilistic such as Bayesian [24] and grey methods [25] are also employed to build future data. However, none can guarantee reliable forecast as the problem in Eq. 1 remains unsolved. A comprehensive review of the literature reveals that most previous research seeks one-to-one relationships among influencing variables and electricity demand.

A common problem is that, due to numerous sources of impact, misleading conclusions might be drawn in univariate relationship models, i.e., between a single variable and demand. In other words, while for example a variable is taken to have a decreasing impact on electricity demand, in reality such decrease might not be due to that variable but other unconsidered or ignored variables.

Misunderstanding about system behaviours causes unreliable projection. Therefore, the only classic solution is to consider all variables concurrently which lie in multi-dimensional regression. On the other hand, multivariate regression in such as that in Eq. 1, is barely practical using hard mathematics as level of complexity exponentially increases with number of variables. Alternatively, the state of the art in demand forecast is based on utilization of artificial neural networks (ANN) [26–28], genetic algorithm (GA) [29], fuzzy and hybrid methods [30–32], and other soft approaches for learning and prediction of demand patterns. In a primary model, load was predicted for specific daily hours based on a combination of load data and temperature, and using a multi-layer perceptron [33]. Smaller size mainly linear problems could be simply modelled using single-layer perceptron [34]. Single hidden layer feed-forward networks are also among widely-used models [35], mainly using standard or enhanced back propagation rule, e.g., using conjugate gradient algorithm (GCA) [1]. However, recurrent models such as Hopfield network [36], and fuzzy cognitive map (FCM) [37] became more popular as more relationship configurations could be obtained. More sophisticated neural structures have evolved to correlate multi variables to electricity demand, a task math models barely accomplish. Overall, neural regression models appeared to be viable alternatives to the stepwise and classic regression models in learning energy consumption patterns [38] being the introduction to forecast. Numerical and neural models provide easier implementation, and more information especially about relationships among variables themselves, e.g., impact of climate change on cost of electricity use [6], in addition to their relationships with electricity demand. Another advantage of neural models is based on efficient utilization of principal component analysis (PCA) [39], exploratory factor analysis (EFA) [40], both on the assumption of linear relationship between observed variables, and other variable reduction techniques. At last, neural networks are applicable to both short and long term projection [41] and in general outperform other methods [42].

An n-node fuzzy cognitive map [37] is a recurrent neural network where each node (so called concept) C_j, $C_j \in \{C_1 \ldots C_n\}$, is the output of all other nodes C_i, $C_i \in \{C_1 \ldots C_n\}$, $i \neq j$, i.e., n interrelated single-layer (n−1)-node perceptron networks. The goal is to train the matrix of edge weights (or FCM adjacency matrix) to hold the FCM at convergence (Eq. 2). In other words, if a neutral (linear with unit slope) activation function is employed, i.e., net weight is directly assigned to output without squashing, a product of the $1 \times n$ state vector of nodes $C = \{C_1 \ldots C_n\}$ at cycle τ ($C^{(\tau)}$) and the $n \times n$ adjacency matrix W, should reveal the same values for the next cycle ($C^{(\tau+1)} = C^{(\tau)}$).

To obtain W as a non-identity matrix, initial weights must start from zero while the perceptron rule (Eq. 3 with learning rate α) applies to all $w_{i,j} \in W$, provided $i \neq j$. Accordingly, the matrix W describes all possible relationships among variables, including input-output relationship as well as relationships among inputs themselves.

$$C^{(\tau+1)} = C^{(\tau)} = C^{(\tau)} \times W^{(\tau)} \tag{2}$$

$$w_{ij}^{(\tau+1)} = w_{ij}^{(\tau)} + \alpha\, C_i^{(\tau)}(C^{(\tau)} \times W_j^{(\tau)} - C_j^{(\tau)}) \tag{3}$$

$$C_j^{(\tau)} = C^{(\tau)} \times W_j^{(\tau)} , \quad w_{jj}^{(\tau)} = 0 \tag{4}$$

Upon sufficient training, matrix W simply provides an (n−1)-dimensional linear regression model for all variables assigned to FCM nodes, i.e., each variable (node C_j) is a result of all variables (nodes C_i, i = 1 … n, i $\neq$ j) which influence on C_j through weights $w_{i,j}$ as shown in Eq. 4. W is also called the relationship model of all variables [43] with wide range of applications especially when it comes to multivariate systems. However, an important issue is that although impacts of variables on a specific output do not need to be independent, perceptron learning is limited to single-trend systems, i.e., extents of impacts of variables on a specific output must not have inverse relationship with one another. In this case, utilization of nonlinear rules such as Hebbian rule is more viable. Another design issue affecting accuracy of forecast and generalization is related to the employed activation function. There are various choices among sigmoid-based, linear and hard limiters depend on the range and format of data, e.g., grey or binary. While there are highly efficient activation models such as cumulative sigmoid activation [37], a new philosophy suggests retaining neutral activation at output (i.e., net weights to appear directly at output) while replacing incoming edge weights with functions rather than crisp values, i.e., natural activation technique [43]. In addition, edge functions, such as polynomial, trigonometric, etc., allow for variables to follow multi-trends.

Nonlinear Hebbian rule facilitates learning of multi-trend behaviours. Of the fundamental ideas in biological learning, Hebb law [44] has become an established learning method in artificial neural networks (NN) when it comes to distinct patterns of inputs, used in classification, memory and retrieval systems. The idea is to strengthen the connection between neurons i and j, if neuron i is near enough to excite neuron j making it more sensitive to such stimuli. Therefore, a learning rate α is multiplied by the values of the input and output neurons to be added to the current weight of the link connecting input to the output. On the other hand, to prevent weights from indefinite growth, a decay factor φ is multiplied by the connecting weight and the value of the memorized output to mimic the phenomenon of forgetting in biological brain. Equation 5 [45] shows the concept of Hebbian rule for weight update from iteration τ to $\tau + 1$ in a typical feed forward neural network.

$$\Delta w_{ij}^{(\tau)} = \alpha \, C_i^{(\tau)} \, C_j^{(\tau)} - \varphi \, C_j^{(\tau)} w_{ij}^{(\tau)} \quad and \quad w_{ij}^{(\tau+1)} = w_{ij}^{(\tau)} + \Delta w_{ij}^{(\tau)} \qquad (5)$$

The basic rule in Eq. 5 needs modifications in recurrent models such as FCM depend on several criteria such as sequence of activation of neurons, and selection of active nodes. Kosko proposed the initial model known as differential Hebbian (DH) model [46]. DH was then improved into balanced differential algorithm (BDA) [47]. In BDA, weight update depends on values of selected FCM nodes which are acting at the same time. However, major advancements were related to non-linear Hebbian learning (NHL) [48] and active Hebbian learning (AHL) [49]. NHL is based on the premise that FCM graph has to be updated synchronously whereby all neurons (state vector of nodes or concepts) are updated at the same time. In contrast, AHL involves the sequence of active concepts and updates new weights of all concepts as influenced by the active concept at any time.

2.3 Electricity Demand as Function of Time

Neural regression uses actual dataset such as 365 × 24 records of diurnal hourly data of temperature, wind speed, humidity, etc. as inputs, and observed electricity demand or scheduled generation as output, to obtain multivariate equations that can be generalized, i.e., to be applied to future known input factors to guess on the amount of output (electricity demand) in future. The problem in Eq. 1 is seemingly resolved, yet the reality is the variables $v_1,\ldots,v_m$ are only the known variables although still subject to measurement errors. More adversely, such multivariate are also subject to forecasting error. For example, exact climate data are not always available. Climate forecast is subject to uncertainty which leads to augmented error in electricity demand forecast as forecast model itself is not error free.

Here, we resort to an entirely different strategy by looking not at influencing variables but electricity demand itself as a function of time, especially when short term forecast is needed. Significance of short term forecast is in management of energy which relates to controlling and scheduling of power systems, power system protection [50], and dynamic demand control in smart grid implementations [51]. The focus varies from minutes to hours depending on desired accuracy, sensitivity of the network, dispatch and settlement policy, e.g., 5 min dispatch and 30 min settlement in Australian.

The stochastic nature of demand as a function of time has been modelled using time series analysis, such as autoregressive moving average (ARMA) [52, 53], autoregressive integrated moving average (ARIMA) [54, 55], Kalman filter for prediction based on multivariate and then update based on observed demand level [26], and cyclic time horizon such as daily, weekly [39, 56], and monthly [1], known as naive forecasting. The weekly patterns are particularly interesting sources for forecasting. In many cases, electricity demand curves during two successive weeks follow similar trends [39], which is even consistent throughout the year.

More people in the world are now following weekly routines of working days and weekends activities, and therefore knowledge of day-of-week and time-of-day improves accuracy of modelling, e.g., in modelling urban transportation routines [57], and electricity demand [31]. This could be a reason for identical weekly usage patterns under similar circumstances such as consistent temperature and other climate variables. Accordingly, available demand data of a particular day-time of the week can be used to guess on the demand level of the same day-time in the following week. AEMO uses a seven days rolling window to forecast the demand, supply, and reserve for daily peak demand, at each NEM region over the next seven days. In particular the estimates include the total generation capacity, the daily peak demand level, net interchange, and reserve, for every region [58]. A comprehensive summary of naive forecast models of electricity demand such as on weekly basis is given in [56].

3   FCM-Based Forecasting

In the remainder of the article, we present three dynamic forecasting models using recurrent structure of fuzzy cognitive map (FCM), with minimal size of training set, and yet reliable forecast accuracy and robustness. At any instance of time, at most three weeks of half hourly time series data samples have been used (mainly within one week rolling window) to forecast up to one day ahead. In the first model (Sect. 3.1) database is updated in parallel (one day (d) demand data (D_d) updates data of the same day in the past week (D_{d-7})). In the second model (Sect. 3.2), fresh half hourly data at current time instance (t_i) denoted as (Dt_i) replaces the last observed data (Dt_{i-1}) in a serial fashion. At last, the final model in Sect. 3.3 utilizes both parallel and serial updating for forecasting with up to one day lead time.

3.1   Forecast Using 7-Days Data, Daily Rolling Window, and Parallel Updating

This section suggests a supervised learning approach for daily demand curves within one week to be used for short term prediction of electricity demand, while a time series smoother is used to refine forecasts in real-time depending on desired frequency. A precise Hebbian learning technique is introduced and implemented to discover patterns of usage behaviour from weekly data with half hour frequency. The database therefore includes 7 records or 7×48 piece of data. Inspired by a naive forecast technique [56] the ideas is that the next amount of demand (during the next half hour) relates to the demand at the same day-time in the past week. In other words, the demand level Dt_i of half hour t_i (e.g., hh:00 to hh:30) in day d of week k is related to Dt_i in day d of week $k-1$, and so is the relationship between transition from half hour t_i to t_{i+1} in the two weeks. This inherently lets the influence of variables such as users occupancy and usage behaviour be inclusive and contribute into accuracy of forecast on the assumption that users most likely follow similar weekly routines as supported by the literature. However, the model is unsupervised so that any new pattern of demand behaviour will have an impact due to plasticity of neural pathways in the system memory.

Another idea is to perform learning on the correlation of all 48 intraday demand levels to one another, in contrast to other methods where focus goes on learning the forward sequence along time. The philosophy is that not only the demand level of a future interval is related to (influenced by) demand level in a past interval as conventionally suggested in time series analysis, but also a past or current demand level might be related to (or influenced by) the demand level in future. This could be generally formulated as: demand level during any time period t_i is related to demand during another period t_j at least within a single day, i.e., t_i and t_j are two intraday intervals. For example, users may opt to postpone turning their air conditioner on from time t_i to t_j due to any reason, such as time-of-use tariff, temperature, or other explanatory variables. Likewise, a task to be performed in future might be brought forth such as doing laundry earlier so that one could sleep earlier. It could be therefore said that the demand levels Dt_i and Dt_j are interactive with mutual impact. On the other hand, the implicit impact of explanatory multivariate on demand levels becomes apparent in their interactions. For example, a negative cause-effect link from node Dt_i to node Dt_j can be interpreted in terms of why an increase in demand during t_i has caused decrease in demand during t_j. The answer can lie in the fact that cooling during hot time t_i still keeps the house cool until and during t_j. On the basis of this idea, in this method we aimed to mutually correlate all 48 intraday demand variables using a recurrent neural structure that is in contrast to feed forward structures (e.g., in [56]). At last, a correlation must be created between latest scheduled demand and observed demand levels inspired by naive forecasting. Such correlation could be made using any of the time series analysis technique available in the literature. Therefore apart from weekly and daily correlations, the next half hourly demand (which makes basis of decision for the next scheduled generation) is predicted from the previous actual demand (last 30 min) just observed.

Figure 1 shows the concept for scheduling generation based on the memorized knowledgebase (i.e., relationship among half-hourly demand levels of each specific day and therefore demand patterns throughout the week), refined by a real-time closed-loop corrective action method (smoother) using last observed demand level and its difference with last forecasted demand level. FCM is used for data mining where an adjacency model of the network suggests relationships among half-hourly demand levels. Accordingly, a 48-node FCM is employed to satisfy the desired accuracy. The 48×48 adjacency matrix (W) is trained to attain FCM convergence as described for Eq. 2, where state vector (C) is a 1×48 vector that is iteratively loaded with observed demand levels ($C_1 \ldots C_{48} = Dt_1 \ldots Dt_{48}$) of each and every of the past seven days ($d_1 \ldots d_7$). The training rule is based on fundamental Hebbian concept for learning and forgetting (Eq. 5), however, both phenomena are merged into one differential rule (Eq. 6) to encourage simultaneous growth and decay of

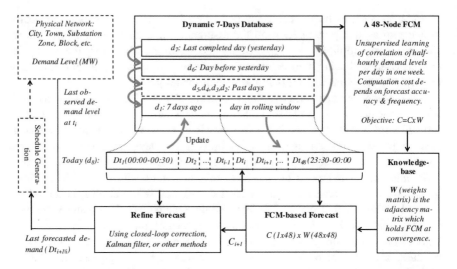

Fig. 1 Forecast expert with parallel updating where d_8 is assumed to have similar behavior with d_1. Forecasts during day d_8 are on the bases of values of $Dt_{i+1} \ldots Dt_{48}$ adapted from d_1, while fresh observations during d_8 ($Dt_1 \ldots Dt_i$) are stored in d_1 for use in the following week. *Bold arrows* show data flow within the database

weights in response to input state vectors. The unsupervised model has fair flexibility to out-of-sample data without scarifying smooth trends.

$$\Delta w_{ij}^{(\tau)} = \alpha \, \Delta C_i^{(\tau)} \Delta C_j^{(\tau)} \quad and \quad w_{ij}^{(\tau+1)} = w_{ij}^{(\tau)} + \Delta w_{ij}^{(\tau)} \tag{6}$$

$$D(t_{i+1|i}) = \rho \, C_{i+1} + (1-\rho)D(t_{i|i}) - \varepsilon \, \Delta D(t_i)\left(1 - D\left(t_{i|i}\right)\right) \tag{7}$$

$$and, \qquad \Delta D(t_i) = D(t_{i|i-1}) - D(t_{i|i})$$

$$D(t_{i+1|i}) = \frac{\left((1-\mu_1)C_{i+1} + (1-\mu_2)D\left(t_{i|i}\right)\right)}{2} - \varepsilon \left(D\left(t_{i|i-1}\right) - D\left(t_{i|i}\right)\right) \tag{8}$$

$$where, \quad \mu_1 = \left|D\left(t_{i|i}\right) - C_i\right|, \quad \mu_2 = \left|D\left(t_{i-1|i-1}\right) - D\left(t_{i|i}\right)\right|$$

$$and, \qquad \varepsilon = \varepsilon_0 \, var\left\{D\left(t_{i-2|i-2}\right) \ldots D\left(t_{i|i}\right)\right\}$$

The database in Fig. 1 is dynamic and frequently updated with freshly observed datum along the rolling window which replaces its counterpart in past week, e.g., Dt_i of today (d_8) replaces Dt_i in seven days ago (d_1). Accordingly, the knowledgebase is also dynamic as training of W continues. Next, the trained FCM will be used for forecasting the demand during the next time period (Dt_{i+1}). Here, the state vector C is consisted of two parts, first from Dt_1 to Dt_i which are the observed demand levels of the current day (d_8), and second from Dt_{i+1} to Dt_{48} adapted from the last similar day (d_1). The product of the state vector and the trained FCM adjacency matrix gives a new vector whose elements C_{i+1} and onwards are expected to reveal forecasted Dt_{i+1} and onwards, respectively. For half hour lead time, only forecasted value of Dt_{i+1} is concerned as revealed by C_{i+1}. The notation $Dt_{i+1|i}$ represents the estimate of demand at time step t_{i+1} given observations up to and including at time t_i. Accordingly, $Dt_{i|i}$ is a posteriori state or observation at t_i. Dt_{i+1} (obtained with respect to demand behaviours of the past 7 days) is further refined in a separate process with respect to the last observed demand level Dt_i and the last forecasting error ΔDt_i. Equation 7 presents the resultant $Dt_{i+1|i}$ where constant ρ represent the assumed similarity of Dt_{i+1} with FCM output C_{i+1} being extracted from its match in past week. ε multiplies by recent forecasting error ΔDt_i and is applied as the gain of a negative feedback to complete the proportional controller. The values of constants ρ, ε are both obtained using a standard annealing technique on the available dataset given in Sect. 4.1

In another attempt, an improvement is made by specifying sources of forecasting error. The week to week error μ_1 accounts for the dissimilarity between last observed Dt_i and the FCM-generated Dt_i (i.e., previously revealed by C_i) which therefore acts against ρ. μ_1 could therefore be modelled as $1-\rho$. Time to time demand error μ_2 is the difference between demand levels Dt_i and Dt_{i-1}, and is related to consistency of usage. The recent forecasting error remains as a corrective feedback with gain ε, which overall results in modification of Eq. 7 to Eq. 8. However, this time the feedback gain ε follows the curve's variance (of last three observations) rather than a fixed value. It is also important to avoid over training as addressed in [56], and therefore FCM cycles (τ_{max}) are kept to minimal in order to increase accuracy of forecast rather based on the observed data, almost similar to a higher gain in a Kalman-based implementation.

3.2 Forecast Using 7-Days Data, Weekly Rolling Window, and Serial Updating

The model described in the previous section is tested using actual dataset. The obtained results are given in Sect. 4.1 along with comparison against similar works in the literature. This section, presents an improved version of the above system that facilitates forecasting up to one day lead time. The structure of the forecast model, although new, is fundamental and solely based on FCM working principles. In this method, forecasting for any shorter lead time (e.g., 1 h, 6 h) could be made even more accurately than the benchmark (one day lead time). A general concept of the forecast model is presented in Fig. 2.

A single 336-node state vector of past week (PW) records the initial dataset being 7 × 48 pieces of observed data in one week, from Dt_1(d:hh:mm) until Dt_{336} (d + 7:hh:mm-00:30). An FCM is trained through application of the differential rule in Eq. 6 on the state vector PW. The adjacency model (W) is derived to hold FCM at convergence. Since W provides a model of mutual relationships among all usage levels in one week, it is suggested that W can be generalized to rebuild the data of

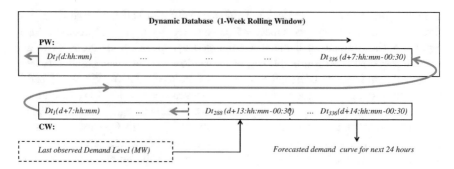

Fig. 2 Schematic diagram of the second forecast expert based on serial updating of data. *Bold arrows* show serial flow of data along and from the current week (CW) to and along the past week (PW)

the following week as also shown in the experiments in Sect. 4.1. With the current observation (let us name it Dt_{288}) occurring at (d + 13:hh:mm-00:30) during the current week (CW), and other known data from first day-time of CW until Dt_{288}, the model must predict the remaining data of CW that is up to one day ahead (Dt_{289} (d + 13:hh:mm) until Dt_{336} (d + 14:hh:mm-00:30)).

The solution is straight forward based on loading known data into respective nodes in the trained FCM and then running for stable values of unknown nodes, i.e., forecasting one day ahead. The forecast trials are then repeated upon every new half-hourly observation while the one-week rolling window (left to right) updates the training set (PW) in a serial fashion. As a result, there will be 336 forecast trials (each with one day lead time) for forecasting seven successive days that can also be repeated for following weeks dynamically.

3.3 Benchmark FCM Forecast Model with Parallel-Serial Database Updating

This section re-examines FCM method using a two dimensional dataset updating strategy. We use merely an unsupervised FCM with no support from other smoothing, regression, or forecast techniques. Following three classic naive forecasting rules (Eqs. 9–11) this method is going to serve as an FCM-based benchmark model (in Sect. 4.2). The database consists of a weekly rolling window, updating itself as well as the past two weeks in serial (by shifting to left) with every half hourly observation. Accordingly, the following 24 h as targeted for forecasting (*Forecast Day*) will have its counterparts in parallel from the past two weeks. The training set is selected from four days as follows. Previous day (Eq. 9), last day of the same type (whether holiday or working day) (Eq. 10), same day in the past week (Eq. 11), and same day in two weeks ago (extension of Eq. 11).

$$D(d : hh : mm) = D(d - 1 : hh : mm) \tag{9}$$

$$D(d_{type} : hh : mm, type) = D(d - 1_{type} : hh : mm, type) \tag{10}$$

$$\begin{aligned} and, \quad type &\in \{weekday, weekend(or\ holiday)\} \\ D(d : hh : mm) &= D(d - 7 : hh : mm) \end{aligned} \tag{11}$$

This selection allows for very small yet most relevant training set which encompasses naive techniques in the literature [56]. It does relate a day to its counterparts in past, while at the same time FCM inference relates all intraday load levels to one another within the day. There are methods in the literature which used limited hours of the previous days for naive forecast, e.g., recent six hours from previous day [39], in order to keep training set as small as possible. Also, most methods employed feed forward (FF) structures which indeed provide lower computation cost than recurrent structures. However, FCM based implementation is still advantageous as it provides a model of mutual relationships among load levels during different hours of day which is a property beyond FF models. On the other hand, to compensate for higher computation cost in recurrent models we suggest PCA to compress the training set in terms of both variables (intraday readings) and samples (past days). The results obtained using this method along with related PCA is given in Sect. 4.2.

4 Results and Discussion

The obtained results presented in Sect. 4.1 are related to the methods in 3.1 and 3.2, while results in Sect. 4.2 are related to the method in Sect. 3.3 of this article. The dataset of historical demand in Victoria, Australia, has been used for the experiments which are publically available on the AEMO website (Australian Energy Market Operator (AEMO) [4].

4.1 Preliminary Results

Starting from 00:00-00:30 on 8 August 2013 the rolling window moves forward to forecast demand levels in the following weeks. As the rolling window moves, freshly observed data replace old counterparts in the past week. Accordingly, upon completion of second week, the database (and therefore knowledgebase) is prepared for forecasting third week's demand levels. Figure 3a–g include the obtained results on seven days. The maximum absolute intraday forecast errors (i.e., each within one day) are 189, 214, 195, 195, 367, 310, 240 MW, with their maximum of 367 MW

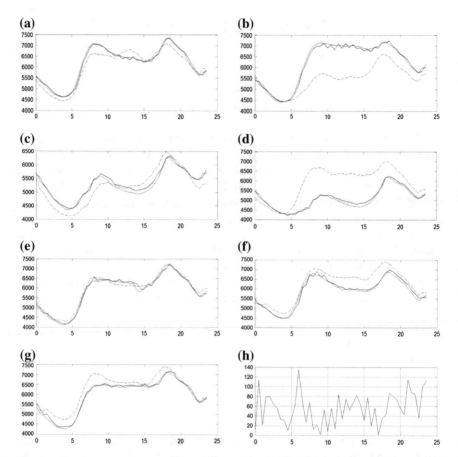

Fig. 3 **a–g** Actual demand levels on 8–14 August 2013 (*solid*), versus actual demand levels on respective days in previous week 1–7 August 2013 (*dotted*). The forecast intraday demand curves are shown with bold-faced perforated lines, **h** seven days average intraday error (with maximum at 6:00 AM)

obtained on the fifth day (Monday), while the average maximum weekly forecast error equals 244 MW. Figure 3h shows the average intraday forecast error (i.e., for the same intervals on the seven days). It is observed that on average, maximum intraday forecast errors occurs at 6:00 AM (around 130 MW) and then at 9:00 PM (around 115 MW) which closely match with peak demand in morning and evening. In the experiments in Fig. 3, optimal values of proportion rate $\rho = 4.892$, and feedback gain $\varepsilon = 0.676$ have been employed while FCM cycles are kept to minimal so that the computation cost for each forecast is as small as around 417 ms on an *Intel* 2.7 GHz machine with 64-b OS and 8 GB RAM. Through the modification of the model, based on week to week and time to time error control, maximum average daily error of 312 MW and average weekly forecast error of 203 MW was obtained on the same dataset, showing slight improvements. Figure 4a–g show the obtained

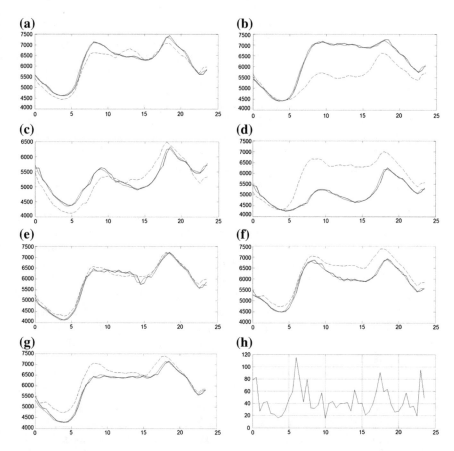

Fig. 4 a–g Similar experiments as presented in Fig. 3, however with control over sources of error, **h** seven-days average intraday error (with maximum at 6:00 AM)

results with 30 min lead time. The maximum daily error is still on Monday, while on average, the maximum intraday error is also at around 6 AM as shown in Fig. 4h.

It must be noted that the surge in error before and after midnight, i.e., the two ends of the error curve, is merely a computation error and has no root in reality. The considerable error at the first point (at 00:00) is due to lack of fresh data and conducting forecast only based on the respective demand level of the same day-time in the previous week. Reversely, the error at the last point (at 23:30) is due to high level of integration between memorized data and the forecast. Best results were obtained with around 4300 FCM cycles per forecast, with maximum daily and mean weekly errors of respectively 291 and 194 MW, and 4.8 s run-time per forecast on the same machine. Maximum intraday forecast error of 115.26 MW is still obtained at 6:00 AM (morning peak demand) while the forecast error related to evening peak demand seems to be slightly harnessed.

The series of experiments (in Figs. 3 and 4), reveal relatively reliable prediction of intraday demand levels with half hour lead time. Maximum intraday forecast error of less than 120 MW was obtained in two successive weeks which seemingly compares well with 325 MW (or adjusted R-squared of 0.7526 for Victoria) in [18] and adjusted R-squared 0.63–0.89 obtained for predicting weekly electricity demand in [59], both using temperature-demand regression models. However, the developed model does not provide predictions of growth in peak and mean demand levels which are provided in [18, 59], that is due to lack of direct influence from climate-related variables.

Another shortcoming is that unfortunately in both experiments the maximum average intraday error occurs at around morning peak (around 06:00 AM) although the forecasting error for the more serious peak in evening is controlled especially using the modified model (Fig. 4). Better forecasts for evening peak as compared against morning peak is obviously due to availability of more intraday observations in the evening. This signifies that FCM training window must be rolling on half-hour basis rather than on daily intervals so that consistent training will be performed prior to each half-hourly forecast.

Accordingly, instead of parallel database updating method (Sect. 3.1) serial updating (Sect. 3.2) was employed expecting more consistent intraday forecast error on and off peak. Figure 5 shows an example of forecasting one day ahead on half-hour basis with 12.73 % averaged absolute error on a typical weekday (at left) being the first day.

Despite serial database updating, it is still realized qualitatively that as more intraday samples are received in the current day, forecast error decreases for the remainder of the day, including the second peak, and until the same hour of the following day (at right). Hence, the main drawback is still related to forecasting the first peak in morning which required more extensive work on this model.

The performance of weekly rolling window for shorter lead time is expected to be better than the parallel window. Therefore, to examine this model, a seven days testing set from 15 to 21 August 2013 was forecasted on half-hour basis and with 12 h lead time. Figure 6 shows the forecasts made exactly at noon and midnight

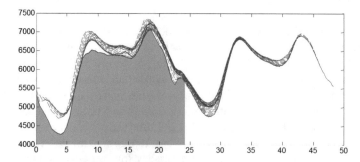

Fig. 5 Forecasting electricity demand (MW) up to one day lead time using weekly rolling window. While the rolling window covers a day (at *left*), the following day (at *right*) is being predicted (on half hour basis)

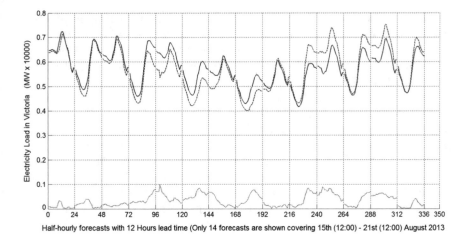

Fig. 6 Forecasting electricity demand (MW) with 12 h lead time using weekly rolling window. Absolute error between the actual load levels (*dotted*) and forecast loads (*solid*) does not exceed 100 MW (shown at the *bottom*). The one week forecast shows MAPE of 5.8717 % purely using FCM

intervals for the following 12 h. The maximum absolute error between the actual and forecast load levels falls below 100 MW with the MAPE of $M = 5.8717$ % (Eq. 12). The obtained error shows better performance than the parallel model, however, it does not compare well against AEMO's MAPE of 2.4 % for Victorian demand on a similar horizon (August 2013 [60]).

$$M = \frac{100\%}{336} \sum_{j=1}^{7} \sum_{i=1}^{48} \left| \frac{Dt_{i|i}(d+j) - Dt_{i|i-1}(d+j)}{Dt_{i|i}(d+j)} \right| , d \, is \, 15, \, Aug \qquad (12)$$

4.2 FCM-Based Benchmark Forecasting Model

The strategies in previous models were incorporated into one, as described in Sect. 3.3. This section describes the related experimental work for daily forecast of seven successive days (15 through 21 August 2013). The results signify that the four-days database and derived FCM-based knowledgebase provide a reliable naive forecasting platform, however, this has to be further analyzed for different types of days. An advantage is about flexibility of FCM graph to expansion so that it can serve as the base for more sophisticated techniques to build upon, where clean data or clear knowledge of other multivariate (e.g., temperature) is available.

Depending on availability of new data, such as temperature observations, new nodes can be added or removed from the graph without disturbing other variables or the ongoing sequence of forecasts. The inference process for tuning graph's adjacency matrix simply adapts to existence of new variables by expanding the matrix at both dimensions. Other than techniques to be integrated into FCM inference, the final outputs (sequence of forecast) can be given to separate modules such as smoothers (e.g., Kalman filter) and other time series analysis algorithms.

Figure 7a–g show the forecast results on the seven test days along with the maximum errors and respective observations given in Table 1. The maximum absolute error of 658.4 MW on one day lead time, and estimated MAPE error of $M = 8.32\%$ has been obtained on the small testing set, which accounts for 91.7 % average accuracy of the model. Exact MAPE value however could be obtained from Eq. 12. To further investigate the properties of the system and interpretation of the obtained results, a PCA was conducted on both training and testing sets. The three weeks dataset (1–21 August) is analyzed where variables are the 48 half hourly

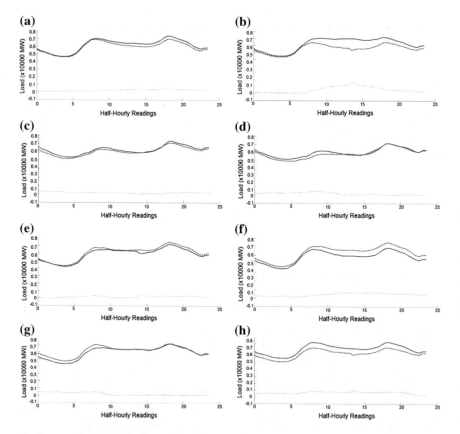

Fig. 7 **a–g** Obtained daily forecast for Thursday 15 through Wednesday 21 Aug, 2013, using FCM-based naive forecasting with 4-days dataset. Actual and forecast intraday load levels are indicated with dotted and solid curves, respectively. Bottom lines show the forecast error. **h** Forecast with revised rules for Friday 16

Table 1 The obtained errors from the experiments in Fig. 7

Test day	Thu 15 Aug	Fri 16 Aug	Sat 17 Aug	Sun 18 Aug	Mon 19 Aug	Tue 20 Aug	Wed 21 Aug
Max. error	444 MW	1667 MW	349 MW	441 MW	447 MW	757 MW	504 MW
	37	28	1	17	29	34	16
Period	18:00–18:30	13:30–14:00	00:00–00:30	08:00–08:30	14:00–14:30	18:30–17:00	07:30–08:00

readings and the samples are the 21 days. The Scree test gives a sharp elbow on the second component which makes analysis straightforward mainly using PC1 and PC2.

Figure 8 shows score plots on PC1–PC2 (a) and PC1–PC3 (b). The scores on PC1–PC3 at least show the distinction between working days and weekends, while in addition, scores on PC1–PC2 plot show clear distinction between Mondays, from the rest of the weekdays. The three midweek days (Tuesdays, Wednesdays, and Thursdays) sit together. Fridays however are on islands which indicate dissimilar behavior to other days as well as to one another that might be due to their busy and diverse schedules. This single fact justifies the highest error of 1667 MW on Friday at around 2 PM. Weekends, although barely grouped together still reveal reliable conclusions. For example, it is well observed that 17 (Saturday) is close to 11 (Sunday) being the last day of the same type (Eq. 10), as well as to 10 (Saturday) in the past week (Eq. 11). However, it is barely related to its preceding day 16 (Friday) which defies the rule in Eq. 9. On the other hand, Wednesdays and Thursdays

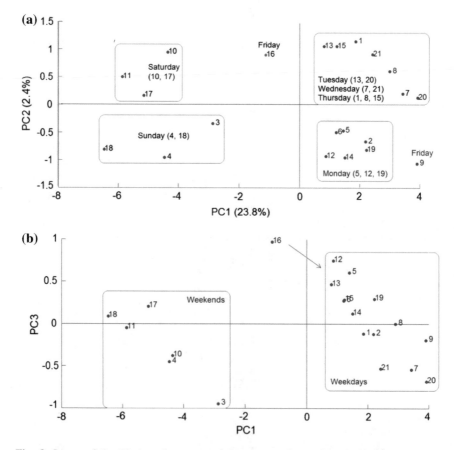

Fig. 8 Scores of the 21 days dataset reveal facts yet to be confirmed using larger dataset throughout the year

support the same rule as they show close behavior with their preceding days. Mondays and Tuesdays are exceptional in that they only follow their counterparts in previous weeks. Accordingly, more counterpart days in series of past weeks can be considered instead of only one week in Eq. 11. Inspired by this analysis, the basic rules in Eq. 9–11, or any additional rule, could be tailored for each individual day. Indeed a more extensive analysis is required such as using larger dataset from different seasons before confirming the naive rules. Another benefit of such analysis is in reduction of redundant samples while using larger dataset. Days which are confirmed to be in one cluster can be represented with limited members of that cluster which leads to compression of dataset (in terms of number of samples) without losing the traits.

Another strategy for dataset compression is based on reduction of correlated variables (half hourly readings) by grouping and then representing groups with their key members that is different from reduction via transformation to PCs space. The loading plots in Fig. 9 show this strategy which results in reduction from 48 to 30 variables, where selection criterion is merely based on the Euclidean distance to the group's centre of shape. The grouping itself is for up to four members per group unless additional members fall within a threshold distance on PC1 to the centre of shape. Another fact is that readings 10–20 (between 5 and 10 AM) are most volatile. The biplots show most Mondays, Fridays, and Sundays are related with 11 AM and onwards (negative on PC2), while other days are more related to other variables (midnight until 11 AM). Extensive PCA is required on yearly data to customize the rules, which eventually lead to a development of a forecasting platform merely using demand curves and no other data.

For example, Fig. 10 shows the intraday loadings for 73 selected days covering the entire year 2013. While most readings fall within interlinked clusters at the left and right extremes, it is confirmed that the 15th–21st (07:30-10:30) and then 39th–42nd (19:30-21:00) readings are notably more volatile and therefore should be treated more independently, while an optimization technique can be incorporated for selection of other representatives from the clusters.

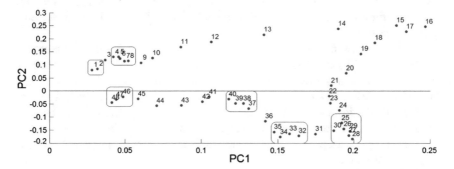

Fig. 9 The loadings on the 48 readings. Grouping allows to reduce readings to 30 instances: 1, 6, 9, 10, 11, 12, 13, 14, 15, 16, 17, 18, 19, 20, 21, 22, 23, 24, 27, 30, 31, 33, 36, 38, 41, 42, 43, 44, 45, 47, half hours

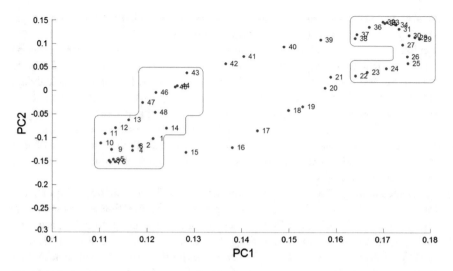

Fig. 10 The loadings on the 48 variables for 73 selected days (on every five days basis) along the year 2013

5 Conclusion

Weekly patterns provide reliable basis for forecasting, however, there are exceptions such as irregular public holidays, most Fridays and Mondays, or days which climate and other variables change tremendously. It is therefore concluded that a highly reliable forecast platform must take account of more system behaviours, upon which forecasting rules intelligently evolve. For example, as a conclusion and proof of concept, the analysis in this section is used to modify the rules for Friday 16th on the assumption that it is close only to its preceding day.

Accordingly, the training set changes from the four-days (as used for other days) to a two-days training set, i.e., only considering the previous day, and the last day of the same type (rules in Eq. 9 and 10). Figure 7h shows the obtained result with reduced error down to 989 MW, still at around 2 PM. The overall system accuracy is increased to 93 %.

Apart from customizing the training set, and utilization of larger yet compressed dataset, the FCM-generated knowledgebase can also be improved in terms of activation and training [37, 43]. Accordingly, the future work involves three steps: to conduct more extensive analysis to confirm a robust naive benchmark based on weekly patterns in one year for forecasting up to one day ahead, to elaborate a relationship model for concurrent analysis of multivariate interactions, and to implement this concept in a small scale, such as individual household level, or statistical local area.

This study is part of the Simulation of Electricity Demand project, at the CSIRO Ecosystem Sciences Urban Systems Program, in Victoria, AU. The authors are grateful to the internal and external peer reviewers, and to Dr. Chi-Hsiang Wang, whose comments improved this submission, and the Australian Energy Market Operator (AEMO) for their reliable electricity demand dataset.

References

1. A. Abraham, B. Nath, A neuro-fuzzy approach for modelling electricity demand in Victoria. Appl. Soft Comput. **1**(2), 127–138 (2001)
2. H. Liao, D. Niebur, Exploring independent component analysis for electric load profiling, in *IJCNN '02. Proceedings of the 2002 International Joint Conference on Neural Networks*, vol. 3 (2002), pp. 2144–2149
3. C.-H. Wang, G. Grozev, S. Seo, Decomposition and statistical analysis for regional electricity demand forecasting. Energy **41**(1), 313–325 (2012)
4. AEMO, Australian Energy Market Operator's Historical demand. (AEMO Web, 2013), http://www.nemweb.com.au/REPORTS/Archive/HistDemand/. Accessed 27 Oct 2013
5. L. Hernández, C. Baladrón, J.M. Aguiar, L. Calavia, B. Carro, A study of the relationship between weather variables and electric power demand inside a smart grid. Sensors **12**, 11571–11591 (2012)
6. K. Pilli-Sihvola, P. Aatola, M. Ollikainen, H. Tuomenvirta, Climate change and electric power consumption—witnessing increasing or decreasing use and costs? Energy Policy **38**, 2409–2419 (2010)
7. Y. Messaoud, H.Y.H. Chen, D.Q. Fuller, The influence of recent climate change on tree height growth differs with species and spatial environment. PLoS One **6**, e14691 (2011)
8. S. Parkpoom, G.P. Harrison, Analyzing the impact of climate change on future electric demandin Thailand. IEEE Trans. Power Syst. **23**, 1441–1448 (2008)
9. U. Wilke, F. Haldi, J.-L. Scartezzini, D. Robinson, A bottom-up stochastic model to predict building occupants' time-dependent activities. Build. Environ. **60**, 254–264 (2013)
10. U. Wilke, Probabilistic bottom-up modelling of occupancy and activities to predict electricity demand in residential buildings, thesis, Solar Energy and Building Physics Laboratory, EPFL, Switzerland (2013)
11. H. Wilhite, A socio-cultural analysis of changing household electricity consumption in india, in *Tackling Long Term Global Energy Problems*, Environment and Policy, ed. by D. Spreng, T. Flüeler, D.L. Goldblatt, J. Minsch (Springer, Dordrecht, 2012), pp. 97–113
12. Z. Atakhanova, P. Howie, Electricity demand in Kazakhstan. Energy Policy **35**(7), 3729–3743 (2007)
13. V. Bianco, O. Manca, S. Nardini, Electricity consumption forecasting in Italy using linear regression models. Energy **34**(9), 1413–1421 (2009)
14. A. Shiu, P.-L. Lam, Electricity consumption and economic growth in China. Energy Policy **32**(1), 47–54 (2004)
15. M. Filippini, Swiss residential demand for electricity by time-of-use. Resour. Energy Econ. **17**(3), 281–290 (1995)
16. J.M. Griffin, the effects of higher prices on electricity consumption. Bell J. Econ. **5**(2), 515–539 (1974). The RAND Corporation
17. J.A. Espey, M. Espey, Turning on the lights: a meta-analysis of residential electricity demand elasticities. J. Agric. Appl. Econ. **36**(1), 65–81 (2004)
18. M.J. Thatcher, Modelling changes to electricity demand load duration curves as a consequence of predicted climate change for Australia. Energy **32**, 1647–1659 (2007)
19. J. Moral-Carcedo, J. Vicéns-Otero, Modelling the non-linear response of Spanish electricity demand to temperature variations. Energy Econ. **27**(3), 477–494 (2005)
20. D.J. Sailor, J.R. Munoz, Sensitivity of electricity and natural gas consumption to climate in the USA—methodology and results for eight states. Energy **22**, 987–998 (1997)

21. D.J. Sailor, Relating residential and commercial sector electricity loads to climate—evaluating state level sensitivities and vulnerabilities. Energy **26**, 645–657 (2001)
22. R. Steadman, A universal expression of apparent temperature. J. Appl. Meteorol. **23**, 1674–1687 (1984)
23. A. Harveya, S.J. Koopmana, Forecasting hourly electricity demand using time-varying splines. J. Am. Stat. Assoc. **88**, 424 (1993)
24. R. Cottet, M. Smith, Bayesian modeling and forecasting of intraday electricity load. J. Am. Stat. Assoc. **98**(464), 839–849 (2003)
25. A.W.L. Yao, S.C. Chi, J.H. Chen, An improved grey-based approach for electricity demand forecasting. Electr. Power Sys. Res. **67**(3), 217–224 (2013)
26. J.T. Connor, A robust neural network filter for electricity demand prediction. J. Forecast. **15**(6), 437–458 (1996)
27. G.A. Darbellay, M. Slama, Forecasting the short-term demand for electricity: do neural networks stand a better chance? Int. J. Forecast. **16**(1), 71–83 (2000)
28. T. Al-Saba, I. El-Amin, Artificial neural networks as applied to long-term demand forecasting, Artif. Intell. Eng. **13**(2), 189–197 (1999)
29. P.-F. Pai, W.-C. Hong, Forecasting regional electricity load based on recurrent support vector machines with genetic algorithms. Electr. Power Syst. Res. **74**(3), 417–425 (2005)
30. W. Bartkiewicz, Neuro-fuzzy approaches to short-term electrical load forecasting, in *Proceedings of the IEEE-INNS-ENNS International Joint Conference on Neural Networks*, vol. 6 (2000), pp. 229–234
31. P.-C., Chang, C.-Y. Fan, J.-J. Lin, Monthly electricity demand forecasting based on a weighted evolving fuzzy neural network approach. Int. J. Electr. Power Energy Syst. **33**(1), 17–27 (2011)
32. S. Roychowdhury, Fuzzy curve fitting using least square principles, in *IEEE International Conference on Systems, Man, and Cybernetics*, vol. 4, Oct 1998 (1998), pp. 4022–4027
33. K.H. Kim, J.K. Park, K.J. Hwang, S.H. Kim, Implementation of hybrid short-term load forecasting system using artificial neural networks and fuzzy expert systems. IEEE Trans. Power Syst. **10**, 1534–1539 (1995)
34. O. Motlagh, Z. Jamaludin, S.H. Tang, W. Khaksar, An agile FCM for real-time modeling of dynamic and real-life systems. Evolving Syst.: Spec. Issue Temporal Aspects Fuzzy Cogn. Maps (2013)
35. G. Zhang, B.E. Patuwo, M.Y. Hu, Forecasting with artificial neural networks: the state of the art. Int. J. Forecast. **14**, 35–62 (1998)
36. M.R. Khan, Č. Ondrůšek, The hopfield model for short-term load prediction, in *2nd Spring International Power Engineering Conference*, UVEE, FEI, Brno University of Technology, Czech Republic (2001), pp. 81–85
37. O. Motlagh, S.H. Tang, W. Khaksar, N. Ismail, An alternative approach to FCM activation for modeling dynamic systems. Appl. Artif. Intell. **26**(8), 733–742 (2012)
38. G.K.F. Tso, K.K.W. Yau, Predicting electricity energy consumption: a comparison of regression analysis, decision tree and neural networks. Energy **32**(9), 1761–1768 (2007)
39. J.W. Taylor, L.M. De Menezes, P.E. McSharry, A comparison of univariate methods for forecasting electricity demand up to a day ahead. Int. J. Forecast. **22**(1), 1–16 (2006)
40. O.G. Santin, Behavioural patterns and user profiles related to energy consumption for heating. Energy Build. **43**(10), 2662–2672 (2011)
41. J.V. Ringwood, D. Bofelli, F.T. Murray, Forecasting electricity demand on short, medium and long time scales using neural networks. J. Intell. Rob. Syst. **31**(1–3), 129–147 (2001)
42. K. Kandananond, Forecasting electricity demand in thailand with an artificial neural network approach. Energies **4**, 1246–1257 (2011)
43. O. Motlagh, S.H. Tang, M.N. Maslan, F.A. Jafar, M.A. Aziz, A novel graph computation technique for multi-dimensional curve fitting. Connection Sci. **25**(2–3), 129–138 (2013)
44. D.O. Hebb, *The Organization of Behavior: A Neuropsychological Theory* (Wiley, New York, 1949)

45. M. Negnevitsky, *Artificial intelligence:a guide to intelligent systems* (Pearson Education Ltd, England, 2005)
46. J.A. Dickerson, B. Kosko, Virtual worlds as fuzzy cognitive maps. Presence **3**(2), 173–189 (1994)
47. A. Vazquez, A balanced differential learning algorithm in fuzzy cognitive maps, Technical report, Departament de Lenguatges I Sistemes Informatics, Universitat Politecnica de Catalunya (UPC) (2002)
48. E.I. Papageorgiou, C.D. Stylios, P.P. Groumpos, Fuzzy cognitive map learning based on nonlinear Hebbian rule, in *Australian Conference on Artificial Intelligence* (2003), pp. 256–268
49. E.I. Papageorgiou, C.D. Stylios, P.P. Groumpos, Active Hebbian learning algorithm to train fuzzy cognitive maps. Int. J. Approx. Reason. **37**(3), 219–247 (2004)
50. S. Soleymani, A.M. Ranjbar, A.R. Shirani, A new structure for electricity market scheduling power electronics, in *PEDES International Conference on Drives and Energy Systems* (2006), pp. 1–5
51. A.J. Conejo, J.M. Morales, L. Baringo, Real-time demand response model. IEEE Trans. Smart Grid **1**(3), 236–242 (2010)
52. S.S. Pappas, L. Ekonomou, D.C. Karamousantas, G.E. Chatzarakis, S.K. Katsikas, P. Liatsis, Electricity demand loads modeling using Auto-Regressive Moving Average (ARMA) models. Energy **33**(9), 1353–1360 (2008)
53. S.S. Pappas, L. Ekonomou, P. Karampelas, D.C. Karamousantas, S.K. Katsikas, G.E. Chatzarakis, P.D. Skafidas, Electricity demand load forecasting of the Hellenic power system using an ARMA model. Electric Power Syst. Res. **80**(3), 256–264 (2010)
54. E. Erdogdu, Electricity demand analysis using cointegration and ARIMA modelling: a case study of Turkey. Energy Policy **35**(2), 1129–1146 (2007)
55. J.W. Taylor, Short-term electricity demand forecasting using double seasonal exponential smoothing. J. Oper. Res. Soc. **54**, 799–805 (2003)
56. H.S. Hippert, D.W. Bunn, R.C. Souza, Large neural networks for electricity load forecasting: are they overfitted? Int. J. Forecast. **21**(3), 425–434 (2005)
57. V. Gogate, R. Dechter, B. Bidyuk, C. Rindt, J. Marca, Modeling transportation routines using hybrid dynamic mixed networks, in *Proceedings of the Twenty-First Conference on Uncertainty in Artificial Intelligence (UAI2005)* (2005), pp. 217–224
58. AEMO, Australian Energy Market Operator's. (AEMO Website, 2013), http://www.aemo.com.au/Electricity/Data/Forecast-Supply-and-Demand/7-Day-Outlook. Accessed 13 Dec 2013
59. S. Howden, S. Crimp, Effect of climate and climate change on electricity demand, in *Modsim 2001 International congress on modelling and simulation* (2000), pp. 655–660
60. AEMO, Australian Energy Market Operator's pre-dispatch demand forecasting for August 2013, published 2 September 2013. (AEMO Website, 2013b), http://www.aemo.com.au/Electricity/Data/PreDispatch-Demand-Fore-casting-Performance. Accessed 9 Dec 2013

Author Biographies

Omid Motlagh is a research engineer with the Commonwealth Scientific and Industrial Research Organisation (CSIRO), in Melbourne, Australia. His research interest is complexity, chaos, and development of machine learning strategies for electricity demand modeling and forecast. He has introduced and applied *natural* and *cumulative* activation to computational geometry, computer graphics, robotics and autonomy, physics-based modelling, pattern recognition, forecast, diagnosis, deduction, and other complex problems. He is a member of the Australian Society for Operations Research.

George Grozev is a principal research scientist and a research group leader, "Buildings, Utilities and Infrastructure" in the Urban Systems Program, CSIRO Ecosystem Sciences, Melbourne, Australia. His background is in operations research and he has extensive experience in energy market modelling and simulation, network theory, and algorithms and software development. During his tenure at CSIRO, he has carried out research and development in the areas of high performance infrastructure, sustainable development, electricity, and gas markets. He is a member of the Australian Society for Operations Research.

Elpiniki I. Papageorgiou is an Assistant Professor at the Department of Computer Engineering of the Technological Educational Institute of Central Greece, in Lamia, Greece. She has been actively involved in several research projects, funded by national and international organizations, including the European Union and the Greek Research Foundation, related with the development of new methodologies based on soft computing and artificial intelligent techniques for solving complex systems.

Development of Artificial Intelligence Based Regional Flood Estimation Techniques for Eastern Australia

Kashif Aziz, Ataur Rahman and Asaad Shamseldin

Abstract This chapter focuses on the development of artificial intelligence based regional flood frequency analysis (RFFA) techniques for Eastern Australia. The techniques considered in this study include artificial neural network (ANN), genetic algorithm based artificial neural network (GAANN), gene-expression programing (GEP) and co-active neuro fuzzy inference system (CANFIS). This uses data from 452 small to medium sized catchments from Eastern Australia. In the development/training of the artificial intelligence based RFFA models, the selected 452 catchments are divided into two groups: (i) training data set, consisting of 362 catchments; and (ii) validation data set, consisting of 90 catchments. It has been shown that in the training of the four artificial intelligence based RFFA models, no model performs the best across all the considered six average recurrence intervals (ARIs) for all the adopted statistical criteria. Overall, the ANN based RFFA model is found to outperform the other three models in the training. Based on an independent validation, the median relative error values for the ANN based RFFA model are found to be in the range of 35–44 % for eastern Australia. The results show that ANN based RFFA model is applicable to eastern Australia.

1 Introduction

Flood is one of the worst natural disasters that cause significant impacts on economy. About 951 people lost their lives and another 1326 were injured in Australia by floods during 1852–2011 [1]. During 2010–11, over 70 % of Queensland state

K. Aziz · A. Rahman (✉)
School of Computing, Engineering and Mathematics, Western Sydney University Australia, Building XB, Kingswood, Locked Bag 1797, Penrith, NSW 2751, Australia
e-mail: a.rahman@westernsydney.edu.au

A. Rahman
Institute for Infrastructure Engineering, Western Sydney University, Penrith, NSW, Australia

A. Shamseldin
University of Auckland, Auckland, New Zealand

© Springer International Publishing Switzerland 2016
S. Shanmuganathan and S. Samarasinghe (eds.), *Artificial Neural Network Modelling*, Studies in Computational Intelligence 628,
DOI 10.1007/978-3-319-28495-8_13

was affected by severe flooding, with the total damage to public infrastructure was estimated to be \$5 billion [2]. Design flood estimate, a probabilistic flood magnitude, is used in engineering design to safeguard water infrastructures and to minimise the overall flood damage. To estimate design floods, at-site flood frequency analysis is the most commonly adopted technique; however, at the locations where streamflow record is unavailable or is of limited length or of poor quality, regional flood frequency analysis (RFFA) is generally adopted to estimate design floods. A RFFA technique attempts to use flood data from a group of homogeneous donor gauged catchments to make flood estimation at ungauged location of interest.

For developing the regional flood prediction equations, the commonly used techniques include the rational method, index flood method and quantile regression technique. These techniques generally adopt a linear method of transforming inputs to outputs. Since hydrologic systems are often non-linear, RFFA techniques based on non-linear methods can be a better alternative to linear methods. Among the non-linear methods, artificial intelligence based techniques have been widely adopted in various water resources engineering problems. However, their application to RFFA problems is quite limited.

This chapter presents the development of artificial intelligence based RFFA methods for eastern Australia. The non-linear techniques considered in this chapter are artificial neural network (ANN), genetic algorithm based artificial neural network (GAANN), gene-expression programing (GEP) and co-active neuro fuzzy inference system (CANFIS).

2 Review of Artificial Intelligence Based Estimation Methods in Hydrology

RFFA essentially consists of two principal steps: (i) formation of regions; and (ii) development of prediction equations. Regions have traditionally been formed based on geographic, political, administrative or physiographic boundaries [3, 4]. Regions have also been formed in catchment characteristics data space using multivariate statistical techniques [5, 6]. Regions can also be formed using a region-of-influence approach where a certain number of catchments based on proximity in geographic or catchment attributes space are pooled together based on an objective function to form an optimum region [7–9].

For developing the regional flood prediction equations, the commonly used techniques include the rational method, index flood method and quantile regression technique (QRT). The rational method has widely been adopted in estimating design floods for small ungauged catchments [4, 10–12]. The index flood method has widely been adopted in many countries, which relies on the identification of homogeneous regions [13–18]. The QRT, proposed by the United States Geological Survey (USGS), has been applied by many researchers using either an Ordinary Least Square (OLS) or Generalized Least Square (GLS) regression techniques [19–27].

Most of the above RFFA methods assume linear relationship between flood statistics and predictor variables in log domain while developing the regional prediction equations. However, most of the hydrologic processes are nonlinear and exhibit a high degree of spatial and temporal variability and a simple log transformation may not guarantee achievement of linearity in modeling. Therefore, there have been applications of artificial intelligence based methods such as artificial neural networks (ANN), genetic algorithm based ANN (GAANN), gene expression programming (GEP) and co-active neuro-fuzzy inference system (CANFIS) in water resources engineering such as rainfall runoff modeling and hydrologic forecasting, but there have been relatively few studies till to-date involving the application of these techniques to RFFA [28–33]. Application of these techniques may help developing new improved RFFA techniques for Australia, which experiences a highly variable rainfall and hydrologic conditions. Unlike regression based approach, the artificial intelligence based techniques do not impose any fixed model structure on the data rather the data itself identifies the model form through use of artificial intelligence.

3 Methodology

Since the first neural model by McCulloch and Pitts [34], there have been developments of hundreds of different models which are considered ANN. The differences in them might be the functions, the accepted values, the topology, the learning algorithms, and the like. Since the function of ANN is to process information, they are used mainly in fields related to information processing. There are a wide variety of ANN that are used to model real neural networks, and study behaviour and control in animals and machines, but also there are ANN which are used for engineering purposes such as pattern recognition, flood forecasting, and data compression.

In this study, the adopted ANN modelling, Lavenberg-Marquardt method was used as the training algorithm to minimize the mean squared error (MSE) between the observed and predicted flood quantiles. The purpose of training an ANN with a set of input and output data is to adjust the weights in the ANN to minimize the MSE between the desired flood quantile and the ANN predicted flood quantile. Three hidden-layered neural networks were selected with 7, 3 and 1 neurons to each of these three layers. Two inputs, catchment area (A) and rainfall intensity with duration equal to time of concentration (t_c) and a given average recurrence interval (ARI) were used in one input layer and one output layer with one output called predicted flood quantile (Q_{pred}). The transfer function used for the hidden layers and the output layer was all hyperbolic tangent sigmoid function. Transfer functions calculate a layer's output from its net input. A maximum training iteration of 20,000 was adopted. All dependent and independent variables were standardized to the range of (0.05, 0.95), so that extreme flood events, which exceeded the range of the training data set could be modelled between the boundaries (0, 1) during testing.

A learning rate of 0.05 was used together with a momentum constant of 0.95. MATLAB was used to perform the ANN training. To select the best performing model the different combinations of hidden layers, algorithm, and number of neurons were observed against the MSE value. In order to obtain the best ANN-based model, the MSE values between the observed and predicted flood quantiles were calculated and the training was undertaken to minimize this error.

The major difference between GA and the classical optimization search techniques is that the GA works with a population of possible solutions; whereas, the classical optimization techniques work with a single solution [35]. GA is based on the Darwinian-type survival of the fittest strategy, whereby potential solutions to a problem compete and mate with each other in order to produce increasingly stronger individuals. Each individual in the population represents a potential solution to the problem that is to be solved and is referred to as a chromosome [36]. An initial population of individuals (also called chromosomes) is created and according to an objective function in focus the fitness values of all chromosomes is evaluated. From this initial population parents are selected who mate together to produce off springs (also called children). The genes of parents and children are mutated. The fittest among parents and children are sent to a new pool. The whole procedure is carried over until any of the two stopping criteria is met i.e. the required number of generations has been reached or convergence has been achieved. Chromosomes are the basic unit of population and represent the possible solution vector; they are assembled from a set of genes that are generally binary digits, integers or real numbers [37, 38].

An initial population is crowded with "n" number of chromosomes where "n" is referred to as the population size. An objective function comprising of feed forward ANN model with complete description of its architecture is defined. It reads training patterns once at the start of model and stores them in memory for applying to each chromosome. The total number of genes l of each chromosome represents the total synaptic weights of ANN model.

$$\{g_1, g_2, \ldots g_l\} = \{w_{(if \to hr)}, w_{(ib \to hr)}, w_{(hf \to or)}, w_{(hb \to or)}\} \qquad (1)$$

where 'w' represents the value of a synaptic weight, subscript 'i' represents a node of input layer, 'h' is a node of hidden layer and 'o' represents the output layer node, 'f' is serial number of node which forwards the information (i.e. $f = 1, 2, 3, \ldots$), 'r' is serial number of node which receives information (i.e. $r = 1, 2, 3, \ldots$), 'ib' represent the bias node of input layer and 'hb' is bias node of hidden layer.

At the start of model, the fitness values of all the chromosomes of population are evaluated by ANN function. The real values stored in the genes of chromosome are read as the respective weights of ANN model. The ANN performs feed forward calculations with the weights read from genes of forwarded chromosome, and calculates MSE. The inverse of MSE is regarded as the fitness value of chromosome. By this way, the fitness values of all chromosomes of initial population are calculated by ANN function.

The selection operator selects two parent chromosomes randomly. The roulette wheel operator with elitism is used in this model. Elitism is a scheme in which the best chromosome of each generation is carried over to the next generation in order to ensure that the best chromosome does not lost during the calculations. The selected parents are mated to produce two children having the same number of genes. The uniform crossover operator is used with a crossover rate of $p_c = 1.0$. In uniform crossover, a toss is done at each gene position of an offspring and depending upon the result of toss, the gene value of 1st parent or 2nd parent is copied to the offspring. The genes of children are then mutated with the swap mutation operator with a mutation rate of $p_m = 0.8$. The mutated children are then evaluated by ANN function to know their fitness values. The fitness values of all the four chromosomes (2 parents & 2 children) are compared and the two chromosomes of highest fitness values are then sent to a new population and the other two are abolished. The evolutionary operators continue this loop of selection, crossover, mutation and replacement until the population size of new pool is same as old pool. One generation cycle completes at this stage and process is repeated until any of two stopping criteria is fulfilled i.e. maximum number of generations are reached or the convergence has been achieved. And the best chromosome which is tracked so far through the number of generations is sent to the ANN function. The genes of best chromosome are read as weights of ANN model and represent the optimised weights of ANN model. With these weights, the model is said to be fully trained. Finally, the train and test sets are simulated by using these weights.

The GAANN is coded in C language and some sub routines of LibGA package [39] for evolutionary operators of GA has been used with alterations to read and process the negative real values.

Gene-expression Programming (GEP) is used to perform a non-parametric symbolic regression. Symbolic regression although is very similar to traditional parametric regression, does not start with a known function relating dependent and independent variables as the latter. GEP programs are encoded as linear strings of fixed length (the genome or chromosomes), which are afterwards expressed as nonlinear entities of different sizes and shapes [40–42].

GEP automatically generates algorithms and expressions for the solution of problems, which are coded as a tree structure with its leaves (terminals) and nodes (functions). The generated candidates (programs) are evaluated against a "fitness function" and the candidates with higher performance are then modified and re-evaluated. This modification evaluation cycle is repeated until an optimum solution is achieved. In GEP a population of individual combined model solutions is created initially in which each individual solution is described by genes (sub-models) which are linked together using a predefined mathematical operation (e.g. addition). In order to create the next generation of model solutions, individual solutions from the current generation are selected according to fitness which is based on the pre-chosen objective function. These selected individual solutions are allowed to evolve using evolutionary dynamics to create the individual solutions of the next generation. This process of creating new generations is repeated until a certain stopping criterion is met [43].

Two important components of the GEP include the chromosomes and the expression trees (ETs). The ETs are the expression of the genetic information encoded in the chromosomes. The process of information decoding from chromosomes to the ETs is called translation, which is based on a kind of code and a set of rules. There exist very simple one to one relationships between the symbols of the chromosome and the functions or terminals they represent in the genetic code. To predict the flood quantiles the set of independent variables (predictor variables) to be used in the individual prediction equation are to be identified. Then a set of functions (e.g. e^x, x^a, $sin(x)$, $cos(x)$, $ln(x)$, $log(x)$, 10^x, etc.) and arithmetic operations $(+, -, /, *)$ are defined. The terminals and the functions form the junctions in the tree of a program.

In GEP, k-expressions (from Karva notation) which are fixed length list of symbols are used to represent an ET. These symbols are called chromosomes, and the list is a gene. The Gene "$sqrt, \times, \pm, a, b, c$ and d" can be represented as ET. The GEP gene contains head and a tail. The symbols that represent both functions and terminals are present in the head while tail only contains terminals. The length of the head of the gene h is selected for each problem while the length of the tail is a function of length of the head of the gene.

In order to obtain the best GEP model, the mean squared error was used as 'fitness function', which was based on the observed and predicted flood quantiles; the training was undertaken to minimize this error.

For the CANFIS model development, model catchments were clustered based on model variables (A, I_{tc_ARI}) into several class values in layer 1 to build up fuzzy rules, and each fuzzy rule was constructed through several parameters of membership function in layer 2. A fuzzy inference system structure was generated from the data using subtractive clustering. This was used in order to establish the rule base relationship between the inputs.

In order to obtain the best CANFIS models, the MSE was used as the 'fitness function', which was based on the observed and predicted flood quantiles; the training was undertaken to minimise this error. Lavenberg-Marquardt (LM) method was used as the training algorithm to minimize the MSE. CANFIS model was trained with a set of input and output data to adjust the weights and to minimize the MSE between the desired outputs and the model outputs. The testing data set was selected randomly to produce a reasonable sample of different catchment types and sizes. Two inputs (A, I_{tc_ARI}) were used in one input layer and one output layer with one output (Q_{pred}).

In the case of CANFIS, the bell membership function and the TSK neuro fuzzy model were used, as this type of fuzzy model best fits the multi-input, single output system [44]. LM algorithm was used for the training of CANFIS model. The stopping criteria for the training of the CANFIS network was set to be a maximum of 1000 epochs and training was set to terminate when the MSE drops to 0.01 threshold value.

The following statistical measures were used to compare various RFFA models [45, 46]:

- Ratio between predicted and observed flood quantiles:

$$\text{Ratio of predicted and observed flood quantile} = \frac{Q_{pred}}{Q_{obs}} \qquad (2)$$

- Relative error (RE):

$$\text{RE}\,(\%) = \text{Abs}\left[\frac{(Q_{pred} - Q_{obs})}{Q_{obs}} \times 100\right] \qquad (3)$$

- Coefficient of efficiency (CE):

$$\text{CE} = 1 - \frac{\sum\limits_{i=1}^{n}(Q_{obs} - Q_{pred})^2}{\sum\limits_{i=1}^{n}(Q_{pred} - \bar{Q})^2} \qquad (4)$$

where Q_{pred} is the flood quantile estimate from the ANNs-based or GEP based RFFA model, Q_{obs} is the at-site flood frequency estimate obtained from LP3 distribution using a Bayesian parameter fitting procedure [47, 48] and $\bar{Q}$ is the mean of Q_{obs}. The median relative error and median ratio values were used to measure the relative accuracy of a model. A Q_{pred}/Q_{obs} ratio closer to 1 indicates a perfect match between the observed and predicted value and a smaller median relative error is desirable for a model. A CE value closer to 1 is the best; however a value greater than 0.5 is acceptable.

4 Data Selection

Eastern Australia was selected as the study area, which includes states of New South Wales (NSW), Victoria (VIC), Queensland (QLD) and Tasmania (TAS). This part of Australia was selected since this has the highest density of streamflow gauging stations with good quality data as compared to other parts of Australia. For RFFA study, streamflow data (annual maximum series) and climatic and catchment characteristics data are needed. A total of 452 catchments, which are rural and are not affected my major regulation were selected for this study. The locations of the selected catchments are shown in Fig. 1. A total of 96, 131, 172 and 53 stations from NSW, VIC, QLD and TAS, respectively were selected.

The catchment sizes of the selected 452 stations range from 1.3 to 1900 km^2 with the median value of 256 km^2. For the stations of NSW, VIC and QLD, the upper limit of catchment size was 1000 km^2, however for Tasmania; there were 4 catchments in the range of 1000 to 1900 km^2. Overall, there are about 12 % catchments in the range of 1 to 50 km^2, about 11 % in the range of 50 to 100 km^2, 53 % in the range of 100 to 500 km^2 and 24 % greater than 1000 km^2.

The annual maximum flood record lengths of the selected stations range from 25 to 75 years (mean: 33 years). The Grubbs and Beck [47] method was adopted in detecting high and low outliers (at the 10 % level of significance) in the annual maximum flood series data. The detected low outliers were treated as censored flows in flood frequency analysis. Only a few stations had a high outlier, which was not removed from the data set as no data error was detected for these high flows.

In estimating the flood quantiles for each of the selected stations, log-Pearson III (LP3) distribution was fitted to the annual maximum flood series using Bayesian method as implemented in FLIKE software [48]. According to Australian Rainfall

Fig. 1 Locations of the selected catchments

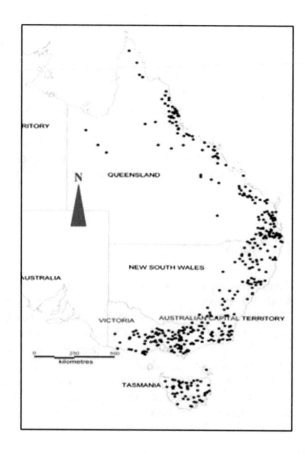

and Runoff (ARR), LP3 is the recommended distribution for at-site flood frequency analysis in Australia [4], and hence it was adopted. In previous applications [27, 49, 50], it has been found that LP3 distribution provide an adequate fit to the Australian annual maximum flood data.

In developing the prediction equations for flood quantiles, initially a total of five explanatory variables were adopted as part of study [45] using the same 452 catchments. These variables are: (i) catchment area expressed in km^2 (A); (ii) design rainfall intensity values in mm/h $I_{tc, ARI}$ (where ARI = 2, 5, 10, 20, 50 and 100 years and t_c = time of concentration (hour), estimated from $t_c = 0.76A^{0.38}$) (iii) mean annual rainfall expressed in mm/y (R); (iv) mean annual areal evapo-transpiration expressed in mm/y (E); (v) main stream slope expressed in m/km (S). In both of these studies where the ANN and Gene Expression programming techniques were used to develop the prediction equation, this was found that two variables (A and $I_{tc, ARI}$) model outperformed other models. Based on this finding, two predictor variables i.e., A and $I_{tc, ARI}$ are selected for this study.

5   Results

At the beginning each of the four artificial intelligence based RFFA models is trained using MATLAB codes (developed as a part of this research) by minimising the mean squared error between the observed and predicted flood quantiles for each of six ARIs (2, 5, 10, 20, 50 and 100 years). This is done using the training data set consisting of 362 catchments.

Table 1 shows the median of the absolute relative error values for the ANN, GAANN, GEP and CANFIS based RFFA models. It can be seen that ANN based RFFA model outperforms the other models with a median RE value of 42.1 % over all the six ARIs. In some cases, the GAANN based RFFA model performs better or equal to the ANN based model i.e. for ARIs of 2, 5, 20 and 100 years; however, for 50 years ARI it shows a very high RE (60 %). In terms of consistency over the ARIs, ANN outperforms the other three models. Both GEP and CANFIS have quite high RE values (GEP = 54.02 %, CANFIS = 59.46 %). Importantly, CANFIS shows very high RE values for 2 years ARI (94.02 %) and 50 years ARI (71.94 %).

Table 1 Median RE (%) values of four artificial intelligence based RFFA models (training)	ARI (years)	ANN	GAANN	GEP	CANFIS
	2	43.75	40.92	73.3	94.02
	5	39.53	39.31	43.91	43.55
	10	39.14	41.01	43.25	45.27
	20	40.38	40.29	54.61	46.07
	50	43.32	60.00	54.22	71.94
	100	46.30	45.28	54.82	55.89
	Overall	42.07	44.47	54.02	59.46

Table 2 Comparison of training and validation results for the ANN based RFFA model

ARI (years)	Training				Validation		
	CE	Q_{pred}/Q_{obs} (median)	RE (%) (median)		CE	Q_{pred}/Q_{obs} (median)	RE (median)
2	0.59	1.03	43.75		0.69	1.04	37.56
5	0.73	1.12	39.53		0.59	0.99	40.39
10	0.64	1.06	39.14		0.63	1.02	44.63
20	0.71	1.10	40.38		0.69	1.04	35.62
50	0.70	1.08	43.32		0.68	1.14	39.09
100	0.64	1.15	46.30		0.40	1.10	44.53
Overall	0.67	1.09	42.07		0.61	1.06	40.30

Overall, in terms of RE value, the ANN is the best performer, followed by the GAANN, GEP and CANFIS.

The CE, median Q_{pred}/Q_{obs} ratio and median relative error values are compared in Table 2 for the training and validation datasets for the ANN based RFFA model. In terms of CE value, the best agreement between the training and validation data sets is found for ARIs of 10, 20 and 50 years, a reasonable degree of agreement is found for ARIs of 2 and 5 years and relatively poor agreement is found for the ARI of 100 years where the CE value for the validation data set is remarkably small. With respect to median Q_{pred}/Q_{obs} ratio value, the best agreement between the training and validation data sets is found for 2 years ARI, a moderate agreement is noticed for 10, 20, 50 and 100 years ARIs and a poor agreement is found for 5 years ARI. However, for 5 years ARI validation data set gives a very good Q_{pred}/Q_{obs} ratio value (0.99). In relation to the median relative error values, the best agreement between the training and validation data sets is found for ARIs of 5 and 100 years, a moderate agreement for ARI of 50 years and poor agreement for ARIs of 2 and 10 years. From these results, it is noted that the ANN based RFFA model shows different degrees of agreement between the training and validation data sets for different ARIs across the three criteria adopted here.

From the results of the training of the four artificial intelligence based RFFA models, it has been found that none of the four models perform the best in all the adopted assessment criteria over the six ARIs. Based on the four different criteria as shown in Table 3, the performances of the four models are assessed in a heuristic

Table 3 Ranking of the four artificial intelligence based RFFA models with respect to training

Criterion	Rank 1	Rank 2	Rank 3	Rank 4
Scatter plot of Q_{obs} Vs Q_{pred}	ANN	GANN	CANFIS	GEP
Median Q_{pred}/Q_{obs}	ANN	GEP	GAANN/CANFIS	#
Median RE	ANN	GAANN	GEP	CANFIS
Median CE	GAANN	ANN	GEP/CANFIS	#
Overall score ANN-15, GAANN-12, GEP-10, CANFIS-7				

manner. In this assessment, a model is ranked based on four different criteria as shown in Table 3. Four different ranks are used, with a relative score ranging from 4 to 1. If a model is ranked 1 for a criterion, it scores 4. For ranks of 2, 3 and 4, scores of 3, 2 and 1, respectively are assigned.

Table 3 shows that the ANN based RFFA model has the highest score of 15, followed by the GANN with a score of 12. The GEP receives a score of 10, while the CANFIS receives only 7 making it the least favourable model in terms of its performance during training. The ANN based model is placed at rank 1 in the 3 out of 4 criteria. Hence, it is decided that the ANN based RFFA model is the best performing artificial intelligence based model in terms of training/calibration of the model.

Table 4 shows the ranking of the four artificial intelligence based RFFA models based on the agreement between the training and validation using three criteria. Four different ranks are used with a relative score ranging from 4 to 1 as mentioned

Table 4 Ranking of the four artificial intelligence based RFFA models with respect to agreement between training and validation

Criterion	Rank 1	Rank 2	Rank 3	Rank 4
Median Q_{pred}/ Q_{obs}	GEP (Best agreement: Q_2, Q_5, Q_{10}, Q_{20} Moderate agreement: Q_{50}, Q_{100} Poor agreement: none)	ANN (Best agreement: Q_2, Q_{10}, Q_{100} Moderate agreement: Q_{20}, Q_{50} Poor agreement: Q_5)	CANFIS (Best agreement: Q_5, Q_{10}, Q_{20}, Q_{50}, Q_{100} Moderate agreement: none Very poor agreement: Q_2)	GAANN (Best agreement: Q_2, Q_{10}, Q_{20}, Q_{100} Moderate agreement: Q_5 Very poor agreement: Q_{50})
Median RE (%)	GEP (Best agreement: Q_2, Q_5, Q_{10}, Q_{20}, Q_{100} Moderate agreement: Q_{50} Poor agreement: none)	ANN (Best agreement: Q_5, Q_{100} Moderate agreement: Q_{50}, Q_{20} Poor agreement: Q_2, Q_{10})	GAANN (Best agreement: Q_{50}, Q_{100} Moderate agreement: Q_{20} Very poor agreement: Q_2, Q_5, Q_{10})	CANFIS (Best agreement: Q_5, Q_{10}, Q_{20}, Q_{50}, Q_{100} Moderate agreement: none Significantly poor agreement: Q_2
Median CE	GAANN (Best agreement: Q_2, Q_5, Q_{20}, Q_{100} Moderate agreement: Q_{10} Poor agreement: Q_{50})	ANN (Best agreement: Q_{10}, Q_{20}, Q_{50} Moderate agreement: Q_2, Q_5 Poor agreement: Q_{100})	CANFIS (Best agreement: Q_{10}, Q_{20}, Q_{50}, Q_{100} Moderate agreement: Q_5 Poor agreement: Q_2)	GEP (Best agreement: Q_5, Q_{20}, Q_{50} Moderate agreement: Q_{10}, Q_{100} Poor agreement: Q_2)

Overall score ANN-9, GEP-9, GAANN-7, CANFIS-5

earlier. It is found that the ANN and GEP based RFFA models both score 9, followed by the GAAANN and CANFIS. Overall, ANN based RFFA model shows the best training/calibration and the CANFIS the least favourable one.

Figure 2 compares the predicted flood quantiles for the selected 90 test catchments from the ANN based RFFA model with the observed flood quantiles for 20 years ARI (Q_{20}). The observed flood quantiles are estimated using an LP3 distribution and Bayesian parameter estimation procedure [48]. It should be noted here that the observed flood quantiles are not free from error; these are subject to data error (such as rating curve extrapolation error), sampling error (due to limited record length of annual maximum flood series data), error due to choice of flood frequency distribution and error due to selection of parameter estimation method. This error undermines the usefulness of the validation statistics (e.g. RE); however, this provides an indication of possible error of the developed RFFA model as far as practical application of the RFFA model is concerned. The ratio Q_{pred}/Q_{obs} and RE values are used for the assessment of models; however, the CE value is not very useful here as the mean of observed flood quantile is not known.

Figure 2 shows a good agreement overall between the predicted and observed flood quantiles; however, there is some over-estimations by the ANN based RFFA model when the observed flood quantiles are smaller than about 50 m³/s. Most of the test catchments are within a narrow range of variability from the 45-degree line except for a few outliers. The plots of predicted and observed flood quantiles for other ARIs showed very similar for ARIs of 2, 5, 10 and 20 years. Results for ARIs of 50 and 100 years exhibited some overestimation by the ANN based RFFA model for smaller to medium discharges.

Figure 3 shows the boxplot of relative error (RE) values of the selected test catchments for ANN based RFFA model for different flood quantiles. It can be seen from Fig. 3 that the median RE values (represented by the thick black lines within the boxes) are located very close to the zero RE line (indicated by 0–0 horizontal

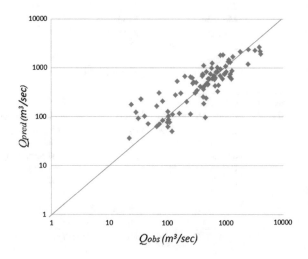

Fig. 2 Comparison of observed and predicted flood quantiles for ANN based RFFA model for Q_{20}

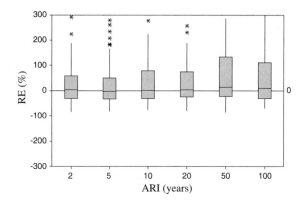

Fig. 3 Boxplot of relative error (RE) values for ANN based RFFA model

line in Fig. 3), in particular for ARIs of 2, 5, 10 and 20 years. However, for ARIs of 50 and 100 years, the median RE values are located above the zero line with ARI of 50 years showing the highest departure, which indicates an overestimation by the ANN based RFFA model. Overall, the ANN based RFFA model produces nearly unbaised estimates of flood quantiles as the median RE values match with the zero RE line quite closely as can be seen in Fig. 3.

In terms of the spread of the RE (represented by the width of the box), ARI of 50 and 100 years present the highest RE band and ARIs of 2 and 5 years present the smallest RE band, followed by ARI of 20 and 10 years. The RE bands for 50 and 100 years ARIs are almost double to RE bands of 2 and 5 years ARIs. This implies that ANN based RFFA model provides the most accurate flood quantile estimates for 2 and 5 years ARIs, and the least accurate flood quantiles for ARIs of 50 and 100 years. Overall. the boxplot in Fig. 3 shows that better results in terms of RE values are achieved for the smaller ARIs (i.e. 2, 5, 10 and 20 years ARIs) as compared to higher ARIs for the ANN based RFFA model. Some outliers (evidenced by notable overestimation with a positive RE) can be seen for all the ARIs, which may need to be examined more closely for data errors or issues regarding the hydrology and physical characteristics of these catchments; if these catchments are deemed to be genuine outliers they should be removed to enhance the ANN based RFFA model; however, this has not been undertaken in this chapter.

Figure 4 shows the boxplot of the Q_{obs}/Q_{pred} ratio values of the selected 90 test catchments for ANN based RFFA model for different ARIs. The median Q_{obs}/Q_{pred} ratio values (represented by the thick black lines within the boxes) are located closer to 1–1 line (the horizontal line in Fig. 4), in particular for ARIs of 2, 5, 10 and 20 years. However, for ARI of 50 years (and to a lesser degree for ARI of 100 years), the median Q_{obs}/Q_{pred} ratio value is clearly located above the 1–1 line.

These results indicate that the ANN based RFFA model generally provides reasonably accurate flood quantiles with the expected Q_{obs}/Q_{pred} ratio value very close to 1.00, although there is a noticeable overestimation for ARI of 50 and 100 years. In terms of the spread of the Q_{obs}/Q_{pred} ratio values, ARI of 2 and 5 years provide the lowest spread followed by ARIs of 20, 10, 100 and 50 years.

Fig. 4 Boxplot of Q_{pred}/Q_{obs} ratio values for ANN based RFFA model

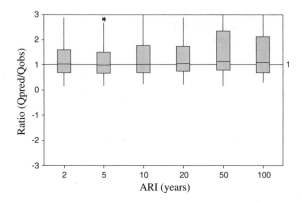

Considering, the RE and Q_{obs}/Q_{pred} ratio values as discussed above, it can be concluded that ANN based RFFA model generally provide unbiased flood estimates for smaller to medium ARIs (2 to 20 years); however, the model slightly overestimates the observed flood quantiles for higher ARIs (50 to 100 years).

6   Conclusion

This chapter presents development and testing of non-linear artificial intelligence based regional flood frequency analysis (RFFA) models. For this purpose, a database of 452 small to medium sized catchments from eastern Australia has been used. Four different artificial intelligence based RFFA models have been considered in this research. It has been found that non-linear artificial intelligence based RFFA techniques can be applied successfully to eastern Australian catchments. Among the four artificial intelligence based models, the ANN based RFFA model has been found to be the best performing model, followed by the GAANN based RFFA model. It has been shown that in the training of the four artificial intelligence based RFFA models, no model performs the best for all the six ARIs over all the adopted criteria. Overall, the ANN based RFFA model is found to outperform the three other models in the training/calibration. Based on independent validation, the median relative error values for the ANN based RFFA model are found to be in the range of 35 to 44 % for eastern Australia.

References

1. D. Carbone, J. Hanson, Floods: 10 of the deadliest in Australian history. Australian Geographic, http://www.australiangeographic.com.au/journal/the-worst-floods-in-australian-history.htm. Accessed 18 July 2013
2. PWC, *Economic impact of Queensland's natural disasters*. (Price Waterhouse Cooper (PWC), Australia, 2011)

3. NERC, *Flood studies report*. (Natural Environment Research Centre (NERC), London, 1975)
4. Institution of Engineers Australia (I.E. Aust.), *Australian rainfall and runoff: A guide to flood estimation*, vol. 1, ed by D.H. Pilgrim (I. E. Aust., Canberra, 1987, 2001)
5. M.C. Acreman, C.D. Sinclair, Classification of drainage basins according to their physical characteristics and application for flood frequency analysis in Scotland. J. Hydrol. **84**(3), 365–380 (1986)
6. R.J. Nathan, T.A. McMahon, Identification of homogeneous regions for the purpose of regionalisation. J. Hydrol. **121**, 217–238 (1990)
7. D.H. Burn, Evaluation of regional flood frequency analysis with a region of influence approach. Water Resour. Res. **26**(10), 2257–2265 (1990)
8. T.R. Kjeldsen, D. Jones, An exploratory analysis of error components in hydrological regression modelling. Water Resour. Res. **45**, W02407 (2009)
9. K. Haddad, A. Rahman, Regional flood frequency analysis in eastern Australia: Bayesian GLS regression-based methods within fixed region and ROI framework—Quantile Regression vs Parameter Regression Technique. J. Hydrol. **430–431**(2012), 142–161 (2012)
10. T.J. Mulvany, On the use of self-registering rain and flood gauges. Inst. Civ. Eng. (Ireland) Trans, **4**(2), 1–8 (1851)
11. G.G.S. Pegram, M. Parak, A review of the regional maximum flood and rational formula using geomorphological information and observed floods, ISSN 0378-4738. Water South Africa **30**(3), 377–392 (2004)
12. A. Rahman, K. Haddad, M. Zaman, G. Kuczera, P.E. Weinmann, Design flood estimation in ungauged catchments: a comparison between the probabilistic rational method and quantile regression technique for NSW. Aust. J. Water Resour. **14**(2), 127–137 (2011)
13. T. Dalrymple, Flood frequency analyses. U.S. Geological Survey Water Supply Paper, 1543-A, 11–51, (1960)
14. J.R.M. Hosking, J.R. Wallis, Some statics useful in regional frequency analysis. Water Resour. Res. **29**(2), 271–281 (1993)
15. B.C. Bates, A. Rahman, R.G. Mein, P.E. Weinmann, Climatic and physical factors that influence the homogeneity of regional floods in south-eastern Australia. Water Resour. Res. **34**(12), 3369–3382 (1998)
16. A. Rahman, B.C. Bates, R.G. Mein, P.E. Weinmann, Regional flood frequency analysis for ungauged basins in south-eastern Australia. Aust. J Water Res. 3(2), 199–207, 1324–1583 (1999)
17. T.R. Kjeldsen, D.A. Jones, Predicting the index flood in ungauged UK catchments: on the link between data-transfer and spatial model error structure. J. Hydrol. **387**(1–2), 1–9 (2010)
18. E. Ishak, K. Haddad, M. Zaman, A. Rahman, Scaling property of regional floods in New South Wales Australia. Nat. Hazards **58**, 1155–1167 (2011)
19. M.A Benson, Evolution of methods for evaluating the occurrence of floods, U.S. Geological Surveying Water Supply Paper, 30, 1580-A (1962)
20. D.M. Thomas, M.A. Benson, Generalization of streamflow characteristics from drainage-basin characteristics, U.S. Geological Survey Water Supply Paper 1975, US Governmental Printing Office, 1970
21. J.R. Stedinger, G.D. Tasker, Regional hydrologic analysis - 1. Ordinary, weighted and generalized least squares compared. Water Resour. Res. **21**, 1421–1432 (1985)
22. G.R. Pandey, V.T.V. Nguyen, A comparative study of regression based methods in regional flood frequency analysis. J. Hydrol. **225**, 92–101 (1999)
23. A. Rahman, A quantile regression technique to estimate design floods for ungauged catchments in South-east Australia. Aust. J. Water Resour. **9**(1), 81–89 (2005)
24. V.W. Griffis, J.R. Stedinger, The use of GLS regression in regional hydrologic analyses. J. Hydrol. **344**, 82–95 (2007)
25. T.B.M.J. Ouarda, K.M. Bâ, C. Diaz-Delgado, C. Cârsteanu, K. Chokmani, H. Gingras, E. Quentin, E. Trujillo, B. Bobée, Intercomparison of regional flood frequency estimation methods at ungauged sites for a Mexican case study. J. Hydrol. **348**, 40–58 (2008)

26. K. Haddad, A. Rahman, Regional flood estimation in New South Wales Australia using generalised least squares quantile regression. J Hydrol Eng ASCE **16**(11), 920–925 (2011)

27. K. Haddad, A. Rahman, J.R. Stedinger, Regional flood frequency analysis using bayesian generalized least squares: a comparison between quantile and parameter regression techniques. Hydrol. Process. **25**, 1–14 (2011)

28. T.M. Daniell, Neural networks—applications in hydrology and water resources engineering. Paper presented at the international hydrology and water resources symposium. Perth, Australia, 2–4 October 1991

29. R.S. Muttiah, R. Srinivasan, P.M. Allen, Prediction of two year peak stream discharges using neural networks. J. Am. Water Resour. Assoc. **33**(3), 625–630 (1997)

30. C. Shu, D.H. Burn, Artificial neural network ensembles and their application in pooled flood frequency analysis. Water Resour. Res. **40**(9), W09301 (2004). doi:10.1029/2003WR002816

31. U.C. Kothyari, Estimation of mean annual flood from ungauged catchments using artificial neural networks, in *Hydrology: Science and Practice for the 21st Century*, vol. 1 (British Hydrological Society, 2004)

32. C.W. Dawson, R.J. Abrahart, A.Y. Shamseldin, R.L. Wilby, Flood estimation at ungauged sites using artificial neural networks. J. Hydrol. **319**, 391–409 (2006)

33. C. Shu, T.B.M.J. Ouarda, Flood frequency analysis at ungauged sites using artificial neural networks in canonical correlation analysis physiographic space. Water Resour. Res. **43**, W07438 (2007). doi:10.1029/2006WR005142

34. W.S. McCulloch, W. Pitts, A logic calculus of the ideas immanent in nervous activity. Bull Math Biophys **5**, 115–133 (1943)

35. A. Jain, S. Srinivasalu, R.K. Bhattacharjya, Determination of an optimal unit pulse response function using real-coded genetic algorithm. J. Hydrol. **303**, 199–214 (2005)

36. A.J.F.V. Rooij, L.C. Jain, R.P. Johnson, *Neural network training using genetic algorithms* (World Scientific Publishing Co. Pty. Ltd., River Edge, 1996), p. 130

37. M. Mitchell, *An Introduction to Genetic Algorithms* (MIT Press, Cambridge, 1996)

38. K. Aziz, A. Rahman, A.Y. Shamseldin, M. Shoaib, Co-Active neuro fuzzy inference system for regional flood estimation in Australia. J Hydrol Environ Res **1**(1), 11–20 (2013)

39. L.C. Arthur, L.W. Roger, in *LibGA for solving combinatorial optimization problems*, ed. by Chambers. Practical Handbook of Genetic Algorithms (CRC Press, Boca Raton, 1995)

40. C. Ferreira, Gene expression programming in problem solving, in *proceedings of the 6th Online World Conference on Soft Computing in Industrial Applications* (invited tutorial), 2001

41. C. Ferreira, Gene expression programming: a new adaptive algorithm for solving problems. Complex Syst. **13**(2), 87–129 (2001)

42. C. Ferreira, *Gene-expression programming; mathematical modeling by an artificial intelligence* (Springer, Berlin, 2006)

43. D.A.K. Fernando, A.Y. Shamseldin, R.J. Abrahart, Using gene expression programming to develop a combined runoff estimate model from conventional rainfall-runoff model outputs. Paper presented at the 18th world IMACS / MODSIM Congress, Cairns, Australia 13–17 July 2009

44. A. Aytek, Co-Active neuro-fuzzy inference system for evapotranspiration modelling. Soft. Comput. **13**(7), 691–700 (2009)

45. K. Aziz, A. Rahman, G. Fang, S. Shreshtha, Application of artificial neural networks in regional flood frequency analysis: a case study for Australia. Stochast Environ. Res. Risk Assess. **28**(3), 541–554 (2013)

46. K. Aziz, A. Rahman, G. Fang, K. Haddad, S. Shrestha Design flood estimation for ungauged catchments: application of artificial neural networks for eastern Australia. In: *World Environment and Water Resources Congress*, ASCE, Providence, Rhodes Island, USA, 2010

47. F.E. Grubbs, G. Beck, Extension of sample sizes and percentage points for significance tests of outlying observations. Technometrics **14**, 847–854 (1972)

48. G. Kuczera, Comprehensive at-site flood frequency analysis using Monte Carlo Bayesian inference. Water Resour. Res. **35**(5), 1551–1557 (1999)

49. K. Haddad, A. Rahman, P.E. Weinmann, G. Kuczera, J.E. Ball, Streamflow data preparation for regional flood frequency analysis: lessons from south-east Australia. Aust. J. Water Resour. **14**(1), 17–32 (2010)
50. K. Haddad, A. Rahman, F. Ling, Regional flood frequency analysis method for Tasmania, Australia: A case study on the comparison of fixed region and region-of-influence approaches. Hydrol. Sci. J. (2014) doi: 10.1080/02626667.2014.950583

Artificial Neural Networks in Precipitation Nowcasting: An Australian Case Study

Benjamin J.E. Schroeter

Abstract Accurate prediction of precipitation is beneficial to many aspects of modern society, such as emergency planning, farming, and public weather forecasting. Prediction on the scale of several kilometres over forecast horizons of 0–6 h (nowcasting) is extrapolated from current weather conditions using radar and satellite observations. However, in Australia, the use of radar for nowcasting is challenging due to sparse radar coverage, particularly in regional areas. Satellite-based methods of precipitation estimation are therefore an appealing alternative; however, the ever-increasing spatial and temporal resolution of satellite data prompts investigation into options that can meet operational performance needs while also managing the large volume of data. In this chapter, the use of Artificial Neural Networks to nowcast precipitation in Australia is explored, and the current limitations of this technique are discussed. The Artificial Neural Network in this study is found to be capable of meeting or exceeding the performance of the industry-standard Hydro-Estimator method using a variety of Machine Learning metrics for the chosen verification scene. Further research is required to determine the optimal configuration of model parameters and generalisation of the model to different times and areas. This may assist Artificial Neural Networks to better reflect seasonal and orographic influences, and to meet operational performance benchmarks.

1 Introduction

Precipitation nowcasting in Australia currently relies on the use of a Radar (Radio Detecting and Ranging) network spanning 60 instruments distributed across the continent. These instruments emit electromagnetic waves in short pulses, which are scattered back to the radar upon intercepting particles of precipitation. Information about the location and magnitude of the precipitation is then interpreted from the

B.J.E. Schroeter (✉)
Bureau of Meteorology, Hobart, Australia
e-mail: b.schroeter@bom.gov.au

© Springer International Publishing Switzerland 2016
S. Shanmuganathan and S. Samarasinghe (eds.), *Artificial Neural Network Modelling*, Studies in Computational Intelligence 628,
DOI 10.1007/978-3-319-28495-8_14

returning signal, determining the reflectivity of the pulse (which measures the volume of precipitation) and its velocity (which measures rate of movement and its trajectory in comparison with the instrument) [1]. From this information, maps of rain rate can be created and types of precipitation inferred (e.g. stratiform or convective) in the local region [2]. These maps are used to inform short-term forecasting of weather, known as nowcasting. Nowcasting specifically refers to weather prediction for a local area for a period of 0–6 h after the time of the observations, detailing the initial conditions as well as the extrapolation of these conditions to forecast the very short-term future weather conditions through numerical weather prediction and other techniques [3]. The Australian Bureau of Meteorology uses nowcasting and numerical weather prediction techniques to estimate precipitation amounts to reach ground level 3–6 h ahead [4].

The majority of Australia's radar instruments are located at or near major population areas, such as the capital cities, or along the coastline (Fig. 1). Coverage in rural areas and the interior of the continent is sparse. Given the size of the Australian continent and the remoteness of much of its interior, achieving continent-wide radar coverage is a significant challenge.

Recent improvements to satellite technology have vastly increased spatial and temporal resolution of imagery that can be used to observe weather in much greater detail [6]. This affords researchers the opportunity to take advantage of comprehensive remotely sensed data as a more cost-effective tool in precipitation modelling. However, increased data resolution also leads to increased data volume, and the need to utilise methods that can process a substantial data load in a short period

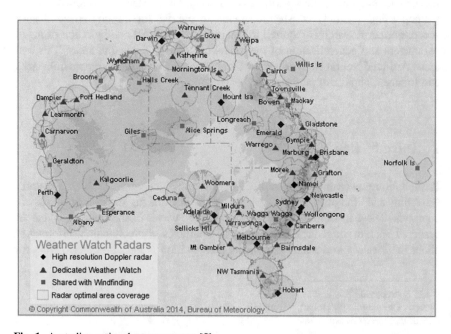

Fig. 1 Australian national coverage map [5]

of time while also mapping the non-linearity of precipitation events. An Artificial Neural Network (ANN) is possibly one such method. In this study, the performance of an ANN is tested against an industry method (Hydro-Estimator) to determine whether ANNs have the potential to meet the required accuracy for precipitation estimation at the time of observation for the Australian continent. This study does not seek to evaluate the ANN method for forecasting purposes at this time.

2　Data, Models and Methods

The data used in this study include a Blended Rainfall Product (rain gauge and Radar), Hydro-Estimator precipitation estimates, and Himawari-8 satellite imagery for a precipitation event over Darwin on January 29, 2015. The model was trained and monitored using two consecutive hourly time periods starting from 0800 and verified using the third (Table 1). Monitoring data was used to observe model evolution during execution and was not used in analysis.

2.1　Blended Rainfall Product

The training data were derived from a Bureau internal blended rainfall product, which represents an optimally interpolated blend of reflectivity data and rainfall gauge measurements [7]. These data cover a spatial domain 256 km surrounding coordinates (-11.082 N, 132.233E) at 2 km resolution and covering a temporal window of the previous 30 min of accumulated precipitation. After preliminary quality control is applied, such as the removal of clutter, barrier effects and real-time adjustment using gauges, the individual data sets are blended to form a single gridded rain field product representing an accumulation over the observation period and expressed as mm/hr. This is achieved through interpolation, which uses a form of Kriging described in Sinclair and Pegram [8].

2.2　Hydro-Estimator

Hydro-Estimator is a proprietary industry model from the US National Oceanographic and Atmospheric Administration (NOAA) that estimates precipitation using infrared satellite data [9]. The model estimates the location and rate of

Table 1 Temporal partitioning of data set

Training	Monitoring	Verification
0800–0900 h	0900–1000 h	1000–1100 h

rainfall by determining below-average brightness temperature values at cloud tops, and adjusts rainfall rate based on the relative temperature of the pixel to its surroundings. This makes it useful for providing rainfall estimates in large, sparsely populated regions, and therefore a potential candidate for rainfall estimation in the vast interior of the Australian continent. Global hourly data from Hydro-Estimator is available through public repositories in ASCII format as integers in the range [0, 256]. The 0.045° resolution data must be converted to a rainfall rate (RR = mm/hr) via the following formula [10]:

$$RR = (value - 2.0) \times 0.3$$

Under this mapping, a value of 0 indicates missing data and a value of 2 necessarily indicates no precipitation.

2.3   Himawari-8

Himawari-8 satellite data consists of 16 Visible (VIS), Near Infrared (NIR) and Infrared (IR) channels with a spectral range of between 0.47 and 13.3 μm at a temporal resolution of 10 min and a spatial resolution of between 0.5 and 2 km (Table 2). Of these, bands 7 through 16 were used in feature selection, in an attempt to train a model that could function without daylight. Using Schmit, et al. [11] for reference, the feature set was cultivated *apriori* to satellite bands in the IR portion

Table 2 Himawari-8 satellite band configuration [22]

Band	Central wavelength (μm)	Resolution (km)	Used in training
1	0.47	1	–
2	0.51	1	–
3	0.64	0.5	–
4	0.86	1	–
5	1.6	2	–
6	2.3	2	–
7	3.9	2	YES
8	6.2	2	YES
9	6.9	2	YES
10	7.3	2	YES
11	8.6	2	YES
12	9.6	2	YES
13	10.4	2	YES
14	11.2	2	YES
15	12.4	2	YES
16	13.3	2	YES

of the instrument's capabilities, largely to avoid the need for daylight and the anomalous error that may be introduced where visible reflectivity contributes to sensor reading (as it does with VIS and NIR).

2.4   Data Homogenisation, Cleansing and Instance Extraction

The three data sources used in this study were of different projections, temporal and spatial scales. As such, it was necessary to homogenise the data for effective model comparison and verification. Hydro-Estimator (equirectangular latitude/longitude) and Himawari-8 (normalised geostationary projection [GEOS]) were reprojected into the blended rainfall product's gnomonic projection and cropped to the same spatial domain. The datasets were then temporally aligned by first summing the radar scenes (30 min) to cover the same temporal window as Hydro-Estimator (1 h), then by averaging a composite of the 6 temporally collocated Himawari-8 scenes (10 min) to cover the same period.

Finally, training instances were extracted from the resulting dataset by transecting the data space along the z-axis through the satellite bands down to the equivalent rain field grid cell (Fig. 2). Invalid or missing data in either the radar or satellite bands were omitted from the training, monitoring and verification datasets. The final data sets totalled approximately 35,000 training instances each and were representative a contiguous 3 × 1 h time period. The Artificial Neural Network was trained on these transect instances using the first hour, monitored on the second and

Fig. 2 Instance transect through the data space

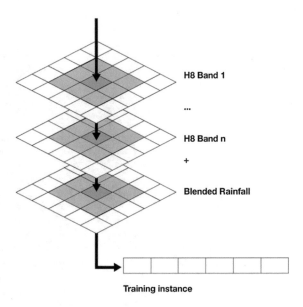

verified on the third. Hydro-Estimator was compared directly with the associated radar observation values in the third hour.

2.5   Artificial Neural Network

The single-layer feed-forward Artificial Neural Network (ANN) used in this study takes the net input signals of the transect training vector, multiplies them by a weight matrix, sums with a bias vector, and passes them through the activation function. The model input layer is made up of the 10 features of the training instance; the hidden layer consists of 30 neurons, and an output layer of a single neuron indicating the amount of precipitation as a regressed value (Fig. 3).

The network was trained with standard back-propagation [12] via Stochastic Gradient Descent (SGD) and attempts to resolve the issues of slow/no convergence, vanishing gradient descent [13] and over-fitting through the following optimisations:

2.5.1   Momentum

Momentum involves the addition of the previous t step's delta term, which effectively assumes that the error gradient will continue to move in the same direction of the model space, the average "downhill direction" [14]. Assuming this to be the case, momentum has the capacity to increase the speed at which convergence is found. The momentum term is simply a weight parameter $0 < m < 1$ applied to the weight delta δ_W (and similarly the bias delta) as follows:

Fig. 3 Illustration of artificial neural network

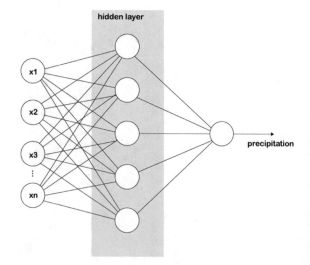

$$\hat{\delta}_W^{(t)} = \delta_W^{(t)} + m\delta_W^{(t-1)}$$

This case study assumes an initial arbitrary momentum term of 0.5 and a momentum adjustment schedule, which saturates to 0.9 at the final epoch.

2.5.2 Learning Rate Decay

Another optimisation to address the convergence problem is the application of a learning rate adjustment schedule or decay (often referred to as an annealed learning rate [14]). In this study, a simple decay-by-factor approach was applied to the learning rate τ via the following formula:

$$\tau_{t+1} = \alpha\tau_t$$

Given the large variance in the input data, an initial learning rate τ of 0.0001 and a decay factor α of 0.9 (that is, the next step uses 90 % of the previous learning rate) were applied.

2.5.3 Rectified Linear Units (ReLU)

The problem of vanishing gradient descent [13] is largely subverted through the use of a different activation function to the standard sigmoid, specifically the use of a linear or near-linear activation function. For this case study, a generic linear activation function with a hard value of 0 on the x-intercept was chosen for the regression task to predict continuous values.

2.5.4 Dropout Regularisation

To prevent the ANN model from over-fitting training data, Hinton, et al. [15] devised a means to prevent the co-adaptation of features by introduction a selective omission of input signals at each layer of the network. This has the effect of regularising the network and is generally applied by generating a dropout vector ξ for each layer l to be pairwise multiplied with the activation and derivative activation signals.

$$\xi_l = \begin{cases} 0 & \text{with probability } \rho \\ \frac{1}{1-\rho} & \text{otherwise} \end{cases}$$

$$\phi(f_{l-1}(x)W_l + b_l) \otimes \xi_l$$
$$\phi'(f_{l-1}(x)W_l + b_l) \otimes \xi_l$$

where ϕ and ϕ' are the activation and derivative activation functions respectively, $f_{i-1}(x)$ are the activations of layer l, W_1 the weight matrix of layer l and b_1 the bi-as vector.

2.5.5 Early Stopping Criterion

The final optimization technique included in the model was to place an early stopping criterion on model execution. This again reduces over-fitting and was implemented by limiting the number of training epochs to ten thousand and model execution to 4 h over the training scene, whichever came first. To reiterate, the model in this study was trained using a precipitation event over Darwin on January 29, 2015.

3 Verification

The primary means to measure rainfall estimates are rain gauges, Radar rainfall estimates, or objective analysis of one or both types of measurement [16] and these estimates are verified through one or several metrics. This study uses the afore-mentioned blended rainfall product as reference for verification and measures performance using the following formulae of measurement:

Mean Error (ME) measures the average forecast error in the difference between observed o and modelled values y with a perfect score being 0. However, it is possible to get a perfect score for a poor forecast, as the metric does not measure the correspondence between forecast and observations.

$$ME = \frac{1}{N} \sum_{i=1}^{N} (y_i - o_i)$$

Mean Absolute Error (MAE) recognises the magnitude of the error for a more general indication of model performance in either direction with a perfect score of 0. Unlike other metrics, the MAE is unambiguous and often featured in forecasting literature [17], however, MAE does not discriminate between over and under estimation.

$$MAE = \frac{1}{N} \sum_{i=1}^{N} |y_i - o_i|$$

The Mean Squared Error (MSE) incorporates both the bias and the variance of the model [18], two quantities a forecasting experiment such as this should seek to minimise towards a perfect score of 0. This metric penalises outliers, encouraging stable homogeneous performance across the verification data set. Additionally, this metric was used as the objective/loss function in model training for this study.

$$MSE = \frac{1}{N} \sum_{i=1}^{N} (y_i - o_i)^2$$

Root Mean Squared Error (RMSE) adds an additional form of measurement, again penalising outliers [19]. The metric is widely used in Machine Learning literature, typically as a training or objective function (perfect score is 0). However, it is considered ambiguous and unreliable in the forecasting space [17, 20, 21], as it is a function of three separate metrics: the sum of squared errors (SSE), the mean of the SSE, and the square root of the mean. This metric may be preferable to encourage conservative forecasting.

$$RMSE = \sqrt{\frac{1}{N} \sum_{i=1}^{N} (y_i - o_i)^2}$$

Pearson's R was used to measure the linear correlation (if any) between the predicted and observed values. Note: this statistic can also be used to measure the correlation between prediction error and observation to observe trends as rainfall rate varies.

$$r = \frac{\sum_{i=1}^{N} (y_i - \bar{y})(o_i - \bar{o})}{\sqrt{\sum_{i=1}^{N} (y_i - \bar{y})^2} \sqrt{\sum_{i=1}^{N} (o_i - \bar{o})^2}}$$

The strength of correlation was categorised on a scale of values between −1 and +1, whereby an absolute value of 1 indicates a perfect correlation and 0 no correlation. The scale formalised by Dancy and Reidy [23] was adapted to interpret the strength of the correlation as indicated in Table 3. Where values of r fall between classifications a nominal distinction was made. For example, a score of $r = 0.35$ was denoted "weak to moderate".

Lastly, these metrics were used to derive a Skill Score, a metric that shows a model's performance relative to some reference value, such as climatology or the output of another model [16]. A set of skill scores was chosen using the verification outputs of Hydro-Estimator (ME, MAE, MSE, RMSE and correlation) for reference.

Table 3 Nominal classes of correlation strength (adapted from [23])	Value	Strength
	1.0	Perfect correlation
	0.7–0.9	Strong
	0.4–0.6	Moderate
	0.1–0.3	Weak
	0.0	No correlation

$$Skill = \frac{score_{estimate} - score_{reference}}{score_{perfect} - score_{reference}}$$

This study considered all of the aforementioned metrics in deriving skill to observe model performance under a range of measurement schemes. Note that a Skill Score > 0 indicates that ANN outperforms Hydro-Estimator on the given metric.

4 Results

In the following figures the blended rainfall product is compared against the estimation technique of interest. It is observed that transect conditions requiring valid data in the Radar; Hydro-Estimator and Himawari-8 datasets has significantly reduced the data available for training and verification as indicated by the uncoloured portions of the images. The industry method Hydro-Estimator used in this study overestimates both the areal coverage and magnitude of precipitation in the target scene (Fig. 4) and fails to capture the complexity of the high-intensity core in the lower left of the rain field. Subjective analysis shows that the technique has captured the general spatial distribution in the area.

There is a weak, positive correlation (r = 0.16) between the predicted and observed values, showing little association between reference observations and Hydro-Estimator estimates (Table 4). A value close to zero for Mean Error (ME) is

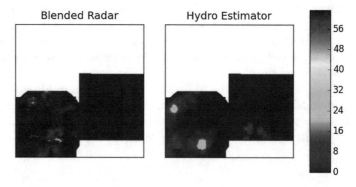

Fig. 4 Hydro-estimator output (mm/hr)

Table 4 Performance of hydro-estimator and ANN (stochastic gradient descent)

Model	ME (mm/hr)	MAE (mm/hr)	MSE (mm/hr)^2	RMSE (mm/hr)	r
Hydro-estimator	0.77	1.11	12.55	3.54	0.16
ANN (SGD)	0.77	1.02	4.29	2.07	0.24

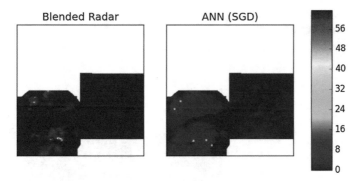

Fig. 5 Artificial neural network (SGD) output (mm//hr)

Table 5 ANN skill score (using hydro-estimator as reference) over each metric

ME	MAE	MSE	RMSE	R
0.00	+0.08	+0.66	+0.41	+0.09

desired, and the Hydro-Estimator score is promising for this metric. However, the score of 12.55 for Mean Square Error (MSE) and 3.54 for Root Mean Squared Error (RMSE) are comparatively high, and are discussed further in the following section.

The Artificial Neural Network was run for four hours (5,187 epochs) and, despite promising initial results, Dropout Regularisation was ultimately disabled due to an adverse effect on model performance and is discussed in the following section. The ANN has similarly overestimated the areal coverage of the precipitation event while producing only pinpoints of intensity proximal to what was observed (Fig. 5). The model performance against observations was largely similar in ME and MAE but significantly improved on MSE and RMSE compared with the Hydro-Estimator output (Table 4). Again it is observed that the missing data in training, monitoring and verification is identical to that of the Hydro-Estimator comparison. This indicates that the missing data of the rain field is the primary driver for conditionally omitting data from the experiment.

The skill of the Artificial Neural Network is shown in Table 5 as a comparison against the Hydro-Estimator model for reference. The difference in skill between the ANN and the Hydro-Estimator is negligible in terms of Mean Error, Mean Absolute Error, and correlation. The Skill Scores for Mean Squared Error (MSE) and Root Mean Squared Error (RMSE) are significantly higher for the ANN.

5   Discussion

Initial results from the ANN appear promising as noted in the respective metrics and Skill Scores; however, precipitation coverage at lower intensities is significantly over-estimated. Upon closer inspection without explicit scaling to match observations

Fig. 6 Artificial neural
network without explicit
scaling (mm/hr)

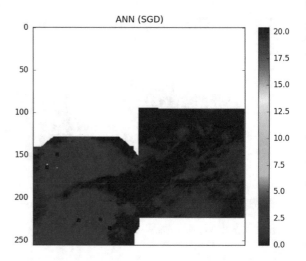

(Fig. 6) this may be an issue of model tuning (parameter configuration on one or
several of the optimisation techniques), the relatively low number of precipitating
cells in the training scene, or the low dimensionality of the problem in general. The
detrimental effect of Dropout Regularisation may also be attributed to the low
dimensionality of the problem as in McMahan, et al. [24], where the feature distribution is comparatively sparser than convolutional or vision-based learning tasks and
this is particularly the case given the noisy nature of the target regression value. In this
instance, Dropout Regularisation adversely affects model performance simply by
reducing the amount of data on which to train.

The ANN model produced favourable ME (0.77), MAE (1.02), MSE (4.29) and
RMSE (2.07) scores, matching or outperforming the Hydro-Estimator under these
measures. However, these MSE and RMSE scores are not unexpected, given that
the model was trained using MSE as an objective/loss function [25]. Considering
that MAE is a more reliable measure of forecast quality [17, 21], the ANN has a
marginal improvement upon the Hydro-Estimator. A slight increase in positive
correlation ($r = 0.24$) was also noted.

In general, neither approach faithfully recreated the distribution and magnitude
of the rain field in the target scene. The punitive nature of a point-based evaluation
also heavily penalises both techniques, as a close estimate off by a few grid cells
would score poorly, albeit still be of use to forecasters. By evaluating a neighbourhood around a cell for verification, it is possible in future to be able to reduce
the spatial fallacies [26] involved with this form of evaluation (locational, atomic,
modifiable area unit problem) to account for the dislocation as well as incorporating
the fact that rain does not necessarily fall immediately below its cloud due to effects
such as wind speed and direction [27].

The feature engineering in this approach is admittedly simplified, using *apriori*
knowledge of the satellite instrument capabilities to thin the number input features
[28]. As such, there remain a number of feature engineering methods available for

future investigation, such as Principle Component Analysis [29] through which the expected loss of data may result in a simpler model space. Further to this, the approach to data set selection and generation could also be improved by incorporating knowledge from both the Machine Learning and Meteorological fields. Provost [30] notes that a balanced dataset with a more equitable range of targets generally improves performance over an imbalanced dataset, such as the scene chosen for this study. Further investigation is warranted with a wider variety of precipitation (light, moderate, intense) covering a greater proportion of the training scene to test this approach. The temporal homogenisation used in this study can also be taken further; including both the satellite scan line time (rather than the file time metadata) and considering the additional minutes required for ice crystals detected at cloud tops to fall as precipitation (possibly) below the cloud [31].

Lastly given the noisy nature of the target regression variable, multi-category classification metrics, such as rainfall range classes of threshold exceedance [32] could also be considered for subsequent investigation.

6 Conclusion

It is clear that both the ANN and the Hydro-Estimator have differing strengths among the metrics used in this study. That the ANN overestimates precipitation area to a greater extent than Hydro-Estimator is perhaps the most significant failing of the use of the Artificial Neural Network method in precipitation estimation, though its accuracy in other metrics appears promising. At this stage, it is unlikely that a simple Artificial Neural Network, such as that provided in this study, could easily replace the Hydro-Estimator model in terms of performance on such an imbalanced training data set. However, given that only a simplified example of an ANN is used here against a fully operational industry model for a small data sample, the results suggest that the ANN is worthy of further investigation. Improved parameterisation, data curation and model training could yield still greater accuracy of rain rate estimation and this case study provides compelling support for further research into the use of Machine Learning techniques such as the Artificial Neural Network for precipitation estimation.

References

1. Australian Bureau of Meteorology. *How Radar Works* (2015), http://www.bom.gov.au/australia/radar/about/what_is_radar.shtml. Accessed 3 March 2015
2. P. May, A. Protat, A. Seed, S. Rennie, X. Wang, C. Cass et al., The use of advanced radar in the Bureau of Meteorology, presented at the International Conference on Radar (Barton, A.C.T., 2013)
3. J. Sun, M. Xue, J.W. Wilson, I. Zawadzki, S.P. Ballard, J. Onvlee-Hooimeyer et al., Use of NWP for nowcasting convective precipitation: recent progress and challenges, Bull. Am. Meteorol. Soc. **95**, 409–426 (2014/03/01 2013)

4. Australian Bureau of Meteorology, *Numerical Prediction Charts—Weather and Waves* (2015), http://www.bom.gov.au/australia/charts/chart_explanation.shtml. Accessed 25 Aug 2015

5. Australian Bureau of Meteorology, *Optimal Radar Coverage Areas* (2015), http://www.bom.gov.au/australia/radar/about/radar_coverage_national.shtml. Accessed 1 March 2015

6. Australian Bureau of Meteorology, *Japan launches new weather satellite* (2014), http://media.bom.gov.au/releases/16/japan-launches-new-weather-satellite/. Accessed 3 March 2014

7. D. Jakob, A. Seed, Spatio-temporal characteristics of blended radar/gauge data and the estimation of design rainfalls, presented at the hydrology and water resources symposium 2014, 2014

8. S. Sinclair, G. Pegram, Combining radar and rain gauge rainfall estimates using conditional merging. Atmos. Sci. Lett. **6**, 19–22 (2005)

9. R.J. Kuligowski, *Hydro-Estimator—Technique Description* (2015), http://www.star.nesdis.noaa.gov/smcd/emb/ff/HEtechnique.php. Accessed 27 Feb 2015

10. National Oceanic and Atmospheric Administration, *STAR Satellite Rainfall Estimates—Hydro-Estimator—Digital Global Data* (2014), http://www.star.nesdis.noaa.gov/smcd/emb/ff/digGlobalData.php. Accessed 30 Jan 2014

11. T.J. Schmit, M.M. Gunshor, W.P. Menzel, J.J. Gurka, J. Li, A.S. Bachmeier, Introducing the next-generation advanced baseline imager on GOES-R. Bull. Am. Meteorol. Soc. **86**, 1079–1096 (2005)

12. D.E. Rumelhart, G.E. Hinton, R.J. Williams, Learning representations by back-propagating errors. Cogn. Model. vol. 5 (1988)

13. S. Hochreiter, Untersuchungen zu dynamischen neuronalen Netzen, Master's thesis, Institut fur Informatik, Technische Universitat, Munchen, 1991

14. A.P. Engelbrecht, *Computational Intelligence: An Introduction* (Wiley, 2007)

15. G.E. Hinton, N. Srivastava, A. Krizhevsky, I. Sutskever, R.R. Salakhutdinov, Improving neural networks by preventing co-adaptation of feature detectors, *arXiv preprint* (2012). arXiv: 1207.0580

16. E.E. Ebert, Methods for verifying satellite precipitation estimates, in *Measuring Precipitation from Space: EURAINSAT and the Future*, ed. by V. Levizzani, P. Bauer, F.J. Turk (Springer, Dordrecht, The Netherlands, 2007), pp. 345–356

17. C.J. Willmott, K. Matsuura, Advantages of the mean absolute error (MAE) over the root mean square error (RMSE) in assessing average model performance. Clim. Res. **30**, 79 (2005)

18. M.D. Schluchter, Mean Square Error, in *Wiley StatsRef: Statistics Reference Online*, (Wiley, 2014)

19. Kaggle Inc., *Root Mean Squared Error (RMSE)* (2015), https://www.kaggle.com/wiki/RootMeanSquaredError. Accessed 17 March 2015

20. T. Chai, R. Draxler, Root mean square error (RMSE) or mean absolute error (MAE)? Geoscientific Model Dev. Discuss. **7**, 1525–1534 (2014)

21. J.S. Armstrong, F. Collopy, Error measures for generalizing about forecasting methods: empirical comparisons. Int. J. Forecast. **8**, 69–80 (1992)

22. Japan Meteorological Agency, *Imager (AHI)* (2015), http://www.data.jma.go.jp/mscweb/en/himawari89/space_segment/spsg_ahi.html. Accessed 2 March 2015

23. C. Dancy, J. Reidy, Statistics without maths for psychology, *IEEE Statistics without maths for psychology,* 2004

24. H.B. McMahan, G. Holt, D. Sculley, M. Young, D. Ebner, J. Grady et al., Ad click prediction: a view from the trenches, in *Proceedings of the 19th ACM SIGKDD international conference on Knowledge discovery and data mining*, 2013, pp. 1222–1230

25. I.J. Goodfellow, D. Warde-Farley, P. Lamblin, V. Dumoulin, M. Mirza, R. Pascanu et al. Pylearn2: a machine learning research library, *arXiv preprint* (2013). arXiv:1308.4214

26. T. Bacastow, *Understanding Spatial Fallacies* (2010), https://www.e-education.psu.edu/sgam/node/214. Accessed 25 May 2010

27. S.S. Chen, J.A. Knaff, F.D. Marks Jr, Effects of vertical wind shear and storm motion on tropical cyclone rainfall asymmetries deduced from TRMM. Mon. Weather Rev. **134**, 3190–3208 (2006)
28. L. Bottou, *Feature Engineering* (2010), http://www.cs.princeton.edu/courses/archive/spring10/cos424/slides/18-feat.pdf. Accessed 20 May 2010
29. I.H. Witten, E. Frank, M.A. Hall, *Data Mining—Practical Machine Learning Tools and Techniques*, 3rd edn. (Morgan Kaufmann, Burlington, MA, 2011)
30. F. Provost, Machine learning from imbalanced data sets 101, in *Proceedings of the AAAI'2000 workshop on imbalanced data sets* (2000), pp. 1–3
31. F. Fabry, I. Zawadzki, Long-term radar observations of the melting layer of precipitation and their interpretation. J. Atmos. Sci. **52**, 838–851 (1995)
32. Centre for Australian Weather and Climate Research, *Forecast Verification: Issues, Methods and FAQ* (2015), http://www.cawcr.gov.au/projects/verification/. Accessed 5 Sept 2015

Construction of PM_x Concentration Surfaces Using Neural Evolutionary Fuzzy Models of Type Semi Physical Class

Alejandro Peña and Jesús Antonio Hernández

Abstract Pollution by particulate matter (PM_x) is the accumulation of tiny particles in the atmosphere due to natural or anthropogenic activities. Particulate matter becomes a pollutant that seriously affects the health of people. In order to reduce its concentration (PM_x), understanding its behavior in space is necessary, overcoming both physical and mathematical limitations. Limitations here refer to little information that a set of monitoring stations provided with regard to air quality and with respect to the dynamics of a pollutant. Furthermore, to the effect that an emission source produces within a certain area (*source apportionment*). Therefore, this work proposes the development of a model for spatial analytical representation of PM_x concentration over time as fuzzy information. The design of the model is inspired by the structure of a *Self-Organizing Map* (SOM). The model consists of an input layer (*n_sources*) and an output layer (*m_stations*) that were determined in shape and size for the study area. Connections between layers are defined by a *Lagrangian backward Gaussian puff tracking model*, which depend on the meteorological dynamics of the area. The model allows the estimation of emissions in *n_sources*, based on the measurement of (PM_x) concentration in the *m_stations* that were considered. The connection weights are adjusted by using evolutionary algorithms. The model showed a series of analytical forecasting maps that describe the spatial temporal behavior of PM_x concentration in terms of the puffs emitted by *n_sources*. The result is a spatial neural evolutionary fuzzy model of type semi-physical class. Its application can support the improvement of air quality in an study area.

A. Peña (✉)
Grupo de Investigación en Modelamiento Computacional y Simulación (GISMOC),
Escuela de Ingeniería de Antioquia, Envigado Colombia,
Km2 + 200 Mts. Variante al Aeropuerto José María Córdova,
Envigado, Antioquia, Colombia
e-mail: pfjapena@gmail.com

J.A. Hernández
Grupo de Investigación en Inteligencia Computacional (GIIC),
Universidad Nacional de Colombia, Sede Medellín, Colombia
e-mail: jahernan@unal.edu.co

© Springer International Publishing Switzerland 2016
S. Shanmuganathan and S. Samarasinghe (eds.), *Artificial Neural
Network Modelling*, Studies in Computational Intelligence 628,
DOI 10.1007/978-3-319-28495-8_15

341

Keywords Particulate matter (PM$_x$) · Self-organizing map (SOM) · Estimation distribution algorithms (EDA) · Forecasting maps · Neural-fuzzy model

1 Introduction

Governments have established various measures in order to improve air quality, but pollution by particulate matter (PM$_x$) is still considered a serious threat for human health [1]. Scientific evidence has shown that prolonged exposure of humans to this pollutant has been associated with an increase in human morbidity and mortality due to the effects generated on health [2–4]. In recent years, particulate matter has been classified as a carcinogenic pollutant [5]. European legislation on air quality for PM$_{10}$ (PM < 10 µm) and PM$_{2.5}$, establishes a daily limit in terms of the concentration for PM$_{10}$ of 50 ug.m^{-3} [1]. Due to this problem, building scale maps that explain the behavior of the PM$_x$ concentration constitutes an essential tool in monitoring compliance, and assessing the implementation of policies to mitigate the effects this pollutant generates on human population [6].

One of the main questions when trying to decrease the effects of particulate matter (PM$_x$) concentration in a study zone is directly related to a better understanding of its spatial behavior throughout time. To solve this question it is necessary to overcome physics and mathematical restrictions. In the physical field, they go from the pollution sort to the impossibility of carrying out massive measurement campaigns. Besides, the little information provided about the spatial behavior of the pollutant by a set of m_stations that monitor the air quality. From the mathematical point of view, the restrictions are associated with the relationship between n punctual sources of emission and m air quality monitoring stations, their location and the analytical spatial representation of the phenomenon [7, 8].

From the point of view of models, these restrictions are associated with the underestimation generated by Chemistry Transport Models (CTM) with respect to the reference concentration in estimating PM$_x$ emissions in space. This is due to the uncertainty associated with the composition of emissions in a particular source, and the formation of secondary pollutants due to the weather dynamics within the study area [9]. Alternatively, empirical models based on geostatistics [10] and the use of regression models [11], have yielded many promising results in the spatial representation of PM$_x$. These models have strong restrictions as the conditions under which they were developed, assume that the mean and variance are stationary, meaning that the relationship between observed measurements and CTM for PM$_x$ do not vary over time and space [12].

This chapter provides a spatial model inspired by the structure of *Self organizing maps* (SOM), to explain the spatiotemporal behavior of PM$_x$ concentration in a study zone. The developed model possesses two layers. A first layer or point layer (input layer), which is composed of a set of n_sources of particle emissions of PM$_x$, while a second layer (output layer) allows describing the spatial behavior of the concentration of PM_x throughout time. The connections of the model are described in terms of a *lagragian gaussian puff model* of type *backward puff tracking*, setting a spatial neural model which allows obtaining a series of forecast maps for PM_x. The analytical representation of these maps is given in terms of the puffs emitted by each of the sources, which are represented by *Gaussian functions*, and setting a neural evolutionary fuzzy model of semi physical class supporting the decision making process in order to mitigate the effects of PM_x concentration within the area of study.

For the analysis and validation of the proposed model two stages were considered. In a first stage, a series of theoretical measurement campaigns using the CALMET/CALPUFF model were carried out. This was done in order to assess the ability of the model with regard to the estimation of emissions from *n_sources* and using the obtained surfaces reflecting the temporary space PM_{10} concentration behavior within an area of study and considering different configurations for *n_sources* of emission and *m_stations* of monitoring air quality. In a second stage, a series of actual measurement campaigns were carried out in order to evaluate the behavior of the model with reference to a number of obtained surfaces that account for the spatiotemporal behavior of PM_{10} concentration within the study area. Departing from the densification the model shows the point performance of PM_{10} concentration in space, according to the weather dynamics that the model incorporates [13].

Thus, the proposed model overcame a group of restrictions imposed by the limited information provided by *m_stations* that monitored air quality and reveals the spatiotemporal behavior of a dispersion phenomenon within an area of study. In particularly the dispersion of particulate matter PM_{10} that is mainly generated by stationary sources. Due to its design, the proposed model can be used to determine the space temporal behavior of PM_{10} concentration in a given study area and thanks to its ability to adapt itself to different environments by learning, and because it incorporates to its structure the weather dynamics governing the dynamics of dispersion of pollutants in an area of study.

1.1 Trends on Pollutants Dispersion Modelling

Dispersion models are models that explain the behavior of pollutants in the atmosphere, and are used by the environmental authorities in order to establish a framework that allows mitigating the generated effects of such pollutants on human health. Regarding the development of pollutant dispersion models two well-defined trends of development can be found in literature: a first trend focusses on the development of models trying to explain the punctual behavior of pollutants in an area of study, while the second trend focusses on the development of models that explain the spatiotemporal behavior of pollutants in space.

Within the first trend, a model based on the Positive Matrix Factorization (PMF) can be found. This has been proposed in order to determine the source apportionment (SA) of different pollutants on points that are spatially distributed in a study zone [14], applying multidimensional patterns of the weather dynamics in that zone. Argyropoulos and Samara [15] presented a model to calculate the SA for PM_x in a study zone by a mass and energy balance (CMB) based on a source-receptor model. This model integrates the weather dynamics of this zone, allowing a unique solution to the conventional problem that presents the chemical mass balance CMB in estimating SA. For the automation of the data management and for estimating the concentration for PM_{10}, the model takes the name of Robotic Chemical Mass Balance (RCMB).In a later work, Givehchi et al. [13] adopted a methodology to determine the sources that generate episodes of high PM_{10} concentrations in desert regions close to Teheran. For this identification, the authors used the HYSPLIT model (*Hybrid Single Particle Lagrangian Integrated Trajectory*), in order to determine the sources in the desert area that contributed more to these pollution episodes. Another work recounts the main source-receptor models that have been used for the SA to determine tracing elements, organic and elemental carbon (PAHS). This work showed in conclusion, the difficulty for these models to identify the origin of pollutants from industrial sources [16]. However, these models have been used for making decisions in order to mitigate the effects of the concentration of certain pollutants generated on specific points within a study area.

In a second development trend, work focuses on the development of models that attempt to explain the spatiotemporal behavior of PM_x concentration in a study zone, using source-receptor models that integrate geospatial concepts from geographical information systems (GIS). In this way, a model was developed by De la Rosa et al. [17] in 2007; they conducted a study of space pollution for PM_{10} in southern Spain by building a series of geochemical maps by interpolating data obtained from a set of 17 stations that monitored air quality and that were spatially distributed in the zone. These maps allowed determining the origin of PM_{10} pollution in cities and sites that have high ecological value in Andalucia, Spain. Afterward, Pisoni et al. [18] proposed a methodology that integrates a chemical contaminant transport model with geostatistics to obtain the temporary concentration maps for PM_{10}.

The construction of these maps was done through the interpolation of a series of specific measures of PM_{10} concentration, allowing evaluating the concentration regularly within the area of study. This tool helped to implement more effective strategies to mitigate the impact generated from this type of pollutant on human health.

Within this trend, and consistently with the objectives pursued by the environmental authorities, to provide information to the public about air quality in the area, Singh et al. [19] proposed a spatial model to estimate hourly concentrations of ozone and particulate matter PM_x by spatial interpolation of a set of stations that monitor air quality and by using techniques of cokriging, which are supported in a deterministic chemical model (CTM). This model was used to estimate the concentrations of these pollutants in areas of difficult access, or in which it is impossible to carry out measurement campaigns. In the same line, other work can be found in which two models based on the principles of computational intelligence such as the Self Organizing Maps (SOM) and the Bayesian Hierarchical Model (BHM) were used to estimate spatial concentrations of benzene, affecting the urban zone in the city of Leipzig, Germany. Both models helped identify periodic spatial variations of benzene in the winter and in late summer [20]. Another study [21] has been highlighted which used neural networks with the *k-means* technique, explaining the spatiotemporal behavior for $PM_{2.5}$ and PM_{10} concentration from time series and describing the punctual behavior of these pollutants according to the weather dynamics within a study zone. This spatial model allowed, through clustering, evidencing the effects of vehicular traffic, anthropometric activities as well as the sea spray on these series.

In order to mitigate the effects that pollution causes, and consistently with the second trend of development, other authors have proposed geospatial models to determine the origin of particulate matter PM_{10} affecting a study area. In a first model, Diaz et al. [22] used a series of factors of spatial diffusion that, departing from a set of monitoring stations, enabled the estimation of the concentration for PM_{10} in neighboring areas in which it is not possible to perform measurement campaigns [22], similar to the other models described above. In this way, Pilla and Borderick [23] propose an information system to spatially measure the exposure of the inhabitants of Dublin to concentrations of particulate matter PM_{10}. For this system, different dispersion models were incorporated, as well as different layers that deal with land use and population density in order to take measures to mitigate the effects of such pollutants, especially on humans when traveling to work.

According to the above, the scientific community shows an observable interest for the development of information systems supporting spatial decision making with the objective to mitigate the effects of particulate matter PM_X concentration within a study zone. Many of these systems incorporate specific source-receptor type dispersion models, which lead to a limitation in explaining a spatiotemporal phenomenon, as happens in the phenomena of dispersion of pollutants in the atmosphere.

2 Methodology

2.1 Neuro Evolutionary Spatial Model

To determine the spatiotemporal behavior of the concentration of particulate matter PM_x within an area of study, the model has a structure inspired on a *Self-organizing map (SOM)* with two layers [24, 25]. The input layer or source layer has n point sources (*n_sources*). The output layer or air quality monitoring stations layer has m cells or monitoring stations (*m_stations*). Both input and output layer are spatially restricted through the limits of the study zone, as shown in Fig. 1.

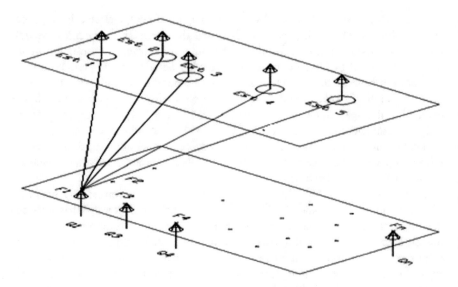

Fig. 1 Geospatial SOM to determine the space-time behavior of PMx concentration in a study zone

The relationship between layers of the SOM, represent the puffs' flow among the sources and the monitoring stations, and is defined by the weather dynamics within the study area and in terms of a *Lagrangian gaussian model* of the type *backward gaussian puff tracking* [26, 27]. For the emission estimation, the proposed model takes the form on the Eq. 1.

$$\Phi_j\left(x_j, y_j, z_j, k\right) = \sum_{j=1}^{ne} \sum_{i=1}^{nf} \sum_{k=1}^{np} \phi\left(x_{o,i,k}, y_{o,i,k}, x_j, y_j\right).G\left(z_{o,i,k}, z_j\right) \tag{1}$$

where:

$\Phi\left(x_j, y_j, z_j, k\right)$: PM$_x$ contribution of each puff to the *j_cell*, from the piling of emitted puffs from each of the *n_sources* in a moment k [1/m^3].

x_j, y_j, z_j: Spatial location of each cell or monitoring station in the output layer, in UTM_x [km], UTM_y [km] and MSL (meter sea level) coordinates.

k, a moment of time.

$\phi\left(x_{o,i,k}, y_{o,i,k}, x_j, y_j\right)$: Form and size of the emitted puffs from each one of the *i_sources*.

$x_{o,i,k}, y_{o,i,k}$: Location of each emitted puff from an *i_source*, in a moment k. In UTM_x [km], UTM_y [km] coordinates.

$G(z, k)$: Effects of *n_reflections* of each pollutant puff on the Earth surface and the thermal inversion layer located in a height H(k). The mixing layer is between both surfaces.

nf: Number of sources (in the sources layer), $i = 1,2,3,........, nf$
ne: Number of monitoring stations (cells number—output layer), $j = 1,2,........, ne$
np: Number of emitted puffs from each i_source, $k = 1,2,3,....,np$.
The form and size of the puffs are mathematically defined in Eq. 2.

$$\varphi\left(x_{o,i,k}, y_{o,i,k}, x_j, y_j\right) = \frac{1}{(2\pi)^{3/2}\sigma_x\sigma_y} Exp\left[-\frac{1}{2}\left[\left(\frac{x_j - x_{o,i,k}}{\sigma_x}\right)^2 + \left(\frac{y_j - y_{o,i,k}}{\sigma_y}\right)^2\right]\right]$$

(2)

where

σ_x, σ_y: Coefficients of turbulent diffusion that determine the form and size of each emitted puff [m].

Mathematically, the concept of reflections and virtual sources is expressed in Eq. 3, [28].

$$G(z,k) = \sum_{n=-\infty}^{\infty}\left[\exp\left(-\frac{(2nH(k) - h_e(k) - z)}{2\sigma_z^2(t)}\right) + \exp\left(-\frac{(2nH(k) + h_e(k) - z)}{2\sigma_z^2(t)}\right)\right]$$

(3)

where

$h_e(k)$: Height of each pollutant puff [m].
z: Calculated Height for the *n_reflections* concept.

2.1.1 Solution Structure (Evolutionary Individual)

In accordance with the spatial structure of the proposed model, the solution structure or evolutionary individual is defined as in Fig. 2:
Where:
Q_i: Pollutant amount inside each emitted puff from the source [g].
BDL: Initial mixing layer high for the study zone [m].
$K_{1,i}$: Number of point sources spatially clustered.
$K_{2,i}$: Opening size of the gases output (m^2).
$K_{3,i}$: Gases output speed (m/s).
For the concentrations computation and following the solution structure, the emissions value is recalculated with the known information of the source, as in Eq. 4.

$$Q_i = Q_i / \left(K_{1,i} * K_{2i} * K_{3,i}\right)$$

(4)

Q_1	Q_2	Q_3	......	Q_n	K_{11}	K_{12}	K_{13}	....	K_{1n}	K_{2n}	K_{3n}	BDL

Fig. 2 Solution structure or evolutionary individual

In Eq. 4, the constants $K_{1,i}$, $K_{2,i}$, $K_{3,i}$, are given according to the adaptation the evolutionary model makes with regard to the known information of a point source during a set of measurement campaigns within the same study zone. Using the CALMET/CALPUFF model [29] through Eq. 5 the value of those constants is obtained.

$$K_{1,i} = K_{1,i}/nf_i \qquad K_{2,i} = K_{2,i}/\phi_{ch,i} \qquad K_{3,i} = K_{3,i}/v_i \qquad (5)$$

where:

nf_i: Number of sources, clustered by companies.
$\phi_{ch,i}$: Chimney size of the point source $\phi_{ch,i} = 2\ m$.
v_i: Gases output speed, $v_i = 50\ m/s$.

From Eq. 4 it can be observed that for a same value of emission, a larger value of pollutant (Q_i) is required, depending of the known information of the source. The fitness function that qualifies the emission pattern (or evolutionary individual quality) regarding the solution is given as in Eq. 6.

$$FA = \frac{K_e}{\frac{1}{2}\left(\sum_{j=1}^{ne}\left(\left(\sum_{i=1}^{nf}\sum_{k=1}^{np} C_{cj}\left(x_{o,i,k}, y_{o,i,k}, z_{o,.i,k}\right)\right) - C_{bj}\left(x_j, y_j, z_j\right)\right)^2\right)} \qquad (6)$$

where:

$C_{bj}\left(x_j y_j, z_j\right)$: Base Concentration, or activation value for j_cell[ug/m^3].

$C_{cj}\left(x_{o,i,k}, y_{o,i,k}, z_{o,i,k}\right)$: Puff piling in j_cell by the connections coming from each of thei_source due to the emission pattern [ug/m^3].

K_e: Proportionality constant scaling the fitness function.

2.1.2 Evolutionary Asynchronous Model (Emission Estimate)

One of the most important elements in the present development of Evolutionary Computation (EC) is the EDA (Estimation Distribution Algorithms), which unlike algorithms for traditional evolution, do not require genetic operators, but statistically based operators for the evolution of the population of individuals or sets of possible solutions [30, 31].

For estimating emissions, the proposed Neuro Evolutionary Fuzzy Spatial Computational Model (SCM) incorporates an EDA, the MAGO (Multi dynamics Algorithm for Global Optimization). MAGO integrates three autonomous dynamics of evolution: Emergent dynamics, Crowd dynamics and Accidental dynamics [32]. For the particular case study, these dynamics will be used sequentially with respect to a single population of individuals, generating three stages in the process of emission estimation at a source and for each step of time k, which comprises a measurement campaign.

This mechanism began with the random generation of a population of individuals (*Stage 0*), which includes a set of random solutions generated by random sampling of the solution space of the problem. In a first stage (*Stage 1*), the algorithm proceeds with the emerging application of population dynamics, generating a fast convergence (SC) in estimation of emissions. The first stage or Emergent dynamics (*Stage 1*) allows correction for calculated PM_x concentrations of the best individuals toward the very best of all. At this stage the new subgroup of individuals is given as follows in Eq. 6:

$$Q_i^{(g)} = Q_i^{(g)} + F^{(g)} \cdot \left(Q_{i,B}^{(g)} - Q_{i,M}^{(g)} \right) \tag{6}$$

where to incorporate the information of relationships between genes of the population, the weighting factor is obtained from the correlation matrix, Eq. 7:

$$F^{(g)} = \frac{S^{(g)}}{\|S^{(g)}\|} \tag{7}$$

where:

$F^{(g)}$: Modification factor between the variables that determine the second substructure of generation g.

$S^{(g)}$: Indicates the covariance matrix of the whole population over the generation g in instant k.

$\|S^g\|$: Represents the diagonal of the correlation matrix.

At the second stage, the Crowd Dynamics [32], a new subgroup of individuals is created by randomly sampling over the hyper-rectangular distribution, defined by Eq. 8:

$$LB^{(i)} = Q_M^{(i)} - \sqrt{diag(S^{(g)})}, UB^{(i)} = Q_M^{(i)} + \sqrt{diag(S^{(g)})} \tag{8}$$

where:

$Q_M^{(i)}$: Represents the average of the whole population of genes that makes up the former genetic structure of the solution individuals, over the generation g for each instant of time k.

$diag(S^{(g)})$: Variance of each individual.

Individuals generated by the Emergent Dynamics at each generation are incorporated into the new population, in terms of an amount of N1, as long as it exceeds the value of the Fitness Function (FF) of their predecessors. The Crowd Dynamics adds N2 individuals (around the actual mean). It should be noted that for each step of time *k*, the SCM will require a number of generations, *g*, which indicate the number of times the population of individuals evolve.

Pseudo code of MAGO
1: j = 0, Generation of the initial population, with auniformrandomlydistribution in the search space.
2: Repeat
3: Evaluate each individual by means of the objective function
4: Calculate the covariance matrix of the populationand the first, second and third dispersion.
5: Calculate the N1, N2 and N3 cardinalities of groups G1, G2 and G3.
6: Select N1 of the best individuals, modify according to the objective function and make them compete. Winners pass to the next generation *j* + *1*.
7: Sampling N2 individuals from the uniform distribution in the hyper rectangle [LB (j), UB (j)], and move on to the next generation j + 1.
8: Sampling N3 individuals from the uniform distribution throughout the whole search space and move to the next generation j + 1.
9: j = j + 1
10: Do until a termination criterion is satisfied.

Fig. 3 MAGO pseudocode

Finally, the population is completed by the Accidental Dynamics, a subgroup of individuals created randomly over the whole searching space, as the initial population at the beginning of a campaign of measurement, but of a dynamic size, N3. The three dynamics can maintain genetic diversity of the population along a measurement campaign. The three dynamics (Stage 1, 2 and 3) are applied in each generation to the population of individuals and for each step of time, k, allowing the population to prepare for estimating emissions at an instant of time k + 1. The population of individuals in a constant number N is kept summing the dynamical cardinalities N1, N2 and N3. In the following Fig. 3, the pseudo code of the MAGO algorithm is presented.

2.2 Case Study

For the analytical representation of the spatiotemporal behavior of the concentration of particulate matter PM_{10}, a study zone was selected. This zone comprises an area of 50 * 50 km^2 and It is characterized by surrounding mountains with heights up to 3200 (*msl*), which significantly modifies the direction and speed of the wind. The computed multiannual wind rose shows that most of the time (31 % calmness) there is no atmospheric movement around the monitoring stations. Meanwhile, the predominant winds come from the north 20 % and from the south 14 % of the time, as is shown in Fig. 4.

For the analysis and validation of the proposed model, two stages were defined that allowed evaluating the mechanism for estimating emissions from *n_sources* of emission and by using a set of *m_stations* of monitoring of air quality, and that are

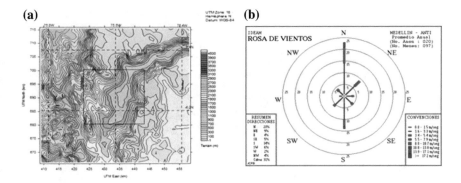

Fig. 4 Study zone. **a** Upper view—level curves. **b** Multiannual wind rose

located in the study area. For the first stage, a total of 24 theoretical campaigns were carried out in order to measure the concentration for PM_{10} in the study area, using the CALMET/CALPUFF model for different configurations of monitoring *m_stations* and *n_sources* [18]. In a second stage, a total of 7 real campaigns were carried out, which allowed to measure the concentration of PM_{10} in the same area, taking into account for this purpose a total of 973 point sources of emission. These were reduced to 403 point sources per cluster (*403_sources*) according to the spatial organization and 17 monitoring stations of air quality (*17_stations*). The emission sources are completely from the industrial and service sector within the study zone. In the second stage, for the background concentration, 137 emission sources were taken into account. These sources correspond with the area sources and fuzzy sources, the later ones coming from the flow of traffic, open air handling and transportation of particulate matter as in quarries, activities related with construction, the service sector, commerce and nourishment in the study zone. To achieve the estimated emissions the first stage counted with a total of 48 h (2 days) per campaign, while for the second stage and given the number of emission sources, a total of 480 h (20 days) per campaign were taken. The meteorological dynamics of the study area that define the connections between the layers of the model was estimated fora height of 10 m to 0.5 km resolution. The layers that make up the model were defined in shape and size for the study area as shown in Fig. 5.

2.3 Model Evaluation

For each measurement campaign considered for each stage, the estimation of emissions at *n_sources* was carried out considering the reference hourly concentrations for PM_{10} measured in each of *m* monitoring stations of air quality (*m_stations*).In its first stage, for each hourly estimation process, the model delivered as a result two digital elevation models (DEM) for PM_{10} by point densification of the output layer (*source apportionment*). Each of these models

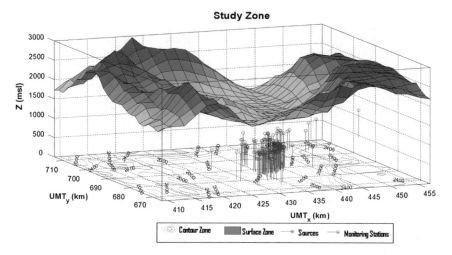

Fig. 5 Spatial location of point sources, area sources, fuzzy sources and monitoring stations (*UTM_x, UTM_y coordinates*)

comprises a total of 10 * 10 points, which are arranged regularly for a first DEM, while for a second DEM, these points are arranged irregularly.

After finishing the point densification of the output layer in the first stage, it continued with the selection of interpolation model taking into account the most relevant features from five spatial interpolation models based on the principles of computational intelligence, which have been used as reference for the representation of spatial phenomena such as:

1. IDW method (Inverse Distance Weight) [33].
2. Cokriging—Radial Basis Neural nets (FBR) [19].
3. Integrated method of Kohonen-IDW maps [21, 34].
4. NURBS method (*Non Uniform Rational Basis Splines*) [7].
5. *Takagi sugeno NURBS* method [33, 35].

After selection of interpolation model, the construction of hourly analytic surfaces for PM_{10} concentration was also carried out for both stages.

For a general evaluation of the proposed model, eight statistic metrics were used according to the fuzzy model developed by Park & Ok-Hyum [36]: Fractional Bias (FB), Geometric Mean Bias (MG), Geometric Mean Variance (VG), Index of Agreement (IOA), Within a Factor of Two (FAC2), Normalized Mean Square Error (NMSE), Index of Agreement (IOA), and Unpaired Accuracy of Peak (UAPC2). They are sorted in the following manner:

1. To measure the discrepancy between the baseline and estimated emissions on each considered *n_sources* of emission (*input layer*), MRE, IOA metrics were used.
2. To measure the discrepancy between the reference of PM_{10} concentrations and the calculated PM_{10} concentrations on each *m_stations* monitoring air quality

spatially located in the study area (*output layer*), the NMSE, UAPC2 and IOA metrics are used.

3. For the selection of the interpolation model, all the arranged metrics by the Park & Ok-Hyum model were used [36]. These metrics allow the assessment of the quality of the interpolation in terms of DEM accounting for the source apportionment in the study area (*output layer*).

4. For a general evaluation of selected interpolation model in both stages, FB, MG, VG, FAC2 and IOA metrics assess the quality of the interpolation in terms of the point densification of the output layer, and in terms of the points of PM_{10} reference concentration, where monitoring stations of air quality are located.

In accordance with the Park & Ok-Hyum model [36], each of the applied metrics assumes a qualitative fuzzy value of Good (G), Fair (F), OverFair (OF), Under Fair (UF) and Poor (P). With respect to the performance indicator of the model, quantitative values are assigned to each quality in the following way: G (8.5), F (5.5), OF (6.0), UF (5.0) and P (2.5), where the maximum score that a model can reach in terms of its behavior of the data is 68 points. According to this score, a model can be ranked in the following descending order with grades from A t to Z, where A represents the best level of performance, while Z represents the lowest level of performance.

3   Results

The results from the proposed model, compared to the estimation of emissions according to each of the measurement campaigns carried out using the CALMET/CALPUFF model in the first stage, are shown in Table 1.

Regarding to the measurement campaign conducted on the days 215–216 for a configuration of *5_sources* and *25_stations*, the average results obtained by spatial interpolation models used as a reference against the hourly densification grids are shown in Table 2.

The results obtained by the interpolation model versus spatial analytical representation of the PM_{10} concentration for both stages, was made taking into account the punctual reference values for PM_{10} measured in each *m_stations* (Point Index), and considering the spatial densification of the output layer(Spatial Index), they are presented in Table 3. The metrics used FB, MG, VG, FAC2 and IOA represent the amount of contaminant present at the interpolation model and in the study area.

The general structure of the concentration surfaces obtained as a result of interpolation and modelling the spatial behavior of the PM_{10} concentration on the study area in the first stage is shown in Fig. 6.

The general structure of the point and spatial correlation diagrams reflecting the spatial behavior of the hourly interpolation model is shown in Fig. 7.

Table 1 General results of the hourly estimation of emissions for each theoretical measurement campaign (first stage)

Configuration			Point index (*m_stations*)			Estimated emissions (*n_sources*)	
Days	Sources	Stations	NMSE	UAPC2	IOA	MRE	IOA
03–05 04/8:00	4	20	0.488	0.012	0.810	0.134	0.854
08–10 08/11:00	4	6	0.104	−0.043	0.702	0.047	0.758
08–10 09/13:00	4	10	0.100	−0.043	0.691	0.047	0.739
08–10 08/16:00	4	15	0.102	−0.048	0.814	0.030	0.818
08–10 09/12:00	4	20	0.061	0.010	0.985	0.042	0.987
32–33 32/1:00	4	4	0.197	−0.155	0.711	0.076	0.665
60–62 61/2:00	4	4	0.168	−0.127	0.758	0.122	0.726
91–92 92:15:00	4	4	0.167	−0.113	0.727	0.045	0.763
121–122 122/18:00	4	4	0.166	−0.137	0.743	0.086	0.771
152–153 152/18:00	4	4	0.191	0.161	0.752	0.129	0.730
169–170 170/3:00	4	4	0.051	0.037	0.858	0.036	0.773
200–201 200/21:00	4	10	0.065	0.016	0.927	0.063	0.874
215–216 215/12:00	5	5	0.048	0.020	0.881	0.061	0.854
215–217 216/12:00	5	10	0.040	0.021	0.860	0.116	0.888
215–217 216/14:00	5	15	0.077	0.029	0.886	0.125	0.917
215–216 215/14:00	5	20	0.074	0.028	0.863	0.050	0.814
215–216 215/8:00	5	25	0.019	−0.026	0.987	0.103	0.911
281–282 281:23:00	4	4	0.074	−0.154	0.702	0.074	0.649 ·
299–300 300/15:00	4	5	0.030	0.030	0.859	0.171	0.873
299–300 299/8:00	4	10	0.055	0.165	0.930	0.009	0.873

(continued)

Table 1 (continued)

Configuration			Point index (*m_stations*)			Estimated emissions (*n_sources*)	
Days	Sources	Stations	NMSE	UAPC2	IOA	MRE	IOA
299–300 300/8:00	4	15	0.063	0.025	0.878	0.050	0.932
299–300 299/16:00	4	20	0.017	0.014	0.951	0.040	0.971
335–336 335/16:00	4	4	0.059	0.019	0.829	0.008	0.774
Mean			0.105	−0.011	0.831	0.072	0.822
Score			G	G	G	G	G
General grade							A

Table 2 Obtained results by the interpolation models proposed in the first stage for a regular point densification (10 * 10 points) and for an irregular point densification (10 * 10 points) from the campaign on the days 215–216 at 12:00

	NURBS(I)		Koh-IDW (R)		TSK(I)		TSK(R)		IDW(R)	
FB	0.0001111	G	−0.0010082	G	0.0165561	G	0.0455780	G	0.0228327	G
NMSE	0.0163013	G	0.0114598	G	0.0121486	G	0.0136514	G	0.0131965	G
MG	0.9942878	G	0.9944282	G	1.0121008	G	1.0427821	G	1.0165692	G
VG	1.0141951	G	1.0121065	G	1.0129690	G	1.0145486	G	1.0138470	G
FAC2	0.5567901	G	0.5333333	G	0.5086420	G	0.4788215	G	0.5148148	G
IOA	0.5238519	G	0.3170034	F	0.2957412	F	0.3788214	F	0.2099060	F
UAPC2	0.2296449	G	0.2926553	G	0.3342298	G	0.3609910	G	0.3685556	G
MRE	−0.0124441	G	−0.0101842	G	0.0076331	G	0.0374228	G	0.0112567	G
Score	68		64.5		64.5		64.5		64.5	
	NURBS(R)		Koh-IDW (I)		Radial(R)		IDW(I)		Radial(I)	
FB	−0.4941252	G	0.0128698	G	0.0007982	G	0.0004590	G	−0.501408	G
NMSE	0.2793502	G	0.0135772	G	0.0105381	G	0.0096290	G	0.2793502	G
MG	0.5966920	G	1.0085556	G	0.9971540	G	0.9982475	G	0.5966920	G
VG	1.3273490	G	1.0142992	G	1.0105806	G	1.0107813	G	1.3273490	G
FAC2	0.5123457	G	0.3333333	F	0.3456790	F	0.2839506	F	0.2812345	F
IOA	0.0662445	P	0.1532171	P	0.1420121	P	0.0095830	P	0.0662445	P
UAPC2	−0.1070951	G	0.3690680	G	0.3637503	G	0.2914025	G	−0.1070951	G
MRE	−0.6852289	G	0.0030310	G	−0.0066364	G	−0.0053863	G	−0.6852289	G
Score	62		58.5		58.5		58.5		58.5	

Table 3 General evaluation of the proposed model against the analytical spatial representation of concentration for PM_{10} in the study zone (first stage)

Days	Sources	Stations	Point index			Spatial index					
			NMSE	UAPC2	IOA	FB	MG	VG	FAC2	IOA	Score
03–05 04/8.00	4	20	0.488	0.012	0.810	0.143	1.264	1.034	0.873	0.825	68
08–10 08/11:00	4	6	0.104	−0.043	0.702	0.276	1.135	1.124	0.943	0.653	68
08–10 09/13:00	4	10	0.100	−0.043	0.691	0.156	1.135	1.105	0.943	0.775	68
08–10 08/16:00	4	15	0.102	−0.048	0.814	0.090	1.104	1.000	0.945	0.832	68
08–10 09/12:00	4	20	0.061	0.010	0.985	0.069	1.102	1.000	0.948	0.844	68
32–33 32/1:00	4	4	0.197	−0.155	0.711	0.460	1.263	1.204	0.868	0.765	68
60–62 61/2:00	4	4	0.168	−0.127	0.758	0.567	1.329	1.286	0.858	0.769	63.92
91–92 92:15:00	4	4	0.167	−0.113	0.727	0.308	1.116	1.124	0.923	0.838	68
121–122 122/18:00	4	4	0.166	−0.137	0.743	0.289	1.174	1.174	0.900	0.862	68
152–153 152/18:00	4	4	0.191	0.161	0.752	0.246	1.281	0.839	0.860	0.840	68
169–170 170/3:00	4	4	0.051	0.037	0.858	0.062	1.063	1.065	0.953	0.873	68
200–201 200/21:00	4	10	0.065	0.016	0.927	0.040	1.147	1.148	0.930	0.874	68
215–216 215/12:00	5	5	0.048	0.020	0.881	0.105	1.126	1.117	0.928	0.839	68

(continued)

Table 3 (continued)

Days	Sources	Stations	Point index			Spatial index						Score
			NMSE	UAPC2	IOA	FB	MG	VG	FAC2	IOA		
215–217 216/12:00	5	10	0.040	0.021	0.860	0.048	1.230	1.191	0.865	0.897		68
215–217 216/14:00	5	15	0.077	0.029	0.886	0.048	1.255	1.185	0.865	0.926		68
215–216 215/14:00	5	20	0.074	0.028	0.863	0.039	1.138	1.150	0.935	0.935		68
215–216 215/8:00	5	25	0.019	−0.026	0.987	0.049	1.265	1.075	0.860	0.947		68
281–282 281:23:00	4	4	0.074	−0.154	0.702	0.178	1.259	1.150	0.870	0.786		65.45
299–300 300/15:00	4	5	0.030	0.030	0.859	0.176	1.284	1.061	0.873	0.812		68
299–300 299/8:00	4	10	0.055	0.165	0.930	0.025	1.016	1.016	0.990	0.848		68
299–300 300/8:00	4	15	0.063	0.025	0.878	0.037	1.118	1.033	0.938	0.851		68
299–300 299/16:00	4	20	0.017	0.014	0.951	0.030	1.155	1.017	0.903	0.882		68
335–336 335/16:00	4	4	0.059	0.019	0.829	0.029	0.972	1.126	0.795	0.755		68
Mean index			0.105	−0.011	0.831	0.151	1.171	1.097	0.903	0.836		67.71
Score			G	G	G	G	G	G	G	G		
General grade												A

(a)   **(b)**

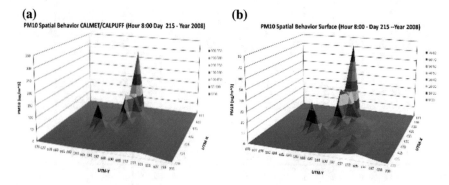

Fig. 6 Spatial behavior of the PM$_{10}$ concentration. Hour 8 day 215 year 2008 (*5_stations, 25_sources, first stage*). **a** CALMET/CALPUFF source apportionment, **b** Concentration surface for PM$_{10}$

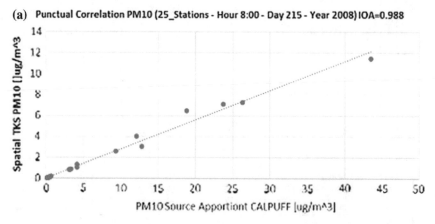

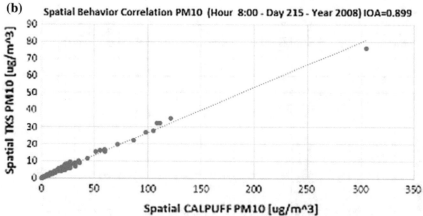

Fig. 7 Hourly correlation diagrams obtained in the first stage (*5_stations, 25_sources*). **a** Point correlation for PM$_{10}$ resulting from the interpolation taking into account *m_stations* that monitor air quality within a study zone per campaign. **b** Spatial correlation resulting from the spatial interpolation taking into account the point densification made by the CALMET/CALPUFF model

The results from the model against the analytical representation of the spatial behavior of the concentration of PM_{10} in the study area during the second stage are shown in Table 4.

The general structure of the concentration surfaces, obtained from the interpolation and exposition of the spatial behavior of the PM_{10} concentration with in the study area during the second stage, are shown in Fig. 8.

The general structure of the correlation diagrams, that are showing the quality of the spatial interpolation resulting from the PM_{10} concentration surfaces in its second stage, are shown in Fig. 9.

4 Discussion

The surfaces which account for the behavior of spatiotemporal PM_{10} concentration for both model validation steps generally showed consistency with the weather dynamics that characterized the study area, as shown in Fig. 10. It is noteworthy that the form and size of the layers of the proposed model were defined in shape and size according to the study area.

According to Table 1, it can be observed with respect to the estimation of emissions in each of the emission sources considered for each campaign, that the proposed model reached correlation values close to an average of 82 %, with a relative error near 7 %. Regarding the calculation of concentrations, the model reached an average rate similar to the estimative (83.1 %), with a NMSE index reaching a value close to 0.105 on average, with an underestimation of the maximum concentration values close to 1 % on average as shown in the UAPC2 index, which shows the quality of the model with regard to both the estimation of emissions, as well as with respect to the calculation of concentrations in each of the monitoring *m_stations* considered for each campaign.

According to Table 2, the interpolation models used as reference for the spatial analytical representation of the PM_{10} concentration in the study area were classified depending on the score achieved by each model and applying the Park & Ok-Hyum scheme.

Thus, high performance interpolators were interpolators which reached scores on average at 65.2, while the average performance interpolators are interpolators which reached lower scores than this value on average at 59.2. It should be noted that a score of 68 points is the maximum score established by the model of Park & Ok-Hyum. This classification was also made taking in account the information such models need for interpolating, or the explanatory power they have regarding to the phenomenon of dispersion of this pollutant in accordance with the principles of a *Lagrangian Gaussian puff model*.

- *Interpolators of High Performance*: According to Table 2, the interpolators based on NURBS (*Non Uniform Rational Basis Splines*) for irregular point grids had

Table 4 Obtained results from the proposed model using the hourly surfaces representing spatial behavior of the PM$_{10}$ concentration in the study zone, during its second stage

Day/hour	Point index			Spatial index						Estimate emissions	
	NMSE	UAPC2	IOA	FB	MG	VG	FAC2	IOA	Score	MRE	IOA
102–121 112/00	0.239	0.264	0.923	−0.020	1.043	1.028	0.066	0.973	68.0	0.276	0.985
122–141 140/12	0.181	0.072	0.815	0.013	1.024	1.003	0.057	0.977	61.5	0.115	0.850
142–161 150/08	0.142	0.061	0.827	−0.052	1.044	1.006	0.068	0.809	64.5	0.292	0.881
162–181 170/12	0.162	0.050	0.839	0.003	1.051	1.035	0.059	0.913	61.5	0.304	0.850
182–201 185/12	0.221	0.077	0.746	0.031	1.017	1.000	0.072	0.858	64.5	0.086	0.791
202–221 205/12	0.132	0.080	0.877	0.026	1.052	1.004	0.056	0.930	59.0	0.167	0.802
222–243 225/12	0.166	0.064	0.877	0.026	1.021	1.005	0.066	0.787	64.5	−0.104	0.737
Mean index	0.177	0.095	0.843	0.004	1.036	1.011	0.064	0.892	63.357	0.162	0.842
Score	G	G	G	G	G	G	G	G		G	G
General grade									A		

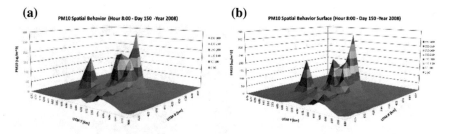

Fig. 8 Spatial behavior of the PM_{10} concentration, Hour 8:00, day 150, year 2008 (*17_stations, 403_sources, first stage*). **a** Source apportionment by the proposed model, **b** obtained surface from the interpolation of the point densification

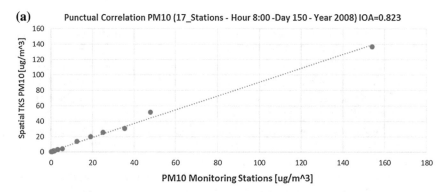

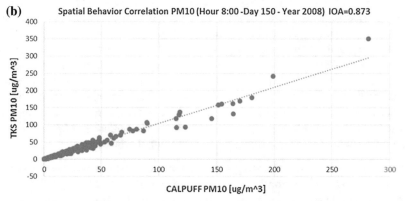

Fig. 9 General diagrams of hourly correlation (*second stage*). **a** Point correlation for PM_{10} resulting from the interpolation and taking into account *m_stations* that monitor the air quality within a study zone, per each measurement campaign. **b** Spatial correlation resulting from the interpolation taking into account the point densification made by the proposed model

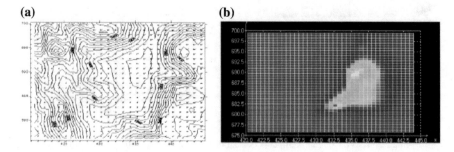

Fig. 10 **a** Prevailing wind field in the study area(CALMET/CALPUFF). **b** Spatial map for PM_{10} concentration according to the spatial distribution of sources and stations, and in accordance with the shape and size of the study area

the best performance indices. That was mainly due to the analytical definition of their basis functions, which are similar to the equations that are defining the puffs, which are traveling on the connections of the proposed model. Primarily because their basis functions are located spatially at points where there is a greater amount of information, which overcomes the limitation of little existing information in other areas of the zone. Another interpolator, which turned out as one of the best performance indices, was the method of *Takagi sugeno* for grids with irregular points. This was due to the properties of this method in the *compression - decompression* of complex surfaces [33, 34, 37].

- *Interpolators of Average Performance:* In Table 2, it can be seen, that the IDW and Koh-IDW methods on irregular point grids showed a poor performance, mainly because these methods are very sensitive to the amount of available information about a phenomenon in an area of study. Geostatistical methods for representing and interpolating complex surfaces base their power on the amount of available information within an area of study, as in the interpolation methods of Kohonen-IDW and IDW for regular grids, and unlike NURBS(R) interpolation methods and Neural Networks with Radial Basis Functions (RBF). These showed acceptable results, mainly because these methods are very sensitive to the saturation of information, especially when there is redundant information, as in a continuous phenomenon, and when it is spatially distributed.

According to the above, the integration of the high performance interpolation methods of *Takagi sugeno* and *Non uniform puffs functions* NURBS (TKSN) will set up an interpolation model based on the physical phenomenon of pollutant dispersion. The rational basis functions will be given in terms of concentration of *puffs* in space or *macropuffs*, which are consistent in shape and size to the equation of puffs that are emitted by the sources. Thus, the overall structure of the proposed interpolation model TKSN is denoted and defined as in Eqs. 9 and 10 [8, 35, 37]:

$$PM_{x,1} = \sum_{i=1}^{np} \emptyset_{i,x} \big(u_{i,x}.PM_x(i,j) + u_{i+1,x}.PM_x(i+1,j)\big) / \sum_{i=1}^{np} \emptyset_{i,x} \qquad (9)$$

$$PM_{x,2} = \sum_{i=1}^{np} \emptyset_{i,x} \big(u_{i,x}.PM_x(i,j+1) + u_{i+1,x}.PM_x(i+1,j+1)\big) / \sum_{i=1}^{np} \emptyset_{i,x} \qquad (10)$$

According to the above equations, the output of the TSKN model is denoted and defined by the Eq. 11, and its structure is shown in Fig. 11.

$$PM_{x,s} = \sum_{i=1}^{np} \emptyset_{i,y} \big(u_{i,y}.PM_{x,1} + u_{i+1,y}.PM_{x,2}\big) / \sum_{i=1}^{np} \emptyset_{i,y} \qquad (11)$$

where:

$PM_{x,S}$: Indicates output system of the study.

$PM_{x,1}$, $PM_{x,2}$: Spatial subsystem in UTM_y [km].

$PM_{y,1}$, $PM_{y,2}$: Spatial subsystem in UTM_y [km].

Fig. 11 TKSN, spatial interpolation model integrating *Takagi sugeno* and NURBS (*Non Uniform Rational Basis Splines*) modelling the estimation of emissions

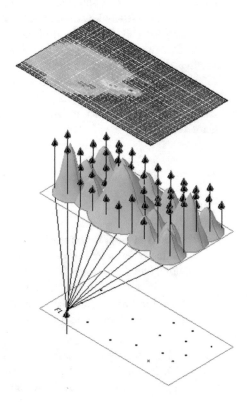

C_{1x}	C_{2x}	...	C_{nmx}	C_{1y}	C_{2y}	...	$C_{nm,y}$	σ_{1x}	σ_{2x}	...	σ_{nmx}	σ_{1y}	σ_{1y}	...	σ_{nmy}
Genotype 1				Genotype 2				Genotype 3				Genotype 4			

Fig. 12 Solution structure, or individual, used for the interpolation of the concentration surfaces for PM_x

$\mu_{i,x}$: Spatial influence of each monitoring station on each j_cell of densification axis in axis UTM_x [km].

$\mu_{i,y}$: Spatial influence of each monitoring station on each j_cell of densification axis in axis UMT_y [km].

$\varnothing_{i,x}$: Macropuff influence on each j_cell, in axis UTM_x [km].

$\varnothing_{i,y}$: Macropuff influence on each j_cell, in axis UTM_y [km].

According to the proposed model for the spatial interpolation, and in agreement with the structure of solution to the problem of estimating emissions, the structure of solution is extended, see Fig. 12, and in accordance with the position of each *macro-puff* used to obtain the spatial concentration surface of PM_{10}:

Where:

Cix, Ciy: Location space (UTM_x [km], UTM_y [km]) of each macropuff.

Ωix, Ωjy: Size of each macropuff [m].

Kix, Kjy: Eccentricity deformation parameters for each macropuff.

i = 1,2,...,nx: Spatial resolution of the concentration on the axis UMT_x depending of the cell densification in the output layer.

j = i, 2,...,ny: Spatial resolution of the concentration on the axis UMT_y depending of the cell densification in the output layer.

According to the point densification of the output layer in terms of the spatiotemporal relationship between sources and sinks, the fitness function, which describes the quality of each individual with respect to the solution of the problem, is denoted and defined as in Eq. 12:

$$FA_{SM} = 2/\sqrt{\sum_{i=1}^{(n+1).(m+1)} \left(PM_{x,d}(i,j) - PM_x(i,j)\right)^2} \qquad (12)$$

where:

FA_{SM} Fitness function, in terms of the inverse of the mean square error for each location of each cell of densification

$PM_{10,}$
$_d$ Digital elevation model (DEM) for PM_{10} concentration, or reference DEM, obtained after the emission estimation process (μg/m3)

PM_{10} DEM for PM_{10}, obtained after the interpolation representation process through adaptive model (μg/m3)

For building concentration surfaces for PM_{10}, the proposed interpolation model uses the MAGO algorithm independently of the version used for the emissions estimation process. According to the structure of solution, the MAGO will identify the highest concentration of *puffs* in space or *macropuffs*, overcoming the limitations imposed by spatial interpolation models of medium and high performance, which are dependent on the spatial distribution of information used for interpolation.

Results from the interpolation model during the first stage for a particular hour for each measurement campaign show that the model reached a score of 67.71 (Grade A) on average, according to the model of Park and Ok-Hyum [36]. This promoted mainly by the average values reached by the MG (1.171) and VG (1.097) indices, which show that the amount of contaminant in the areas of concentration for PM_{10}, generally have the same amount of contaminant in PM_{10} present in the study area. From a point view, the proposed model reached an average IOA of 0.836 and the FAC2 equals to 0.903, with a slight underestimation of concentration for PM_{10} as shown in the FB index, which took a value of 0.151. These indices show that the surface indicates in its outcome the same spatial PM_{10} concentration value in each of the points obtained as a result of the point densification, and each of the reference points (*m_stations*) that were used to estimate the emissions. It should be noted that the performance of the interpolation model, in terms of the IOA, was favored by a larger number of stations that monitored the air quality within the study area.

Table 4 shows the behavior of the proposed modeling the second phase of validation. The model achieved the score A with 63.35, this promoted mainly by the IOA achieved by the estimation model (0842), in calculating the concentrations (0.843), and there presentation of the spatial behavior of the concentration as a result of the point densification for PM_{10} of the study area (0.892). As in the previous case, MG (1.036) and VG (1.011) indices reached values close to unity, which shows the similarity between the amount of contaminant in the study area and the amount of contaminant on the concentration surface, even though the model shows a small underestimation of the concentration, as shown by the point index MRE (0.162) and the FB spatial index (0.004).

According to the above, it can be observed the good performance achieved by the proposed model with the analytical representation of the spatiotemporal behavior of the PM_{10} concentration in the two stages of validation, integrated into a single model the calculation of PM_{10} concentration, the dynamics of the dispersion phenomenon of pollutants and the pollutant cloud within a study area, setting a neural evolutionary fuzzy model of the type semi physical class, which allows to explain the spatial phenomenon of pollution by particulate matter PM_x in a study zone.

5 Conclusions

The proposed Neural Evolutionary Fuzzy model overcomes the limitations imposed by the lack of spatial information that a set of *m_stations* that monitor the air quality, yields with regard to the behavior of a dispersion phenomenon. Thanks to its adaptability, it can be used to describe the behavior of particulate matter concentration (PM$_x$) in any study area, if information about the meteorological dynamics of the study area is available.

This model obtains analytic surfaces for PM$_x$ concentration, takes a series of rational basis functions or *macropuffs*, which are obtained because of the clustering of puffs in space. The equation defining each *macropuff*, used by the interpolation, is determined by the equation that defines each *puff* emitted by each source, constituting a semi physical adaptive model; these puffs depend on the pollutant dispersion.

This model incorporates an asynchronous evolutionary tool, this due to the fact that for each instant of time during a measurement campaign the algorithm processes an emission source estimation and then it continues with the clustering of *puffs* in space to achieve the PM$_x$ concentration surfaces according to the TSKN model structure of interpolation.

Point densification of the study zone is achieved through spatial continuity that generates the background concentration for PM$_x$ due to the effect of the emission of pollutant puffs for each of the sources considered. The representation of each of the sources of emission that forms the structure of solution as a stochastic process, and its mapping on the output layer of the proposed model, allowed to obtain a series of forecasting maps that show the spatial behavior of PM$_{10}$ concentration throughout the duration of a measurement campaign. Thus, manipulating each of the genes of the solution structure establishes a set of actions to mitigate the effects that concentration can cause within an area of study.

The proposed model can be extended by evolution, to determine the spatio-temporal behavior of particulate matter PM$_x$ for different particles sizes as PM$_{2.5}$, PM$_{1.0}$, since the rate of deposit of the particles is defined by the Eq. 9, which in turn define the fitness function of the integrated model to estimate emissions.

References

1. EEA European Environmental Agency, *Air Pollution in Europe 1990–2004, European Report* (EEA—European Environmental Agency, Brueles, Europe, 2007)
2. B. Brunekreef, S. Holgate, Air pollution in health. Lancet **360**(9341), 1233–1242 (2002)
3. D. Dockery, C. pope, X. Xu, J. Spengler, J. Ware, J. Fay and E. Speizer, An association betwenn air pollution and mortality in six U.S. cities. N. Engl. J. Med. **329**(24), 1753–1759 (1993)
4. C. Pope, D. Dockery, J. Shwartz, Review of epidemiological evidence of health effects of particulate air pollution. Inhal. Toxicol. **7**, 1–18 (1995)
5. D. Loomis, Y. Grosse, B. Lauby-Secretan, F.E. Chisssassi, V. Bouvard, T. Benbrahim, N. Guha, R. Baan, H. Mattock and S. Straif, The carcinogenity of outdoor air pollution. Lancet Oncol. **14**(13), 1262–1263 (2013)
6. R. Beelem, G. Hoek, D. Vienneau, M. Eeftens, K. Dimakoupolou, X. Pedeli, Development of NO2 and NOx land use regression models for estimating air pollution exposure in 36 study areas in Europe—the ESCAPE project. Atmos. Environ. **72**, 16–23 (2013)
7. P.A. Peña, J.A. Hernández, V.M. Toro, Evolutionary inverse lagrangian puff model. Environ. Model. Softw. **25**(12), 1890–1893 (2010a)

8. P.A. Peña, J.A. Hernández, V.M. Toro, Evolutionary inverse modelling for PM10 polltutant dispersion, in *Soft Computing Methods for Practical Environmental Solutions: Techniques and Studies* (IGI-Global, 2010b), pp. 293–313
9. R. Stern, P. Builtjes, M. Shaap, R. Timmermans, R. Vautard, A. Hodzic, M. Memmesheimer, H. Fldmann, E. Renner, R. Wolke, A. Kerschbaumer, A model inter-comparison study focussing on epiodes with elevated Pm10 concentrations. Atmos. Environ. **42**(19), 4567–4588 (2008)
10. M. Brauer, M. Ammann, R. Burnett, A. Cohen, F. Dentener, M. Ezzati, S. Henderson, M. Krzyzanowski, R. Martin, R. Van Dingenen, A. Van Donkelaar, G. Thurston, Exposure assessment for estimation of the global durden of desease attributable to outdoor air pollution. Environ. Sci. Technol. **46**(2), 652–660 (2011)
11. C. Lloyd, P. Atkinson, Increased accuracy of geostatistical prediction of nitrogen dioxide in the United Kingdom with secondary data. Int. J. Appl. Earth Obs. Geoinf. **5**(4), 293–305 (2004)
12. N. Hamm, A. Finley, M. Shaap, A. Stein, A spatially varying coefficient model for mapping PM10 air quality at European Scale. Atmos. Environ. **102**, 393–405 (2015)
13. R. Givehchi, M. Arhami, M. Tajrishy, Contribution of the middle eastern dust source areas to PM10 levels in urban receptors: case study of Theran, Iran. Atmos. Environ. **75**, 287–295 (2013)
14. C. Yiu-Chung, O. Hawas, D. Hawker, P. Vowles, D. Cohen, E. Stelcer, R. Simpson, G. Golding, E. Christensen, Using multiple type composition data and wind data in PMF analysis to apportion and locate sources of air pollutants. Atmos. Environ. **45**, 439–449 (2011)
15. G. Argyropoulos, C. Samara, Development and application of a robotic chemical mass balance model for source apportionment of atmospheric particulate matter. Environ. Model Softw. **26**, 469–481 (2011)
16. M. Adewale, M. Taiwo, R. Harrison, S. Zongbo, A review of receptor modelling of industrially emitted particulate matter. Atmos. Environ. **97**, 109–120 (2014)
17. J. de la Rosa, A. Sánchez, A. Alastuey, X. Querol, Y. Gonzalez-Castanedo, R. Fernández-Camacho, A. Stein, Using PM10 geochemical maps for defining the origin of atmospheric pollution in Andalusia (southern Spain). Atmos. Environ. **44**, 4595–4605 (2010)
18. E. Pisoni, C. Carnevale, M. Volta, Sensitivity to spatial resolution of modeling systems designing air quality control policies. Environ. Model. Softw. **25**, 66–73 (2010)
19. V. Singh, C. Carnevale, G. Finzi, E. Pisoni, M. Volta, A cokriging based approach to reconstruct air pollution mpas, processing measurement station concentrations and deterministic model simulations, Environ. Model. Softw. **26**, 778–786 (2011)
20. K. Strebel, G. Espinosa, F. Giralt, A. Kindler, R. Rallo, M. Richter, U. Schlink, Modelling airborne benzene in space an time with self organizing maps and Bayesian techniques. Environ. Model Softw. **41**, 151–162 (2013)
21. M. Elangasinghe, N. Singhal, K. Dirks, J. Salmond, S. Samarasinghe, Complex time series analysis of PM10 and PM2.5 for a coastal site using artificial neural network modelling and k-means clustering. Atmos. Environ. **94**, 106–116 (2014)
22. M. Diaz de Quijano, D. Joly, D. Gilbert, N. Bernard, A more cost effective geomatic approach to modelling PM10 dispersion across Europe. Appl. Geogr. **55**, 108–116 (2014)
23. F. Pilla, B. Broderick, A GIS model for personal exposure to PM10 for Dublin commuters, Sustain. Cities Soc. **15**, 1–10 (2015)
24. P. Isazi, Redes de Neuronas Artificiales—Un enfoque práctico (2004)
25. C. Coello, Introduccion a la Computación Evolutiva, Centro de Investigacones y Estudios Avanzados - CINVESTAV, delta.cs.cinvestav.mx/~ coello/comevol/apuntes.pdf.gz, 2014
26. P. Israelsson, K. Do, E. Adams, A comparison of three lagrangian approaches for extending near field mixing calculations. Environ. Model Softw. **21**, 1631–1649 (2006). doi:10.1016/j.envsoft.2005.07.008
27. P.A. Peña, R.J. Hernández, EDA's classfier systems for identifying patterns of particulate matter (PMx) dispersion in a Study Area, in *Lagrangran Modelling of the Atmmosphere* (Grindelwald, Switzerland, 2011)

28. F. Martín Ll., C. Gonzalez and I. e. a. Palomino, SICAH, Sistema Informático para el Control y Prevención de la Contaminación Atmósferica en Huelva, Madrid: Centro de Investigaciones Energeticas, Medioambientales y Tecnológicas, CIEMAT (2002)

29. J. Scire, F. Robe, M. Fernan, R. Yamartino, A useros guide for the CALMET/CALPUFF model (versión 5.0), (Earth Tech. Inc., Concord, 2000)

30. P. Larrañaga, P. Lozano, *Estimation of Distribution Algorithms* (Kluwer, Massachissetts, 2002)

31. J. Lozano, P. Larrañaga, E. Bengoetxea, Towards a new evolutionary computation: advances on estimation of distribution algorithms. (Springer, 2006)

32. J. Hernández, J. Ospina, A multi dynamics algorithm for global optimization, Math. Comput. Model. **52**(7), 1271–1278 (2010). *doi:*10.1016/j.mcm.2010.03.024

33. P.A. Peña, R.J. Hernández, Compression of Free Surfaces Base on the Evolutionary Optimization of a NURBS—Takagi Sugeno, in *23rd. ISPE International Conference on CAD/CAM, Robotics & Factories of the Future* (Bogota, 2007b), ISBN: 978-958-978-597-3

34. P.A. Peña, R.J. Hernández, C. Parra, Modelo Evolutivo Integrado para la Interpolación/Descomposición de Modelos DIgitales de Elevación, in *2da. Conferencia Ibérica de Sitemas y Tecnologías de la Información* (Oporto, Portugal, 2007a), ISBN:978-972-8830-88-5

35. P.A. Peña, R.J. Hernández, P.R. Jímenez, Construction of Concentration Surfaces PMx using Fuzzy Neural models of Semiphysical Class, in *Spatial Statistics* (Columbus, Ohio, 2013)

36. O.-H. Park, M.-G. Seok, Selection of appropiate model to predict plume dispersion in coastal areas. Atmos. Environ. **41**, 6095–6101 (2007). doi:10.1016/j.atmosenv.2007.04.010

37. P.A. Peña, R.J. Hernández, M. Toro, Asynchronous evolutionary modelling for PM10 spatial characterization, in *Proceedings of the 18th. World IMACS Congess and MODSIM 09 International Congress on Modelling and Simulations* (Sidney, Australia, 2009)

Application of Artificial Neural Network in Social Media Data Analysis: A Case of Lodging Business in Philadelphia

Thai Le, Phillip Pardo and William Claster

Abstract Artificial Neural Network (ANN) is an area of extensive research. The ANN has been shown to have utility in a wide range of applications. In this chapter, we demonstrate practical applications of ANN in analyzing social media data in order to gain insight into competitive analysis in the field tourism. We have leveraged the use of an ANN architecture in creating a Self-Organizing Map (SOM) to cluster all the textual conversational topics being shared through thousands of management tweets of more than ten upper class hotels in Philadelphia. By doing so, we are able not only to picture the overall strategies being practiced by those hotels, but also to indicate the differences in approaching online media among them through very lucid and informative presentations. We also carry out predictive analysis as an effort to forecast the occupancy rate of luxury and upper upscale group of hotels in Philadelphia by implementing Neural Network based time series analysis with Twitter data and Google Trend as overlay data. As a result, hotel managers can take into account which events in the life of the city will have deepest impact. In short, with the use of ANN and other complementary tools, it becomes possible for hotel and tourism managers to monitor the real-time flow of social media data in order to conduct competitive analysis over very short timeframes.

Keywords Artificial neural networks (ANNs) · Hospitality · Social Media analysis · Kohonen · Forecasting · Competitive Analysis · Lodging · Hotel Occupancy

T. Le (✉) · P. Pardo · W. Claster
Ritsumeikan Asia Pacific University, Jumonji Baru 1-1, Beppu, Oita, Japan
e-mail: le.thai.jp@ieee.org

P. Pardo
e-mail: pardorit@apu.ac.jp

W. Claster
e-mail: wclaster@apu.ac.jp

© Springer International Publishing Switzerland 2016
S. Shanmuganathan and S. Samarasinghe (eds.), *Artificial Neural Network Modelling*, Studies in Computational Intelligence 628,
DOI 10.1007/978-3-319-28495-8_16

1 Introduction

Starwood Hotels and Resorts was one of the very first hotels to realize the critical role of social media data, and to leverage the information to support customers in their travel decisions [1]. Gradually, not only have more and more tourism businesses become actively involved in online activities on different social network channels such as Twitter and Facebook, but many of them are also considering social media data as a valuable and timely information source of input for various decision making processes [2]. Ironically, regardless of the prevalent adoption of social media usage in the tourism industry (e.g. [3–7]), there is still a lack of comprehensive guidelines on how online social data can be interpreted to gain competitive knowledge in the hospitality industry. In this chapter, we introduce approaches based on using unsupervised ANN, namely self-organizing map (SOM) based methods to analyze Twitter and Google Trends data of two different groups of hotels, namely luxury and upper upscale, in Philadelphia during the period between 2011 and 2014. First, we look at how related data is collected and pre-processed. Then, the chapter shares and discusses the implementation of SOMs, which are trained by ANN, in analyzing textual contents of hotel's management tweets. An application for using ANN in predicting the occupancy rate of the two groups of hotels with different overlay data is subsequently examined.

2 Data Collection and Pre-Processing

Social media data is scattered throughout the Internet in many forms, but most are in the form of micro-blogs, which found on various online social networks such as Facebook, Twitter, etc. Because of this, in this research, data was mainly collected from two sources: Tweets data from Twitter and search queries data from Google Trend, which were also effectively employed in various related researches (e.g. [8–12]). Moreover, data regarding Philadelphia hotels' average occupancy rate between January 2008 and May 2014 is provided from Smith Travel Research Inc. (STR). Regarding the Twitter data, we collected thousands of tweets posted by the public Twitter accounts of eight different hotels in Philadelphia, which we categorized into two groups: luxury and upper upscale. The Google Trend data is normalized query volume of keywords worldwide, which in this case were the names of the examined hotels as in Table 1.

After collecting all the data, we proceed to the pre-processing procedure, in which the data is cleaned up to ensure a sound subsequent analysis. In particular, all the duplicated tweets, English stop-words (the, a, an, etc.), numeric figures, Philadelphia's different entities (PA, Philly, etc.), and hyperlinks are filtered and eliminated from the dataset. Regarding AKA Rittenhouse hotel, since only a corporate Twitter account is found, solely tweets concerned with location in Philadelphia are selected for the purpose of this research.

Table 1 Input query string used to collect google trend data and tweets data collection results

Hotel names	Input query string on google trend	Number of retrieved user's tweets
Hyatt at the Bellevue	Hyat1t at The Bellevue	352/353 (99.7 %)
Windsor Suites	Windsor suites philadelphia	3035/3037 (99.9 %)
Le Meridien Philadelphia	lemeridienphiladelphia	694/694 (100 %)
Kimpton Hotel Palomar Philadelphia	hotelpalomarphiladelphia + palomarphiladelphia	2006/2011 (99.7 %)
Sofitel Philadelphia	sofitelphiladelphia	3151/6868 (46 %)
Four Seasons Hotel Philadelphia	four seasons philadelphia	3149/9081 (35 %)
The Latham Hotel	thelatham hotel + the lathamphiladelphia	40/40 (100 %)
AKA Rittenhouse Hotel	Aka_Rittenhouse philadelphia	2924/2929 (99.8 %)

3   Neural Network Facilitated Self-Organized Map

Since we postulate that the tweet's contents represent the marketing strategies used in approaching customers through online channels, understanding the relationship of different keywords being used in the tweets can help us to gain viable competitive knowledge from the different hotels' manager's perspective. However, since each of the keywords belongs not only to one but several documents, or tweets, the collected tweets can be considered as a huge sparse matrix of several thousand-dimensional vectors. Hence, an algorithmic approach enabling unsupervised clustering is needed to transform this matrix into meaningful visual expositions. In this section, we share an application of an unsupervised ANN that is able to cluster multi-dimensional data into a two-dimensional informative map. The map is called Self-Organizing-Map (SOM) proposed Kohonen [13], which is recognized to be a very effective analysis tool in data clustering facilitated by an unsupervised ANN learning algorithm [14–17].

In order to clearly picture the use of SOM in the analysis of social media data in the field of tourism and hospitality, results of such a neural network training process on the management tweets of the Four Season Philadelphia and Sofitel Philadelphia hotels, which belong to two different hotel ranks listed as luxury and upper upscale respectively, are introduced as follows in Fig. 1.

As we can clearly see, the above generated SOM map (Fig. 1) contains a total of 1029 nodes representing 1029 neurons, classifying over 9716 terms from 3149 management tweets posted by the Four Season hotel into 20 clusters. Each of the clusters is pictured by different colors, which has a algorithmically generated central

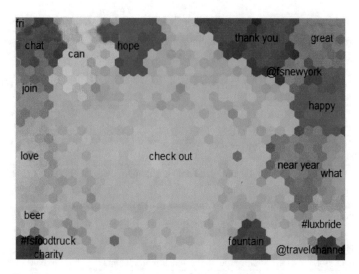

Fig. 1 Self-organizing-map (SOM) for four season Philadelphia, 2011–14

concept expressed as the cluster name. In the same way that color on a world map color only serves as a visual demarcation between countries, the color in these visualizations do not express any qualitative or quantitative attributes of the data. However, the position, as well as the size of each of the clusters does have meaning. To illustrate, the most prevalent cluster "check out" located at the center of the map containing terms such as "culture", "events", and "local", shows a marketing effort as it suggests various available activities occurring locally in Philadelphia. Specifically, the cluster "#fsfoodtruck" and "charity" located in a position adjoining each other doubtless describe an effort to market the corporation's food-truck campaign around September 2014 to raise donation for the Children's Hospital in Philadelphia. Above that is the "beer" cluster, which probably points to a beer festival that occurred in the city around that time. A portion of the map is covered by the "new year" concept, which is located right next to "happy", "#luxbride", and "what" cluster. If we look at the details of the messages which include these top keywords, we can find terms such as "resolution", "wedding", "weddingplanning", etc., which in fact describe a marketing promotion for the hotel on the social networks of a wedding package during the New Year period. Additionally, located at the top left corner of the map are three aggregated clusters namely "fri", "chat", "join" that help us to learn about the publicity surrounding of regular speaking events featuring some famous regional editors in fashion and lifestyle. Through the lens of the Kohonen map, it is possible to gain a deeper insight into the hotel's marketing strategies and campaigns during this time, and also to picture the hotel's different emphases in approaching customers via online social media.

In comparison with the previous SOM (Fig. 1) the one found in Fig. 2 for Sofitel Philadelphia's hotel management tweets, shows several noticeable differences. Other than the centered "holiday" cluster, the "#conciergechoice" motif seems to be

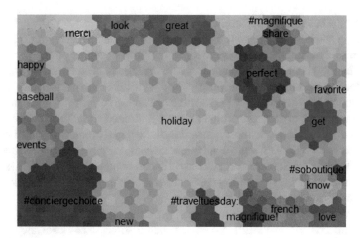

Fig. 2 Self-organizing-map (SOM) for Sofitel Philadelphia, 2011–14

the largest among them. The fact that this concept was surrounded by the "new" and "events" clusters implies that the hotel enjoys giving out advice and suggestions about new activities and events to its customers. If we look deeper into the "new" cluster, some related phrases can be found such as "new native American voices exhibit", "new app for iphone", and "new and coming-soon restaurants". In addition, the "baseball" cluster located next to "events" also implies that the hotel marketers seem to be leveraging the popularity of news regarding local baseball games to make their entity prevalent on the social network. Despite the fact that there are many events being promoted by the hotel's tweets, most of them relate to "baseball", and "battle", hence the algorithm creates a separate grouping for this concept. Noticeably, many concepts shown on the map are described by keywords in French such as "merci" or "magnifique". This also suggests that the Sofitel Philadelphia concentrates on advertising one of its unique features, which is the combination of French style and American living.

4 Time Series Prediction with Neural Network

Because of its great impact on various aspects in the field of tourism, hotel occupancy forecasting has always been a main focus of hotel managers. Hotel occupancy is not only a metric for the internal assessment of a single hotel, but it is also an indicator of the changing patterns of customers within a specific geographic area such as city, or country. In other words, hotel occupancy rates reflect the consumer's cognition and behavior, which are also recognized to possess a linkage with social media data. Because of this, we propose forecasting the hotel occupancy rate of these two groups of hotels in Philadelphia using the retrieved data from the Twitter social network. Our goal was to find if there are strong quantitative relationships between management tweets and hotel occupancy rate. Moreover, this forecasting model also makes use of Google Trend data to partly reflect the contribution of consumer's online behavior in the hotel occupancy rate.

Within the scope of this chapter, implementations of time series forecasting facilitated by a Multilayer Perceptron ANN with different overlay data using the Weka data mining software [18] is suggested. Part of the collected data (occupancy rate, management tweets, and Google trends) is filtered so that the respective time period is aligned with each other. Then, it is divided into training and testing dataset with a ratio of 7:3. Different evaluation metrics namely Mean Absolute Error (MAE), Root Mean Squared Error (RMSE), Mean Squared Error (MSE) are used to fit the trained model on the testing dataset to compare the respective results with the performance of Linear Regression.

Regarding the analysis on the luxury group of hotels (Table 2), Weka's Multilayers Perceptron (MLP) provided a better forecast of occupancy than did Linear Regression and this is partly due to MLP flexibility in being able to tune multiple parameters, in particular the learning rate (−L) and momentum (−M). This improvement showed up across all the evaluation measures (MAE, MAPE, RMSE, and MSE). However, the introduction of social media data: Tweets, Google Trend, or both, did not improve the forecasts.

In the analysis on the upper upscale group of hotel we observed a different scenario (Table 3). Even though the MLP model outperformed linear regression in the analysis without overlay data, social media data showed itself to be significant. Specifically, management tweets greatly improved the forecasting results with linear regression model, and the inclusion of both management tweets and Google trends data shows slight enhancement in the case of MLP model.

Table 2 Prediction analysis result on the luxury group of hotels data using MLP and linear regression

	No overlay data	Tweets overlay	Trend overlay	Tweets + trend overlay
Linear regression	MAE 6.1587	MAE: 6.8693	MAE: 7.1971	MAE: 6.8693
	MAPE: 8.0194	MAPE: 8.9465	MAPE: 9.3409	MAPE: 8.9465
	RMSE: 8.0204	RMSE: 9.2922	RMSE: 9.8314	RMSE:9.2922
	MSE: 64.3276	MSE: 86.3453	MSE: 96.6559	MSE: 86.3453
Multilayer perceptron (−L 0.3 −M 0.2)	**MAE: 3.8041**	MAE: 5.0625	MAE: 4.939	MAE: 3.899
	MAPE: 5.06	MAPE: 6.6659	MAPE: 6.495	MAPE: 5.2032
	RMSE: 4.3529	RMSE: 5.9638	RMSE: 6.1949	RMSE: 5.4229
	MSE: 18.9478	MSE: 35.5664	MSE: 38.3763	MSE: 29.4084
Multilayer perceptron (−L 0.2 −M 0.001)	MAE: 4.0597	**MAE: 3.8467**	**MAE: 3.5707**	**MAE: 3.9372**
	MAPE: 5.3961	**MAPE: 5.1286**	**MAPE: 4.7686**	**MAPE: 5.262**
	RMSE: 4.7331	**RMSE: 4.4672**	**RMSE: 4.5605**	**RMSE: 4.8992**
	MSE: 22.4024	**MSE: 19.9558**	**MSE: 20.7982**	**MSE: 24.0018**

Table 3 Prediction analysis result on the upper upscale group of hotel data using MLP ANN and linear regression

	No overlay data	Tweets overlay	Trend overlay	Tweets + trend overlay
Linear regression	MAE: 5.662	**MAE: 2.5313**	MAE: 5.662	MAE: 5.662
	RMSE: 6.4643	**RMSE: 3.3175**	RMSE: 6.4643	RMSE: 6.4643
	MSE: 41.7875	**MSE: 11.0061**	MSE: 41.7875	MSE: 41.7875
Multilayer perceptron (−L 0.1 −M 0.4)	**MAE: 3.5747**	MAE: 5.496	MAE: 4.1679	MAE: 3.6409
	RMSE: 4.5417	RMSE: 6.1738	RMSE: 5.004	RMSE: 4.676
	MSE: 20.627	MSE: 38.1156	MSE: 25.0399	MSE: 21.8645
Multilayer perceptron (−L 0.1 −M 0.5)	MAE: 4.1775	MAE: 5.3354	MAE: 4.9444	**MAE: 3.2727**
	RMSE: 4.8488	RMSE: 6.0356	RMSE: 5.7739	**RMSE: 4.4111**
	MSE: 23.5109	MSE: 36.4279	MSE: 33.3382	**MSE: 19.4582**

In overall, it is noticeable that either linear regression or MLP ANN has its own advantages with different input data of the two groups of hotels in Philadelphia. However, the ANN model is shown to be more adaptable, since its parameters are easily altered accordingly with different scenario.

5   Conclusion

In this chapter, we have shown that ANN facilitated SOM is powerful in analyzing social media data to gain competitive knowledge in the field of tourism. Because of the flexibility of the algorithm, the introduced methodology can be easily customized and employed in different business scenarios. The capability of MLP ANN is also demonstrated through its application in time series forecasting, hotel occupancy prediction particularly, with different overlay data. ANN is without doubt a very dynamic and supportive in business analytics, and its potentials will be surely pushed beyond any recognizable boundaries in the present context.

References

1. L.H. Lanz, B.W. Fishhof, R. Lee, How are hotels embracing social media in 2010—examples of how to begin engaging. HVS sales and marketing services
2. C.K. Anderson, The impact of social media on lodging performance. Cornell Hospitality Rep. **12**(15), 4–11 (2012)
3. W. Claster, Q.T. Le, P. Pardo, Implication of social media data in business analysis: a case in lodging industry, Proceedings of the APUGSM 2015 conference, (Japan, 2015), p. 61
4. I. Blal, M.C. Sturman, The differential effects of the quality and quantity of online reviews on hotel room sales. (Cornell Hospitality Quarterly, 2014)

5. H. Choi, P. Liu, Reading tea leaves in the tourism industry: a case study in the Gulf oil spill. (March 24, 2011)
6. G. Seth, Analyzing the effects of social media on the hospitality industry. UNLV Theses/Dissertations/Professional Papers/Capstones, paper 1346 (2012)
7. W. Claster, P. Pardo, M. Cooper, K. Tajeddini, Tourism, travel and tweets: algorithmic text analysis methodologies in tourism. Middle East J. Manage. **1**(1), 81–99 (November 2013)
8. W. He, S. Zha, L. Li, Social media competitive analysis and text mining: a case study in the pizza industry. Int. J. Inf. Manage. **33**(3), 464–472 (June 2013)
9. X. Yang, B. Pan, J.A. Evans, B. Lv, Forecasting Chinese tourist volume with search engine data, tourism management, vol. 46, pp. 386–397 (IFebruary 2015). ISSN 0261-5177
10. L. Dey et al., Acquiring competitive intelligence from social media. *Proceedings of the 2011 Joint Workshop on Multilingual OCR and Analytics for Noisy Unstructured Text Data*, ACM (2011)
11. G. Barbier, H. Liu, Data mining in social media. Social Network Data Analytics, (2011), pp. 327–352
12. Z. Xiang, B. Pan, Travel queries on cities in the United States: implications for search engine marketing for tourist destinations, Tourism Manage. **32**(1), 88–97 (February 2011). ISSN 0261-5177
13. T. Kohonen, Self-organized formation of topologically correct feature maps. Bio. Cybern. **43** (1), 59–69 (1982)
14. D. Isa, V.P. Kallimani, L.H. Lee, Using the self-organizing map for clustering of text documents. Expert Syst. Appl. **36**(5), 9584–9591 (July 2009). ISSN 0957-4174
15. W. Claster, D. Hung, S. Shanmuganathan, Unsupervised Artificial Neural Nets for Modeling Movie Sentiment, *2nd International Conference on Computational Intelligence, Communication Systems and Networks*, 2010
16. W.B. Claster, M. Cooper, Y. Isoda, P. Sallis, Thailand—tourism and conflict: modeling tourism sentiment from twitter tweets using naïve bayes and unsupervised artificial neural nets. CIMSim2010, Computational intelligence, modelling and simulation, 2010, pp. 89–94
17. Y.C. Liu, M. Liu, X.L. Wang, Application of self-organizing maps in text clustering: a review, applications of self-organizing maps, ed. by Dr. M. Johnsson, (2012). ISBN: 978-953-51-0862-7
18. M. Hall, E. Frank, G. Holmes, B. Pfahringer, P. Reutemann, I.H. Witten, The WEKA data mining software: an update. SIGKDD Explor. **11**(1), (2009)

Sentiment Analysis on Morphologically Rich Languages: An Artificial Neural Network (ANN) Approach

Nishantha Medagoda

Abstract The extraction and analysis of human feelings, emotions and experiences contained in a text is commonly known as sentiment analysis and opinion mining. This research domain has several challenging tasks as well as commercial interest. The major tasks in the area of study are, identifying the subjectivity of the opinionated sentence or clause of the sentence and then classifying the opinionated text as positive or negative. In this chapter we present an investigation of machine learning approaches mainly the application of an artificial neural network (ANN) to classifying sentiments of reader reviews on news articles written in Sinhala, one of the morphologically rich languages in Asia. Sentiment analysis provides the polarity of a comment suggesting the reader's view on a topic. We trained from a set of reader comments which were manually annotated as positive or negative and then evaluated the ANN architectures for their ability to classify new comments. The primary interest in this experiment was the exploration of selecting appropriate Adjectives and Adverbs for the classification of sentiment in a given language. The experiment was conducted in different weighting schemes by examining binary features to complex weightings for generating the polarity scores of adjectives and adverbs. We trained and evaluated several ANN architectures with supervised learning for sentiment classification. A number of problems had to be dealt with in this experiment and they are: the unavailability of the main part of speech, adjective and adverb and the sample size of the training set. Despite the issues, our approach achieved significant results for sentence level sentiment prediction in both positive and negative classification.

Keywords Sentiment analysis · Artificial Neural Network (ANN)

N. Medagoda (✉)
Auckland University of Technology, Auckland, New Zealand
e-mail: nmedagod@aut.ac.nz

© Springer International Publishing Switzerland 2016
S. Shanmuganathan and S. Samarasinghe (eds.), *Artificial Neural Network Modelling*, Studies in Computational Intelligence 628,
DOI 10.1007/978-3-319-28495-8_17

377

1 Introduction

Text mining is a sub area of the study of natural language processing that relates to in a way understanding and generating high quality information from human languages, such as English, French, Japanese, and Hindi etc. The understanding of a given language is not only of the spoken language but in the use of written scripts as well. Text mining is more suited to the written text of documents including the unstructured textual information about, facts and opinions. Opinions are subjective expressions of human thoughts, emotions and feelings. The research area of analyzing the opinions contained in texts is popularly known as opinion mining and it is basically about two methods that are run in a sequence [1]. The first identifies the subjectivity of the opinionated sentence or clause of the sentence and the second classifies the opinionated text as positive or negative. The former method is known as subjectivity classification and the latter one is referred to as sentiment analysis.

The modern world contains an unimaginably vast amount of digital information which is growing faster than ever before [2]. As the result of recent technological advances everyone is benefited, today anyone can gather and analyse considerably high volumes of data for extracting useful information on matters that are of interest to the individual concerned.

In general, data collected for any research or simple analysis can be broadly divided into two main categories; (a) Quantitative Data (b) Qualitative Data. This qualitative text data can be related to thoughts, opinions and experience of a respondent. The data for a particular topic can be collected in a number of ways. The most traditional and popular method, in past decade has been paper based questionnaires [3]. But today, in this technologically sophisticated world, the Internet is used not only to collect the data but to analyse in a very efficient manner as well. In the paper based questionnaire method, qualitative text data is acquired from open-ended questions. It is apparent that presently more people specially web surfers express their views, opinions or experience on politics, products, services and many other things in the web itself [4]. This has increased rapidly with the introduction of social networks in the later part of the 20th century. Identifying such information systematically using opinion mining software tools saves time and money significantly when compared with alternative methods, such as surveys or market research. Another advantage of such an approach is that mining opinions using a systematic approach reveals precise information hidden in these views [5].

The benefits of opinion mining are several fold and they are;

Customers interested in finding specific information on a certain product or service, can use an opinion mining system to gather he relevant information without having to read the verbal comments of the clients who had used the same product or service.

Governments as well as political parties benefit enormously from the use of data mining approaches to predict election results based on comments given by the public using the social networks as the means for voicing their opinions.

Manufacturers or merchants interested in determining the success of a new version of a product or service, based on its popularity or identifying the demographics of those who like or dislike the special features of the commodity, could establish customer profiles before launching a new advertising campaign. Identifying this kind of information systematically by using opinion mining tools saves time and money by comparison with the use of time consuming surveys or market research. In addition, the results are likely to be more accurate and reliable since the data has been created by real customers in ideal situations without forcing responses from them.

Traditionally, the opinion or comments are collected using questions where the researcher has allocated some space in the questionnaire to write views. These questions are defined as open-ended questions and information other than that of a quantitative nature is included in the responses [3]. Today the most popular and freely available source for collecting such information, is the World Wide Web. Blogs, review sites and micro blogs, all provide a good understanding of the perceptions of customers of the products and services of interest [6]. The social networks are the newly invented repositories for customer comments. Social networks especially contain a wealth of human feelings and expressions on a vast number of topics such as politics, products, services and actions taken by the governments or state institutions [5]. The content of the social webs is dynamic and rapidly changing to reflect the societal and sentimental fluctuations of contributors as well as the use of a language. Even though contents of the social media are messy and highly heterogeneous they represent a valuable source of information of attitudes, interests and expectations of the citizens and consumers everywhere [7]. With the dramatic increase in the Internet usage the blog pages are also growing rapidly. Unlike the comments or opinions, in social webs the blogger content tends to be longer and the language is more structured. Blogs are used as a source of opinions in many of the studies related to sentiment analysis [8]. Review sites are useful for any consumer who is looking for others' comments on a certain product or service. A large and growing body of user generated reviews is available on the Internet [6].

The core of the sentiment analysis is first identifying the subjectivity of a sentence containing an opinion followed by judging the polarity (sentiment) of the view expressed. These tasks are carried out using machine learning algorithms as well as many other methods. The choice of the specific learning algorithm used is a critical step. The methods of determining the semantic orientation used for identifying the polarity of the sentence are categorized into two approaches: supervised and unsupervised classification techniques. The evidence of initial attempts on the application of an unsupervised approach by Pang et al. in 2002 and supervised method by Peter Turney can be seen in Kobayasi et al. [9]. Supervised classification algorithm is one of the learning algorithms most frequently used in text classification systems [10]. In supervised classification two sets of opinion are required namely, training and testing data sets. The training data set is used to train the classifier to learn the variation of the characteristics of the sentence or document and the test data is used to validate the performance of the classification algorithm. The supervised machine learning techniques, such as Naïve Bayes, support vector machines (SVM) and maximum entropy, are the most popular ones and they have been proven to be the most successful in sentiment classification [6].

In this research, we investigated the application of supervised classification algorithm using an Artificial Neural Network (ANN) for categorising the sentiment of an opinion. In addition to examining the applicability of ANN, we investigated the usage of adjectives and adverbs as important features of classification algorithm.

This article is organised as follows. Previous related work on this theme and related topics by other researchers is explained in Sect. 2. Then we discussed the methodology including feature selection, learning algorithms and evaluation in Sect. 3. The results of the case study in Sinhala with accuracy measurements are presented in Sect. 4. Finally, Sect. 5 concludes the paper with some interesting initial results and anticipations relating to future directions.

2   Previous Related Work

Currently many researchers pay attention for sentiment classification using machine learning algorithms. Among these classification algorithms Naïve Bayes and SVM are more common and give higher accuracy [4]. However, there are several studies that have used neural networks for classification of sentiments into positive or negative. Backpropagation neural networks (BPNN) and modified backpropagation networks (MBPNN) were initially proposed for text classification [11, 12]. A neural network based sentiment classification index was implemented combining the machine learning techniques and information retrieval methods by Long et al. [13] to detect the harmful negative blogger comments quickly and effectively. In that study for each sentiment four orientation indexes were used as the input for the BPNN approach. These indexes included two types of point wise mutual information, latent semantic index and the other stated as index SO-A, For a word SO-A is defined as the association with a positive paradigm (a set of positive words) minus the strength of its association with a negative paradigm (a set of negative words). The language specific features such as adjectives, adverbs, nouns and verbs were considered to construct the input vector. The vocabulary size that is the dimension of the feature vectors range from 35 to 48. Authors mentioned that dimensionality is the major problem as the textual data will grow exponentially. In the paper, the use of tf-idf and n-gram features in the classification has been mentioned but there was no evidence of it in the way how they were used. The results on the use of such features were not made available either. Different data sets were tested to evaluate the performance of the proposed algorithm which had given F values in the range of 0.4–0.8. The calculation complexity of the proposed indexes was the major drawback of the proposed method.

Cicero et al. [14] proposed a sentiment classification method for short texts using deep convolution neural network. In their approach the authors suggested character-to sentence-level information to perform sentiment analysis of short texts. The neural network architecture jointly used character-level, word-level and sentence-level representations to perform sentiment analysis. Authors claimed the novelty of the approach as use of two convolution layers which allow the handling of the words and sentences. In the first layer network transforms the words into real-valued feature vectors that capture the morphological, syntactic and semantic information of the words. The network was evaluated for performance with two corpuses, and each sentence had five classes. Backpropagation neural network architecture was proposed for the work with the network trained by minimizing a negative likelihood over the training set. Classification accuracies achieved were in the range of 79.4–85.7 in different algorithm scenarios.

A recursive neural network (RNN) model that learns compositional vector representations for phrases and sentences of arbitrary syntactic type and length was introduced by Richard et al. [15] to capture the compositional meaning of longer phrases, preventing them from a deeper understanding of languages. The compositionality (an ability to learn vector representations for various types of phrases and sentences of arbitrary length), was modelled by matrix-vector representation. Based on the authors' statement, the vector captures the meaning of that constituent. The matrix captures how it modifies the meaning of the other word that it combines with. In the first approach, the experiment was carried out using the adjective-adverb pair with the aim of predicting fine gained sentiment distributions. The publicly available data set was used to extract the adjective-adverb pairs. The network was trained by computing via backpropagation. The approach gave 79 % of accuracy.

3 Neural Network Based Sentiment Classification Methodology

In this section, we outline the neural network based sentiment classification methodology used in the paper. Our proposed method consists of five steps and they are; (1) data pre-processing, (2) feature selection, (3) feature indexing, (4) classification and finally (5) evaluation.

3.1 Step 1: Data Set and Pre-processing

We conducted the experiment on non-English text data, collected from one of the morphologically rich language in Asia, Sinhala. It is an official language in Sri Lanka with over more than 22 million speaker population in the country. Sinhala belongs to the Indo-Arian branch of the Indo-European language family. It is a SOV (subject-object-verb) word order language with its own scripts. Modern Sinhala scripts consist of 18 vowels and 43 consonants. We performed an experiment for testing the application of an ANN approach to Sentiment analysis using a set of Sinhala opinions provided for newspaper articles. 2084 comments were extracted from the text data from a leading online newspaper called "Lankadeepa" (http://lankadeepa.lk/). As this is a domain independent classification, a sample

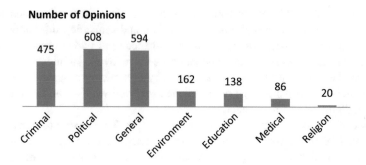

Fig. 1 Opinion sample distribution

consisting of different news articles was chosen, the domains included in the sample being; Political, Criminal, Education, Religion, Medical and General. The sample consists of general and political discussions rather than the Medical and Religion related news (Fig. 1). The data set contains 44,426 words with an average comment length of 21 words.

In the initial step of pre-processing stage, we cleaned the text data by removing the words with spelling mistakes and all punctuation marks with numerals. With an aim of applying supervised (Neural Network) classification techniques to analysing the comments, the collected comments were classified into sentiment categories manually, and the categories are: positive, negative and neutral. The classification was done manually by native speakers. If a comment expressed was in supportive of the news article then it was classified as positive and if the expression was opposed to the contents of the article, the comment was classified as a negative one. On the other hand, if the comment was in the form of an objective issue then it was coded as neutral.

The coding is done by individuals who are native speakers of the language. Three language users coded the opinions, and the common agreement of opinions was included in the sample in the first pass. Opinions differently coded were reconsidered and the researcher coded those based on his understanding. Finally, based on the categorisation, the tested sample comprised of 745 positive (P), 838 Negative (N) and 500 neutral (N) opinions.

In the next step, we removed all the function (stop) words. Function (grammatical) words are words which have little meaning but essential to maintain the grammatical relationships with other words [16]. Function words also known as stop words include prepositions, pronouns, auxiliary verbs, conjunctions, grammatical articles or particles. For a given language, the set of function words is closed and freely available. In text analysis, these words are dropped in order to reduce the dimension of the feature vector set. Since these function words carry less importance to the meaning, it is reasonable to remove them all. But in this research, the stop words relating to negation were not removed, such as no, not and can't. The sense of these words affects the total meaning hence the sentiment scale of the comments.

3.2   Step 2: Feature Selection

Sentiment classification methods based on vector models always require a vector which represents the opinion or review. The components of the vector are known as features, they are the special characteristics of an opinion. These features may be a language specific characteristic or most of the cases they provide statistical information retrieved from the opinion. The most widely used approach in document classification is referred to as a bag of words methods where each term presenting the document is considered as a separate feature [17]. A relevant set of features always provide useful information to discriminate the different opinions. On the other hand irrelevant, redundant or noisy features decrease the accuracy also, increasing the computational complexity. Therefore, the performance of sentiment classification can be enhanced by applying effective and efficient feature selection methodologies.

Feature selection methods are classified as filter, wrapper and embedded methods. Filter feature section methods are less complex in terms of computational processing, simple and they give considerably good levels of accuracies. Some of filter feature selection methods are; Information gain, Gain ratio, chi-square, gini index, odds ratio, document frequency and mutual information.

In this experiment, document term frequencies are used to select the most relevant terms that can represent an opinion. But in sentiment classification, it is important to consider respective language features to select most relevant terms in addition to the term frequencies. In this experiment part of speech (POS) is considered to select the language specific terms. For the purpose of sentiment classification, the most suitable POS categories are Adjectives and Adverbs. Adjectives always describe the nouns and adverbs add the manner of the verbs. Therefore, Adjectives and Adverbs are most important language units (parts of speech) when analysing sentiments in any language [18, 19]. Hence, initially, we calculated the term frequencies of all words contained in the experiment data set. Then, filtered out the adjectives and adverbs in the list using a predefined set of adjectives and adverbs for the Sinhala language. The predefined adjective and adverb list is retrieved from a 10 million words corpus of the target language Sinhala. The list consists of 7503 adjectives and 671 adverbs. It was also found that more that 70 % of the opinions out of 2083 include at least one adjective. But the number of opinion includes at least one adverbs was less than 15 %.

Initially, the frequency distribution of adverbs and adjectives is investigated to select the most relevant word list to construct the feature vector. This investigation is done on the opinions after removing the stop words explained in step2. The frequency distribution of adjectives and adverbs is given in the Fig. 2.

The minimum frequency of 1 indicates that the adjective/adverb is less important to the concept of the comment being talked about. On the other hand, a word with maximum frequency is highly correlated with the concept. It is also noted that it follows the Zip's rule which suggests that 80 % of the words are with less frequencies and 20 % of them show higher frequency counts [20]. The list of adjective/adverb for

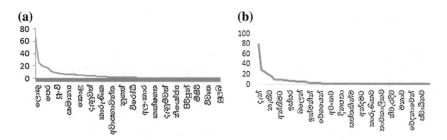

Fig. 2 Adjectives and Adverbs distributions (**a** *left*) Adjectives (**b** *right*) Adverbs

further analysis was selected by removing the adjective/adverb with fewer occurrences. The threshold for this selection was decided by considering the Zip's law and the graph of adjective/adverb versus frequencies (Fig. 2). It was concluded to use the words with frequencies greater than 3 as the adjective/adverb list for the rest of analysis. This selection contributes to 83 % of the total adjectives and adverbs.

In this selection, 128 adjectives were selected out of 328 different types' adjectives. This is approximately 49 % and the total adjectives considered for the classification is 83 %. The similar statistics for the adjectives are 37 and 87 % respectively. This reveals that as quantity for both groups are sufficiently contributed to the feature vector however, these quantities poorly represent the type or quality of each adjective and adverb group.

We also investigated the number of opinions that had at least one adjective included within this selected frequency greater than 3. Only 37 % included at least one adjective and 10 % of the opinion has at least one adverb. But the estimate for combined features was 42 %.

3.3   Step 3. Feature Indexing

The next step was to calculate the weightages of selected adjective/adverb for the feature vector. In this work, document frequency and inverse document frequency (tf-idf) weights were tested on the above mentioned data set. Where tf denotes the term frequency for the opinion which is simply the number of times a given term appears in that opinion. This value is normalized to avoid the bias to long opinions and to give the exact importance for the same word in a shorter one. And it is calculated using the following equation

$$tf_{i,j} = \frac{n_{i,j}}{\sum_k n_{k,j}}$$

where $n_{i,j}$ is number of times the term t_i appears in the opinion d_j and the denominator is the sum of all the words in the opinion d_j.

The inverse document frequency (idf) is a measure of the general importance of the term. Idf is obtained by dividing the number of all opinion by the number of opinions containing the term. Then the logarithm of the quotient is calculated as

$$idf_i = log \frac{|D|}{|\{j : t_i \in d_j\}|}$$

where $|D|$ total number of opinionsconsidered and $|\{j : ti \in dj\}|$ is the number of opinions where the term ti appears. The division-by-zero occurs when the term t_i is not present in the opinions. To avoid this one can change the denominator to $1 + |\{j : t_j \in d_j\}|$.

Finally;

$$(tf - idf)_{i,j} = tf_{i,j} \times idf_i$$

The other weighing alternative to the above is binary representation. The binary vector for each and every opinion is constructed by considering the presence and absence of highly frequent adjective or adverb. As for the tf-idf the length of the binary vector is the length of the word (adjective or adverb) with the highest frequency. We classified the opinions as either positive, negative or neutral using both vector representation of tf-idf and binary.

3.4 Step 4A: Classification using Artificial Neural Networks

A neural network is a massively parallel distributed processors made up of single units, which has a natural propensity for storing knowledge and making it available for use [21]. These processors known as neurons are structured in different manners that are known as network architectures. In general three fundamental architectures can be identified namely; Single layer Feedforward networks, Multilayer Feed forward networks and Recurrent networks (Fig. 3). In the single layer architecture, we have an input layer of source nodes that are connected to the output layer. Here network is feedforward or acyclic. In the second type there are many layers between the input and output layers that are commonly known as hidden layers. A recurrent neural network has at least one feedback loop.

In addition to the above classification neural nets can be classified into two categories and they are; Feed forward and feedback networks. Feed forward has a higher accuracy in classification tasks [21, 22]. Therefore, we deployed feed forward neural network in the sentiment classification. The back-propagation networks (BPN) is the best among the feed forward neural network. The iterative gradient algorithm in BPN minimizes the mean square error between the actual output of a multilayer feedforward perceptron and the desired output. The parameter of the

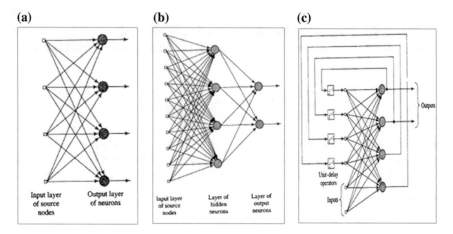

Fig. 3 Different types of neural network architectures (**a** *left*) single layer (**b** *mid*) multi-layer (**c** *right*) recurrent

BPN, such as, the number of layers could be one or two [23]. The backpropagation algorithm runs in two passes; forward pass and backward pass. The purpose of the forward pass is to obtain the activation value, and the backward pass is to adjust the weights and biases based on the difference between the desired and the actual network outputs. These two passes will be continued iteratively until the network converges [11].

3.5 Step 4B-Training the neural network

The experimental opinion set was trained with different neural network architectures. The sample was divided into several training and test samples in the rage of 60–90 %. In addition, we trained and tested the network using 10 % cross validation. Each training set was obtained after randomly sorting the original opinion data set. The main parameters of BPN, such as, the number of hidden neurons, training iterations, and training rates are all tested arbitrarily. The best performance combination is selected after completing the evaluation.

3.6 Step 5. Evaluation

In general, text categorization algorithms are evaluated using Precision, Recall and F-measures in addition to classification accuracies. These standard measures have significantly higher correlation with human judgments [24]. These are first defined for the simple case where a text categorization system returns the categories.

Table 1 Precision and recall contingency table

	Relevant	Non-relevant
Retrieved	True positives (t_p)	False positives (f_p)
Not retrieved	False negatives (f_n)	True negatives (t_n)

Precision (P) is the fraction of retrieved documents that are relevant

$$Precision = \frac{\#(relevent\ items\ retrived)}{\#(retrieved\ items)} = P(relevant|retrived)$$

Recall (R) is the fraction of relevant documents that are retrieved

$$Precision = \frac{\#(relevent\ items\ retrived)}{\#(relevent\ items)} = P(retrived|relevant)$$

These notions can be made clear by examining the following contingency (Table 1);

Then;

$$P = \frac{tp}{(tp + fp)}$$

$$R = \frac{tp}{(tp + fn)}$$

The measures of precision and recall concentrate the evaluation on the return of true positives, asking what percentage of the relevant documents has been found and how many false positives have also been returned.

A single measure that trades off precision versus recall is the F measure, which is the weighted harmonic mean of precision and recall. F score is a measure of a test's accuracy. There are different weights that can be calculated for F measure. The balance F measure equally weights precision and recall and it is commonly written as F1

$$F_1 = \frac{2PR}{(P + R)}$$

Then, the F1 score can be interpreted as a weighted average of the precision and recall, where an F1 score reaches its best value at 1 and worst score at 0.

4 Experimental Results

In the first attempt, we experimented the sentiment classification using binary vectors. That is the input feature vector for the neural network is a binary vector of adjectives and adverbs. We tested several network architectures using the representation.

The experimental scenarios and the classification accuracies are given the Table 2. The test cases include testing the different language features such as adjectives and adverbs considering each case individually and combining both. In addition, the neural networks were trained to find the ideal number of hidden layers. The cases which obtained highest accuracies are presented in the Table 2.

Based on Table 2, the highest classification accuracy is shown when the feature vector consists of only adverbs. But out of 2083 opinions only 208 include the adverbs of frequency greater than 3. On the other hand, the best test case with respect to Precision, Recall and F measure are shown in Table 3.

Table 3 reveals that opinions with adjectives and adverbs give the highest F measure when the neural network was set up with no hidden layers.

The experiment was continued with different approaches as the results of binary representation did not given acceptable performance. Next we experimented with feature vectors where the weights of the features are tf-idf measures as explained in Sect. 3.

Table 2 Classification accuracies in binary vector

Feature	Number of features	Number of opinions	Number of hidden layers	Accuracy
Adjectives	127	704	1	45.3901
Adverbs	25	208	1	53.6585
Adjectives and Adverbs	178	740	1	49.3243

Table 3 F measures in binary vector

Feature	Number of features	Number of opinions	Number of hidden layers	P	R	F
Adjectives	127	704	a	0.459	0.482	0.445
Adverbs	25	208	1	0.37	0.463	0.411
Adjectives and Adverbs	178	740	0	0.48	0.48	0.453

Table 4 Classification accuracies for tf-idf measures

Feature	Number of features	Number of opinions	Number of hidden layers	Accuracy
Adjectives	127	781	0	42.3816
Adverbs	25	217	a	48.3871
Adverbs	25	217	1	48.3871
Adjectives and Adverbs	153	872	0	43.1193
Adjectives and Adverbs	153	872	2	43.0046

The best performance showed in the classification (Table 4), if the feature vector constructed was only by adverbs than the other cases of adjectives or both. It is noted that these accuracies are shown in both cases where the number of hidden layers were set to 'a' and 1. Where 'a' is the default number of hidden layers used in WEKA and it is equal to 'a' = (number of attributes + number of classes)/2. In these cases the value is 14. The precision and recall with F measures for these cases are given the Table 5.

Based onto Table 5 the highest F value is shown in the case where adjectives and adverbs were combined in the feature vector, however, as in the case of binary vector use and here as well the number of hidden layers remained same.

Due to this poor performance given by different classification settings authors decided to carry out the experiment by dropping the "neutral" class. The approach is empirically defensible as most of the sentiment classification experiments for English and other major languages have been conducted without the "neutral" class [25, 26]. In similar situations, authors have considered the above classification accuracies and built the input vector for the neural network including both adjectives and adverbs. In addition, used the 10 fold cross validation for training the network. The classification accuracy measures with precision, recall and F values are given in Table 6.

Table 5 F measures in tf-idf vector

Feature	Number of features	Number of opinions	Number of hidden layers	P	R	F
Adjectives	127	781	0	0.408	0.424	0.415
Adverbs	25	217	a	0.431	0.484	0.396
Adverbs	25	217	1	0.45	0.484	0.329
Adjectives and Adverbs	153	872	0	0.425	0.431	0.423
Adjectives and Adverbs	153	872	2	0.361	0.43	0.374

Table 6 Classification accuracies for positive and negative classes for 10-fold cross validation

Feature	Number of features	Number of opinions	Number of hidden layers	Accuracy	P	R	F
Adjectives and Adverbs	153	703	a	53.3428	0.525	0.533	0.486
Adjectives and Adverbs	153	703	0	52.9161	0.524	0.529	0.521
Adjectives and Adverbs	153	703	1	53.3428	0.525	0.533	0.498
Adjectives and Adverbs	153	703	2	54.7653	0.545	0.548	0.505
Adjectives and Adverbs	153	703	3	52.7738	0.523	0.528	0.52

The classification accuracies achieved with the proposed neural network classification algorithm is 55 % giving a F value 0.51. Two hidden layers were utilised in the neural network architecture for training in this case. A significant improvement was shown over the classification with "neural" "class". On the other hand, of all the cases experimented, the highest F measure was resulted from a network with no hidden layers in the training algorithm. Another experiment was carried out for different training strategies by dividing the sample into 80 % for training and rest for the testing.

Based on Table 6, the highest classification accuracy obtained by the network with no hidden layers also showed the highest F value. Another fact noticed in this experiment was that in 80 % training sample classification accuracy (Table 7) dropped by 0.86 % when compared with 10-fold cross validation test.

To understand the classification errors in both cases, the confusion matrices for both cases were examined. In the investigation, authors found that the error of the classification largely occurred in positive cases than those in the negatives (Tables 8 and 9).

From the above analyses, it can be concluded that the neural network training for sentiment classification gives better performance in 10-fold cross validation training with two hidden layers in terms of classification accuracy.

Table 7 Classification accuracies for positive and negative class of 80 % training sample

Feature	Number of features	Number of opinions	Number of hidden layers	Accuracy	P	R	F
Adjectives and Adverbs	153	703	a	49.6454	0.485	0.496	0.412
Adjectives and Adverbs	153	703	0	53.9007	0.55	0.539	0.508
Adjectives and Adverbs	153	703	1	50.3546	0.523	0.504	0.396
Adjectives and Adverbs	153	703	2	51.773	0.588	0.518	0.39
Adjectives and Adverbs	153	703	3	49.6454	0.488	0.496	0.425

Table 8 Confusion matrix 10-fold cross validation

	Negative	Positive
Negative	306	67
Positive	251	79

Table 9 Confusion matrix 80 % sample

	Negative	Positive
Negative	56	15
Positive	50	20

5 Discussion and Conclusion

In this experiment an attempt was made to investigate the rate of success with neural network based sentiment classification for morphological rich languages. The method proposed examined the basic single word vector space model as the input for the neural network. The single word feature vector comprises adjective-adverb language features. The backpropagation architecture with different hidden layers were tested for the classification of the sentiments in the comments given by "Lankadeepa" newspaper reviewers. The Maximum accuracy obtained was around 53 % by the proposed approach and authors postulate several reasons for the average classification accuracies. One reason for this average accuracy may be the lack of structural linguistic information represented by the model. The vector used in this approach includes a single word and it does not represent the contextual information. The only form for linguistic information applied was the part of speech of the word whether it is adjective or adverb. The model can be improved immediately by introducing the bigrams or trigrams instead of single word use. The bigrams deals with collocations of words in a comment that will add the importance of two adjacent words to the polarity of the comment. It is further viewed that the combination of both unigrams (single word) and bigram have a greater chance of increasing the accuracy over the baseline accuracies.

As mentioned in the introduction, Sinhala language considered in this experiment is morphologically rich and the polarity of the words also changes with the word formation. As an example the word ගොඳ (good) has several morphological forms. These morphological forms carry different polarity strength in different contexts. As an example the polarity of word ගොඳම (best) is much higher than the word ගොඳ (good). Both words functioned as adjectives and the difference between them was not distinguished by the neural network model. Authors suggest that incorporating different weight calculation such as sentiment score for these morphologically derived words could be a better solution.

Another enhancement for the existing modal is applying the feature selection procedures such as information gain or chi-square. The current feature selection mechanism based on the word frequencies and some highly correlated words with the sentiment might have been filtered out in the model tested in this research. On the other hand, words with less or no contribution to the polarity of the sentiment might have been included in the feature set.

Based on Tables 6 and 7 in Sect. 4, the network architecture deployed for the sentiment classification did not affect the classification accuracies significantly. In addition, the accuracies are approximately equal for different number of hidden layers. Although the above complexities and weaknesses in the proposed classification algorithm gave around 53 % accuracy, the experiment shows that neural network based sentiment classification for morphologically rich languages is plausible and with further improvements suggested the classification success rate could be further increased.

References

1. B. Liu, Sentiment analysis and subjectivity, in Appear in Hand Book of Natural Language Processing, ed by N. Indurkhya, F.J. Damerau (2010), pp. 3–8
2. V. Turner, J.F. Gantz, D. Reinsel, S. Minton, The digital universe of opportunities: rich data and the increasing value of the internet of things, Framingham (2014)
3. R.W. Nishantha Medagoda, An application of document clustering for categorizing open-ended survey responses, Colombo (2011)
4. S.S.W. Nishantha Medagoda, A comparative analysis of opinion mining and sentiment classification in non-english languages, in *International Conference on Advances in ICT for Emerging Regions*, Colombo, Sri Lanka (2013)
5. B.H. Kasthuriarachchy, K. De Zoysa, H.L. Premaratne, A reviw of domain adaption for opinion detection and sentiment classification, in *The international Conference on Advances in ICT for Emerging Regions*, Colombo (2012)
6. G. Vinodhini, R.M. Chandrasekaran, Sentiment analysis and opinion mining: a survey, in *International Journal of Advanced Research in Computer Science and Software Engineering*, pp. 282–292 (2012)
7. C. Rodriguez, J. Grivolla, J. Codina, A hybrid framework for scalable opinion mining in social media: detecting polarities and attitude targets, in *Proceedings of the Workshop on Semantic Analysis in Social Media* (2012)
8. C.-H. C. Teng-kai, Blogger-centric contextual advertising, in *8th ACM conference on Information and knowledge management* (2009)
9. Kobayasi N, K. Inui, Y. Matsumoto, "Opinion mining from web documents: extraction and structurization," in *Transaction of the Japanese Society for Artificial Intelligence* (2007)
10. S.B. Kotsiantis, I.D. Zaharakis, P.E. Pintelas, Machine Learning: a review of classification and combining techniques. Artif. Intell. Rev. **26**, 159–190 (2006)
11. L.-S. Chena, C.-H. Liub, H.-J. Chiua, A neural network based approach for sentiment classification in the blogosphere. J. Inform. **5**, 313–322 (2011)
12. Z.-B. Xu, C.-H. Li, B. Yu, Latent semantic analysis for text categorization using neural network. Knowl. Based Syst. **21**, 900–904 (2008)
13. L.-S. Chen, H.-J. Chiu, Developing a Neural Network based Index for Sentiment Classification, in *International MultiConference of Engineers and Computer Scientists*, Hong Kong (2009)
14. C.N. dos Santos, M. Gatti, Deep convolutional neural networks for sentiment analysis of short texts, in *25th International Conference on Computational Linguistics*, Dublin, Ireland (2014)
15. R. Socher, B. Huval, C.D. Manning, A.Y. Ng, Semantic compositionality through recursive matrix-vector spaces, Stroudsburg, PA, USA (2012)
16. H.Z. Jingyi Zhang, Improving function word alignment with frequency and syntactic information, in *Twenty-Third International Joint Conference on*, Beijing (2013)
17. K.K. Bharti, P.K. Singh, A Three Stage unsupervised dimension reduction method for text clustering, J. Comput. Sci. **5**, 156–169 (2014)
18. M. Hu, B. Liu, Mining and summarizing customer reviews, in *Proceedings of the tenth ACM SIGKDD international conference on Knowledge discovery and data mining*, New York, NY, USA (2004)
19. F. Benamara, C. Cesarano, D. Reforgiato, Sentiment analysis: adjectives and adverbs are better that adjectives alone, in *International Conference on Weblogs and Social Media (ICWSM)*, Boulder, Colorado (2007)
20. S. T. Piantadosi, Zipf's word frequency law in natural language: a critical review and future directions. Psychon. Bull. Rev. 1112–1130 (2014)
21. S. Haykin, *Neural Networks A comprehensive Foundation* (Pearson Education, Ontario, 2001)
22. W. Zhou, L. Chen, S. Dong, L. Dong, Feed-forward neural networks for accurate and robust traffic classification, vol. 4 (2012)

23. L.-F. Chen, C.-T. Su, M.-H. Chen, A neural-network approach for defect recognition in TFT-LCD photolithography process. IEEE Trans. Electron. Packag. Manuf. 1–8 (2009)
24. C. Manning, H. Schütze, Text Categorization, in *Foundations of Statistical Natural Language Processing*, Cambridge, MA, MIT Press (1999)
25. A.L. Maas, R.E. Daly, P.T. Pham, D. Huang, A.Y. Ng, C. Potts, Learning word vectors for sentiment analysis, Stroudsburg, PA, USA (2011)
26. T. Wilson, J. Wiebe, P. Hoffmann, Recognizing contextual polarity in phrase-level sentiment analysis, Stroudsburg, PA, USA (2005)

Predicting Stock Price Movements with News Sentiment: An Artificial Neural Network Approach

Kin-Yip Ho and Wanbin (Walter) Wang

Abstract Behavioural finance suggests that emotions, moods and sentiments in response to news play a significant role in the decision-making process of investors. In particular, research in behavioural finance apparently indicates that news sentiment is significantly related to stock price movements. Using news sentiment analytics from the unique database RavenPack Dow Jones News Analytics, this study develops an Artificial Neural Network (ANN) model to predict the stock price movements of Google Inc. (NASDAQ:GOOG) and test its potential profitability with out-of-sample prediction.

Keywords Technical analysis · News sentiment · Artificial neural network

JEL classification C45 · C53 · G14 · G17

1 Introduction

Technical analysis is the study of past price movements with the aim of forecasting potential future price movements. Market participants who use technical analysis often exploit primary market data, such as historical prices, volume and trends, to develop trading rules, models and even technical trading systems. These systems comprise a set of trading strategies and rules that generate trading signals, such as buy and sell signals, in the market. Several studies [1–5] examine the profitability of trading strategies, which include moving average, momentum and contrarian strategies. In particular, Park and Irwin [1] show that out of 95 modern studies on technical trading strategies, 56 of them provide statistically significant evidence that technical analysis generates positive results. Han et al. [2] demonstrate that a relatively

K.-Y. Ho · W. (Walter) Wang (✉)
Research School of Finance Actuarial Studies and Statistics,
The Australian National University, Canberra 2601, Australia
e-mail: walter.wang@anu.edu.au

© Springer International Publishing Switzerland 2016
S. Shanmuganathan and S. Samarasinghe (eds.), *Artificial Neural Network Modelling*, Studies in Computational Intelligence 628,
DOI 10.1007/978-3-319-28495-8_18

straightforward application of a moving average timing strategy outperform the passive buy-and-hold strategy. Bessembinder and Chan [3] suggest that technical trading rules have varying degrees of success across different international stock markets; in general, these rules tend to be more successful in the emerging markets. Fernandez-Rodríguez et al. [4] examine the profitability of a simple technical trading rule based on the Artificial Neural Networks (ANNs) and conclude that the ANN trading rule is mostly superior to a passive buy-and-hold trading strategy during "bear" market and "stable" market episodes.

Most of the existing research on technical trading rules and strategies focuses on objective and unambiguous rules based on historical market information without considering investor sentiment. Behavioral finance shows that information flows in markets play a significant role in human decision-making, and financial decisions are significantly driven by mood and sentiment [6]. In particular, research in behavioural finance apparently indicates that news sentiment is significantly related to stock price movements [7–12]. For instance, Antweiler and Frank [8] suggest that internet messages have a significant impact on stock returns and disagreement among the posted messages is associated with increased trading volumes. Schmeling [9] finds that sentiment negatively forecasts aggregate stock market returns on average across countries. Moreover, Schmeling [9] suggests that the impact of sentiment on stock returns is higher for countries that have less market integrity and are more susceptible to market overreaction and herding. Wang et al. [12] show evidence that whilst news volume does not Granger-cause stock price change, news sentiment does Granger-cause stock price change. In general, these papers suggest that the impact of sentiment on stock markets cannot be ignored.

In this paper, we combine a trading strategy based on the ANN model with news sentiment to build our ANN model for predicting the stock price movements of Google Inc. (NASDAQ:GOOG). GOOG is an American public corporation specializing in internet-related services and products that enhance the ways people connect with information [13]. Its primary source of revenue comes from delivering online advertising that are relevant to consumers and cost-effective for advertisers. Founded by Larry Page and Sergey Brin as a privately held company in 1998, GOOG became a public corporation after the initial public offering (IPO) on August 19, 2004. In the past decade, its share has grown by more than 1500 %. As of December 31, 2014, Google had 53,600 full-time employees. Its current range of services includes web search, email, mapping, office productivity, and video sharing services. We focus on GOOG for the following reasons: one, as a major stock on NASDAQ, GOOG is one of the few that has relatively straight forward transaction data because it is a non-dividend-paying stock. As noted on Google's Investor Relations website, Google has "never declared or paid a cash dividend nor do we expect to pay any cash dividends in the foreseeable future" [13]; two, since its IPO 11 years ago, GOOG is considered one of the best performers in the stock market, as its stock price has risen by more than 15 times over the past decade and only 13 stocks in the S&P500 index have outperformed GOOG; three, GOOG has a very high volume of outstanding shares (over 300 million shares with an average daily trading volume of 2.4 million shares) and high stock price (over $600 in September 2015), making it unlikely the subject of price manipulation [13]; four, as a frequently traded share with a large market capitalization exceeding US$400 billion, news directly related to GOOG is frequently reported in various major media outlets. These news releases are a rich source of data to examine the impact of news sentiment on GOOG's price movements.

To quantify the sentiment associated with each news release, we use the dataset obtained from the RavenPack News Analytics Dow Jones Edition (RavenPack). RavenPack systematically tracks and analyzes information on more than 2,200 government organizations, 138,000 key geographical locations, 150 major currencies, 80 traded commodities and over 30,000 companies. It is a comprehensive database covering more than 1,200 types of firm-specific and macroeconomic news events. Among its many benefits, RavenPack delivers sentiment analysis and event data that are most likely to affect financial markets and trading around the world—all in a matter of milliseconds. It continuously analyses relevant information from major real-time newswires and trustworthy sources such as Dow Jones Newswires, regional editions of the Wall Street Journal and Barron's and Internet sources including financial sites, blogs, and local and regional newspapers, to produce real-time news sentiment scores. All relevant news articles about entities are classified and quantified according to their sentiment, relevance, topic, novelty and market effect. In terms of the sentiment, RavenPack uses a proprietary computational linguistic analysis algorithm to quantify the positive and negative perceptions on facts and opinions reported in the news textual content. The core of the algorithm can be divided into two steps. In the first step, RavenPack builds up a historical database of words, phrases, combinations and other word-level definitions that have affected the target company, market or asset class. Subsequently, in the second stage, the text in the specific news story is compared with the historical database, and the sentiments score is generated accordingly. In this paper, we use the sentiment score to classify news type.

The remainder of the paper is organized as follows. In the second section, we discuss the artificial neural networks (ANN) used to generate predictions. In Sect. 3, we discuss the datasets used in this paper. The empirical results of using ANN to predict stock price movements of Google Inc. are discussed in Sect. 4. The last section concludes this paper.

2 Methodology

In the past decade, the availability of datasets has increased tremendously in various fields including finance. Therefore, the empirical applications of data mining techniques, such as classification, clustering and association, have become increasingly important [14]. In particular, there is a burgeoning strand of literature on the applications of data mining techniques to the analysis of stock price movements [15]. This strand of literature suggests that the Artificial Neural Network (ANN) model is fast becoming one of the leading data mining techniques in the field of stock market prediction [16–25]. Chang et al. [22] suggest that ANN can be employed to enhance the accuracy of stock price forecasting. De Oliveira et al. [25] also show that the ANN model is a feasible alternative to conventional techniques of predicting the trends and behavior of stocks in the Brazilian market.

The structure of the ANN model mimics the human brain and nervous system [26–28]. Most neural networks contain three types of layers: input, hidden, and output layers. Each neuron in a hidden layer receives the input data attributes from every single neuron in an input layer, and the attributes are added through applied weights and converted to an output value by an activation function. Subsequently, the output is passed to neurons in the next layer, providing a feed-forward path to the output layer.

Probabilistic Neural Network (PNN) is one of the most widely implemented neural network topologies [29]. PNN is developed based on the classical Bayesian classifier, whose goal is to statistically minimize the risk of misclassifications. Based on the concept of posterior probability, which assumes that the probability density function of the population from which the data are drawn is known a priori, the decision rule of the PNN is to classify a sample to the class with the maximum posterior probability. The PNN then uses a training set to obtain the desired statistical Bayesian information.

In this paper, the PNN is implemented by using the MATLAB Neural Network Toolbox, with the network structures specified according to the default settings [30]. More specifically, the PNN creates a two-layer network structure. The first layer has radial basis network neurons, and calculates its weighted inputs by vector distance between its weight vector and the input vector, multiplied by the bias. The second layer has competitive transfer function neurons, and calculates its weighted input with dot product weight function and its network inputs with the sum of network inputs.

3 Datasets

We use the daily prices of Google Inc. (ticker symbol "NASDAQ:GOOG") obtained from SIRCA Thomson Reuters Tick History (TRTH). The news dataset for GOOG is obtained from RavenPack News Analytics Dow Jones Edition (RavenPack), which provides sentiment analysis for the news articles relevant to GOOG. For each news article, RavenPack provides the following key information: the date and time each news article is released, a unique firm identifier, and several variables that measure the relevance, content, sentiment and form of the article. For instance, the "Relevance" score, which ranges from 0 to 100, indicates the extent to which the underlying news story is directly relevant to GOOG; a score of 100 indicates the article is highly relevant. The "Event Sentiment Score" (ESS) measures the sentiment associated with the news article. ESS ranges from 0 to 100, where 0 indicates extremely negative news, 50 indicates neutral news, and 100 indicates extremely positive news. To compute the ESS, RavenPack uses a proprietary computational linguistic analysis algorithm to quantify positive and negative perceptions on facts and opinions reported in the news textual content.

We construct the Daily Sentiment Score (DSS) for GOOG using the Relevance Score and ESS based on the formula provided below. The period that we use to calculate the DSS in day $i-1$ is the 24-h period before the market opens in day i:

$$DSS_{i-1} = \sum_{\text{all news about the given firm in 24 h before market open in the day } i} I(\text{Relevance} = 100) * (\text{ESS} - 50)$$

$$(1)$$

Figure 1 shows the Daily Sentiment Score for GOOG from January 1, 2013 to June 30, 2015.

In this paper, we use daily opening and closing prices of GOOG from January 1, 2013 to June 30, 2015 to test the predictive accuracy of the PNN model. The prices are obtained from SIRCA's Thomson Reuters Tick History (TRTH) database. Figure 2 shows the daily closing prices of GOOG. In 2014, the price experienced a big jump, which is due to the Google 2-for-1 stock split on April 3, 2014. As a

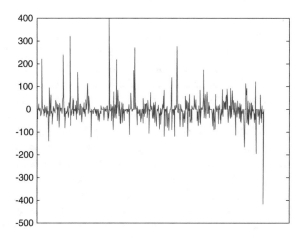

Fig. 1 Daily sentiment score of the NASDAQ:GOOG (Jan 2013–Jun 2015)

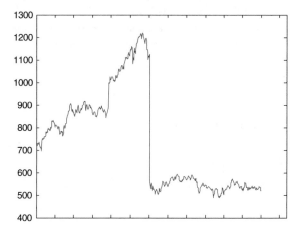

Fig. 2 Daily closing prices of the NASDAQ:GOOG (Jan 2013–Jun 2015)

result of the stock split, GOOG's shareholders received two shares (Class A and Class C) for every one share that they owned. The main difference between these two classes is that Class A confers voting rights whereas Class C does not.

To compute the returns of GOOG, we calculate the difference between the natural logarithm of the daily opening and closing stock prices and multiply the difference by 100 as the stock return in the day i.

$$\text{Stock\_return}_i = 100 * \ln\left(P_{iopen}/P_{iclose}\right) \tag{2}$$

P_{iopen} is the opening price of the GOOG in day i and P_{iclose} is the closingprice of the GOOG in day i. As we use daily opening and closing stock price to calculate the stock returns, the stock split of GOOG does not affect our estimation results.

4 Empirical Results

In this paper, the stock price movement is an "up" ("down") in day i if the closing price of GOOG in day i is larger (smaller) than the opining price of GOOG in day i. We use the "up" and "down" movements of the stock prices as our training patterns. The PNN model is trained on the training data and subsequently tested to assess its performance on the testing data. Basically, the process of training or learning helps us obtain the optimum neural network weights by minimizing the model error, which is the difference between the actual output and the desired one. In this paper, we employ data from January 1, 2013 to December 31, 2014 as the training set and data from January 1, 2015 to June 30, 2015 as the testing set.

Given these sets, this paper uses stock price returns in the last three trading days (i.e. $\text{Stock\_return}_{i-3}, \text{Stock\_return}_{i-2}, \text{Stock\_return}_{i-1}$) and DSS in the last trading day (i.e. DSS_{i-1}) as input features in the PNN model. Based on this approach, 4 indices are obtained. In other words, there are 4 input nodes and one output node. Table 1 shows the basic statistics of these inputs (the close-to-open return and the daily sentiment score of GOOG).

For the application of binary classification in the PNN model, sensitivity and specificity are used to assess the performance of the model. In this paper, we define the pattern as "price up" and "price down". Then we calculate the following variables:

Table 1 The basic statistics of the close-to-open return and the daily sentiment score of GOOG

	Mean	Highest	Lowest
Close-to-open return	−0.03 %	4.03 %	−5.48 %
Daily sentiment score	−2.39	396	−418

Table 2 Performance characteristics of the PNN prediction model

	Sensitivity	Specificity	Prediction rate
PNN	52.83 %	55.71 %	54.47 %

$$\text{Sensitivity} = \frac{\text{Number of true ``up''}}{\text{Number of true ``up''} + \text{Number of false ``down''}}$$

$$\text{Specificity} = \frac{\text{Number of true ``down''}}{\text{Number of true ``down''} + \text{Number of false ``up''}}$$

$$\text{Prediction rate} = \frac{\text{Number of true ``up''} + \text{Number of true ``down''}}{\text{Number of prediction days}}$$

Table 2 shows the performance characteristics of our PNN prediction model. The sensitivity is 52.83 % and the specificity is 55.71 %. According to the random walk directional forecast, the stock price has a fifty-fifty chance of closing higher or lower than the opening price [4, 31]. As can be seen, the total prediction rate of the PNN model is 54.47 %, which implies that the PNN model can perform better than a random walk directional forecast.

5 Concluding Remarks

Many papers on trading strategies build trading rules based on historical data such as stock price and volume. In this paper, we use sentiment scores of news articles related to GOOG to develop an ANN model to predict its stock price movements. More specifically, by defining an "up" ("down") in day i as the closing price in day i being larger (smaller) than the opening price in day i, our empirical results provide better predictive accuracy than a random walk directional forecast. Our model provides a potentially profitable trading strategy with the following rules: if the model predicts an "up" movement, we should buy the stock at the stock market open and sell the stock at the stock market close; in contrast, if the model predicts a "down" movement, we should sell the stock at the stock market open and buy the stock at the stock market close.

References

1. C.H. Park, S.H. Irwin, What do we know about the profitability of technical analysis? J. Econ. Surv. **21**(4), 786–826 (2007)
2. Y. Han, K. Yang, G. Zhou, A new anomaly: The cross-sectional profitability of technical analysis. J. Financ. Quant. Anal. **48**(05), 1433–1461 (2013)

3. H. Bessembinder, K. Chan, The profitability of technical trading rules in the Asian stock markets. Pac. Basin Fin. J. **3**(2), 257–284 (1995)
4. F. Fernandez-Rodrıguez, C. Gonzalez-Martel, S. Sosvilla-Rivero, On the profitability of technical trading rules based on artificial neural networks: Evidence from the Madrid stock market. Econ. Lett. **69**(1), 89–94 (2000)
5. P.-H. Hsu, C.-M. Kuan, Reexamining the profitability of technical analysis with data snooping checks. J. Financ. Econometrics **3**(4), 606–628 (2005)
6. J.R. Nofsinger, Social mood and financial economics. J. Behav. Finance **6**(3), 144–160 (2005)
7. R. Neal, S.M. Wheatley, Do measures of investor sentiment predict returns? J. Financ. Quant. Anal. **33**(04), 523–547 (1998)
8. W. Antweiler, M.Z. Frank, Is all that talk just noise? The information content of internet stock message boards. J. Finance **59**(3), 1259–1294 (2004)
9. M. Schmeling, Investor sentiment and stock returns: Some international evidence. J. Empir. Finance **16**(3), 394–408 (2009)
10. T. Lux, Sentiment dynamics and stock returns: the case of the German stock market. Empirical Economics **41**(3), 663–679 (2011)
11. S.-L. Chung, C.-H. Hung, C.-Y. Yeh, When does investor sentiment predict stock returns? J. Empirical Finance **19**(2), 217–240 (2012)
12. W.W. Wang, K.-Y. Ho, W.-M.R. Liu, K.T. Wang, The relation between news events and stock price jump: an analysis based on neural network, in *20th International Congress on Modelling and Simulation (MODSIM2013)*, Adelaide, Australia (2013)
13. Google Inc, Investor Relations (2015)
14. S.-H. Liao, P.-H. Chu, P.-Y. Hsiao, Data mining techniques and applications—A decade review from 2000 to 2011. Expert Syst. Appl. **39**(12), 11303–11311 (2012)
15. J. Patel, S. Shah, P. Thakkar, K. Kotecha, Predicting stock and stock price index movement using trend deterministic data preparation and machine learning techniques. Expert Syst. Appl. **42**(1), 259–268 (2015)
16. K.-J. Kim, I. Han, Genetic algorithms approach to feature discretization in artificial neural networks for the prediction of stock price index. Expert Syst. Appl. **19**(2), 125–132 (2000)
17. Q. Cao, K.B. Leggio, M.J. Schniederjans, A comparison between Fama and French's model and artificial neural networks in predicting the Chinese stock market. Comput. Oper. Res. **32** (10), 2499–2512 (2005)
18. M.R. Hassan, B. Nath, M. Kirley, A fusion model of HMM, ANN and GA for stock market forecasting. Expert Syst. Appl. **33**(1), 171–180 (2007)
19. E. Guresen, G. Kayakutlu, T.U. Daim, Using artificial neural network models in stock market index prediction. Expert Syst. Appl. **38**(8), 10389–10397 (2011)
20. Y. Kara, M.A. Boyacioglu, Ö.K. Baykan, Predicting direction of stock price index movement using artificial neural networks and support vector machines: The sample of the Istanbul Stock Exchange. Expert Syst. Appl. **38**(5), 5311–5319 (2011)
21. J.-Z. Wang, J.-J. Wang, Z.-G. Zhang, S.-P. Guo, Forecasting stock indices with back propagation neural network. Expert Syst. Appl. **38**(11), 14346–14355 (2011)
22. P.-C. Chang, Wang D-d, Zhou C-l, A novel model by evolving partially connected neural network for stock price trend forecasting. Expert Syst. Appl. **39**(1), 611–620 (2012)
23. Preethi G, Santhi B (2012) Stock market forecasting techniques: A survey. J. Theor. Appl. Inf. Technol **46**(1)
24. J.L. Ticknor, A Bayesian regularized artificial neural network for stock market forecasting. Expert Syst. Appl. **40**(14), 5501–5506 (2013)
25. F.A. de Oliveira, C.N. Nobre, L.E. Zárate, Applying Artificial Neural Networks to prediction of stock price and improvement of the directional prediction index—Case study of PETR4, Petrobras, Brazil. Expert Syst. Appl. **40**(18), 7596–7606 (2013)
26. T. Hill, L. Marquez, M. O'Connor, W. Remus, Artificial neural network models for forecasting and decision making. Int. J. Forecast. **10**(1), 5–15 (1994)
27. G. Zhang, B.E. Patuwo, M.Y. Hu, Forecasting with artificial neural networks: The state of the art. Int. J. Forecast. **14**(1), 35–62 (1998)

28. A. Bahrammirzaee, A comparative survey of artificial intelligence applications in finance: Artificial neural networks, expert system and hybrid intelligent systems. Neural Comput. Appl. **19**(8), 1165–1195 (2010)
29. D.F. Specht, Probabilistic neural networks. Neural Networks **3**(1), 109–118 (1990)
30. M.H. Beale, M.T. Hagan, H.B. Demuth, Neural network toolbox user's guide (2015)
31. Y. Hashimoto, T. Ito, T. Ohnishi, M. Takayasu, H. Takayasu, T. Watanabe, Random walk or a run: Market microstructure analysis of foreign exchange rate movements based on conditional probability. Quant. Finance **12**(6), 893–905 (2012)

Modelling Mode Choice of Individual in Linked Trips with Artificial Neural Networks and Fuzzy Representation

Nagesh Shukla, Jun Ma, Rohan Wickramasuriya, Nam Huynh and Pascal Perez

Abstract Traditional mode choice models consider travel modes of an individual in a consecutive trip to be independent. However, a persons choice of the travel mode of a trip is likely to be affected by the mode choice of the previous trips, particularly when it comes to car driving. Furthermore, traditional travel mode choice models involve discrete choice models, which are largely derived from expert knowledge, to build rules or heuristics. Their approach relies heavily on a predefined specific model structure (utility model) and constraining it to hold across an entire series of historical observations. These studies also assumed that the travel diaries of individuals in travel survey data is complete, which seldom occurs. Therefore, in this chapter, we propose a data-driven methodology with artificial neural networks (ANNs) and fuzzy sets (to better represent historical knowledge in an intuitive way) to model travel mode choices. The proposed methodology models and analyses travel mode choice of an individual trip and its influence on consecutive trips of individuals. The methodology is tested using the Household Travel Survey (HTS) data of Sydney metropolitan area and its performance is compared with the state-of-the-art approaches such as decision trees. Experimental results indicate that the proposed methodology with ANN and fuzzy sets can effectively improve the accuracy of travel mode choice prediction.

N. Shukla (✉) · J. Ma · R. Wickramasuriya · N. Huynh · P. Perez
SMART Infrastructure Facility, Faculty of Engineering and Information Sciences,
University of Wollongong, Wollongong, NSW 2522, Australia
e-mail: nshukla@uow.edu.au

J. Ma
e-mail: jma@uow.edu.au

R. Wickramasuriya
e-mail: rohan@uow.edu.au

N. Huynh
e-mail: nhuynh@uow.edu.au

P. Perez
e-mail: pascal@uow.edu.au

© Springer International Publishing Switzerland 2016
S. Shanmuganathan and S. Samarasinghe (eds.), *Artificial Neural Network Modelling*, Studies in Computational Intelligence 628,
DOI 10.1007/978-3-319-28495-8_19

Key words Travel mode choice · Consecutive trips · Artificial neural networks · Fuzzy sets

1 Introduction

Travel mode choice is an important aspect of travel behavior, and also one of the four steps in transportation demand estimation for urban planning [20]. It refers to the procedure of assigning available travel modes (e.g. car, walk, bus, and train) to each trip of an individual based on the person's demographic and social characteristics and the environmental conditions. Travel mode choice has received significant research attention from academia, industry and governmental management agencies. Current studies on travel mode choice mainly based on discrete choice methods which rely heavily on some prerequisites such as specific model structure, complete historical travel diaries, and expert knowledge, to build rules or heuristics for mode choice analysis. However, real data used for travel mode choice analysis seldom fulfill these requisites. Hence, the models built on top of them may well fit the data available but have limited usefulness for real world applications.

Using machine learning techniques for travel mode choice modeling have drawn a significant attention in recent years. Methods based on machine learning techniques have been reported to have better performance/accuracy in prediction of travel mode choices over conventional statistical models. These methods are mainly data-driven i.e. they are built directly from data without (or minimal) prerequisites in terms of predefined mathematical model. However, majority of existing machine learning methods in literature assign mode to each of an individuals trips as independent procedure without considering the inner interaction of the linked trips. As trips made by an individual is closely related to his/her travel purpose and previous trips (if any), therefore, linked trips need to be considered when developing a machine-learning based models.

Traditionally, mode choice using discrete choice methods tried to link trips in a travel diary (set of trips performed by an individual in a period) by defining an explicit utility model as the measurement or evaluation criterion for travel mode assignment. Such methods required the access to the complete historical travel diaries of individuals in that period. As travel diaries of individuals may vary in many aspects such as number of trips, traveling time period, travel purposes, as well as trip distances; it is hard to apply a common utility model to different types of travel diaries. Particularly, in a given focal period such as morning peak, input data of trips may mainly comprise of partial set of trips in travel diaries, which leads to difficulty in analyzing travel diaries and assigning appropriate modes for the trips in other time period.

Modeling travel mode choice are largely based on near continuous variables in travel survey data. Although data about these variables is crucial for travel mode choice analysis; however, it is too detailed from the point of view of modeling and affects the performance of a model. Research indicates that categorizing this data can effectively improve the predictive performance of a model. To implement this, fuzzy representation is an appropriate technique because it can effectively harness the domain and common knowledge to aid model building. For example, we can use a "morning-peak trip" to describe trips in a given time period without drilling to the detailed time, or use a "short-distance trip" to summarize trips in a short time period or with a short distances.

In this chapter, we focus on travel mode choice modeling approach which considers consecutive (a.k.a linked) pairs of trips of an individual under the assumption that the individuals previous trip mode affects the following trip mode choice. Particularly, we considered linked trips of an individual and developed a machine learning based travel mode choice approach and used fuzzy representation for data discretisation.

The remainder of the chapter is organised as follows. Section 2 provides an overview the related models and techniques in travel mode choice literature. Section 3 analyses the linked trips. Section 4 provides the details of the presented method. Section 5 describes the experiments conducted on a real travel survey (the Household Travel Survey (HTS) of Sydney metropolitan area) and compares the performance of the presented method. Finally, we summarise the main contributions and future work in Sect. 6.

2   Related Works

Discrete choice models are primarily used in travel mode choice studies. Such models include the Dogit and Logit models [12, 13], the multinomial logit (MNL) models [19] and the nested logit models [10]. Arguments about their limitation include (i) a specific model structure needs to be defined in advance, which ignores the partial relationships between explanatory variables and travel modes for different population subgroups; (ii) they lack the ability to model complex non-linear systems, which represent complex relationships involved in human decision making; and, (iii) they check only for conditions that hold across an entire sample of observations in the training dataset and patterns cannot be extracted from different observation subgroups [31].

In past two decades, machine learning has emerged as a superior approach in travel mode choice research by which travel mode choice can be better predicted while alleviating aforementioned shortcomings [8, 27, 28, 31]. For example, Xie et al. [33] report that Artificial Neural Networks (ANN) achieved better results compared to MNL based on a comparative study using work-related travel data. Similarly, Rasmidatta ([26]) compared the nested logit models and the ANN models for long distance travel mode choice, and illustrated a better performance of ANN over other models. Furthermore, there are other studies that compare and contrast the performance of machine learning techniques with other traditional statistical

Table 1 Classification of state-of-the-art approaches in mode choice

Trip type	Data Type			
	Discrete choice models		Machine learning models	
	Crisp data	Crisp & Fuzzy representation	Crisp data	Crisp & Fuzzy representation
Independent trips	[4, 6, 9, 10, 12, 15, 19, 23, 25]	[11]	[8, 15, 27, 28, 31]	[32]
Linked individual trips	[21]	This study	[5]	This study
Linked household trips	[21]		Future work	Future work

techniques and propose to use machine learning techniques such as ANN and Decision Tree (DT) for travel mode choice prediction [8, 15, 24, 27, 28].

Park et al. [25] proposed the use nested model over multinomial logit model and probability distribution function model for access mode choice behaviors to the water transportation mode in Bangkok. Nakayama et al. [23] have proposed a semi-dynamic traffic assignment model with travel mode choice (logit) model between public transit and car. Bhat and Dubey [4] have proposed a new multinomial probit-based model formulation for integrated choice and latent variable (ICLV) models and compared its advantages against traditional logit kernel-based ICLV formulation. Bliemer and Rose [6] have proposed experimental design strategies to determine reliable and robust parameter estimates for multinomial logit models. Karlaftis and Vlahogianni [16] have compared the use of neural networks and statistical methods for their appropriate applicability in transportation research domain. One of the conclusions from the review is that the neural networks are regarded as constraint free and flexible when dealing with nonlinearities and missing data.

Based on the types of trips focused and data used, we summarize the existing research studies in Table 1. Table 1 indicates that the majority of existing research focuses on independent trips and crisp (or original survey) data. Research on linked trips with fuzzy representation has not been reported, which is the main focus of this study.

3   Linked Trip and Its Influence on Travel Modes

Travel mode choice can be understood as the travel mode to which a traveler pre-commit for a given activity (shopping, work, school). Majority of the traditional literature focuses on an independent trip, where each trip is considered as an independent event. This indicates the assumption that an individual has to make independent decision about the travel mode regardless of whether these trips are part of a tour (e.g. home-work-shopping-home). However, due to complexity of patterns of trips performed by an individual in a time period, the assumption of each trip to be independent is inappropriate. Cirillo and Axhausen [9] suggested that individuals maintain their mode during a tour, especially if they use an individual vehicle (car, motorcycle or bicycle). Hence, there is strong dependency between travel modes adopted in consecutive trips.

In this study, we focus on linked trips (a.k.a. consecutive trips) which comprises of a chain of trips made by an individual trip maker. Our selection of linked trips is inspired by the fact that the real household travel survey data are largely incomplete and it mostly characterizes the travel behavior in different time periods of the day. These considerations are different from previous research on travel mode choice. Traditionally, trip-based, activity-based, and tour-based models have considered the trip maker as the central subject. Hence, survey data needs the travel makers entire history of travel diaries when building those models. However, we noted that trip records in household travel survey data are seldom complete. Majority of persons trip records often have incomplete travel sections i.e. isolate trips. Another reason for our considering of linked trips is that transportation planning is often based on common travel periods of the whole community rather than individuals i.e. travel mode is affected by the time periods. For example, during peak hours, many Sydney residents drive to the nearby bus or train stations and then use the public transportation in order to avoid the traffic congestion. Therefore, we are trying to discover the influence of the travel mode used in previous trip on the following trip during given a time period through the linked trips.

Formally, let O and D be the origin and destination of a travel $p_{O \rightarrow D}$, a trip is a section between locations s and t in the travel and denoted by $p_{s \rightarrow t}$. Hence, a travel with $n + 1$ trips from O to D is denoted by

$$p_{O \rightarrow D} = p_{O \rightarrow 1} \cup p_{1 \rightarrow 2} \cup \cdots \cup p_{n \rightarrow D}$$

For convenience, we write $t_0 = p_{O \rightarrow 1}, t_n = p_{n \rightarrow D}, i = 1, \ldots, n - 1, i = 1, \ldots, n - 1$ and trip t_{i+1} is called the linked trip of t_i, $i = 0, \ldots, n - 1$. The travel mode assigned to trip t_i is m_i, $i = 0, \ldots, n$.

Generally, it is not necessary to assume the origin and destination of a travel to be the same location; and locations $O, 1, \ldots, n, D$ can be different. For example, Fig. 1 shows some examples of possible linked trips.

4 Methodology

This section details the modeling methodology adopted in this Chapter for the travel mode choice of an individual using data collected for a household travel survey. As mentioned in Sect. 2, travel mode choice has been studied largely using discrete choice models that possess known limitations. From the viewpoint of machine learning, travel model choice modeling is able to be treated as a classification

Fig. 1 Examples of linked
trips

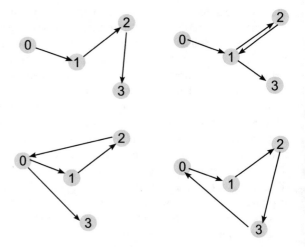

problem, where the classification attribute is the assigned trip mode. This leads us to
explore existing classification methods in machine learning to overcome those
known limitations in discrete choice models. In this study, we mainly use the
artificial neural networks (ANN) and the decision trees (DT) because they are the
widely-used classification methods. We have also employed fuzzy sets to represent
common knowledge about travel time periods, incomes, and other attributes in the
survey data. The first part of this section presents some preliminaries of ANN and
DT followed by a description of linked trips and fuzzy representation of travel
survey data.

4.1 Preliminaries

McCulloch and Pitts [18] developed the concept of artificial neurons to study
cognitive processes. Following this, ANNs have been applied for variety of classification and pattern recognition problems. In general, an ANN consists of a set of
interconnected processing nodes called neurons which are used to estimate mapping
between explanatory variables (attributes) and the responses. Each neuron combines its inputs into a single output value based on an activation function.
A commonly used neural network consists of an input layer, an output layer, and a
hidden layer. Multi-layer feed-forward (MLFF) neural networks are the mostly used
structure. It has been recognized that using multiple hidden layers can improve the
ability of an ANN to model complex linear as well as non-linear relationships.
Hence, we employ MLFF ANN for this study.

Figure 2 illustrates a typical MLFF ANN with one hidden layer where
$x_1, x_2, \ldots, x_n$ represent the set of n explanatory variables as inputs and $y_1, y_2, \ldots, y_o$
represent the set of o responses as outputs. $w_{i,j}$ $(i = 1, \ldots, n; j = 1, \ldots, h)$ represents

Fig. 2 A typical MLFF ANN topology

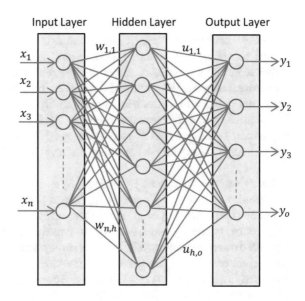

the weight associated with the edge between node n_i^I in input layer and node n_j^H in hidden layer; and $u_{j,k}$ ($j = 1, \ldots, h$; $k = 1, \ldots, o$) represents weight associated with the edge between node n_j^H in hidden layer and node n_k^O in output layer.

When an observation $(x_1, \ldots, x_n; y_1, \ldots, y_n)$ is presented to the network, each explanatory variable is multiplied by appropriate weights and summed up together with respective bias value to adjust magnitude of the output. The resulting value is fed into transfer function to produce output for a particular hidden node. Formally,

$$H_j = \varphi\left(\sum_i^n w_{i,j} x_i + b_j^H\right), j = 1, 2, \ldots, h \tag{1}$$

where H_j represents the output of the j-th hidden node, b_j^H represents the bias value for the j-th node in hidden layer and $\psi(\cdot)$ represents a predetermined transfer function.

Similarly, the value $\bar{y}_k$ of the k-th output node is represented as:

$$\bar{y}_k = \psi\left(\sum_j^h u_{j,k} H_j + b_k^O\right), k = 1, 2, \ldots, o \tag{2}$$

where b_k^O represents bias value for the k-th node in output layer.

Various algorithms have been proposed for training ANN [14, 22]. These algorithms calibrate $w_{i,j}, u_{j,k}, b_j^H, b_k^O$ to accurately predict or produce the responses. We consider the widely-used back-propagation algorithms such as scaled conjugate gradient [22] and Levenberg-Marquardt optimization [14] for ANN training.

DT is a type of rule-based modeling method which maps observations (explanatory variables) to their response through branching criteria. DT resembles human reasoning and produces a white-box model for decision making. In DT, leaves represent a class label and each node represents a condition or rule for tree branching. DT presents several advantages such as robustness to noise, low computational cost for model generation and ability to deal with redundant variables. DT algorithm and its variants such as C4.5 or J4.8 and Classification and Regression Technique (CART) have been used for classification in variety of areas [30]. An application of these algorithms to travel mode choices was reported by Xie et al. [31].

Fuzzy set was introduced by Zadeh [33] as a tool for processing uncertainty in real application systems which involve human perceptions of vague concepts such as young person and big heap. Since then, fuzzy set has been successfully used in engineering, control systems, and decision making [17]. Recently, it has been used in travel demand modeling [32]. In travel survey data, many attributes such as travel cost, travel cost, travel time as well as individual demographic characteristics are measured in more detailed measurements. However, this kind of data is hard to be used for travel mode choice modeling. Reasons for this include (1) the detailed measurements are by no means accurate, for example, an individual cannot tell when exactly he or she started a trip; (2) the detailed measurements prevent a modeler to find useful pattern, e.g., useful pattern about peak hours will be buried in too much detailed patterns; and (3) the detailed measurements increase the difficulties to build and tune a robust model. Machine learning research indicates categorization is an efficient data pre-processing strategy for building models with better performance. Because fuzzy set or fuzzy representation is a powerful tool for summarizing information for categorization purpose, we argue that fuzzy sets can better describe a persons choice of a specific travel mode.

Fuzzy sets can provide better description of and insight into a specific travel mode choice. Generally speaking, a travel mode choice can be described as an *IF-THEN* expression such that:

> *IF* the depart time is 06:30 and the travel distance
> is 20.5 km, *THEN* the travel mode is car-driving.

Although this kind of description is accurate from the modelling perspective and data accuracy, it lacks insight into the real travel behaviour pattern particularly when there are so many similar expressions. Using fuzzy sets, we can provide an intuitive and better understandable expression such as:

> *IF* the depart time is early morning and the
> trip is long, *THEN* the travel mode is car-driving.

Hence, we can combine multiple expressions into an easily understandable expression and obtain insight into the mode choice pattern.

Fuzzy set offers a way to describe the uncertainty in the travel survey dataset. Two types of uncertainties exist in the collected data for travel mode choice analysis. The first type of uncertainty comes from the surveyors questionnaire and responses design; and the other type exists in an individual responses. Question responses in a survey need to be designed with distinguishable and clear meanings. However, it is very common that the provided response options do not fit the real opinion or actual behavior of a responder. Therefore, the responder may either choose from provided options or write down self answers when filling the survey form. Secondly, due to the differences in responders' perception of same question, uncertainty is inevitably added to the collected data. Fuzzy sets can better handle these two kinds of uncertainties. For example, it is hard for any traveler to remember exact start and finish time for different trips in a day, however, they can clearly remember if the trip is started in the morning or afternoon or night. In this situation, asking the traveler to choose from given time periods or to provide exact starting time is not a good option. It would be better to ask them to indicate the starting time using fuzzy terms like early morning, morning or lunch time.

Typically, a fuzzy set is defined as Definition 1. Based on this definition, we can describe uncertainty and common sense knowledge for the travel mode choice analysis. Details of operations and algorithms of fuzzy sets can be found in Klir and Yuan [17] and are not included and discussed here.

Definition 1 Let X be a set of objects (X is called a universe of discourse). A fuzzy set $\tilde{A}$ on X is defined by an associated membership function $\mu_{\tilde{A}}$ such that for any $x \in X$, $\mu_{\tilde{A}}(x) \in [0, 1]$. $\mu_{\tilde{A}}(x)$ is called the membership degree of element x in fuzzy set $\tilde{A}$.

4.2　Linked Trip Representation in Household Travel Survey

In this section, we present the data processing of the household travel survey and the outcome is used as inputs for ANN and DT. Considering the focus of our approach, this process mainly includes extracting linked trips from survey data and representing them based on transportation needs.

Household travel surveys are increasingly used in most of the metropolitan cities to understand the current travel behavior and demand for transport planning. For instance, London Travel Demand Survey covers 8000 households annually [29]; Sydney Household Travel Survey covers 3500 household annually to gather travel related information (source: http://www.bts.nsw.gov.au/HTS/HTS/default.aspx). These travel surveys record socio-economic characteristics, demographic characteristics, household attributes, travel diary, travel purpose, departing and arriving times, and travel modes, among others. These records are used by planners to design or change existing transport plans. This Chapter used the Sydney Household Travel Survey to model the travel mode choices of an individual given other attributes.

In the following, let S be a survey dataset of trips made by L travelers. The m-th trip made by the traveler l is represented by (x^{lm}, y^{lm}), $m \in \{1, 2, ..., M_l\}$,

$l \in \{1, 2, ..., L\}$, where $x^{lm} = (x_1^{lm}, x_2^{lm}, ..., x_n^{lm})$ are values of n explanatory attributes and $y^{lm} = (y_1^{lm}, y_2^{lm}, ..., y_o^{lm})$ are Boolean values of o decision variable (corresponding to possible travel modes) such that only one of them is true $(= 1)$.

To describe linked trips, we introduced an additional set of explanatory variables, which is the travel mode adopted by the individual in front trip, apart from x^{lm} to model the travel mode choices. The additional variable set is represented as x'_{lm}, formally,

$$x'_{lm} = \begin{cases} 0 & m = 1 \\ Y^{l(m-1)} & m \in \{2, ..., M_l\} \end{cases} \tag{3}$$

Since the first trip (i.e., $m = 1$) does not have front trip mode choice (y^{l0}), we use a dummy vector $\mathbf{0} = (0, ..., 0)$ to represent that. In other words, we treat the first trip of an individual to be independent and rest of the trips to be dependent on their front trip.

The modified travel survey dataset includes x^{lm} and x'_{lm} as explanatory variables to model the responses in y^{lm} for all $l \in \{1, 2, ..., L\}$, $m \in \{1, 2, ..., M_l\}$.

4.3 Attributes Fuzzy Representation

There are mainly three typical value types in the explanatory variables i.e. categorical (*ordinal* or *nominal*), continuous, and date-time. Research indicates that categorizing continuous and date-time values can generally improve model performance. To do this, we follow the fuzzy set technique and knowledge from transport needs.

Based on the features of fuzzy sets, we introduce several fuzzy representations to replace two selected variables which are *depart_time* and *household_income*.

In the survey dataset, the *depart_time* variable is recorded in minute-basis from 00:00 to 23:59 for a day. For convenience, we let $X = \{0, ..., 1439\}$ be the minutes apart from the time 00:00 and define fuzzy sets to describe different depart time.

Following Transport for New South Wales (T4NSW) technical documentation [7], four fuzzy sets are defined for *depart time* over the 24 h, which are morning peak" (M), evening peak" (E), inter-peak" (L), and evening/night period" (N).

In our method, morning peak" starts from approximately 6:00AM ($x = 360$) and spans to 10:00AM ($x = 600$), which has 1 h extension before and after the TfNSWs definition (7:00–9:00AM). The membership function μ_M is given by

$$\mu_M(x) = \begin{cases} e^{-\frac{(x-450)^2}{2 \times 60^2}} & x < 450 \\ 1.0 & 450 \leq x \leq 510 \\ e^{-\frac{(x-510)^2}{2 \times 60^2}} & 510 < x \leq 1439 \end{cases} \tag{4}$$

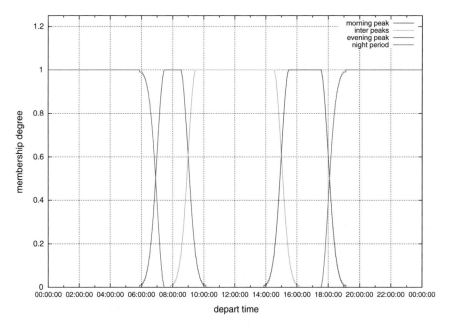

Fig. 3 Fuzzy sets for "depart time"

where x is the minutes from 00:00. By this definition, the morning peak starts gradually from around 6:00AM, reaches the peak between 7:15AM ($x = 450$) and 8:45AM ($x = 510$) and then ends at about 10:00AM. Similarly, the other three fuzzy sets (for evening peak", inter-peak, evening/night period) are defined as following (Fig. 3).

$$\mu_L(x) = \begin{cases} e^{-\frac{(x-570)^2}{2 \times 60^2}} & x < 570 \\ 1.0 & 570 \leq x \leq 870 \\ e^{-\frac{(x-870)^2}{2 \times 60^2}} & 870 < x \leq 1439 \end{cases} \tag{5}$$

$$\mu_E(x) = \begin{cases} e^{-\frac{(x-930)^2}{2 \times 60^2}} & x < 930 \\ 1.0 & 930 \leq x \leq 1050 \\ e^{-\frac{(x-1050)^2}{2 \times 60^2}} & 1050 < x \leq 1439 \end{cases} \tag{6}$$

$$\mu_N(x) = \begin{cases} 1.0 - e^{-\frac{(x-450)^2}{2 \times 60^2}} & x < 450 \\ 0.0 & 450 \leq x \leq 1050 \\ 1.0 - e^{-\frac{(x-1050)^2}{2 \times 60^2}} & 1050 < x \leq 1439 \end{cases} \tag{7}$$

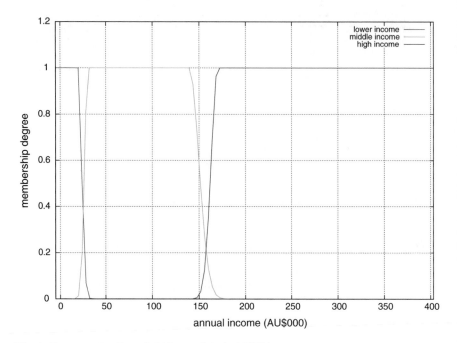

Fig. 4 Fuzzy sets for "household income" (unit AU$000)

In the survey dataset, the variable *household_income* indicates the annual approximate household income which ranges from −AU$5005.74 to AU$402741. Due to the spread of the range, it is hard to get insight of the influence of household income on travel mode choice. We introduced three fuzzy sets to depict understandable concepts which are consistent with individuals ordinary experience on household income levels. The three fuzzy sets are low income (LI), middle income (MI) and high income (HI), which are shown in Fig. 4.

The three fuzzy sets are defined based on the information from the Australian Bureau of Statistics (ABS) and the Australian Taxation Office (ATO). By ABS, the low income household are located in the second and the third income deciles, the high income households are located in the 9-th and 10-th deciles, and the other are middle income households [1]. By ATO income tax rate, the low income threshold is about AU$37000 [3] and family tax offset threshold is about AU$150000 [2]. Hence, we defined the three fuzzy sets based on aforementioned information.

The fuzzy set "low income" is defined as following:

$$\mu_{LI}(x) = \begin{cases} 1.0 & x < 19600 \\ e^{-\frac{(x-19600)^2}{2 \times 3600^2}} & x \geq 19600 \end{cases} \tag{8}$$

where x is the total household income. Similarly, the other two fuzzy sets are defined as

$$
\mu_{MI}(x) = \begin{cases} e^{-\frac{(x-30300)^2}{2\times3600^2}} & x < 30300 \\ 1.0 & 30300 \le x \le 139000 \\ e^{-\frac{(x-139000)^2}{2\times10100^2}} & x > 139000 \end{cases} \tag{9}
$$

$$
\mu_{HI}(x) = \begin{cases} e^{-\frac{(x-172000)^2}{2\times7000^2}} & x < 172000 \\ 1.0 & x \ge 172000 \end{cases} \tag{10}
$$

5 Case Study

5.1 Dataset Description

The Sydney Household Travel Survey (HTS) data is the largest and most comprehensive source of information on personal travel patterns for the Sydney Greater Metropolitan Area (GMA), which covers Sydney, the Illawarra Statistical Divisions and the Newcastle Statistical Subdivision. The data is collected through face-to-face interviews with approximately 3000–3500 households each year (out of 50000 households in the Sydney GMA randomly invited to participate in the survey). Details recorded include (but are not limited to) departure time, travel time, travel mode, purpose, origin and destination of each of the trips that each person in a household makes over 24 h on a representative day of the year. Socio-demographic attributes of households and individuals are also collected.

The presented method has been implemented and tested on a 100,000-trip sample which is randomly selected from the HTS data. We partitioned the sample into three subsets i.e. a training set (30 %), a testing set (35 %) and a validation set (35 %). The three partitions are used for all the experiments conducted in this Chapter.

6 Experiments Design

Before conducting the test, we identified 17 variables based on statistical analysis of their correlations with the travel modes. The identified variables are listed in Table 3. Among these variables, the variables *depart_time* and *household_income* are replaced by the fuzzy sets defined above when conducting the test for fuzzy settings. Further, we use previous trip's mode, pre_mode_new, (as mentioned in Sect. 4.3) to test the scenarios when considering dependent/linked/consecutive trips of an individual in a day (Table 2).

Table 2 Attributes used for testing

Attribute ID	Attribute name	Attribute description
1	*tripno*	Trip order (m)
2	*day_no*	Weekday of a travel
3	*hf*	Household type
4	*occupancy*	Household occupancy
5	*veh_parked_here*	Number of Vehicles Parked at dwelling location
6	*hh_income*	Household income
7	*licence_num*	Number of driving licenses in a household
8	*student_sum*	Number of students in a household
9	*work_athome_sum*	Number of home based working person
10	*resident_num*	Household size
11	*pers_num_trips*	Number of trips made by an individual in a day
12	*purpos11*	Travel purpose
13	*road_dist_xy*	Trip distance
14	*depart_time*	Trip departure time
15	*arrive_time_tune*	Trip ending time slot (30 min interval)
16	*tmode_new*	Travel mode
17	*pre_mode_new*	Travel mode of trip in front

Table 3 Experiments based on DT, ANN

Experiment	Empirical settings		PCI (%)	
ID	Fuzzy sets	Linked trip	DT	ANN
1	N	N	64.71	68.1
2	Y	N	67.67	68.7
3	N	Y	85.63	85.9
4	Y	Y	86.17	86.8

Total 4 experiments (shown in Table 3) have been conducted based on different empirical settings (on DT and ANN) which are:

Experiment 1 We use *travel mode* as decision variable and the others as explanatory variables. Under this setting, we test independent trip mode choice model and use this result as a benchmark for the following tests.

Experiment 2 Replacing the explanatory variables *hh_income* and *depart_time* by their fuzzy sets in Experiment 1. Under this setting, we test the performance of fuzzy sets in travel mode choice modeling.

Experiment 3 We add attribute *pre_mode_new* as an additional exploratory attribute to experiment 1 and test the performance of travel mode choice modeling based on linked trips.

Experiment 4 We add attribute *pre_mode_new* as an additional exploratory attribute to experiment 2 and test the performance of linked trips modeling based on consecutive trip under fuzzy set settings.

7   Result and Analysis

To measure the performance, we use root mean square error (RMSE) value for mode share and percentage of records correctly identified (PCI) for each experiment. The RMSE of mode share (for R modes) for a classifier is:

$$RMS = \sqrt{\sum_{r=1}^{R}(P_r O_r)^2} \tag{11}$$

where, P_r and O_r are the percentage of instances of r-th mode in prediction and original dataset.

TABLE 3 gives the empirical settings and PCI of experiments. Some observations from this experimentation are:

1. Using dependent (or linked) trips achieves higher PCI. For example, the PCI of experiment 1, 2 for both ANN and DT increases significantly from 64.71 to 85.63 % in DT and 69.30 to 84.7 % in ANN.
2. Using fuzzy sets as categorization strategy gives higher PCI to ANN and DT. Experiments 1 and 2 for DT and 3 and 4 for ANN justify the use of Fuzzy sets.
3. ANN performs better than the DT for all the experiments.

Based on the experiments, we can claim that our method improves the PCI of travel mode choice. Table 4 illustrates the mode shares predicted by proposed approach considering ANN and DT with fuzzy sets and linked trips and it is compared with the original mode shares from the HTS data. It illustrates that the mode shares from proposed approach are consistent with that from the HTS data. Particularly, we noted that the mode share for public transport and bicycle are very

Table 4 Mode shares for ann prediction

Travel modes	HTS data (%)	DT prediction (%)	ANN prediction (%)
Car_driver	40.95	43.50	43.11
Car_passenger	20.65	30.76	19.05
Public_transport	8.37	7.54	7.74
Walk	29.26	17.68	29.55
Bicycle	0.77	0.53	0.53

similar to those of the HTS data; and the other modes have slightly varies. The mode share indicates that the public transport and bicycle modes have strong correlation between linked trips; however, the other modes may be affected by many aspects such as travel purpose. This needs more study.

8 Conclusion and Future Work

This Chapter describes a novel methodology for travel mode choices based on linked trips and fuzzy representation. The proposed method considers (i) common sense knowledge by using fuzzy sets to categorize over-detailed attributes; and, (ii) using the linked trips based model that uses travel mode of trips in front as one predictive variable. We use two typical classification techniques, i.e., the ANN and DT, to test the methodology on a real-world household travel survey dataset. The results from various analysis suggests that the use of fuzzy representation and linked trips for mode choice achieves higher performances in both individual mode choice and mode share distribution aspects.

In future, this work can be extended to include other explanatory variables, new fuzzy representation, and linking the individuals in the household to achieve higher modeling performances. Moreover, considering individual trips are affected by household characters, such as household activity and household composition, our next step work will include linked household trips as shown in Table 1.

References

1. Australian Bureau of Statistics, *Household Income and Income Distribution, Australia, 2009-2010* (Canberra, Australia, 2012)
2. Australian Taxation Office, Parent, spouse's parent or individual relative tax offset calculator (2012), Retrieved on 07 May 2013
3. Australian Taxation Office, Household Assistance Package—Tax Reforms (2013), Retrieved on 07 May 2013
4. C.R. Bhat, S.K. Dubey, A new estimation approach to integrate latent psychological constructs in choice modelling. Transp. Res. Part B: Methodol. **67**, 68–85 (2014)
5. J.P. Biagioni, P.M. Szczurek, P.C. Nelson, A. Mohammadian, Tour-based mode choice modeling: Using an ensemble of conditional and unconditional data mining classifiers, in *Transportation Research Board 88th Annual Meeting*, Washington, 11–15 January 2008
6. M.C.J. Bliemer, J.M. Rose, Experimental design influences on stated choice outputs: An empirical study in air travel choice. Transp. Res. Part A: Policy Pract. **45**(1), 63–79
7. Bureau of Transport Statistics, *Sydney Strategic Travel Model (STM): Modelling future travel patterns* (Transport for New South Wales, Sydney, 2011)
8. G.E. Cantarella, S. De Luca, Modeling transportation mode choice through artificial neural networks, in *Fourth International Symposium on Uncertainty Modeling and Analysis, ISUMA 2003*, pp. 84–90, 24–24 Sept 2003. doi: 10.1109/ISUMA.2003.1236145

9. C. Cirillo, K.W. Axhausen, Mode choice of complex tours: A panel analysis, Arbeitsberichte Verkehrs- und Raumplanung, vol. 142 (Zurich: Institut fr Verkehrsplanung, Transporttechnik, Strassenund Eisenbahnbau (IVT), ETH Zurich, 2002)

10. A. Daly, S. Zachary, Improved multiple choice models, in *Identifying and measuring the determinants of mode choice*, ed. by D. Hensher, Q. Dalvi (Teakfield, London, 1979), pp. 335–357

11. M. Dell'Orco, G. Circella, D. Sassanelli, A hybrid approach to combine fuzziness and randomness in travel choice prediction. Eur. J. Oper. Res. **185**(2007), 648–658 (2007)

12. M.J.I. Gaudry, Dogit and logit models of travel mode choice in Montreal. Can. J. Econ. **13**, 268–279 (1980)

13. M.J.I. Gaudry, M.G. Degenais, The dogit model. Transp. Res. Part B: Methodol. **13**, 105–111 (1979)

14. Hagan, M.T., and M. Menhaj (1994) Training feed-forward networks with the Marquardt algorithm, IEEE Trans. Neural Networks, **5**(6), 989–993

15. D.A. Hensher, T.T. Ton, A comparison of the predictive potential of artificial neural networks and nested logit models for commuter mode choice. Transp. Res. Part E: Log. Transp. Rev. **36** (3), 155–172 (2000)

16. M.G. Karlaftis, E.I. Vlahogianni, Statistical methods versus neural networks in transportation research: Differences, similarities and some insights. Transp. Res. Part C: Emerg. Technol., **19** (3), 387–399

17. G. J. Klir, B. Yuan. *Fuzzy Sets and Fuzzy Logic: Theory and Applications* (Prentice Hall, 1995)

18. W. McCulloch, W. Pitts, A logical calculus of the ideas immanent in nervous activity. Bull. Math. Biophys. **7**, 115–133 (1943)

19. D. McFadden, in *Conditional Logit Analysis of Qualitative Choice Behavior, Frontiers in Econometrics*, ed. by P. Zarembka (Academic Press, New York, 1973)

20. M.G. McNally, in *The Four Step Model*, ed. by D.A. Hensher, K.J. Button Handbook of Transport Modeling (Pergamon Publishing, 2007)

21. E.J. Miller, M.J. Roorda, J.A. Carrasco, A tour-based model of travel mode choice. Transportation **32**(4), 399–422 (2005). doi:10.1007/s11116-004-7962-3

22. M.F. Moller, A scaled conjugate gradient algorithm for fast supervised learning. Neural Networks **6**, 525–533 (1993)

23. Sho-ichiro Nakayama, Jun-ichi Takayama, Junya Nakai, Kazuki Nagao, Semi-dynamic traffic assignment model with mode and route choices under stochastic travel times. J. Adv. Transp. **46**(3), 269–281 (2012)

24. P. Nijkamp, A. Reggiani, T. Tritapepe, Modelling inter-urban transport flows in Italy: a comparison between neural network analysis and logit analysis. Transp. Res. Part C: Emerg. Technol. **4**(6), 323–338 (1996)

25. Dongjoo Park, Seungjae Lee, Chansung Kim, Changho Choi, Chungwon Lee, Access mode choice behaviors of water transportation: a case of Bangkok. J.Adv. Transp. **44**(1), 19–33 (2010)

26. I. Rasmidatta, Mode Choice Models for Long Distance Travel in USA, PhD Thesis, University of Texas at Arlington, USA, 2006

27. A. Reggiani, T. Tritapepe, Neural Networks and Logit Models Applied to Commuters Mobility in the Metropolitan Area of Milan, in *Neural Networks in Transport Systems*, ed. by V. Himanen, P. Nijkamp, A. Reggiani (Ashgate, Aldershot, 2000), pp. 111–129

28. D. Shmueli, I. Salomon, D. Shefer, Neural network analysis of travel behavior: evaluating tools for prediction. Transp. Res. Part C: Emerg. Technol. **4**(3), 151–166 (1996)

29. TfL (Transport for London) (2011) Travel in London. Supplementary Report: London Travel Demand. Survey (LTDS), Retrieved on Dec 7, 2013

30. X. Wu, V. Kumar, J.R. Quinlan, J. Ghosh, Q. Yang, H. Motoda, G.J. McLachlan, A. Ng, B. Liu, P.S. Yu, Z. Zhou, M. Steinbach, D.J. Hand, D. Steinberg, Top 10 algorithms in data mining. Knowl. Inf. Syst. **14**(1), 1–37 (2008)
31. C. Xie, J. Lu, E. Parkany, Work travel mode choice modeling using data mining: decision trees and neural networks. Transp. Res. Rec. J. Transp. Res. Board (1854), 50–61
32. G. Yaldi, M.A.P. Taylor, W. Yue, Examining the possibility of fuzzy set theory application in travel demand modelling. J. East. Asia Soc. Transp. Stud. **8**, 579–592 (2010)
33. L.A. Zadeh, Fuzzy Sets. Inf. Control **8**, 338–353 (1965)

Artificial Neural Network (ANN) Pricing Model for Natural Rubber Products Based on Climate Dependencies

Reza Septiawan, Arief Rufiyanto, Sardjono Trihatmo, Budi Sulistya, Erik Madyo Putro and Subana Shanmuganathan

Abstract International Rubber Study Group report in [1] points out that the world natural rubber consumption continues to increase at an average of 9 per cent per year. Especially, the demands of natural rubber tire industry in developed countries such as the USA, Germany, China and Japan have increased steadily. Tropical countries, such as Indonesia, Malaysia, Thailand and Vietnam, members of the Association of Natural Rubber Producing Countries (ANRPC) accounted for about 92 per cent of the global production of natural rubber in 2010. The market price of natural rubber fluctuates reflecting the variations in supply capacity of these production countries. Therefore, knowledge on the natural rubber supply from these countries is significant in order to have an accurate pricing model of natural rubbers. Moreover, the supply of natural rubber is determined by the climatic conditions in these countries. Rubber trees grow and produce best in warm with an ideal temperature between 21–35 °C, an annual rainfall of 200-300 cm and moistly conditions. In this context, the chapter looks at the dependencies of natural rubber market price especially, the climatic conditions in the production countries, and derives at a natural rubber pricing model to provide farmer information regarding the prediction of market price using an artificial neural network (ANN) based prediction approach.

Keywords Climate monitoring · Pricing model · Natural rubber · Artificial neural network (ANN)

R. Septiawan (✉) · A. Rufiyanto · S. Trihatmo · B. Sulistya · E.M. Putro
Center of Information Communication Technology, Indonesian Agency for the Assessment
Application of Technology (BPPT), Jakarta, Indonesia
e-mail: reza.septiawan@bppt.go.id

S. Shanmuganathan (✉)
Auckland University of Technology (AUT), Auckland, New Zealand
e-mail: subana.shanmuganathan@aut.ac.nz

© Springer International Publishing Switzerland 2016
S. Shanmuganathan and S. Samarasinghe (eds.), *Artificial Neural
Network Modelling*, Studies in Computational Intelligence 628,
DOI 10.1007/978-3-319-28495-8_20

1 Introduction

Based on data relating to natural rubber production and consumption in 2013 [2], there were six countries in the world that produced about 92 % of the global natural rubber requirement. In some of these natural rubber producing countries natural rubber production is less than they really need. These countries still need to import natural rubber from other producing countries. China produced 856,000 tons of natural rubber in 2013, but China consumed around 4,150,000 tons natural rubber (Table 1).

Rubber trees grow and produce the best in warm and moistly climate, i.e., temperatures between 21–35 °C with an annual rainfall of 200–300 cm. However, the production capacity of natural rubber fluctuates with the rubber seed type, age of the rubber tree and the rubber processing method used.

Furthermore, of the world's natural rubber producing countries, Thailand has a production capacity of 34 %, Indonesia 30 % and Malaysia 12 %. Nevertheless, the biggest harvesting area for national rubber is in Indonesia, approximately 3.4 million Ha, but the natural rubber (NR) production of Indonesia is approximately lower than that of Thailand. In Thailand the production capacity is 1.5 times higher than that of Indonesia. The productivity of rubber trees in Indonesia is approximately 800 kg/Ha while in Thailand the productivity is approximately 1.7 ton/Ha. Two of the main causes for the low productivity of natural rubber in Indonesia is the type of rubber seed and the most natural rubber trees in Indonesia are already old. Replantation of natural rubber trees is necessary to boost the productivity in Indonesia and other countries. China initiated replantation of natural rubber a few years ago, in an attempt to meet its domestic supply.

Another aspect for the natural rubber price fluctuation is the global price of natural rubber, which mainly follows from the supply demand condition in the global market. The demand side fluctuates with the global economic conditions, such as decreasing demand from the automotive industry due to economic crises. The supply side is derived from the production parameter of natural rubber, such as the climate change (temperature, precipitation), seed type, and age of rubber trees.

Table 1 Production and consumption of natural rubbers in NR-production countries [2]

No	Country	Production (1000 to ns)	Consumption (1000 to ns)	Surplus (1000 to ns)
1	Thailand	4014	519	3495
2	Indonesia	3180	603	2477
3	Vietnam	950	154	796
4	China	856	4150	-3294
5	Malaysia	820	456	364

In this paper, the initial results of a study performed to investigate the relation-ships between natural rubber price and its production environments are elaborated. The parameters used to reflect the natural environment are: harvesting areas, production capacity, temperature, precipitation, vapour and cloud. Climate data is derived from [3], while production data is derived from [4].

2 Methodology

The main objective of this activity is to investigate the potential relationships between environmental and climate parameter data with the production capacity of NR-plantation, and the potential relationships between the production capacity of NR-plantation and the market price of NR products. The indirect relationships between environment and climate data and the market price of the NR products will be derived thereafter.

A prediction system that can provide early information to farmers about the possibility of market price fluctuations based on current or historical environmental data is developed. Initially, using statistical regression techniques appropriate dependent variables are selected. Then an ANN model in WEKA for predicting possible price changes is investigated. WEKA stands for Waikato Environment for Knowledge Analysis [5], which is open source software issued under GNU License. Weka is a collection of machine learning algorithms developed for data mining tasks. WEKA contains tools for ANN prediction and data visualization as well.

Primary and secondary data regarding the climate parameters is necessary in order to develop this Pricing Model related to Natural Rubber products. The parameters needed for this model development are:

1. Data concerning primary and secondary dependent variable parameters of 5 NR-production countries, including temperature, precipitation, vapour, and cloud coverage.
2. Data concerning primary and secondary parameters of production capacity of five NR-production countries, including production capacity, harvesting area, yield and market price of NR-products.

A conceptual system of a prediction/iterative model to analyze the effects of climate change on the productivity of NR- products in Thailand, Indonesia, Malaysia, Vietnam, India and China is derived from climate and productivity data. The climate data is obtained from Tylle [3] and NOAA [6] and the productivity data is obtained from FAO statistics and UNCTAD data [4, 7–10]. The prediction model is based on macroclimate i.e., regional scale historical data spanning over a year.

3 Natural Rubber (NR) and Synthetic Rubber (SR)

Rubbers are categorized into two types, namely natural rubbers (NRs) and synthetic rubbers (SRs). Natural rubber is a naturally occurring substance obtained from the exudations of certain tropical plants while synthetic rubber is artificially synthesized from petrochemical products. Synthetic rubber (SR) attempts to take the place of natural rubber (NR) since rubber is a significant commodity in our life. NR is recognized to be more elastic, easy to process, can be formed with low heat and has a high cracking resistance. Rubber is a polymer of isoprene units, which at normal temperatures irregular in shape (called latex). With decrease in temperature the rubber will inflate and restore its shape (elastic). Therefore, during the production process, some anticoagulant (a solution of soda-Na2CO3), ammonia (NH3) and sodium-sulfite (Na2SO3) are used to keep NR as a liquid before it is coagulated using formic acid in order to form rubber sheets.

The types of natural rubber are [7] (Fig. 1):

1. Latex concentrates: made from freshly tapped field latex and uncoagulated. This is the material obtained from the rubber tree *Hevea brasiliensis*
2. Dry/solid rubber: made from coagulated field latex rubber (rubber sheets, white and pale crepes) or made from remilled rubber sheets. Sometimes the dry rubber is visually graded in sheets or crepes and classified by instruments (block rubber)

Natural rubber tapped from rubber trees will coagulate within a few hours after tapping due to field coagulum agents. Farmers put additional preservatives in latex such as ammonia to prevent the coagulation. Natural rubber has become an important industrial polymer due to its advantages because of the following:

1. Physical resistance: excellent resilience, tear strength, abrasion resistance, impact strength, cut growth resistance.

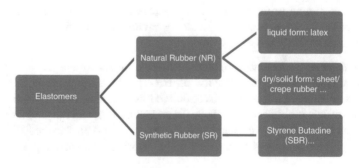

Fig. 1 Rubber product types: Synthetic Rubber and Natural Rubber [7]

Fig. 2 Natural Rubber plantation in Indonesia

2. Environmental resistance: excellent water resistance, low temperature flexibility, resistance to alcohols and acids

The limitations of natural rubber are poor ozone, sunlight, gasoline, and hydrocarbon solvent resistance (Fig. 2).

Rubber trees grow and produce best in warm and moist climate between 21–35 °C with an annual rainfall of 200–300 cm. Rubber trees can grow up to over 40 meters in the wild. However, when the tree is under cultivation it is not allowed to grow higher than 25 m. A Rubber tree can last for over 100 years by nature. However, it is usually replanted after 25–35 years, when latex yield decreases. The production capacity of natural rubber fluctuates with the rubber seed type, age of the rubber trees and the rubber processing. On the whole, in this paper the production capacity relationship to the environment and climate data is investigated and thereafter a pricing model related to the production capacity is derived.

Electromagnetic Chamber (EMC) in BPPT is used to investigate the difference in microwave absorption capability between NR and SR samples from Center of Material Technology in BPPT. The measurement testbed is given in Fig. 3. The absorption capability of NR and SR is measured in the frequency range of 1–3 GHz generated by a signal generator and transmitted through a directional

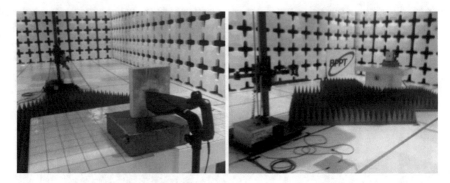

Fig. 3 Testbed of microwave absorption capability of NR and SR in EMC lab of BPPT

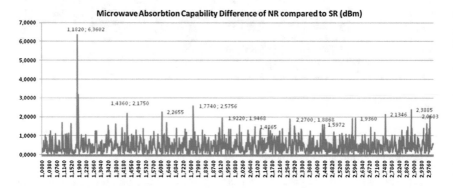

Fig. 4 Microwave absorption capability of NR compared to SR between 1 GHz and 3 GHz

antenna as in Fig. 3. The received signal is measured by an Electromagnetic Interference (EMI) receiver.

In the frequency range between 1–3 GHz, the absorption capability of NR and SR samples is shown in Fig. 4. The absorption capability of NR is higher than the SR sample especially in frequencies: 1.1820 GHz.

The climate data used in this paper is from [3] and earth monitoring station of NOAA in Indonesia [6]. The parameters used are temperature, precipitation, vaporation, and cloud cover. The temperature values are annual means related to the difference between the minimum and maximum temperatures. The cloud cover is given in the percentage of annual cloud cover. The precipitation is given in the total precipitation in milimetres, while the vapour pressure is given in hecta-Pascals.

The temperature of NR-producing countries are relatively warm and moistly with climate in the range of 21–35 °C, except China which has temperatures lower than the ideal climate to grow rubber tree around 8° Celsius (see Fig. 5).

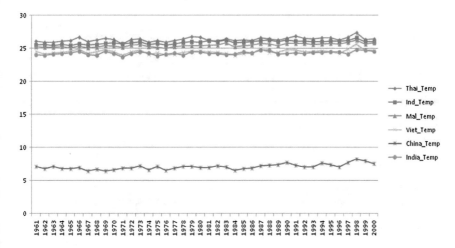

Fig. 5 Annual temperatures of NR-production countries

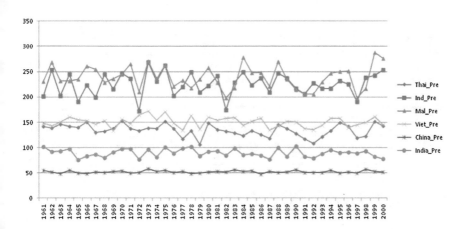

Fig. 6 Annual precipitation of NR-production countries

The precipitation of NR-producing countries are relatively below, the ideal values of precipitation are between 200–300 cm except for Indonesia and Malaysia, which have a precipitation value in the range of ideal values of precipitation for rubber trees. Thailand, Vietnam, India and China have precipitation values lower than the ideal value (see Fig. 6).

In addition, the cloud coverage parameter is considered. Most NR production countries ha values in between 50 and 70 % cloud coverage. Except India which has the lowest cloud coverage (40 %). Malaysia has the highest cloud coverage around 70 % (see Fig. 7).

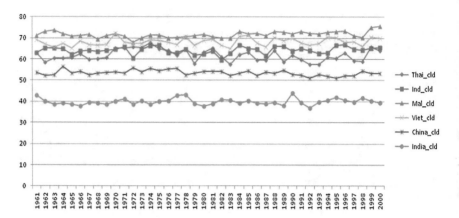

Fig. 7 Annual cloud coverage of NR-production countries

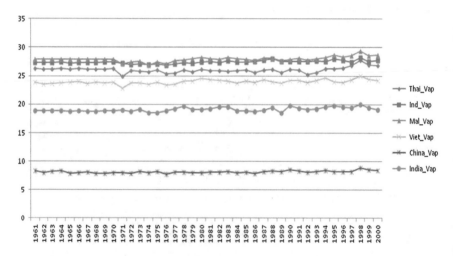

Fig. 8 Annual vapour values of NR-production countries

The vapour level of most NR-production countries is between 24 to 27 Hpa, only India and China have vapouration values below 20 Hpa. India has vapour values around 20 Hpa and China has vapour values below 10 Hpa (Fig. 8).

The above climate data is used as dynamic values for the production and pricing model.

4   Production Data and Price

During the last decade there has been a change in the estate involvement in Natural Rubber (NR) industry. In order to assure natural rubber supply, there has been an increase in smallholder farmer (who owns less than 2–3 ha rubber trees area) involvement in the NR industry. The smallholding NR production of the world increased by an annual rate of approximately 3 % from 5270 thousands of tons in 1998 and 6310 thousands of tons in 2003 [11].

There has been a change of productivity during the last decade. For instance, in the 1990s the Indonesian productivity was 680 kg/Ha for smallholdings, the Thailand Productivity was between 500–800 kg/Ha for smallholdings and 900 kg/Ha in Estates. In 2013, the productivity of rubber trees in Indonesia was approximately 800 kg/Ha. In contrary, Thailand has increased its productivity to approximately 1.7 tons/Ha in 2013 [11] (Table 2).

Indonesia has a NR harvested area approximately 3.5 million Ha, of which 85 % is smallholdings and the majority of the trees are older than 30 years. Therefore, the NR productivity of Indonesia is lower than that of Thailand. The global price of NR was USD 4.6/kg in 2011 and it dropped to USD 1.5/kg in 2015. The NR price keeps decreasing because of higher NR supply in the global market. When the NR price was high in 2011, most of the International Rubber Council members increased their NR production. Beginning in 2013 the NR market experienced an oversupply hence, the price began to decrease.

In the Fig. 9 the production capacity of NR-countries is given. The figure shows a high increase of production in Thailand, while a moderate increase of production in other countries such as Indonesia, China, India and Vietnam. Malaysian production capacity decreased until the end of 1999 but then began to increase by year 2000.

If we consider the harvesting area of each country as given in Fig. 10, the harvesting area is getting higher in Indonesia and Thailand, but in the other

Table 2 NR Costs structure (in 1990s, cents/kg) [11]		Indonesia	Thailand
		Smallholdings	Smallholdings
	Yields (kg/ha)	680	800
	Direct costs		
	Management and labour:	44.9	18.5
	Materials:		
	Tapping and collection	3.7	2.5
	Fertilizers	-	
	Weedicides	6.3	6.5
	Pest control	-	
	Other	1.8	3.3
	Transportation:	2.0	2.3
	Capital costs		
	Planting investments	14.6	20.6

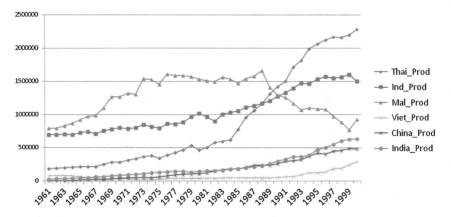

Fig. 9 Productivity of NR in 6 countries

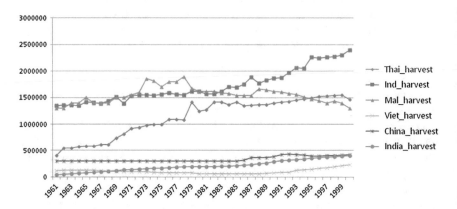

Fig. 10 Harvested area in NR-production countries

countries harvesting areas remain constant or even getting lesser as in Malaysia. The harvesting area in Malaysia is converted into some palm plantation.

Polynomial Regression

In order to investigate the trend of production capacity, harvest area and yield (kg/Ha) of each NR-production country, the following polynomial regression model is used:

$$y = a_0 + a_1 x_i + a_2 x_i^3 + \cdots + a_m x_i^m + \varepsilon_i;$$
$$i = 1, 2, \ldots, n$$

With ε_i is the remaining error, and in this paper the coefficient of determination (R^2 of R *squared*) is used to measure how well the data fits the polynomial model in order to predict the future outcomes and for testing hypotheses. It provides a

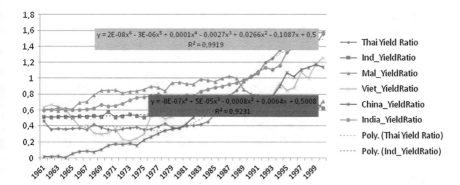

Fig. 11 Yield Ratio in NR-production countries

standard of measurement on how well the observed outcomes are replicated by the model especially, the proportion of total variation within the outcomes explained by the model. The coefficient of determination ranges from 0 to 1. The better the regression fits the data in comparison, the closer the value of R^2 to 1. In addition, a correlation value is used to tell how different parameters are related to each other, with values between -1 and 1, where 0 is no relation, and 1 is a very strong relationship. Square roots are used to have more influence of the extreme values on the results.

Figure 11 shows the trend of yield ratio (kg/ha) coverage in NR-production countries.

The trend of Thailand yield ratio is:

$$y = 2E - 08x^6 - 3E - 06x^5 + 0,0001x^4 - 0,0027x^3 + 0,0266x^2 - 0,1087x + 0,5$$

with the R square value of $R^2 = 0,9919$.

The trend of Indonesia yield ratio is:

$$y = -8E - 07x^4 + 5E - 05x^3 - 0,0008x^2 + 0,0064x + 0,5008$$

with the R square value of $R^2 = 0,9231$

In addition, the relationship between the environmental, climate data and the productivity of natural rubber based on annually production data in Thailand and Indonesia during year 1961–2000 is given in Table 3. The R square is better if it is close to value of 1, while the value of significance_F is better if it is less than 0.05 (similar to value of P).

The environment and climate data is more significant in Indonesia for the production capacity of natural rubber (significance F is 1.69E-09) when compared with that of Thailand (significance F is 0.004442). The Rsquare values is 0.72854 for

Table 3 Productivity varies
by Environment Climate Data

	Thailand	Indonesia
R Square	0.343487	0.72854
Observations	40	40
significance_F	**0.004442**	**1.69E-09**
P_value_Temp	0.126025	**7.39E-06**
P_Value_Pre	0.611101	0.034334
P_Value_cld	0.682699	0.05754
P_Value_Vap	**0.089082**	0.202731

Table 4 Price varies by
environment climate data

	Thailand	Indonesia
R Square	0.569235	0.754618
Observations	10	10
significance_F	0.295101	**0.086096**
P_value_Temp	0.929699	0.226642
P_Value_Pre	**0.063328**	**0.097048**
P_Value_cld	0.215085	0.281241
P_Value_Vap	0.7577	**0.033284**

Indonesia in which the temperature parameter is seen to be the most significant climate parameter related to the production capacity of natural rubber followed by the precipitation level of rainfall. While in Thailand the vapour level is seen to be the most significant parameter.

Furthermore, the prices of natural rubber in Thailand and Indonesia are observed in relation to the environment and climate data (Table 4). The price of NR in Thailand varies less with the changes in the independent data (significance F-0.295101) except for the precipitation (p-value 0.063328), while the price of NR in Indonesia varies more with the changes in the independent data (significance F 0.086096) especially the vapour (p-value 0.033284) and precipitation (p-value is 0.097408).

Furthermore, the relationship of the price and the productivity of the six Natural Rubber production countries is given in Table 5.

Table 5 shows that natural rubber price in Thailand and Indonesia relies significantly on the other NR-production countries (significance F value is 4,32853E-07 and 4.30769E-09 respectively). Interestingly Indonesian NR-price relates more to the Malaysian NR-production capacity, while Thailand NR-price relates more to Indonesian and China NR-production capacities.

The regression line for Thailand NR-price in relation to the Production capacities of other NR-countries is given as follow.

$$Pr_{Thai} = 0.00017 * Prod_{Thai} + 0.00030 * Prod_{Ind} + 0.00037 * Prod_{Mal}$$
$$+ 0.00034 * Prod_{Viet} + 0.00059 * Prod_{CHina}$$

Table 5 Price Prediction varies by Production Capacity of NR-production countries

	Indonesia	Thailand
R Square	**0.937876227**	0.888469505
Observations	22	22
significance_F	**4.30769E-09**	**4.32853E-07**
Coefficient_Thai_Prod	−0.000172368	−0.00165138
Coefficient_Ind_Prod	0.000305791	0.002254715
Coefficient_Mal_Prod	0.000374071	0.00025635
Coefficient_Viet_Prod	0.000343564	−0.001119253
Coefficient_China_Prod	0.000593683	0.006153726
P_Value_Thai_Prod	0.104570761	**0.005929221**
P_Value_Ind_Prod	**0.062850698**	**0.011934348**
P_Value_Mal_Prod	**0.082306757**	0.81007049
P_Value_Viet_Prod	0.426965085	0.616490397
P_Value_China_Prod	0.231043334	**0.024535483**

Based on this analysis the pricing model derived using statistical regression and ANN (in WEKA) show promise. Thailand's NR-production capacity is related more to the vapourisation values of the climate data in Thailand. Furthermore Thailand's NR-price is related more to the precipitation values within Thailand climate data and production capacity of Indonesia and China. In ANN results, the NR-price in Thailand is derived from these parameters using Multilayer perceptron. The results is given in Appendix A.

The correlation coefficient is about 99.93 % with the Root mean squared error of 9.542 and mean absolute error of 6.1406, with the best selected attributes being precipitation and vapourisation within Thailand climate data (Fig. 12).

Furthermore, additional parameters were considered to improve the performance of prediction of Thailand NR-price. The results are given in Appendix B. The use of more parameters in prediction of Thailand NR-price provides a better result especially, the improvement is by 10^3 in error values (see Table 6).

Furthermore, Indonesia's NR-production capacity is more related to the temperature within the country's climate data. Indonesia's NR-price is more related to the precipitation and vapourisation within the country's climate data and the production capacity of Malaysia. Form the ANN results (WEKA), the NR-price in Indonesia was derived from these parameters using a Multilayer perceptron. The results are given in Appendix C (Fig. 13).

Furthermore, additional parameters were considered to improve the performance in the prediction of Indonesian NR-price. The results are given in Appendix D.

The use of more parameters in the prediction of Indonesian NR-price provides a better result especially the improvement is by 10^3 in error values (Table 7).

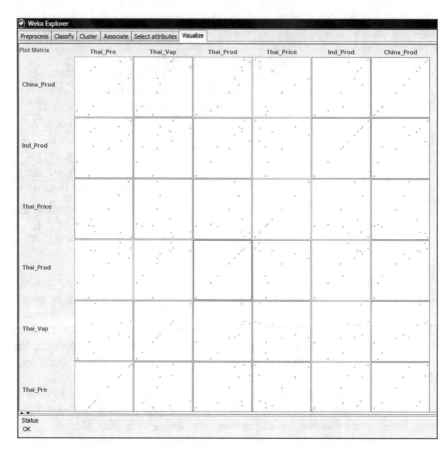

Fig. 12 Relationship of Thailand NR-price, Thailand climate data and NR-production countries

Table 6 Coefficient of significance using 6 and 11 parameters of environmental climate data and productivity data related to Thailand NR-price

Thailand NR-Price	6 parameters	11 parameters
Correlation coefficient	0.99930	1.00000
Mean absolute error	6.14060	0.00720
Root mean squared error	9.54200	0.01020
Relative absolute error	3.0767 %	0.0036 %
Root relative squared error	3.8935 %	0.0042 %

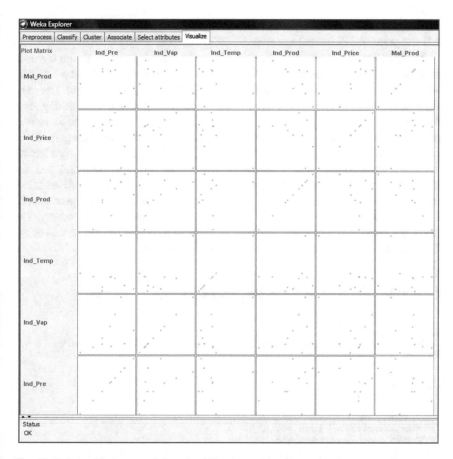

Fig. 13 Relationship between Indonesian NR-price and its climate data in

Tabel 7 Coefficient of significance using 6 and 11 parameters of environmental climate data and productivity data related to Indonesian NR-price	Indonesian NR-price	6 parameters	11 parameters
	Correlation coefficient	0.99700	1.00000
	Mean absolute error	3.67650	0.07420
	Root mean squared error	5.90550	0.08350
	Relative absolute error	7.6992 %	0.1554 %
	Root relative squared error	9.2210 %	0.1303 %

5   Conclusion

It is shown that the price of NR in a NR-producing country is related to its own productivity and other countries productivities as well. The relationships are specified in the corresponding coefficients in the respective regression equations and the corresponding weights of the ANN prediction model. The higher the coefficients the more the significance in the influence of the country's productivity and the price of NR in that country.

Furthermore, the production capacity of each country is related to the environment and climate in that country. The relationships between production capacity and environment as well as climate vary among countries. In Thailand, the most significant independent parameters related to the production capacity are; vapourisation and precipitation. Meanwhile, the NR-price in Thailand is mostly related to the production capacity of Indonesia and China. The pricing model is useful when identifying the potential competitors, for example, NR production capacities in Indonesia and China influence the NR-price in Thailand and the influence is more than that of NR-production capacity in Malaysia. In addition, information regarding the most significant climate variable in the NR-productivity in a country indicates the robustness of NR-trees and plantation in that country. The lower the significance exerted by a climate variable the more the robustness of the NR-tree plantation to the changes in the climate variable.

Appendix A: WEKA Run Data and Result

```
=== Run information ===
Scheme:weka.classifiers.functions.MultilayerPerceptron -L 0.3 -M 0.2 -N 500 -V 0 -S 0 -E 20 -H a
Attributes:   6
            Thai_Pre
            Thai_Vap
            Thai_Prod
            Thai_Price
            Ind_Prod
            China_Prod
Test mode:evaluate on training data
=== Classifier model (full training set) ===
Linear Node 0
    Inputs    Weights
    Threshold   1.749902301630781
    Node 1    -2.4013997901314834
    Node 2    -3.5162050764531148
    Node 3    0.06081154842183814
Sigmoid Node 1
    Inputs    Weights
    Threshold   0.5236236579674346
    Attrib Thai_Pre    -1.8415687438970416
    Attrib Thai_Vap    -1.387438278729828
    Attrib Thai_Prod   -0.3276634761924621
    Attrib Ind_Prod    0.6024382775884405
    Attrib China_Prod  -2.4850489513900356
Sigmoid Node 2
    Inputs    Weights
    Threshold   -3.9635874842464256
    Attrib Thai_Pre    0.3167431758460005
    Attrib Thai_Vap    1.3769004172430026
    Attrib Thai_Prod   1.461297417788155
    Attrib Ind_Prod    0.09336331236180025
    Attrib China_Prod  3.09311785098321
Sigmoid Node 3
    Inputs    Weights
    Threshold   -0.8457331231854387
    Attrib Thai_Pre    -0.8010123058348094
    Attrib Thai_Vap    -0.19957764103345418
    Attrib Thai_Prod   -0.5981736564609482
    Attrib Ind_Prod    -0.5373350160373417
    Attrib China_Prod  -0.18433659889855544
```

Appendix B: WEKA Run Data and Result

```
=== Run information ===
Scheme:weka.classifiers.functions.MultilayerPerceptron -L 0.3 -M 0.2 -N 500 -V 0 -S 0 -E 20 -H a
Attributes:  11
            Thai_Pre
            Thai_Vap
            Thai_Prod
            Thai_Price
            Ind_Prod
            China_Prod
            Thai_Temp
            Thai_cld
            Mal_Prod
            Viet_Prod
            India_Prod
Test mode:evaluate on training data
=== Classifier model (full training set) ===
Linear Node 0
   Inputs    Weights
   Threshold    1.9138957757497645
   Node 1    -1.6452064793932293
   Node 2    -0.9392051406150806
   Node 3    -1.693330715713067
   Node 4    -0.2924948895574329
   Node 5    -1.2750101529740436
Sigmoid Node 1
   Inputs    Weights
   Threshold    0.10261401652715471
   Attrib Thai_Pre    -0.3706483151649728
   Attrib Thai_Vap    0.3079386208779187
   Attrib Thai_Prod    -0.19909232245824043
   Attrib Ind_Prod    -0.10101365603842809
   Attrib China_Prod    -0.7501478450595847
   Attrib Thai_Temp    -0.022826674663922665
   Attrib Thai_cld    0.41677330152507075
   Attrib Mal_Prod    -1.2185659909573794
   Attrib Viet_Prod    1.1115351280851677
   Attrib India_Prod    0.1013896275120634
Sigmoid Node 2
   Inputs    Weights
   Threshold    -0.27725269461675683
   Attrib Thai_Pre    -0.5219136736012591
   Attrib Thai_Vap    0.11568300116923473
   Attrib Thai_Prod    -0.34502069652616585
   Attrib Ind_Prod    -0.1660373873786111
   Attrib China_Prod    -0.7089527729826143
   Attrib Thai_Temp    0.2563604970526755
   Attrib Thai_cld    -0.044679662376493186
   Attrib Mal_Prod    -0.5905218969564233
   Attrib Viet_Prod    0.5167595904466141
   Attrib India_Prod    -0.21709496110404175
Sigmoid Node 3
   Inputs    Weights
   Threshold    0.18994896844661185
   Attrib Thai_Pre    -0.36152949612739
   Attrib Thai_Vap    0.23915007114901676
   Attrib Thai_Prod    -0.2041082472011238
   Attrib Ind_Prod    -0.01317691718789573
   Attrib China_Prod    -0.8876609081714663
   Attrib Thai_Temp    -0.0521448217271642
   Attrib Thai_cld    0.2717446217766461
   Attrib Mal_Prod    -1.216021421164287
   Attrib Viet_Prod    1.170050420024951
   Attrib India_Prod    0.03616020356150899
Sigmoid Node 4
   Inputs    Weights
   Threshold    -0.6186530548104157
   Attrib Thai_Pre    -0.342263287342173
   Attrib Thai_Vap    0.17778395806416508
   Attrib Thai_Prod    -0.185248136427864
   Attrib Ind_Prod    -0.206559041979244
   Attrib China_Prod    -0.2197511842802674
```

```
   Attrib Thai_Temp   0.3818885339740087
   Attrib Thai_cld   0.12924991668705613
   Attrib Mal_Prod   -0.1266407071166203
   Attrib Viet_Prod   0.3356237616116768
   Attrib India_Prod   0.01109119333972902
Sigmoid Node 5
   Inputs   Weights
   Threshold   -0.12921596947624328
   Attrib Thai_Pre   -0.7577985615146368
   Attrib Thai_Vap   -0.06868575732703393
   Attrib Thai_Prod   -0.6182231637754222
   Attrib Ind_Prod   -0.251095203414586
   Attrib China_Prod   -1.0610622695111707
   Attrib Thai_Temp   -0.018400445427221933
   Attrib Thai_cld   -0.09345302657090276
   Attrib Mal_Prod   -0.5617194399754916
   Attrib Viet_Prod   0.3449369784930535
       Attrib India_Prod   -0.4754360443333911
```

Appendix C: WEKA Run Data and Result

```
=== Run information ===
Scheme:weka.classifiers.functions.MultilayerPerceptron -L 0.3 -M 0.2 -N 500 -V 0 -S 0 -E 20 -H a
Attributes:   6
         Ind_Pre
         Ind_Vap
         Ind_Temp
         Ind_Prod
         Ind_Price
         Mal_Prod
=== Classifier model (full training set) ===
Linear Node 0
   Inputs   Weights
   Threshold   -2.02235331049152
   Node 1   2.4809061645508796
   Node 2   -0.13654828302806996
   Node 3   1.8841614129059234
Sigmoid Node 1
   Inputs   Weights
   Threshold   0.34106473198010806
   Attrib Ind_Pre   -0.18777061222657074
   Attrib Ind_Vap   -0.6047803356302281
   Attrib Ind_Temp   -1.3125870419669445
   Attrib Ind_Prod   1.3976929605351807
   Attrib Mal_Prod   2.1265815725425137
Sigmoid Node 2
   Inputs   Weights
   Threshold   -0.689717384941266
   Attrib Ind_Pre   -0.20693673657319617
   Attrib Ind_Vap   0.18349079473081775
   Attrib Ind_Temp   0.07607980536693383
   Attrib Ind_Prod   -0.18131622948707016
   Attrib Mal_Prod   0.24775362765881342
Sigmoid Node 3
   Inputs   Weights
   Threshold   -1.6721322027039116
   Attrib Ind_Pre   1.3272990265105595
   Attrib Ind_Vap   1.661845752213534
   Attrib Ind_Temp   -0.010500786990556022
   Attrib Ind_Prod   -0.30691946961240074
   Attrib Mal_Prod   1.8278096406392794
```

Appendix D: WEKA Run Data and Result

=== Run information ===
Scheme:weka.classifiers.functions.MultilayerPerceptron -L 0.3 -M 0.2 -N 500 -V 0 -S 0 -E 20 -H a
Attributes: 11
       Ind_Pre
       Ind_Vap
       Ind_Temp
       Ind_Prod
       Ind_Price
       Thai_Prod
       Ind_cld
       Mal_Prod
       Viet_Prod
       China_Prod
       India_Prod
=== Classifier model (full training set) ===
Linear Node 0
    Inputs   Weights
    Threshold   -0.6513225543454542
    Node 1   1.1341620537033892
    Node 2   0.7100565252747373
    Node 3   -1.8428244625217802
    Node 4   0.15735546322323843
    Node 5   0.8449830639796779
Sigmoid Node 1
    Inputs   Weights
    Threshold   -0.16855868151638012
    Attrib Ind_Pre   0.23486987830478206
    Attrib Ind_Vap   0.0076257946352456234
    Attrib Ind_Temp   -0.37150298477277954
    Attrib Ind_Prod   -0.16708931970590685
    Attrib Thai_Prod   0.5223904427589126
    Attrib Ind_cld   0.31660664077245154
    Attrib Mal_Prod   0.381185526330691
    Attrib Viet_Prod   -0.7246949102508935
    Attrib China_Prod   0.5263371860286197
    Attrib India_Prod   -0.21326265437105965
Sigmoid Node 2
    Inputs   Weights
    Threshold   -0.4732383000987289
    Attrib Ind_Pre   0.3060357521447042
    Attrib Ind_Vap   0.3213220198702666
    Attrib Ind_Temp   -0.1241555049673523
    Attrib Ind_Prod   -0.1863697704782177
    Attrib Thai_Prod   0.22790124048102864
    Attrib Ind_cld   0.27337097072653005
    Attrib Mal_Prod   0.161581936977685
    Attrib Viet_Prod   -0.340007212517958
    Attrib China_Prod   0.4092507836225177
    Attrib India_Prod   -0.12241711652997499
Sigmoid Node 3
    Inputs   Weights
    Threshold   -0.8360763192281531
    Attrib Ind_Pre   0.4278974745390614
    Attrib Ind_Vap   0.4788776090293571
    Attrib Ind_Temp   1.3899457822502905
    Attrib Ind_Prod   0.30332159568932826
    Attrib Thai_Prod   -0.5224153744397598
    Attrib Ind_cld   0.027180794016561517
    Attrib Mal_Prod   -0.8766581275217855
    Attrib Viet_Prod   0.9071280468671425
    Attrib China_Prod   -0.2934209766250628
    Attrib India_Prod   0.6792461840043068
Sigmoid Node 4
    Inputs   Weights
    Threshold   -0.4761430261598761
    Attrib Ind_Pre   -0.0044173148824384195
    Attrib Ind_Vap   0.1402459988780109
    Attrib Ind_Temp   -0.09251126926640348
    Attrib Ind_Prod   -0.09309501276609462
    Attrib Thai_Prod   0.06156634301797498
    Attrib Ind_cld   0.20517104957525495

```
Attrib Mal_Prod   0.07936814492592777
Attrib Viet_Prod   -0.001560017990104226
Attrib China_Prod   0.1639482981113313
Attrib India_Prod   -0.10321447156373058
Sigmoid Node 5
  Inputs   Weights
  Threshold   -0.595529466656197
  Attrib Ind_Pre   0.324461687674693
  Attrib Ind_Vap   0.5106428966777481
  Attrib Ind_Temp   -0.16092521814003055
  Attrib Ind_Prod   -0.2738550140163375
  Attrib Thai_Prod   0.19813519850530337
  Attrib Ind_cld   0.26012859991215737
  Attrib Mal_Prod   0.08228000798350896
  Attrib Viet_Prod   -0.2944262794736895
  Attrib China_Prod   0.5231689224559062
  Attrib India_Prod   -0.18499440086063007
```

References

1. http://www.rubberstudy.com/
2. Global and China Natural Rubber Industry Report, 2013–2015 by ResearchInChina
3. http://www.cru.uea.ac.uk/~timm/cty/obs/TYN_CY_1_1.html
4. http://faostat.fao.org/site/567/default.aspx#ancor
5. www.cs.waikato.ac.nz/ml/weka/
6. NOAA, www.noaa.gov
7. UNCTAD secretariat (adapted from H. Long, Engineering Properties of Elastomers, The Roofing Industry Educational Institute)
8. UNCTAD secretariat (Links: USDA, NRCS. 2005. The PLANTS Database, Version 3.5. Data compiled from various sources by Mark W. Skinner. National Plant Data Center, Baton Rouge, LA 70874-4490 USA)
9. UNCTAD secretariat (Data: International Rubber Study Group)
10. http://www.unctad.info/en/Infocomm/Agricultural_Products/Caoutchouc/Crop/Natural-rubber-production-
11. C. Barlow, S. Jaysuriya, C. Suan Tan, The World Rubber Industry (Routledge, London)
12. http://ptm.bppt.go.id

A Hybrid Artificial Neural Network (ANN) Approach to Spatial and Non-spatial Attribute Data Mining: A Case Study Experience

Subana Shanmuganathan

Abstract A hybrid artificial neural network (ANN) approach consisting of self-organising map (SOM) and machine learning techniques (top-down induction decision tree/TDIDT) to characterising land areas of interest is investigated using New Zealand's grape wine regions as a case study. The SOM technique is used for clustering map image pixels meanwhile, the TDIDT is used for extracting knowledge from SOM cluster membership. The contemporary methods used for such integrated analysis of both spatial and non-spatial data incorporated into a geographical information system (GIS), are summarised. Recent approaches to characterise wine regions (viticulture zoning) are based on either a single or composite (multi-attribute) index, formulated generally using digital data (vector and raster) representing the variability in environmental and viticulture related factors(wine label ratings and price range) over different spatial and temporal scales. Meanwhile, the world's current wine regions, already well-developed, were initially articulated based on either grapevine growth phenology (growing degree days/GDD, frost days, average/minimum temperature, berry ripening temperature range) or wine style/rating/taste attributes. For both approaches, comprehensive knowledge on local viticulture, land area, wine quality and taste attributes is a *sine qua non*. It makes the characterisation of *newworld* vineyards or new sites (potential vineyards), with insufficient knowledge on local viticulture/environment an impossible task. For such instances and in other similar not so well-known domains, the SOM-TDIDT approach provides a means to select ideal features (discerning attributes) for characterising, in this case, within New Zealand's wine regions or even within vineyards also scientifically validating the currently used factors regardless of present day scale and resolution related issues.

Keywords Self-organising map · Image data clustering · Unsupervised ANNs

S. Shanmuganathan (✉)
Auckland University of Technology (AUT), Auckland, New Zealand
e-mail: subana@gmail.com

© Springer International Publishing Switzerland 2016   443
S. Shanmuganathan and S. Samarasinghe (eds.), *Artificial Neural Network Modelling*, Studies in Computational Intelligence 628,
DOI 10.1007/978-3-319-28495-8_21

1   Introduction

A hybrid artificial neural network (ANN) approach to data mining spatial and non-spatial attributes using information from disparate digital data sources incorporated into a GIS for visualising spatial patterns, correlations and trends in the data is becoming widespread in many application domains. For example, historic census [1], healthcare [2, 3] and socio-economics [4] are some of the disciplines and the publications demonstrated how such an integrated analysis of spatial and non-spatial data had aided the researchers in understanding the influences of many complex issues relating to the chosen topic in a spatial context. The major functions that enhance such integrated analysis of multiple attribute data in a spatial context within a GIS environment, are summarised. Subsequently, contemporary methodologies used for characterising the wine regions of some *old* as well as *new world* wine producing countries are outlined. The methodology section describes the hybrid ANN approach, consisting of SOM clustering and machine learning i.e., top-down induction decision tree/TDIDT knowledge extraction methods, investigated in this research. The interim results of this work demonstrate how the SOM-TDIDT approach could be applied to analysing spatial and non-spatial attribute data (on land related factors) at different spatial scales. The approach is especially useful when analysing issues in less known problem domains i.e., zoning/characterising of some wider implications of certain observed/potential change, i.e., climate change, for decision-making purposes on potential future change scenarios without the need for high resolution satellite imagery as required in contemporary approaches.

1.1   Integrated Analysis of Spatial and Non-spatial Attributes

Both simple and complex spatial data analysis methods are efficient when there is sufficient knowledge in the problem domain as well as the options/solutions being considered. Such methods do not facilitate an explorative analysis of multi-sourced spatial data sets hence, new approaches are urgently required to analyse high volume data in less known problem domains i.e., land areas, to learn about their suitability for different purposes, such as, new vineyard site selection, resources management. The commonly used simple methods applied to integrated analysis of spatial attribute data can be grouped into four main categories and they are: (1) retrieval/classification/measurement, (2) overlay, (3) neighbourhood, and (4) connectivity of network functions [5, 6]. It then appears, contemporary GIS and their functions enable the integration, manipulation, visualisation and analysis of geo-coded data with ease. They enable analysts to pre-process digital map layers that consist of attribute data on various landscape features, observations and measurements. However, they do not enable the explorative analysis of spatial data at different resolutions and formats i.e., vector and raster.

1.2 Artificial Neural Networks (ANNs) and Their Applications

ANNs are biologically inspired approaches to incorporating heuristics into conventional algorithmic computing. The latter needs a clear understanding of the problem addressed and the solutions considered to resolving the problem. Hence, many of real world problems, such as image/character/pattern recognition, cannot be solved using algorithmic computing approaches. In an attempt to overcome the conventional computing issues, W S McCulloch and W Pitts first introduced a mathematical model of a neuron in 1943 [7]. They continued their work and explored network paradigms for pattern recognition despite rotation angle, translation, and scale factor issues. Most of their work involved simple neuron models and these neural network systems were generally referred to as perceptrons (for more details see Chap. 1).

ANNs are sometimes described as a collection of mathematical techniques that can be used for signal processing, forecasting, and clustering. Based on this explanation they are referred as non-linear, multi-layered, parallel regression techniques [8]. ANN modelling is like fitting a line, plane or hyper plane through a set of data points. A line, plane or hyper plane can be fitted through any data set to define relationships that may exist between (what the user chooses to be) the inputs and the outputs; or it can be fitted for identifying a representation of the data on a smaller scale. The human brain is much more complicated and many of its cognitive functions are still unknown and unlikely to be discovered in the near future. However, the following are the main characteristics considered as common functions in real and artificial networks: 1. Learning and adaptation, 2. Generalisation, 3. Massive parallelism, 4. Robustness, 5. Associative storage of information, and finally 6. Spatiotemporal information processing. ANNs are generally defined using four parameters and they are: (1) type of neuron, ANN architecture, learning, and recall algorithms and see Chap. 1 for more details on ANN components.

Intrigued by the potentials of ANNs, professionals from almost all fields are finding methods by way of creating new hybrid models of possible combinations of symbolic and subsymbolic paradigms (Artificial Intelligence, Expert Systems), many of them with Fuzzy techniques and conventional, i.e., statistical, to suit a variety of applications within their own disciplines.

1.3 Self-organising Maps (SOMs)

A SOM is a two-layered feed-forward artificial neural network. It uses an unsupervised learning algorithm to perform non-linear non-parametric regression. Using a process called self-organisation, the network configures itself in such a way that the output layer gradually evolves into a display of topology preserving representation of the input data. In this output display, similar data are clustered near each other (Fig. 3). The topology preserving mapping of the SOM algorithm is useful in projecting multi-dimensional data sets onto low, usually one- or two- dimensional planes. Thus, SOM displays can be used for clustering and visualisation of multi-dimensional data sets that are difficult to analyse using conventional i.e., statistical, methods [9, 10].

1.4   Clustering in Spatial Data Mining

Increasingly, new algorithms are being introduced for clustering spatial data to optimise the efficiency of the clustering process [11]. The problem domains recently being investigated in the spatial data analysis include; improving cluster quality in large volumes of high dimensional data sets [12], noise removal [13], uncertainty, data pre-processing and reduction of running time consumed [14].

1.5   Top-Down Induction Decision Tree (TDIDT)

TDIDT algorithm is considered to be a powerful tool for extracting data classifi-cation rules from decision trees since the mid-1960s. The TDIDT algorithm has formed the basis for many classification systems, such as, Iterative Dichotomiser 3 (ID3) and C4.5 (statistical classifier), the latter is an extension to the ID3. Decision rules can be transformed into a form of a decision tree using TDIDT methods by repeatedly splitting the values of attributes also referred to as recursive partitioning. The TDIDT algorithm (Fig. 1) is based on a set of instances used for training. Each instance can be described by the values of a set of categorical attributes relating to a member of a universe of objects. With that introduction to spatial data analysis and clustering, the next section looks at a few major approaches to characterising and zoning methods adopted in viticulture in the *old world* (France, Spain, Italy, Germany, Portugal, Austria, Greece, Lebanon, Israel, Croatia, Georgia, Romania,

---

IF all the instances in the training set belong to the same class
THEN return the value of the class
ELSE
(a) Select an attribute A to split on+
(b) Sort the instances in the training set into subsets, one
for each value of attribute A
(c) Return a tree with one branch for each non-empty subset,
each branch having a descendant subtree or a class
value produced by applying the algorithm recursively
+ Never select an attribute twice in the same branch

Fig. 1 The basic TDIDT Algorithm [15], p. 48

Hungary and Switzerland) and the more recent *new world* (U.S., New Zealand, Argentina, Chile, Australia and South Africa) wine producing countries.

2 Viticulture Characterising and Zoning

The empirical studies from literature on this very topic and related themes reveal that the majority of the traditional viticulture zoning systems still in use relate to a "*French notion*" of the 19th century's *terroir* concept, which is believed to have originated from Latin many centuries ago. The zoning systems of some generally referred to as *old world* countries, were initially introduced solely to regulate the wine industry especially, to protect winemaker livelihood in already then established wine regions. These old systems and their derivatives portray significantly less reference to geography. However, in the current context, many different environmental factors are used to zoning a wine region to manage a nation's vineyards, wineries and the production of different wine *styles* labelled based on their aroma, flavour, taste compounds and vintage (wine of a particular season's yield from a particular vineyard). Both vineyard zoning and wine labelling are useful when implementing regulatory measures over wine marketing strategies with *designated origins* controlled by the respective state institutions, as well as in some instances for irrigation and other natural resources management purposes [16].

2.1 Viticulture Zoning Approaches

The simplest viticulture zoning approach could be developed with a single factor/attribute based on its spatial distribution [17]. However, the various approaches in use vary based on how they aggregate/disaggregate spatial data to match the demarcating registration boundaries. In general, a demarcating approach involves the assigning of a boundary to a single pre-existing quantifiable variable with historically or customarily accepted spatial distribution. The spatial distribution data of a variable representing any one particular aspect such as, climate, grapevine or wine style/vintage quality, and the respective boundaries at which spatial distribution data of different variables are available. Hence, in more complex approaches with multiple factors, the task of assigning spatial distribution data to the zoning boundaries becomes confound. In the meantime, many describe the well-known French notion *terroir* as a commonsense approach. The *terroir* concept is explained as an attempt to convey semi-mystical group of forces of an area that produced the wine as *special* and linking it to that region and its local attributes, such as atmospheric, environmental and cultural moreover, the wine making practices [18].

Australia has a modern "three tier system" to define its wine regions. In this system, on the op is the "state", then mid-way through "zone" and finally at the bottom "region". In some cases, in addition to the three tiers there are sub-regions defined by the Wine and Brandy Corporation's Geographical Indications Committee (GIC) to further subdivide the bottom tier. For instance, Australia's Adelaide Hills wine region within the Southern part of the country's Lofty zone, has two sub regions, namely, Piccadilly Valley and Lenswood. Similarly, High Eden is defined as a sub region of the Barossa Valley. Interestingly, a new region only declared in 1997 consisting of roughly an area of 15 by 30 km stretch with 6,800 hectares of vineyards, got sub regions within it. Even though GIC has not officially declared any sub regions as such, there are six of them defined unofficially and they are; Seaview, Willunga Plains, Sellicks Foothills, McLaren Flat, McLaren, Vale/Tatachilla and finally Blewitt Springs. Winemakers use the sub regions to describe vineyard blocks based on the difference in a single factor and that is: the temperature experienced during grape berry ripening season. Some describe the approach as a marketing ploy to capitalise on the areas that have already established popularity. It enables the McLaren's less known producers to charge higher prices, which could in turn be reinvested to improving vineyards as the region has a higher price expectation. In fact, within this region, Tatachilla is stated to be a good area [19].

In [20] the authors elaborated on Cabernet Sauvignon phenology from observations of cropping and fruit characteristics at six vineyard sites in Hawke's Bay (New Zealand). The study was based on observations made over three seasons from which a numerical model was developed to characterise the environmental conditions of a vineyard site. The variables used in that model were; air temperature in October and January, seasonal rainfall, rooting depth, and gravel percentage as well as clay-to-silt ratio in topsoil. A "site index" or "SI" calculated using the variables stated above was described to be significantly correlated with soil temperature and volumetric soil moisture content, the latter in turn described to be closely linked to clay-to-silt ratio, air temperature and rainfall. Similarly, vegetative growth, canopy characteristics, precocity of *veraison*, total anthocyanins, total soluble solids (TSS) and malic acid concentrations in grapes were found to be significantly correlated with SI values in the six sites over two seasons. The study concluded SI correlations between particular viticulture variables to be stronger with five climatic indices for the sites studied and described the SI index as a potential gauge for use in vineyard zoning and site selection evaluation processes.

In [21] a vineyard register called VINGIS was developed using the following four sets of variables:

- Agrometeorology (frequency of winter frost damage, spring, fall frost damage),
- Soil (Soil type, Soil forming rock, pH and lime state, physical soil kind, water management features, Humus level, thickness of the production layer of soil. The area's homogeneity concerning the soil type),
- Water management (water management of the area based on site observation), degree of erosion,
- The lie of the land, elevation (slope degree and aspect, elevation above sea level on hill and mountainside, emergence from the environment on the plain and flat areas, relief, area surface on hill and mountainside, relief, area surface on plain

and flat areas, environment proximity of woods, degree of built-up areas), area utilization, road conditions.

The highly complex models of viticulture zoning so far developed basically have attempted to use multiple key variables at the *meso* or micro scale and these models could be classified into two major approaches and they are:

- formulate a complex index by adding appropriately weighted vital factors considered as most contributory to grapevine growth and berry ripening. Such highly complex models seem to have built to deal with *meso* scale data projected onto a national map.
- Study the correlation between selected key variables and most of them at the micro scale (precision viticulture) within a vineyard and projected onto a national map are discussed.

"Precision viticulture", is a refined domain that uses highly advanced technologies to process satellite imagery at multiple resolutions as used in other domains, such as urban planning [22]. The major constraint encountered when incorporating GIS data into precision viticulture systems is that finding processes to aggregate spatial information where resolution and precision are not the same and in most of the cases geostatistics and kriging techniques are used to transform all the data into grids of manageable size. This allows for classical analysis or queries applied to the data in different layers as in [23]. The authors of that study modelled the correlations between spatial variability in vigour, elevation, sugar content and soil resistivity using heterogeneous precision viticulture data by applying uncertainty theories (fuzzy sets, possibility theory and Choqiet integral) all at the micro scale. With the use of high performance computing the methods could be extended to *meso* scale involving several vineyards.

2.2 Remote Sensing in Viticulture

More recent advances in remote sensing and access to satellite imagery have led to the use of airborne multispectral and hyper spectral imagery in precision viticulture with greater flexibility especially, in yield mapping integrated to soil, vegetation vigour or disease properties [24–26]. Literature reviewed reveals the various approaches investigated for satellite imagery classification in the recent past. In [23, 27] fuzzy logic has been applied to satellite imagery pixel analysis and delineation of vine parcels by segmentation of high resolution (aerial images).

Contemporary precision viticulture studies as well use increasingly high resolution aerial imagery and micro climate/environmental data acquired using networks of wireless/wired sensors/probes to identify the different zones within a vineyard block [28]. For example in [29, 30], to understand the complex dynamic relationships between site, soil, water, growth phenology, vine variety and wine quality, in order to manage the vineyard daily operations efficiently, key variable interactions, such as, effective soil water storage, plant rooting depth, onset of water stress, and daily vine water use, were studied. In recent times, since the beginning of the last decade the technology has become an important tool in vineyard water management especially, in countries like Australia, where salinity and water scarcity are major problems as far as agriculture and viticulture are concerned.

3 The Methodology

In this research, all available thematic (feature) map data in digital formats related to viticulture (Table 1) is processed and analysed in a GIS environment using ArcGIS 10.0 supported by ESRI (www.esri.com). The successive processing steps of the approach are outlined in Fig. 2. Initially, all feature layers (vector/polygon maps) obtained from Landcare Research web portal (www.landcareresearch.co.nz/resources/data/lris) are converted into raster maps within an ArcMap project. Secondly, pixel (point attribute) data is extracted from all raster layers for implementing clustering with SOM and then rule extraction using TDIDT method with WEKA and IBM's Clementine sw. SOM based clustering was performed on 437,888 pixels along with their attribute data in Table 1.

Table 1 Attributes used for image pixel clustering for charaterising/zoning NZ wine regions/vineyards

Climate variables	Land form variables	Soil variables
1. Mean annual temperature: strongly influences plant productivity	1. Elevation:	1. Drainage: influences the oxygen availability in upper soil layers
2. Mean minimum winter Temperature: influences plant survival	2. Slope: Major driver of drainage, soil rejuvenation and microclimate	2. Acid soluble phosphorous: indicates a key soil nutrient
3. Mean annual solar radiation: determines potential productivity	3. Aspect: refers to the compass direction the slope faces (north, south, east, or west). Depending on the region's climate, different slopes are selected for the greatest benefit of vineyard production	3. Exchange calcium: both a nutrient and a determinant of soil weathering
		4. Induration (hardness): determines soil resistance to weathering
4. Monthly water balance ratio: indicates average site "wetness"		5. Age: separates recent, fertile soils from older less fertile soils
5. Annual water deficit: gives an indication of soil dryness, it is calculated using mean of daily temperature, daily solar radiation and rainfall (Leathwick, Morgan, Wilson, Rutledge, McLeod, & Johnston, 2002)	4. Hill shade	6. Chemical limitation of plant growth: indicates the presence of salinity of ultramafic substances

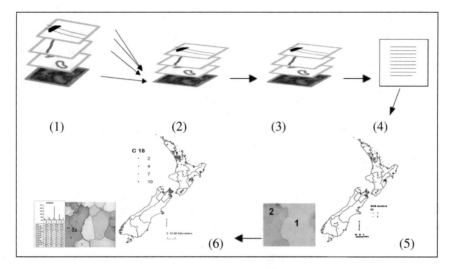

Fig. 2 A schematic diagram showing the methodology. Spatial attribute data on different layers (either vector or raster format) of the land areas are overlaid initially (steps *1* and *2*). Subsequently, vector layers are converted into raster format (*3*) and then point pixel data are extracted from this multi-layered map (*4*) for pixel clustering, knowledge extraction and analysis (*5* and *6*)

3.1 SOM Training Algorithm

A SOM consists of a regular, usually two-dimensional grid of neurons. Each neuron i of the SOM is represented by a weight model vector,

$$m_i = [m_{i1,\ldots\ldots,min}]T,$$

Where, n is equal to the dimension of the input vectors.

The set of weight vectors is called the codebook. Distances (Euclidean) between x and all the prototype vectors are computed. In the basic SOM algorithm, the topological relations and the number of neurons are fixed from the beginning. The number of neurons may vary from a few dozen up to several thousands. It determines the granularity of the mapping, which in turn affects the accuracy and generalisation capacity of the SOM. During an iterative training, the SOM forms an "*elastic net*" that folds onto the "*cloud*" dictated by the input data (Fig. 3). The net

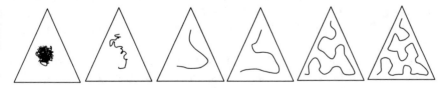

Fig. 3 2D to 1D mapping by a SOM, from [32]

tends to approximate the probability density of the data; the codebook vectors tend
to drift to places where the data is dense, while there would be only a few codebook
vectors in places where data is sparse. The component planes of the map as well
tend to reflect the relationships between the data attributes enhancing the detection
of patterns within the input data [31, 32] in this case, the land attribute data, and
they are studied using graphs and by overlaying the different cluster membership on
New Zealand maps. At each training step, one sample vector x is randomly chosen
from the input data set and the distances (such as the similarities) between the
vector x and all codebook vectors are computed. The best matching unit
(BMU) denoted here by c, would be the map unit whose weight vector is closest to
x: $\|x - m_c\| = \min \{\|x - m_i\|\}$.

After finding the BMU, the weight vectors are updated. The BMU and its
topological neighbours are moved closer to the input vector in the input space. The
update rule for the weight vector of unit i is:

$$m_{i(t)} + \alpha_{(t)} \left[x(t) - m_{i(t)}\right], \quad i\varepsilon N_{c(t)}$$

$$m_{i(t+1)} = m_{i(t)}, \quad i\varepsilon N_{c(t)}$$

Where, t denotes time. $Nc(t)$ is the non-increasing neighbourhood function
around the winner unit c and $0 < \alpha(t) < 1$ is a learning coefficient, a decreasing
function of time [10].

A commercial software package "Viscovery" supported by Eudapics (www.
viscovery.net) was used for clustering the pixels using Kohonen's SOM algorithm.
In the SOM Ward clustering process, similar pixels are grouped together in clusters.
Subsequently, new knowledge in the form of "*Rules*" are extracted by analysing the
different cluster profiles (pixel data) to enhance the identification of attributes for
zoning land areas, such as wine regions or vineyards.

3.2   WEKA's JRip Classifier

JRip classifier in WEKA is an implementation of the RIPPER rule learner created
by William W Cohen [33]. In JRip (RIPPER) classes are examined in increasing
size and an initial set of rules for each class is generated. Incremental reduced error
JRip (RIPPER) proceeds are used by treating all the examples of a particular
judgment in the training data as a class when finding a set of rules that cover all the
members of that class. By repeating the process with all the classes, different sets of
rules are generated for each class [34]. WEKA's JRip classifier model consists of
collection rules and some statistics about those rules (e.g., coverage/no coverage,
true/false positives/negatives).

As anticipated, by evaluating the SOM cluster profiles (values and graphs) as
well as JRip and J48 classifier rules created using SOM membership as the classifier, interesting patterns within and among the clusters and their attributes have

been revealed and they are discussed in the next section. The SOM clustering gives a means to extract information for the selection of meaningful features/attributes for characterising/zoning land areas in this case, NZ wine regions/*terroirs*/vineyards at different scales.

4 Results and Discussion

Initially, all 15 layers, vector as well as raster (Table 1) were overlaid on a New Zealand map and all vector layers are converted into raster layers. Subsequently, from this multi-layered map, 437,888 pixels were extracted with values for New Zealand vineyards. Then pixels were initially clustered into two and then further into four and finally into 18 clusters. The SOM clusters of both (1 and 2 –> 1a and 1c, 1b, similarly 2 –> 2a and 2b) and finally C1–C18, were analysed using their cluster profiles i.e., attribute ranges, graphs, association rules and also by overlaying the SOM cluster membership over New Zealand maps to see the spatial distribution of each cluster. Subsequently, to evaluate the use of the approach at different scales, the SOM cluster profile and rules generated (for the SOM cluster profiles) were analysed at the (1) national with all NZ wine regions, (2) regional or *meso*, using individual regional i.e., Marlborough and Auckland vineyard pixel details separately and finally, (3) at the micro (vineyard) using a few vineyards with Kumeu pixels [35].

4.1 Two-Cluster SOM

At the first level, two SOM clusters were created which showed the major discriminating features in the first level clustering C1 and C2 (Figs. 4 and 5) and they are discussed.

In cluster 1, vineyard pixels 270,584 out of the total 270,798, belong to the South Island vineyards. Similarly, 137,926 out of total 167,090 of cluster 2, lie in the North Island vineyards (Fig. 5). Upper South island has both 1 and 2 clusters, Marlborough region on the top right of the South island mostly consists of cluster 1 and on the top left Nelson got cluster 2 pixels. The former has high annual water deficiency (7.64–312.2 with an average of 215.9) whereas, the latter has low (0.42– 216 with an average of 149.9). The following are the major observations in the second level clustering labelled as 1 a and c together and 1 b, and 2 a and 2b (Figs. 6, 7 and 8):

- Elevation, average minimum temperature, induration (hardness), exchange calcium, acid soluble phosphorous, soil age, drainage, monthly water balance ratio and annual water deficit show very high variability that can be used to characterise the vineyards. Annual water deficit gives an indication of soil

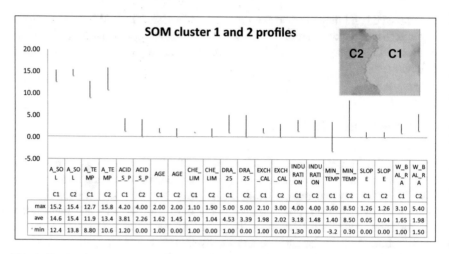

SOM cluster 1 and 2 profiles

	A_SOL	A_SOL	A_TEMP	A_TEMP	ACID_S_P	ACID_S_P	AGE	AGE	CHE_LIM	CHE_LIM	DRA_25	DRA_25	EXCH_CAL	EXCH_CAL	INDU RATION	INDU RATION	MIN_TEMP	MIN_TEMP	SLOPE	SLOPE	W_B AL_R A	W_B AL_R A
	C1	C2	C1	C2	C1	C2	C1	C2	C1	C2	C1	C2	C1	C2	C1	C2	C1	C2	C1	C2	C1	C2
max	15.2	15.4	12.7	15.8	4.20	4.00	2.00	2.00	1.10	1.90	5.00	5.00	2.10	3.00	4.00	4.00	3.60	8.50	1.26	1.26	3.10	5.40
ave	14.6	15.4	11.9	13.4	3.81	2.26	1.62	1.45	1.00	1.04	4.53	3.39	1.98	2.02	3.18	1.48	1.40	8.50	0.05	0.04	1.65	1.98
min	12.4	13.8	8.80	10.6	1.20	0.00	1.00	0.00	1.00	0.00	1.00	0.00	1.00	0.00	1.30	0.00	-3.2	0.30	0.00	0.00	1.00	1.50

Fig. 4 Graph of two cluster SOM showing the range of the first levels clusters. All attribute ranges show considerable variability in clusters 1 and 2. At this first level SOM clustering soil age, chemical limitation of plant growth and monthly water balance have same average however, their ranges vary to an extent. Slope for both 1 and 2 clusters remain the same at 0.4–0.05 (ranging between 0 and 1.26)

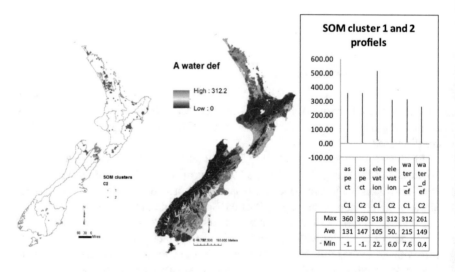

SOM cluster 1 and 2 profiels

	as pect	as pect	ele vat ion	ele vat ion	wa ter _d ef	wa ter _d ef
	C1	C2	C1	C2	C1	C2
Max	360	360	518	312	312	261
Ave	131	147	105	50.	215	149
Min	-1.	-1.	22.	6.0	7.6	0.4

Fig. 5 *Left* and *middle* SOM cluster 1 and 2 pixels overlaid on New Zealand and annual water deficiency maps. *Right* graph showing SOM 1 and 2 cluster profiles of aspect and elevation key: A water def: annual water deficiency

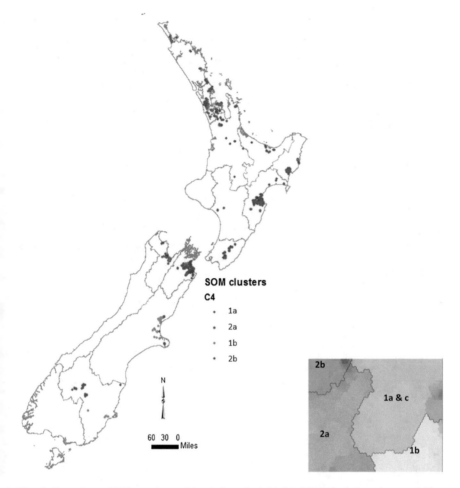

Fig. 6 Four cluster SOM map (second level clustering) (*right*). SOM pixel clustering overlaid on New Zealand map (*left*). Cluster 1b pixels are found in Canterbury and coastal Otago, whereas, cluster 2b pixels are in Auckland and Nelson. Cluster 2a pixels are in mid North Island i.e., Gisborne and Hawk's Bay alone and none of 2a pixels lie in the Central Otago region

dryness, it is calculated using mean of daily temperature, daily solar radiation and rainfall [36].

- Canterbury and coastal Otago vineyard pixels (1b) get separated (with 62.37 m average elevation, 1.09 °C average minimum temperature and 4.88 average drainage) from cluster 1a&c pixels.
- Most of northern North Island and Nelson vineyard pixels are separated into 2b with major differences in average elevation (93.84 m), average minimum temperature (4.59 °C), average induration (2.28), average acid soluble phosphate, age, monthly water balance ratio and annual water deficiency.

C18 no	pixel count	Ele**	A Temp	A mi T**	A Sol R	In-dura	Ex C*	ASI P**	Che Age limit	Slop	Dra ge**	Wat Ba R	Water deficit**
C1	50313	**82.33**	12.39	2.40	14.9	**3.32**	2.00	**4.19**	1.00	1.97	0.06	4.81	1.40	**261.48**
C3	31141	**111.17**	12.20	2.01	14.9	**2.12**	2.00	**3.29**	1.00	2.00	0.07	2.90	1.50	**247.07**
C9	1433	**81.52**	12.40	**2.40**	14.9	3.32	2.00	4.20	1.00	1.98	0.04	4.80	1.40	261.73
C13	16064	**108.54**	12.11	**1.87**	14.8	2.09	2.00	3.32	1.00	2.00	0.05	2.87	1.52	239.95
C8	39187	167.00	11.90	1.00	15.2	3.90	1.90	3.90	1.00	1.60	0.03	5.00	2.10	136.28
C14	14302	167.00	11.90	1.00	15.2	3.90	1.90	3.90	1.00	1.60	0.03	5.00	2.10	136.28
C15	8945	247.64	10.20	2.40	13.9	2.09	2.10	3.00	1.10	1.98	**0.09**	4.57	1.00	**305.53**
C16	2909	**385.65**	10.14	0.78	14.2	3.37	1.62	3.10	1.00	1.99	**0.25**	4.49	1.64	170.18
C5	78678	**54.14**	11.70	**1.20**	14.1	3.29	2.00	3.90	1.00	1.00	0.03	**4.99**	1.70	**213.31**
C6	14929	**105.75**	11.24	**0.52**	13.9	3.38	2.09	3.65	1.00	1.99	0.05	**4.34**	1.73	**181.64**

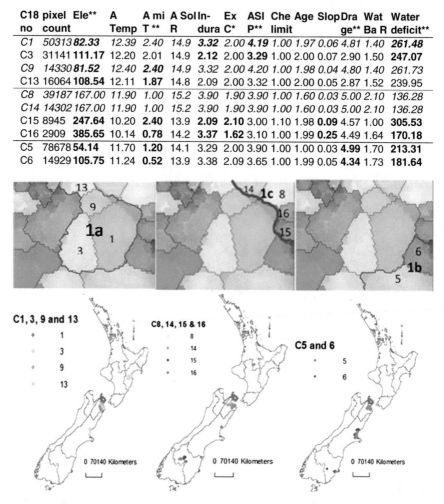

Fig. 7 SOM cluster profiles (*top*), SOM maps (*middle*), New Zealand maps (bottom) with SOM cluster pixel information overlaid showing the spatial spread and profile of 18 SOM clusters (sub clusters of C1 a b and c)

- Drainage is high in 1a, b and c when compared with those of 2 a and b.
- Average minimum temperature variability in clusters is becoming more conspicuous at this level.
- Cluster 1b pixels are located in the coastal areas of Marlborough and Otago regions with the lowest average minimum temperature i.e., 1.09 °C. Induration of this cluster is the highest of all clusters.

4.2 SOM Clusters (18-Cluster) and JRip Classifier Rules

The two cluster SOM is further divided into 18 clusters to look at the attribute variability in wine regions of New Zealand in detail.

Clusters 11 (Auckland, Bay of Plenty, Gisborne, Hawk's Bay, Northland and Waikato) and 17 (Auckland, Northland and Waikato are) of the 18 cluster SOM are found in upper North Island with higher annual average (13.0–15.0 °C) and minimum (3.5–8.5 °C) temperatures (Fig. 8). They also have similar range in acid

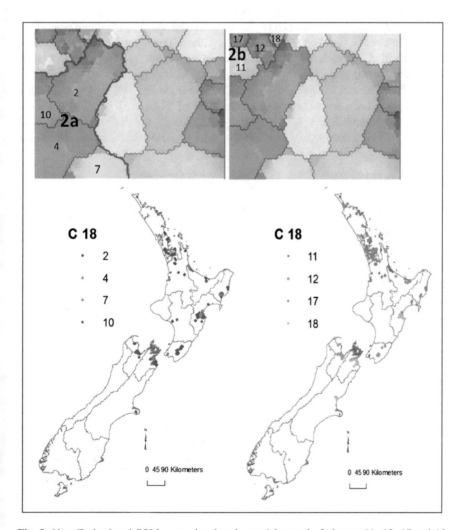

Fig. 8 New Zealand and SOM maps showing the spatial spread of clusters 11, 12, 17 and 18 (original C2b of the two-cluster SOM) in the 18 cluster SOM (*left*). New Zealand map superimposed with clusters 2, 4 7 and 10 (original C2a of the two-cluster SOM) in the 18 cluster SOM) (*right*)

soluble phosphorous (14–15), exchange cation (0–1.6), age (1.0–2.0), slope (0–0.3) and water balance ratio (2.0–3.0) but vary in drainage and induration. Cluster 12 pixels (Fig. 8) are mainly seen in North Island and upper South Island i.e., a few pixels (90) in Auckland, Bay of Plenty, Marlborough, Nelson, Northland, Waikato and Wellington.

Based on SOM clustering patterns, cluster 1 (at the 18 cluster level) consists of 50,313 pixels all belonging to Marlborough region's vineyards alone and no other. Similarly, cluster 14 consists of 14,302 pixel belonging to Marlborough alone (Fig. 9). This shows the diverse as well as some unique nature of the *terroirs* (or vineyards) that produce the world famous premium wine labels [37].

In the next stage, JRip classifier rules were created using 1,395 group summaries (group maximum, average and minimum values Fig. 9) of the 796 vineyard polygons-18 SOM clusters-11 regions and SOM clustering (1–18) as the target variable (Figs. 10 and 11).

The JRip rules extracted from vineyard, wine region and SOM group minimum, average and maximum values gave an accuracy of 94.6915 % when evaluated at tenfold cross-validation (Fig. 12). This means using the JRip rules the SOM clustering of the vineyards (based on attributes studied) could be predicted at a 94 % accuracy level.

4.3   SOM Clustering and Rules Extracted for Marlborough Region

Marlborough wine region pixels are clustered and analysed to see the SOM-TDIDT method approach at the *meso* scale. The cluster profiles are analysed using the rules

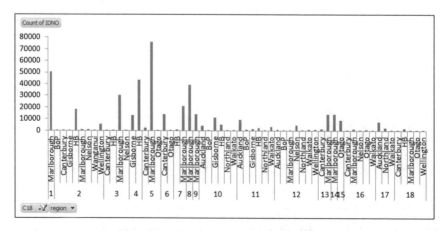

Fig. 9 Graphs showing the total number of pixels in different regions in each of the 18 SOM clusters. Marlborough region vineyard pixels are found in all clusters but 4, 6, 10, 11, 15 and 17. Meanwhile, Clusters 1, 8, 9 and 14 consist of only Marlborough vineyard pixels

SOM cluster	Min aspect	Max aspect
1	0	275.49
2	8.9	359.93
3	-1	282.71
5	-1	360
7	-1	359.97
8	-1	264.43
9	232.55	360
12	3.82	354.15
13	240.76	359.99
14	237.31	359.99
16	0.23	359.86
18	2.05	352.73

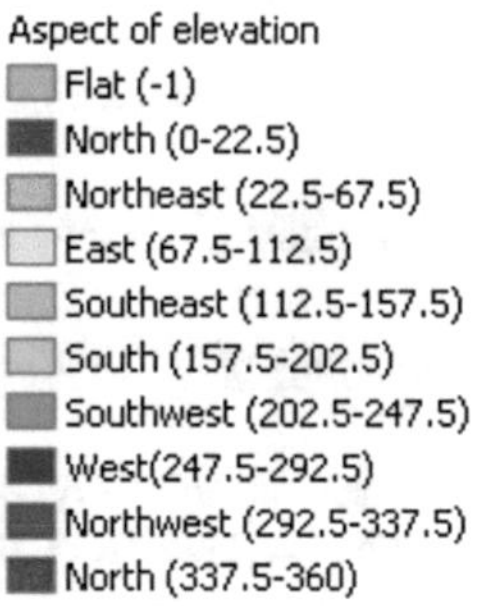

Aspect of elevation
- Flat (-1)
- North (0-22.5)
- Northeast (22.5-67.5)
- East (67.5-112.5)
- Southeast (112.5-157.5)
- South (157.5-202.5)
- Southwest (202.5-247.5)
- West(247.5-292.5)
- Northwest (292.5-337.5)
- North (337.5-360)

Fig. 10 Minimum and maximum aspects of Marlborough cluster pixels. SOM clusters 8 and 14 are found at 167 m elevation,

generated for the cluster profiles. The SOM cluster profiles of the Marlborough wine region (Figs. 13, 14, 15, 16 and 17) clearly distinguish the diverse and unique characters of the region's *terroirs* that give the well-known characteristic flavours of New Zealand Sauvignon blanc [38]. In order to further study the similarities in the Marlborough vineyard *terroirs*, pixel clusters of the region were analysed using C5 algorithm (for the TDIDT method). The following are the findings from the rules and pixel overlays on New Zealand maps:

- C7 has the lowest elevation and almost very high annual water deficiency 260.
- C8 and C14 have the same averages for all attributes except for aspect. C8 pixels are in flatland or with north, northwest, southwest or south aspect (Fig. 10). C14 in the opposite aspect.
- C8 and C14 have the highest drainage (5.00) of all Marlborough clusters 18 SOM clusters.
- Along with C8 and C14, C5 has the highest drainage 5.00. Furthermore, C5 also has the third lowest average temperature.

Decision tree models are useful when developing a classification system to predict or classify future observations based on a set of decision rules. For the C5 algorithm, a data set that is divided into classes relating to the research/a topic (types of land area) is required to build rules that can be used to classify old or new cases with maximum accuracy. For example, a tree constructed could be used to classify the new *terroirs* based on the land attributes. However, for areas where no such prior classification is available SOM clustering could be used to create a new

=== **Classifier model (full training set)** === **JRIP rules:**
1. (Min of W_BAL_RA <= 1)   => C18=15 (16.0/0.0) Otago
2. (Min of A_SOL >= 15.2) and (Min of ASPECT >= 237.31)
                                         =>C18=14 (26.0/1.0)Marlborough** (dark yellow)
3. (Min of A_SOL >= 15.2) and (Min of ELE_25 >= 167)
                                         => C18=8 (30.0/0.0) Marlborough** (dark blue)
4. <u>(Average of ELE_25 >= 284.12) and (Min of A_TEMP <= 11.2)</u>
                                                          <u>=> C18=16 (35.0/0.0)</u>
5. <u>(Average of A_TEMP <= 10.9) and (Min of INDURATION >= 3.6)</u>
                                                          <u>=> C18=16 (2.0/0.0)</u>
6. (region = Canterbury) and (Max of INDURATION >= 3.6) => C18=6 (35.0/0.0)
7. (Min of A_TEMP <= 9.8)   => C18=6 (4.0/0.0)
8. (Min of ELE_25 <= 7)   => C18=7 (52.0/0.0)
9. (Average of INDURATION <= 0)   => C18=18 (53.0/0.0)
10. (Min of ACID_S_P >= 4) and (Average of ASPECT >= 279.33) => C18=9 (61.0/0.0)
    Marlborough* (red)
11. (Min of ASPECT >= 240.76) and (Min of MIN_TEMP <= 2) and (Max of A_TEMP
    >= 12.2)   => C18=13 (64.0/0.0)
12. (Average of MIN_TEMP <= 0.8) and (Min of ASPECT >= 285.04)
                                                          => C18=13 (3.0/0.0)
13. (Average of A_TEMP <= 11.7) and (Min of ELE_25 <= 85) => C18=5 (68.0/0.0)
14. (Max of ELE_25_2 >= 197)   => C18=12 (50.0/6.0)
15. (region = Nelson)   => C18=12 (42.0/18.0)
16. (region = Wellington) and (Average of A_TEMP <= 12.675) => C18=12 (14.0/6.0)
17. (Min of W_BAL_RA >= 3.2)   => C18=12 (3.0/1.0)
18. (Min of ACID_S_P >= 4)   => C18=1 (81.0/1.0) Marlborough* (light blue)
19. (Min of W_BAL_RA <= 1.5)   => C18=3 (75.0/0.0)
20. (Min of INDURATION >= 3.9) and (Max of ASPECT <= 127.99)=> C18=3 (17.0/2.0)
21. (Min of EXCH_CAL <= 1.1) and (Average of ELE_25 >= 90.46) => C18=17
    (104.0/0.0)
22. (Min of EXCH_CAL <= 1.5) and (Max of ASPECT >= 357.68) => C18=17 (6.0/0.0)
23. (Average of A_TEMP >= 15.784198)   => C18=17 (2.0/0.0)
24. (Average of DRA_25 <= 1.9) and (Min of ELE_25 <= 48)=> C18=10 (124.0/0.0)
25. (Min of A_TEMP <= 13.2)   => C18=2 (115.0/6.0)
26. (Average of ELE_25 >= 31)   => C18=11 (148.0/0.0)
27. =>C18=4 (164.0/0.0)

    Number of Rules: 27   Time taken to build model: 0.92seonds

Fig. 11 WEKA JRip rules extracted from NZ vineyards blocks (795 polygons) (11) in 1394 groups and their minimum, average and maximum values for all 15 attributes using 18 SOM cluster value as the class (C1–C18)

class variable based on the attributes themselves. With such a new class variable, using the C5.0 node, either a decision tree or a rule set can be developed. Similar to Ripper algorithm, C5 rules are developed by splitting the data based on the attribute that provides the maximum information gain at each level. For C5 model, the "target" or the class field must be categorical [39]. In this case, the SOM cluster

```
=== Stratified cross-validation
=== Summary ===
Correctly Classified Instances        1320        94.6915 %
Incorrectly Classified Instances      74          5.3085 %
Kappa statistics                      0.9427
Mean absolute error                   0.0074
Root mean squared error               0.0709
Relative absolute error               7.1766 %
Root relative squared error           31.237 %
Total Number of Instances             1394
=== Detailed Accuracy By Class ===
        TP Rate   FP Rate   Precision   Recall   F-Measure   ROCArea   Class
        0.967     0.002     0.978       0.967    0.972       0.982     3
        0.956     0.002     0.97        0.956    0.963       0.991     5
        0.981     0         1           0.981    0.99        0.99      7
        0.955     0.002     0.955       0.955    0.955       0.991     13
        0.988     0.003     0.952       0.988    0.97        0.992     1
        0.967     0         1           0.967    0.983       0.992     9
        0.919     0.001     0.971       0.919    0.944       0.958     16
        0.843     0.022     0.808       0.843    0.825       0.982     2
        0.964     0         1           0.964    0.981       0.996     18
        0.734     0.016     0.734       0.734    0.734       0.972     12
        0.986     0.002     0.98        0.986    0.983       0.992     11
        1         0.001     0.968       1        0.984       1         8
        0.96      0.001     0.96        0.96     0.96        0.979     14
        1         0         1           1        1           1         4
        1         0.002     0.984       1        0.992       0.999     10
        0.878     0.004     0.878       0.878    0.878       0.958     6
        0.956     0.001     0.991       0.956    0.973       0.975     17
        1         0         1           1        1           1         15
Weig Avg. 0.947   0.004     0.948       0.947    0.947       0.988
=== Confusion Matrix ===
  a  b  c  d  e  f  g  h   i  j  k   l  m  n   o   p   q   r  <-- classified as
 88  0  0  0  0  0  0  2   0  1  0   0  0  0   0   0   0   0 |  a = 3
  0 65  0  2  1  0  0  0   0  0  0   0  0  0   0   0   0   0 |  b = 5
  0  0 52  0  1  0  0  0   0  0  0   0  0  0   0   0   0   0 |  c = 7
  1  0  0 64  0  0  1  0   0  0  0   0  0  0   1   0   0   0 |  d = 13
  0  0  0  0 80  0  0  1   0  0  0   0  0  0   0   0   0   0 |  e = 1
  0  0  0  0  2 59  0  0   0  0  0   0  0  0   0   0   0   0 |  f = 9
  0  0  0  0  0  0 34  0   0  0  0   0  0  0   3   0   0   0 |  g = 16
  0  0  0  0  0  0  0 118  0 20  1   0  0  0   1   0   0   0 |  h = 2
  0  0  0  0  0  0  0  2  53  0  0   0  0  0   0   0   0   0 |  i = 18
  0  0  0  1  0  0  0 19   0 58  1   0  0  0   0   0   0   0 |  j = 12
  0  0  0  0  0  0  0  0   0  0 146  0  0  0   1   0   1   0 |  k = 11
  0  0  0  0  0  0  0  0   0  0  0  30  0  0   0   0   0   0 |  l = 8
  0  0  0  0  0  0  0  0   0  0  0   1 24  0   0   0   0   0 |  m = 14
  0  0  0  0  0  0  0  0   0  0  0   0  0 164  0   0   0   0 |  n = 4
  0  0  0  0  0  0  0  0   0  0  0   0  0  0 124  0   0   0 |  o = 10
  0  2  0  0  0  0  1  2   0  0  0   0  0  0   0  36   0   0 |  p = 6
  1  0  0  0  0  0  0  1   0  0  1   0  1  0   1   0 109   0 |  q = 17
  0  0  0  0  0  0  0  0   0  0  0   0  0  0   0   0   0  16 |  r = 15
```

Fig. 12 Confusion matrix and validation of WEKA JRip rules (Fig. 11) created for NZ vineyards blocks (795 polygons) in 1394 groups and their minimum, average and maximum values for all 15 attributes using 18 SOM cluster membership

membership is used as the target and the rules developed give more insight into the *terroirs* of Marlborough.

The C5 rules created using cluster minimum, average and maximum values of Marlborough pixels alone and the maps show the minimum average temperature (> or <11.9 °C in Fig. 15 left) as the major discerning attribute in the 18 SOM clustering (Fig. 14 top). The other attributes are minimum drainage (> or <3.1 in Fig. 14 bottom left), minimum age (> /<1), maximum acid soluble P (<= 3 or > 3 in Fig. 14 bottom right), minimum/maximum elevation (Fig. 15 right) and average/maximum aspect.

Marlborough 18-SOM clusters 1, 2, 3, 7, 9 and 13 consist of minimum average temperature > 11.9 °C (rules underlined in Fig. 13) and all of them are located closer to the coastal area of the region except for some patches of 3 (Fig. 15). Meanwhile, clusters 8, 12 and 18, consist of vineyards located in Marlborough coastal area but <11.9 °C and at higher elevations. Cluster 8 is at elevation <=167 m (rule no 14 in Fig. 13), cluster 12 with maximum elevation > 167 m (rule no 2 in Fig. 13) and cluster 18 with minimum elevation > 149 m (rule no 6 in Fig. 13). This shows that the approach (clustering using SOM techniques and then generating rules) to unravel new knowledge on attributes and their ranges related to the land area could be carried out successfully, in this case, to distinguish *terroirs* among and within a country's wine regions.

Similarly, based on rule no. 1 and 15 (Fig. 13 marked with*), the critical point for a second major attribute, the minimum average temperature is > 11.9 °C and a third attribute, minimum of drainage with the critical point being 3.1.

Clusters 1 and 9 show minimum of drainage > 3.1, the former with average of aspect <= 264.4° (min 0° and max 275.49°) and the latter with average of aspect > 264.4° (min 232.55° and max 360°) next to some of the cluster 1 pixels (table and maps in Fig. 7). Meanwhile, clusters 13, 2, 3 and 7 with minimum average drainage <= 3.1 are further divided based on minimum soil age; cluster 7 has <1 suggesting the area as consisting of recent fertile soil. The rest 13, 2 and 3 (in Fig. 13) vary in minimum elevation, average and maximum aspect (Figs. 15 and 17)

For the Marlborough SOM clusters 12, 14, 16, 18, 5 and 8 with > 11.9 °C minimum average temperature, soil age seems to be the second main discerning attribute, and cluster 18 has soil age <1 (rule no 6 with # in Fig. 13) areas with new fertile soil. The rest with > 1 soil age are further divided based on minimum average temperature and maximum or minimum elevation (Fig. 15).

Cluster 14 and 8 *terroirs* (Fig. 13) found in the inland areas are with minimum annual temperature > 11.2 °C and maximum elevation <= 167 m but complement each other in aspect, for cluster 8 it is min −1 max 264.43° whereas for cluster 14, it is min 237.31 max 359.99 (rules 4 and 14 in Fig. 11c). Cluster 5 is a huge area with minimum elevation <= 149 (rule no 12 ## in Fig. 13).

1. Rule 1 for 1 (81; 1.0) if Min of A_TEMP > 11.9 and Min of DRA_25 > 3.1 and Average of ASPECT <= 264.4 then 1 +
2. Rule 1 for 12 (8; 1.0) if Min of A_TEMP <= 11.9 and Min of AGE > 1 and **Min of A_TEMP > 11.2 and Max of ELE_25 > 167** then 12
3. Rule 1 for 13 (62; 1.0) if Min of A_TEMP > 11.9 and Min of DRA_25 <= 3.1 and Min of AGE > 1 and Average of ASPECT > 229.925 and Min of ELE_25 <= 111 then 13 ++
4. Rule 1 for 14 (25; 1.0) if Min of A_TEMP <= 11.9 and Min of AGE > 1 and **Min of A_TEMP > 11.2 and Max of ELE_25 <= 167** and Average of ASPECT > 229. 925 Then 14 +++
5. Rule 1 for 16 (22; 1.0) if Min of A_TEMP <= 11.9 and Min of AGE > 1 and **Min of A_TEMP <= 11.2** then 16
6. Rule 1 for 18 (20; 1.0) if Min of A_TEMP <= 11.9 and Min of AGE <= 1 and **Min of ELE_25 > 149** then 18 #
7. Rule 1 for 2 (2; 1.0) if Min of A_TEMP > 11.9 and Min of DRA_25 <= 3.1 and Min of AGE > 1 and Average of ASPECT <= 229.925 and Max of ACID_S_P <= 3 and Max of ASPECT <= 136.56 and Min of ELE_25 <= 93 then 2 ***
8. Rule 2 for 2 (8; 1.0) if Min of A_TEMP > 11.9 and Min of DRA_25 <= 3.1 and Min of AGE > 1 and Average of ASPECT <= 229.925 and Max of ACID_S_P <= 3 and Max of ASPECT > 136.56 then 2 ***
9. Rule 3 for 2 (14; 1.0) if Min of A_TEMP > 11.9 and Min of DRA_25 <= 3.1 and Min of AGE > 1 and Average of ASPECT > 229.925 and Min of ELE_25 > 111 then 2 ***
10. Rule 1 for 3 (4; 1.0) if Min of A_TEMP > 11.9 and Min of DRA_25 <= 3.1 and Min of AGE > 1 and Average of ASPECT <= 229.925 and Max of ACID_S_P <= 3 and Max of ASPECT <= 136.56 and Min of ELE_25 > 93 then 3 ***
11. Rule 2 for 3 (72; 1.0) if Min of A_TEMP > 11.9 and Min of DRA_25 <= 3.1 and Min of AGE > 1 and Average of ASPECT <= 229.925 and Max of ACID_S_P > 3 then 3 ***
12. Rule 1 for 5 (56; 1.0) if Min of A_TEMP <= 11.9 and Min of AGE <= 1 and **Min of ELE_25 <= 149** then 5 ##
13. Rule 1 for 7 (43; 1.0) if Min of A_TEMP > 11.9 and Min of DRA_25 <= 3.1 and Min of AGE <= 1 then 7**
14. Rule 1 for 8 (30; 1.0) if Min of A_TEMP <= 11.9 and Min of AGE > 1 and **Min of A_TEMP > 11.2 and Max of ELE_25 <= 167** and Average of ASPECT <= 229.925 then 8 +++
15. Rule 1 for 9 (61; 1.0) if Min of A_TEMP > 11.9 and Min of DRA_25 > 3.1 and Average of ASPECT > 264.4 then 9 *

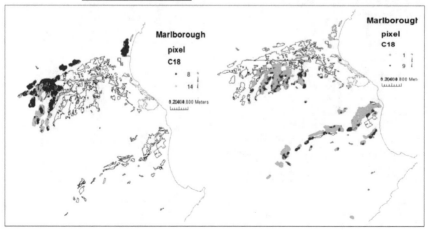

Fig. 13 C5 algorithm rules generated using Clementine software for the Marlborough region pixels. The maps showing cluster 8 and 14 (*bottom left*) as well as 1 and 9 (*bottom right*) of the 18 cluster SOM relating to Marlborough vineyard block pixel attributes. The C8 and C14 have common features, minimum annual temperature > 11.9 °C and Min of DRA> 3.1 but vary in aspect (rules no 1 and 15)

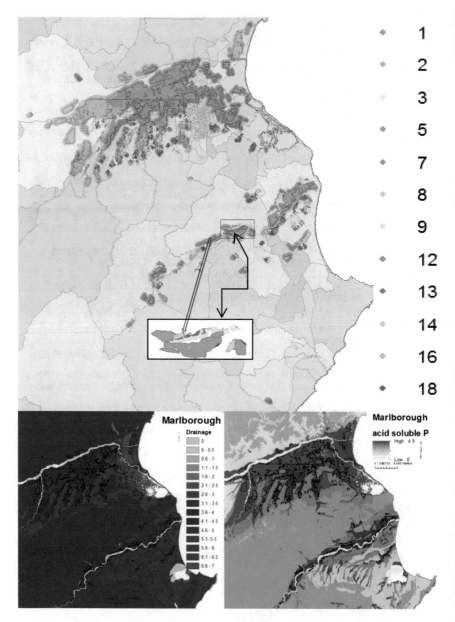

Fig. 14 Marlborough census wards and vineyards (*top*) located mostly in areas with average high drainage i.e., >4.65 (*bottom left*) and higher acid soluble P (*bottom right*)

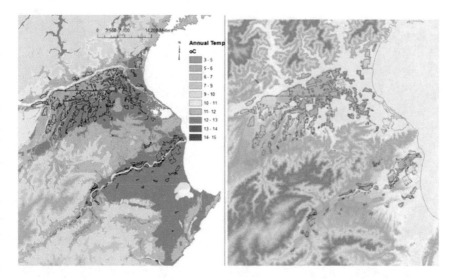

Fig. 15 Average temperature (left) and digital elevation (right). The C5 rules created using cluster minimum, average and maximum values of Marlborough pixels alone, and the maps show the minimum average temperature (> or <11.9 °C in Fig. 13 rules) as the major discerning attribute within the 18 SOM clusters

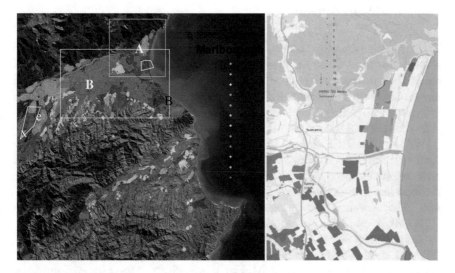

Fig. 16 *Left* Marlborough pixels displayed over satellite imagery show the unique features regarding landscape, terrain and vegetation. ex: some blocks identified for vineyard expansion b (*right*): section A (shown in the *left* image) showing the location related details

◆ 1
◆ 2
◆ 3
◆ 5
◆ 7
◆ 8
◆ 9
◆ 12
◆ 13
◆ 14
◆ 16
◆ 18

Fig. 17 Section B of Fig. 16 (*left*) showing the location and plot related details

4.4   Pixel Clustering and Rules Developed for Within Auckland Region

The section looks at the clusters, their profiles and rules extracted from Auckland pixels alone to evaluate the approach among and within vineyards (at the *meso* scale). For this second regional level evaluation, 22,462 individual Auckland vineyard pixel details were used. The cluster profiles were studied using JRip and J48 rules (with WEKA functions) generated to analyse the similarities within *terroirs* of Auckland wine region (Fig. 18).

WEKA rules (JRIP and J48 in Figs. 19 and 20 respectively) were created using Auckland cluster pixels with SOM clusters 10, 11, 12 and 17 as output classes. The main attributes discerning the pixels are (Fig. 19): average elevation, average drainage and average hill shade with prediction rate 99 % at tenfold cross validation. The elevation of clusters 12, 10 and 17 are at > = 214 m with (1098.0/0.0), <= 40 m with 4140.0/0.0 and > = 92 m with 7767.0/0.0 pixels.

The two main SOM clusters (of C18) 11* with 9,454 and 17** with 7,767 pixels (Fig. 18), containing the majority of the 22,462 Auckland pixels are spread across the western and eastern parts of Auckland respectively. The main attribute separating the two clusters are elevation, minimum temperature and induration. Cluster 17 pixels in the north eastern and Waikiki Island are at slightly higher elevation > = 92 m with minimum temperature 6.47 °C and induration 3.12 as compared to cluster 11 <= 40 m, 4.85 °C and 1.99 for the respective attributes.

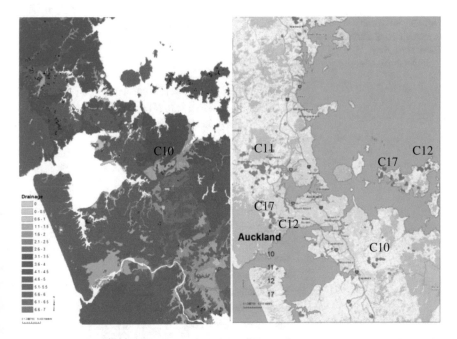

Fig. 18 Maps showing drainage (*left*) and Auckland vineyard pixels (*right*). As per WEKA rules elevation and drainage are the contributing factors in the SOM clustering for the Auckland region. In the drainage map cluster 10 pixels have <= 1.9 and the rest have >1.9

Meanwhile, the other SOM clusters 10 and 12 reflect the minor areas of both extremes found in the Auckland region. Most of the southern Auckland pixels belong to cluster 10 with an average elevation 33.19 m, the lowest for the region and with poor drainage (1.67) when compared with the other Auckland clusters (Fig. 18).

Cluster 12 pixels of the Auckland region are at an average elevation 227.69 m, the highest for the region, and have induration 4 and monthly water balance 3.46 the highest values for the attributes for this region.

From WEKA's J48 results (Fig. 20), the main attribute distinguishing cluster 11 and 17 is elevation; for C11 it is <=68 m whereas, for C17 it is > 68 < 144 m. C12 consists of areas with elevation > 144 m. Most of cluster 11 pixels have hill-shade <=202 (9,432 pixels).

The ability to distinguish land attributes from available digital data shows that the approach can be used to establish meaningful zones in this case *terroirs* at the national and regional scales. The results so far discussed using the same 18 SOM clustering (membership) show promise for zoning of land areas at the national and regional (*meso*) scales.

For the micro scale analysis, Kumeu sub regional pixels were used alone to study the application of the approach among and within vineyards [35]. Kumeu

Fig. 19 J48 results for
Auckland pixel attributes and
SOM clustering
(membership) data

Test mode:10-fold cross-validation

=== Classifier model (full training set) ===

JRIP rules:

===========

(ELE_25 >= 214) => C18=12 (1098.0/0.0)

(ELE_25 <= 40) => C18=10 (4140.0/0.0)

(ELE_25 >= 92) => C18=17 (7767.0/0.0)**

(HILLSHAD >= 219) and (ELE_25 <= 49) => C18=17
(3.0/0.0)

                    => C18=11 (9454.0/0.0)*

Number of Rules : 5

Time taken to build model: 1.28seconds

Stratified cross-validation

=== Summary ===

Correctly Classified Instances	22456	99.9733 %
Incorrectly Classified Instances	6	0.0267 %
Kappa statistic	0.9996	
Mean absolute error	0.0001	
Root mean squared error	0.0103	
Relative absolute error	0.0355 %	
Root relative squared error	2.5105 %	
Total Number of Instances	22462	

=== Detailed Accuracy By Class ===

	TP Rate	FP Rate	Precision	Recall	F-Mea	ROC Area	Class
	1	0	0.999	1	1	1	17
	1	0	1	1	1	1	11
	1	0	1	1	1	1	10
	1	0	1	1	1	1	12
WA	1	0	1	1	1	1	

=== Confusion Matrix ===

a	b	c	d		<-- classified as
7768	2	0	0	|	a = 17
4	9450	0	0	|	b = 11
0	0	4140	0	|	c = 10
0	0	0	1098	|	d = 12

wine region is in the north western part of Auckland and some of the vineyards in
this suburb are famous for their unique wine labels. More insights regarding the
land attributes and their spatial distribution for the success of the *terroirs* were
established in [35].

Fig. 20 J48 results for Auckland pixel attribute and SOM cluster (target) data

Test mode:10-fold cross-validation

=== Classifier model (full training set) ===

J48 pruned tree

ELE_25 <= 68

| DRA_25 <= 1.9: 10 (4140.0)

| DRA_25 > 1.9

| | HILLSHAD <= 202: 11 (9432.0)

| | HILLSHAD > 202

| | | A_SOL <= 15: 11 (20.0)*

| | | A_SOL > 15

| | | | HILLSHAD <= 214: 11 (2.0

| | | | HILLSHAD > 214: 17 (3.0)

ELE_25 > 68

| ELE_25 <= 144: 17 (7767.0)**

| ELE_25 > 144: 12 (1098.0)

Number of Leaves : 7

Size of the tree :   13

Time taken to build model: 0.41 seconds

=== Stratified cross-validation ===

=== Summary ===

Correctly Classified Instances	22459 99.9866 %
Incorrectly Classified Instances	3 0.0134 %
Kappa statistic	0.9998
Mean absolute error	0.0001
Root mean squared error	0.0082
Relative absolute error	0.0245 %
Root relative squared error	2.001 %
Total Number of Instances	22462

=== Detailed Accuracy By Class ===

TP R	FP Rate	Preci	Recall	F-Mea	ROC	Class
1	0	1	1	1	1	17
1	0	1	1	1	1	11
1	0	1	1	1	1	10
1	0	1	1	1	1	12
WA 1	0	1	1	1	1	

=== Confusion Matrix ===

a   b   c   d   <-- classified as

7767   3   0   0| a = 17

0 9454   0   0| b = 11

0   0 4140   0| c = 10

0   0   0 1098| d = 12

5 Conclusions

Literature reviewed for this research reveals that the well-known *centuries-old* as well as contemporary approaches to zoning wine regions or characterising *terroirs* have been developed based on extensive knowledge on the independent factors (local atmospheric, environmental, cultural) and wine quality. Moreover, the old approaches were developed without much reference to geographic location related information. The need for extensive knowledge makes the zoning of wine regions of recent origin and assess new cites for potential vineyards a challenging task. In view of this fact, a new approach consisting of SOM based pixel clustering and TDIDT method within a GIS environment was investigated. The results of this research showed that the SOM-TDIDT approach provides a means to overcome the major impediment when selecting features with unique attributes (along with their cardinality) that can be used for characterising/zoning less known land areas of interest, in this case new *terroirs* within New Zealand wine regions.

The SOM-TDIDT approach as well can be used to overcome resolution and scale related issues that often cause constraints in similar GIS studies when integrating multi-sourced GIS data. The SOM clustering approach to discerning attributes produced promising results at different scales as well. For Marlborough minimum average temperature (> or <11.9 °C) was found to be the major discerning attribute. This also reiterates the current situation there; frost in September determines the crop of the Marlborough region. In addition, minimum drainage (> or <3.1), age, acid soluble P (<=3 or >3), elevation and aspect can be used for characterising the vineyards within this wine region. Similarly, for Auckland, elevation <= 40 / > = 92 / > = 214), drainage (<= 1.9 or > 1.9) and annual solar radiation, along with hill shade (related to elevation) were found to be useful factors for characterising the vineyards of this region. The solar radiation and average and minimum annual temperatures are used as growing degree days (GDD) in the currently used traditional factors. Meanwhile, when Kumeu pixels were used alone in an earlier study [35] to look at the success of the approach at the micro scale, age (<=1-fertile soils /2-older less fertile soils), elevation (> 92 /approx. 40 m), annual average temperature <= 14/> 15) along with aspect and hill shade(the latter two are related to elevation) were are seen as meaningful attributes for characterising among and even within vineyards. The Kumeu results also revealed the approach as capable of detecting the variability in aspect and hill shade arising from the subtle variability in the elevation within vineyards.

References

1. G. Chi, J. Zhu, Spatial regression models for demographic analysis. Popul. Res. Policy Rev. **2008**(27), 17–42 (2008). doi:10.1007/s11113-007-9051-8
2. L. Bissonnette, K. Wilson, S. Bel, T.I. Shah, Neighbourhoods and potential access to health care: the role of spatial and aspatial factors. Health Place **18**(4), 841–853 (2012)

3. T. Wei, S. Tedders, J. Tian, An exploratory spatial data analysis of low birth weight prevalence in Georgia. Appl. Geogr. **32**(2), 195–207 (2012)
4. L. Xiaonian, Z. Yi, F. Zhang, X. Liu, The geographic information platform of new socialist countryside comprehensive services, in *Procedia Environmental Sciences*, vol. 11 (2011), pp. 3–10
5. J.K. Berry, Fundamental operations in computer-assisted map analysis. Int. J. Geogr. Inf. Syst. **1**, 119–136 (1987)
6. S. Aronoff, *Geographic Information Systems: A Management Perspective* (WDL Publ., Ottawa, 1989), p. 294
7. W. Pitts, W.S. McCulloch, "How we know universals the perception of auditory and visual forms. Bull. Math. Biophys. **9**(1947), 127–147 (1947)
8. G.J. Deboeck, T.K. Kohonen, *Visual Explorations in Finance with Self-Organizing Maps (Springer Finance)* (Springer, London, 1998), 285 pp. ISBN 978-1-4471-3913-3
9. J. Vesanto, E. Alhoniemi, Clustering of the self-organizing map. IEEE Trans. Neural Netw. **11** (3) (2000)
10. J. Vesanto, SOM-Based data visualization methods. Intell. Data Anal. J. (1999)
11. R. Chauhan, H. Kaur, M. Alam, Data clustering method for discovering clusters in spatial cancer databases. Int. J. Comput. Appl. (0975–8887) **10**(6), 9–14 (2010)
12. Y. Qian, K. Zhang, GraphZip: a fast and automatic compression method for spatial data clustering spatial data clustering, in *SAC '04, March 14–17, 2004, Nicosia, Cyprus (p. 5). Nicosia, 2004 ACM 1-58113-812-1/03/04.*, Cyprus (2004)
13. M. Ester, H.-P. Kriegel, Jörg S, X. Xu, A density-based algorithm for discovering clusters in large spatial databases with noise, in ed. by E. Simoudis, J. Han, U.M. Fayyad, *Published in Proceedings of 2nd International Conference on Knowledge Discovery and Data Mining (KDD-96)* (1996)
14. B. Li, L. Shi, J. Liu, Research on Spatial Data Mining Based on Uncertainty in Government GIS, in *2010 Seventh International Conference on Fuzzy Systems and Knowledge Discovery (FSKD 2010), 10–12 August 2010*, Yantai, China (2010)
15. M. Bramer, Chapter 3 using decision trees for classification, in *Principles of Data Mining*, London, Springer, ISBN-10: 1846287650 | ISBN-13: 978-1846287657, 2007, pp. 41–50
16. S. Shanmuganathan, Viticultural zoning for the identification and characterisation of New Zealand "Terroirs" using cartographic data, in *Proceedings of GeoCart'2010 and ICA Symposium on Cartography, 1–3 Sept 2010* (Auckland University, Auckland, 2010)
17. E. Vaudour, A.B. Shaw, A worldwide perspective on viticultural zoning. S. Afr. J. Enol. Vitic. **26**(2), 106–115 (2005)
18. Wine fight club, He knows his claret from his Beaujolais…, 2012. www.lazyballerina.com/Winefightclub/winefightclubJul07.pdf. Accessed 2 Oct 2012
19. J. Hook, James Hook Wine, 12 Aug 2008. http://jameshookwine.blogspot.co.nz/2008/08/he-knows-his-claret-from-his-beaujouis.html. Accessed 20 March 2015
20. T. Tesic, D.J. Woolley, E.W. Hewett, D.J. Martin, Environmental effects on cv Cabernet Sauvignon (*Vitis vinifera* L.) grown in Hawke's Bay, New Zealand.: 2. Development of a site index, in *Australian Journal of Grape and Wine Research published by Australian Society of Viticulture and Oenology(ASVO)*, Vol 8 Issue 1, pp. 27–35 (2010)
21. L. Martinovich, Z. Katona, K. Szenteleki, E.P. Boto, A. Szabo, C. Horvath, Updating the Evaluation of Hungarian Wine Producing Fields Using the National GIS Register (VINGIS), 2010. http://www.oiv2007.hu/documents/viticulture/Hungarian_wine_GIS_register_VINGIS_OIV_jav_POSTER.pdf. Accessed 5 July 2010
22. C. Kurtz, N. Passat, P. Ganc, Arski, A. Puissant, Multi-resolution region-based clustering for urban analysis. Int. J. Remote Sens. **31**(22), 5941–5973 (2010)
23. J.N. Paoli, B. Tisseyre, O. Strauss, J. Roger, Combination of heterogeneous data sets in precision viticulture. *Vineyards & Sciences*, 2010. http://www.sferis.com/articles/050412_JnPaoli.pdf. Accessed 5 July 2010

24. M. Ferreiro-Arm'an, J.P.D. Costa, S. Homayouni, J. Mart'ın-Herrero, Hyperspectral image analysis for precision viticulture, in *International Conference on Image Analysis and Recognition(ICIAR 2006)*. 2006., Póvoa de Varzim, Portugal, 2006

25. R.G.V. Bramley, J. Ouzman, P.K. Boss, Variation in vine vigour, grape yield and vineyard soils and topography as indicators of variation in the chemical composition of grapes, wine and wine sensory attributes. Aust. J. Grape Wine Res. **17**, 217–229 (2011)

26. R.G.V. Bramley, M.C.T. Trought, J.P. Praat, Vineyard variability in Marlborough, New Zealand: characterising variation in vineyard performance and options for the implementation of Precision Viticulture. Aust. J. Grape Wine Res. **17**(1), 72–78 (2011). doi:10.1111/j.1755-0238.2010.00119.x

27. P.P.D. Costa, F. Michelet, C. Germain, O. Laviall, G. Grenier, Delineation of vine parcels by segmentation of high resolution remote sensed images. Precis. Agric. **2007**(8), 95–110 (2007)

28. J.A. Taylor, New technologies and opportunities for Australian viticulture, Chapter 3 in precision viticulture and digital terroir: investigations into the application of information technology in Australian vineyards. PhD Thesis. http://www.digitalterroirs.com/. The University of Sydney, 2004

29. L. Smith, P. Whigham, Spatial aspects of vineyard management and wine grape production, in *Presented at SIRC 99—The 11th Annual Colloquium of the Spatial Information Research Centre December 13–15th 1999*. (University of Otago, Dunedin, New Zealand, 1999)

30. P. Buss, M. Dalton, Olden, R. Guy, Precision management in viticulture—an overview of an Australian integrated approach.," in *Universit a Degli Studi Di Teramo, Integrated Soil And Water Management for Orchard Development (Fao Land and Water Bulletin International Seminar on The Role and Importance of Integrated Soil and Water Management for Orchard Development*, Mosciano S. Angelo (Italy), 9–10 May 2004 (2004)

31. E.L. Koua, Using self-organizing maps for information visualization and knowledge discovery in complex geospatial datasets, in *Proceedings of the 21st International Cartographic Conference (ICC) 10–16 Aug 2003*, Durban, South Africa (2003)

32. T. Kohonen, *Self-Organizing Maps*, 3rd ed. (Information Sciences, Berlin, 2001)

33. W.W. Cohen, "Fast effective rule induction," in *Proceedings of the Twelfth International Conference on Machine Learning* (Morgan, Kaufmann., 1995)

34. A. Rajput, R.P. Aharwal, M. Dubey, S.P. Saxena, M. Raghuvanshi, J48 and JRIP rules for e-governance data. Int. J. Comput. Sci. Secur. (IJCSS) **5**(2), 201–207 (2011)

35. S. Shanmuganathan, J. Whalley, Pixel clustering in spatial data mining; an example study with Kumeu wine region in New Zealand, in *20th International Congress on Modelling and Simulation, Adelaide, Australia, 1–6 December 2013* www.mssanz.org.au/modsim2013 (2013), pp. 810–816

36. J. Leathwick, F. Morgan, G. Wilson, D. Rutledge, M. McLeod, K. Johnston, *"Land Environments of New Zealand: A Technical Guide", Ministry for the Environment 2002* (Landcare Research New Zealand Limited, Private Bag 3127, Hamilton, 2002)

37. W.V. Parr, D. Valentin, J.A. Green, C. Dacremont, Evaluation of French and New Zealand Sauvignon wines by experienced French wine assessors. Food Qual. Prefer. **21**(2010), 56–64 (2010)

38. W.V. Parr, J.A. Green, K.G. White, R.R. Sherlock, The distinctive flavour of New Zealand Sauvignon blanc: Sensory characterisation by wine professionals. Food Qual. Prefer. **18** (2007), 849–861 (2007)

39. IBM SPSS Modeler Information Center, http://pic.dhe.ibm.com/infocenter/spssmodl/v15r0m0/index.jsp. http://pic.dhe.ibm.com/infocenter/spssmodl/v15r0m0/index.jsp. Accessed 20 Aug 2013

Printed in the United States
By Bookmasters